Microwave Propagation in Ferrimagnetics

Microwave Propagation in Ferrimagnetics

M. S. Sodha and N. C. Srivastava
Indian Institute of Technology
New Delhi, India

SPRINGER SCIENCE+BUSINESS MEDIA, LLC

Library of Congress Cataloging in Publication Data

Main entry under title:

Microwave propagation in ferrimagnetics.

Bibliography: p.
Includes index.
1. Ferrites (Magnetic materials) 2. Microwaves. 3. Ferrimagnetism. I. Sodha, M. S., 1932- . II. Srivastava, N. C.

QC766.F3M5 538'.45 81-15364
ISBN 978-1-4757-5841-2 ISBN 978-1-4757-5839-9 (eBook) AACR2
DOI 10.1007/978-1-4757-5839-9

Originally published by Plenum Press, New York in 1981

Preface

During the last three decades, interest in the field of interaction of microwaves with ferrimagnetics has steadily increased. Investigations in this field have led to the development of a number of devices used for a variety of applications. The initial emphasis of the investigators was on the microwave behavior of ferrimagnetics placed in cavities and metallic waveguides and associated devices. This work has been presented in various books, monographs, and reviews written during the sixties. In recent years, interest in microwave propagation in ferrimagnetics has shifted from loaded waveguides to relatively new areas, e.g., magnetostatic and magnetoelastic waves in layered structures, microwave propagation in ferrimagnetic strip lines and microstrips, etc. Such investigations are important from the viewpoint of devices such as delay lines, filters, convolvers, guided wave amplifiers, striplines, and microstrip phase shifters, circulators, edge guided mode isolators, etc. As such, we felt the need for a text (meant for graduate students starting work in these areas as well as practicing electrical engineers and applied physicists) which presents a coherent account of the various aspects of propagation of microwaves (electromagnetic as well as magnetoelastic) in biased ferrimagnetics and discusses the relatively recent developments in the theory and operation of the aforementioned devices, and this book is the result.

A biased ferrimagnetic is, in the mathematical sense, a complicated medium, electromagnetically as well as elastically. Hence, the mathematical nature of the presentation is unavoidable but sufficient details are given in the derivation of important results. Although the reader is assumed to have a basic knowledge of electromagnetic theory, elastic field theory, and solid state magnetism, seven appendices on the structure and properties of common ferrimagnetics, the magnetic moments of atoms and ions, coordinate transformation, elastic field theory, precessional mode frequency of ellipsoids, Poynting's theorem and partially magnetized ferrimagnetics have been included to make the book self-contained.

Chapters 1 to 3 present the basic concepts of electromagnetic and magnetoelastic wave propagation in biased ferrimagnetics. The response of a ferrimagnetic to external electromagnetic and elastic fields is discussed in Chapter 1. The permeability tensor has been obtained for magnetically saturated ferrimagnetic, considering the various interactions present in a real ferrimagnetic. The coupled magnetoelastic equations have also been developed. Chapter 2 discusses electromagnetic and magnetoelastic plane wave propagation in ferrimagnetics. Chapter 3 presents an analysis of electromagnetic and magnetoelastic wave propagation across plane interfaces; these studies provide a physical insight into the relevant phenomena and serve as a background for the appreciation of more complex wave propagation phenomena in ferrimagnetics (which is of practical importance). In Chapter 4 we discuss the propagation of magnetostatic waves in layered planar structures; the analysis and experimental results pertaining to a number of practical configurations are presented. The applications to delay lines, filters, convolvers, and magnetostatic wave amplifiers are also discussed. Chapter 5 considers magnetoelastic guided wave propagation in layered planar structures and its applications. Gaussian units have been used throughout the book.

We wish to thank our colleague Prof. A.K. Ghatak for providing the inspiration to write this book. We are also grateful to Dr. I. Rattan for a critical review of the manuscript, to Mr. S. S. Gupta and Mr. S. N. Bajpai for helpful discussions, and to Mr. T. N. Gupta for his kind and ungrudging assistance in the preparation of the manuscript. The financial support of NSF (USA) for the original work reported in this book is gratefully acknowledged.

Delhi
June 1981

M. S. Sodha
N. C. Srivastava†

Contents

Notation xiii

Chapter 1. Introduction 1

1.1. Ferrimagnetics 1
1.2. Magnetic Dipole and Magnetization 2
1.3. A Dipole in a Magnetic Field 4
1.4. The Permeability Tensor 8
1.5. Damping 10
1.6. Exchange Interaction 18
1.7. Magnetocrystalline Anisotropy 22
1.7.1. Cubic Ferrimagnetics 24
1.7.2. Hexagonal Magnetoplumbites 30
1.7.2.1. Uniaxial Magnetoplumbite 30
1.7.2.2. Planar Magnetoplumbite 31
1.8. Dipole–Dipole Interaction: Demagnetization 33
1.9. Magnetoelastic Interaction 44

Chapter 2. Plane Waves in Ferrimagnetics 47

PART A. ELECTROMAGNETIC WAVES 47

2.1. Electromagnetic Field Equations 47
2.2. Exchange-Free Plane Waves 50
2.2.1. The Propagation Constant 50
2.2.2. Special Cases 52
2.2.2.1. Lossless Isotropic Ferrimagnetics 52
2.2.2.2. Lossless Hexagonal Planar Magnetoplumbites 61
2.2.3. Lossless Ferrimagnetics: The General Case 63
2.2.3.1. Mode Classification 63
2.2.3.2. Propagating and Nonpropagating Regions 64
2.2.4. Effect of Damping 67
2.2.5. Comparison with Waves in Magnetoplasmas 70
2.3. The Magnetostatic Approximation 71
2.4. Exchange-Dominated Plane Waves 74

PART B. MAGNETOELASTIC WAVES 79

2.5. Magnetoelastic Field Equations 79
2.6. Exchange-Free Plane Waves 80
2.6.1. The Dispersion Relation 81
2.6.2. Special Cases 83
2.6.3. General Case 86
2.6.3.1. Dispersion Curves 86
2.6.3.2. Refractive Index Diagrams 88
2.6.3.3. Polarization of Modes 91
2.7. Effect of Exchange 93
2.8. Effect of Anisotropy 95

Chapter 3. Wave Propagation Across Interfaces 97

PART A. ELECTROMAGNETIC WAVE PROPAGATION 97

3.1. Ferrimagnetic Half Space 97
3.1.1. Reflection and Transmission Coefficients 98
3.1.2. Regions of Ordinary and Total Reflection 100
3.1.3. Phase Shift in Total Reflection 105
3.1.4. Reflection from Lossy Ferrimagnetics 111
3.1.5. Energy Flow in Total Reflection: Lateral Shift 115
3.1.6. Ray Model for Attenuated Total Reflection 122
3.2. Ferrimagnetic Slab 127

PART B. MAGNETOELASTIC WAVES 133

3.3. Half Space Problem 133

Chapter 4. Magnetostatic Waves in Layered Planar Structures 143

4.1. Ferrimagnetic Half Space 144
4.1.1. Transverse Magnetization 145
4.1.2. Arbitrary Magnetization 150
4.1.3. Experimental Results 155
4.2. Isotropic Ferrimagnetic Slab 160
4.2.1. Magnetization in the Slab Plane 160
4.2.1.1. Transverse Propagation: Magnetostatic Analysis 161
4.2.1.2. Transverse Propagation: Electromagnetic Analysis 167
4.2.1.3. General Magnetostatic Analysis 170
4.2.1.4. Experimental Results and Applications 174
4.2.2. Normal Magnetization 180
4.2.2.1. Magnetostatic Field Analysis 181
4.2.2.2. Energy Distribution and Power Flow 184
4.2.2.3. Zig-Zag Ray Model 187

4.3. Metal-Backed Ferrimagnetic Slab 194
4.3.1. Magnetization in Slab Plane 194
4.3.1.1. Transverse Propagation: Magnetostatic Analysis 195
4.3.1.2. Transverse Propagation: Electromagnetic Analysis . . . 197
4.3.1.3. Arbitrary Propagation: Magnetostatic Analysis 201
4.3.1.4. Effect of Finite Conductivity of Metal 203
4.3.1.5. Experimental Results 206
4.3.2. Normal Magnetization 208
4.4. Ferrimagnetic–Dielectric–Metal Structure 209
4.4.1. Transverse Magnetization in the Film Plane 211
4.4.2. Normal Magnetization 221
4.4.2.1. Dispersion Characteristics 221
4.4.2.2. Energy and Power 223
4.4.3. Arbitrary Magnetization 227
4.5. Metal–Ferrimagnetic–Metal Structure 235
4.5.1. Transverse Magnetization 235
4.5.2. Normal Magnetization 240
4.6. Double Magnetic Layered Structures 242
4.6.1. Air Gap Between Two Ferrimagnetic Substrates 242
4.6.2. Ferrimagnetic Layer Over Another Ferrimagnetic Substrate . . 246
4.7. Periodic Structures 248
4.8. Effect of Magnetocrystalline Anisotropy 252
4.8.1. YIG 256
4.8.2. Zn_2Y Planar Magnetoplumbite 262
4.9. Interaction of Magnetic Waves with Drifting Carriers 266
4.9.1. Physical Explanation of Amplification 266
4.9.2. Air–Ferrimagnetic–Semiconductor Composite 267
4.9.3. Metal–Dielectric–Semiconductor–YIG Structure 273
4.10. Magnetostatic Wave Oscillators 278
4.11. Transduction of Magnetostatic Waves 281

Chapter 5. Magnetoelastic Waves in Layered Planar Structures . . 291

5.1. Transverse Magnetization 292
5.1.1. Ferrimagnetic Half Space 292
5.1.2. Nonmagnetic Layer Over Ferrimagnetic Substrate 297
5.1.3. Ferrimagnetic Slab 303
5.1.4. Ferrimagnetic Layered Nonmagnetic Substrate 304
5.2. Longitudinal Magnetization 308
5.2.1. The Dispersion Relation 308
5.2.2. Numerical Solution of the Dispersion Relation 312
5.2.3. Energy Loss for Leaky Waves 316
5.2.4. Effect of Magnetic Loss 319
5.2.5. Other Studies 319
5.3. Normal Magnetization 321
5.3.1. The Dispersion Equation 321
5.3.2. Dispersion Characteristics 323

5.4. Experimental Results and Devices 325
5.4.1. Excitation and Propagation Loss 325
5.4.2. Magnetoelastic Surface Wave Delay 326
5.4.3. Magnetoelastic Surface Wave Isolator 328
5.4.4. Magnetoelastic Surface Wave Oscillator 328
5.4.5. Magnetoelastic Surface Wave Convolution 331
5.4.6. Other Studies 335

Appendix A. Structure and Properties of Common Ferrimagnetics 337

A.1. Ferrimagnetic Garnet 337
A.1.1. Structure 337
A.1.2. Magnetization 338
A.1.2.1. Pure Garnet 338
A.1.2.2. Mixed Garnets 340
A.1.2.3. Substituted Garnets 340
A.1.3. Magnetocrystalline Anisotropy 341
A.1.4. Linewidth 341
A.1.5. Other Properties 341
A.2. Spinel Ferrites 341
A.2.1. Crystal Structure 341
A.2.2. Magnetization 342
A.2.3. Magnetocrystalline Anisotropy 343
A.2.4. Linewidth 343
A.3. Hexagonal Magnetoplumbites 343
A.3.1. M-Type 343
A.3.2. W-Type 344
A.3.3. Y-Type 344
A.3.4. Z-Type 344
A.3.5. U-Type 345

Appendix B. Magnetic Moment of Atoms and Ions 347

Appendix C. Coordinate Transformations 353

C.1. Eulerian Angles and Rotation Matrices 353
C.2. Transformation of Maxwell's Equations 357
C.3. The Transformed Permeability Tensor 359
C.4. Transformation of Anisotropy Field 360
C.5. Transformation of Magnetoelastic Interaction Energy 361

Appendix D. Basic Elastic Wave Theory 363

D.1. Strain 363
D.2. Stress 365

D.3. Hooke's Law: Stiffness and Compliance 365
D.4. Damping 367
D.5. Boundary Conditions 367
D.6. Elastic Energy 367
D.7. Elastic Field Equations 368
D.8. Elastic Plane Waves in an Infinite Medium 368

Appendix E. Uniform Precessional Mode Frequency for Small Ellipsoids 371

Appendix F. Poynting's Theorem 373

Appendix G. Partially Magnetized Ferrimagnetics 377

References 381

Author Index 395

Subject Index 399

Notation

$\boldsymbol{A}$	see equation (D.33)
$A, A_{\pm}, A_{a}, A_{d}, A_{m}$	arbitrary constants
A_{11}–A_{33}	see equation (D.35)
$A(k_z - k_R)$	see equation (5.46)
a	subscript for air region
a	semiaxis of the ellipsoid
a_0	electron orbit radius
a_1	interdipole spacing
$a_{15}, a_{25}, a_{35}, a_{45}, a_{55}, a_{1\beta}, a_{2\beta}, a_{3\beta}, a_{4\beta}, a_{5\beta}$	elements of $a_{\alpha\beta}$, see equations (5.44), (5.45), and (5.54)
a_f	$= \sqrt{\varepsilon_f}$
$a_{\alpha\beta}$	determinant of dispersion equation coefficients
$\mathbf{B}$	instantaneous magnetic induction
$B, B_{\pm}, B_{a}, B_{d}, B_{m}$	arbitrary constants
$\mathbf{B}_0$	static component of $\mathbf{B}$
$\mathbf{B}_i$	instantaneous internal magnetic induction
B_r	see equation (2.93)
B_z^{in}, B_z^{out}	magnetic induction inside and outside the sample, respectively
$\mathbf{b}$	rf magnetic induction
b	semiaxis of the ellipsoid
b_1, b_2	first and second magnetoelastic constants, respectively
b_x, b_y	cartesian components of $\mathbf{b}$
$\boldsymbol{C}$	stiffness tensor
C	arbitrary constant
C_{11}–C_{44}	elements of $\boldsymbol{C}$
C_{JK}	components of stiffness matrix

c_0	speed of light in vacuum
c	semiaxis of the ellipsoid
c_{ijkl}	elastic stiffness coefficients
c_l, c_t	speed of longitudinal and transverse elastic waves, respectively
$\mathbf{D}$	instantaneous electric displacement
D	lateral shift of rays after reflection, arbitrary constant, see equation (4.225).
$D_\pm$	see equation (3.81)
D_0	see equation (1.32)
D_1	see equation (1.76)
D_2	see equation (1.88)
D_3	see equations (1.89) and (2.18)
D_{ex}	see equation (1.48)
D_L	see equation (4.111)
D_{me}	see equation (3.86)
D_r, D_t	arbitrary constants
D_v	see equation (4.150)
D_τ	see equation (4.141)
$\mathscr{D}$	dispersion relation, see equations (5.43), (5.45) and (5.54)
d	thickness of dielectric region
d	subscript/superscript for dielectric region
$\mathbf{E}$	instantaneous electric field
$\boldsymbol{\mathscr{E}}$	real part of $\mathbf{E}$
$\mathscr{E}_0$	amplitude of $\boldsymbol{\mathscr{E}}$
E_x, E_y, E_z	cartesian components of $\mathbf{E}$
e	charge of electron
e	superscript for nonmagnetic medium
$\mathbf{F}$	force experienced by a dipole
$\mathbf{F}^b$	body force/unit volume
$\mathbf{F}^{me}$	force density due to magnetoelastic stress
F_1, F_2, F_3	see equation (3.14)
$F(\omega, \beta)$	see equation (4.224)
$F_T(\omega, \beta)$	see equation (4.223)
F'_T	see equation (4.240)
F'_{Ti}, F'_{Tr}	see equation (4.231)
$f = \omega/2\pi$	precessional frequency of $\mathbf{h}$
f	superscript/subscript for ferrimagnetic region
f_0	$= \omega_0/2\pi$
$f_I, f_{II}, f_{III}, f'_2$, f'_3, f_2, f_σ	characteristic frequencies in the dispersion relation
Δf	change in oscillator frequency
g	Landé g-factor

g_1	see equation (3.59)
$\mathbf{H}$	instantaneous magnetic field
$\mathbf{H}_0, H_0$	dc magnetic field (biasing field)
H^{ω}	see equation (1.47)
$\mathbf{H}_A$	anisotropy field
$\mathbf{H}_{A0}$	dc component of $\mathbf{H}_A$
$\mathbf{H}_i$	internal magnetic field
H_{bot}	cut-off magnetic field between surface and bulk waves
H_c	coercive field
H_{cr}	$=(\omega/\gamma - 4\pi M_0)$
H_a	see equation (1.85)
H_a^u	see equation (1.82)
H_d	demagnetizing field
H_e	external dc magnetic field
H_{ji}^{ω}	component of Weiss field
$\mathbf{H}_{me}$	see equation (1.109)
$\boldsymbol{\mathcal{H}}$	real part of $\mathbf{H}$
ΔH	linewidth
$\mathbf{h}$	rf magnetic field
$\mathbf{h}_A$	time varying component of $\mathbf{H}_A$
$\mathbf{h}_i$	$\mathbf{h}$ inside the ellipsoid
$\mathbf{h}_{me}$	time-varying component of $\mathbf{H}_{me}$, see equation (2.85)
h_x, h_y, h_z	cartesian components of $\mathbf{h}$
h_{Ax}, h_{Ay}, h_{Az}	cartesian components of $\mathbf{h}_A$
$\hbar = h/2\pi$	h is Planck's constant
I_0	uniform current microstrip
i, j, k, l $(=1, 2, 3, 4)$	integer subscripts
i, r, t	superscripts for incident, reflected, and transmitted waves, respectively
i	circulating electron current
$\mathbf{J}$	total angular momentum of a multielectron shell
J_{ov}	overlap integral for exchange interaction
J_c	current density
$J_y(x)$	magnitude of surface current distribution
j	$=\sqrt{-1}$
$\mathbf{k}$	wavevector
k_0	see equations (1.50) and (4.53)
k_1, k_2	first- and second-order anisotropy constants
k_B	Boltzmann constant
k_R	wavenumber of leaky waves
k_{R0}	wavenumber of Rayleigh waves
k_l, k_t	see equation (2.93)

k_S	see equation (4.204)
$k_t^{(n)}$	see equation (3.60b)
$k_x^{(q)}$	four values of k_x^2
k_x, k_y, k_z	cartesian components of β
Δk	$= k_R - k_{R0}$
$\mathbf{L}$	total orbital angular momentum
$\mathbf{L}^e$	equilibrium position of particles
L	path length between transducers, coil width, length of YIG sample
L_p	attenuation length
$\mathcal{L}$	instantaneous position of particles
$\mathbf{l}$	orbital angular momentum
l	length of microstripline, quantum number
$\mathbf{M}$	magnetization at absolute zero
$\mathbf{M}(\mathbf{r})$	nonuniform magnetization at any point
M_l	see equation (4.233)
M_r	remanent magnetization
$\mathbf{m}$	rf magnetization
m	integer, mass of electron
m	superscript/subscript for ferrimagnetic medium
m^*	effective mass of charge carriers
m_l, m_s	quantum numbers
$m_n^{(q)}$	normalized eigenvectors
m_{nl}	nonlinear component of m at frequency 2ω
$\boldsymbol{N}$	demagnetizing tensor
N	number of magnetic dipoles/unit volume
N_{ij}	elements of $\boldsymbol{N}$
N_v	see equation (4.150)
N_τ	see equation (4.141)
N_x, N_y, N_z	demagnetizing factors
n	positive integer, charge carrier concentration
n_a	see equation (4.158)
n_e	electron density
n_L, n_W	positive integers
n_S	see equation (4.161)
n_x, n_y, n_z	cartesian components of $\hat{n}$
$\hat{n}$	outward normal to a surface, unit vector in the direction of phase propagation, principal quantum number
$\hat{n}_r$	unit vector in the direction of polarization
$\mathbf{P}$	Poynting vector, time averaged power flow/unit area, see equation (C.19)
P	power density
P_0	see equation (C.9)

P_a	see equation (4.160)
P_{ac}	ac component of **P**
P_c	average power delivered to load
P_{dc}	dc component of **P**
P_i	see equation (3.33)
$P_{me}^{(n)}$	see equation (3.73b)
P_n	see equation (4.193)
P_r	see equation (3.33)
P_x	Poynting vector
P_x^t	total power lost
P_{-x}^t	see equation (3.20)
P_{-y}^r	see equation (3.21)
$\bar{\mathbf{P}}$	instant power flow/unit area
P_T	total power radiated
p	see equation (4.129)
p_h	see equation (4.21)
p_α	see equation (4.25)
p, q, r, s $(=1, 2, 3, 4)$	subscripts
p_x, p_y, p_z	see equation (1.73)
Q_1, Q_2	see equation (4.240)
Q_5	arbitrary constant
Q_a	see equation (4.100)
Q_n	see equation (4.193)
Q_q	weight factors
Q_β	arbitrary constants
q	see equation (3.8)
q_α	see equation (4.25)
R	radius of cylinder, reflection coefficient, coordinates transformation matrix, see equations (C.7) and (C.16)
$\|R\|$	absolute value of reflection coefficient
R'	see equation (3.82)
R_-	reflection coefficient when incident and reflected waves are interchanged
$R_1, R_2^\pm$	see equations (3.16) and (C.3)
R_i	input resistance of microstrip
R_g	see equation (4.45)
R_m	$=2R_i/l$
R_x, R_y	see equation (3.43)
$\mathbf{r}$	position vector
r	superscript/subscript for a reflected wave
$\hat{r}$	unit vector in r-direction
$\boldsymbol{S}$	strain tensor

$\mathbf{S}$	spin angular momentum
S_1, S_2	amplitudes of $\mathcal{S}_1$ and $\mathcal{S}_2$, respectively, see equations (3.90) and (3.92)
S_{11}–S_{44}	see equation (D.21)
S_{ij}	elements of $\boldsymbol{S}$
S_I	see equation (D.6)
S_J	components of $\boldsymbol{S}$
S_i	$\equiv \sin^2 \psi_i$
S_f	$\equiv \sin^2 \psi_f$
$\mathcal{S}_1, \mathcal{S}_2$	strength of magnetoelastic surface wave signals
$\boldsymbol{s}$	compliance matrix
s	surface, subscript free space region
s	$= \pm 1$
$\boldsymbol{\mathcal{S}}$	spin of electron
$\boldsymbol{T}$	stress tensor
$\boldsymbol{T}^{me}$	effective stress tensor
$\mathbf{T}^{ex}$	exchange torque
$\mathbf{T}_d$	torque exerted on a dipole
$\mathbf{T}_n$	normal stress field
$\mathbf{T}_x, \mathbf{T}_y$	stress field in x- and y-directions, respectively
T	transmission coefficient, relaxation time,
T	subscript/superscript, see equations (4.226) and (4.239)
T'	see equation (2.111)
T_-	transmission coefficient when incident and reflected waves are interchanged
$\lvert T_\pm \rvert$	absolute value of $T_\pm$
T_{1133}–T_{3333}	see equation (C.29)
T_1, T_5, T_6	cartesian components of $\mathbf{T}_x$
T_2, T_4, T_6	cartesian components of $\mathbf{T}_y$
T_1^{me}–T_6^{me}	see equation (1.113)
T_c	compensation temperature, critical temperature
T_I	see equation (4.15)
T_I^{me}	see equation (1.110)
T_N	Néel temperature
$T_{p,q,r,s}$	see equation (C.23)
T_x, T_y	see equation (C.43)
T_{xy}, T_{yy}, T_{zy}	cartesian components of T_y
t	thickness of dielectric film
t	superscript/subscript for transmitted waves
t_{eff}	see equation (4.192)
U	potential energy of a magnetic dipole in external magnetic field, see equation (3.92)
U_e	elastic energy density

U_{ex}	exchange energy between two ions
U_{me}	magnetoelastic interaction energy
$\mathbf{u}$	elastic displacement field
$\mathbf{u}_n^{(q)}$	normalized eigenvectors
u_l, u_s	longitudinal and shear components of $\mathbf{u}$
u	time-averaged energy density of harmonic fields
u_0^m	arbitrary constant
u_x, u_y, u_z	cartesian components of $\mathbf{u}$
$\tilde{u}$	instant energy density of harmonic fields
V	volume, average volume available/magnetic dipole, see equation (3.92)
V_C	magnitude of convolved signal
v	wave velocity, particle velocity
v_0	orbital speed of electrons, drift velocity of charge carriers
v_R	velocity of Rayleigh waves
v_e	average velocity of energy flow
v_g	group velocity
$v_g^{\pm}$	v_g for RCP and LCP waves, respectively
v_{ph}	phase velocity
$v_{ph}^{\pm}$	v_{ph} of RCP and LCP waves, respectively
Δv	change in v
W	subscript, width of YIG sample, acoustic beam width
X, Y, Z	cartesian coordinates system
x_1, x_2	surfaces of corrugated ferrimagnetic slab
$x_{\pm}$	$= \sqrt{(\mu \pm \kappa)}$
$\hat{x}_3$	unit vector in x_3-direction
$\hat{x}, \hat{y}, \hat{z}$	unit vector along X-, Y- and Z-axes, respectively
$\partial/\partial x, \partial/\partial y, \partial/\partial z$	cartesian components of gradient operator ∇
$\partial^2/\partial x^2, \partial^2/\partial y^2, \partial^2/\partial z^2$	cartesian components of Laplace operator ∇^2
∇	gradient operator
∇^2	Laplace operator
∇_{Ij}	see equation (D.8)
$\nabla_{\mathbf{k}}$	gradient operator with respect to components of $\mathbf{k}$
∇_{iJ}	see equation (D.13)
$\nabla_{\boldsymbol{\alpha}}$	gradient operator with respect to components of $\boldsymbol{\alpha}$
$\boldsymbol{\alpha}$	direction cosines of magnetization vector
$\boldsymbol{\alpha}(\mathbf{r})$	unit vector in the direction of manetization
α	dimensionless, positive parameter
α, β $(=1,2,3,4,5, =1,2,3,4)$	subscripts

$\alpha_{\pm}$	loss factor for RCP and LCP modes, respectively
α'	angle of axis rotation
α_a	see equation (4.158)
α_e	attenuation coefficient of E-mode
$\alpha_{eff}^{\pm}$	see equation (3.31)
α_m	see equation (4.184)
α_n	see equation (4.191)
α'_n	see equation (4.193)
α_p	attenuation factor
α_s	see equation (4.64)
α_v	see equation (4.73)
α_y	$\equiv \cos\theta$
β, β'	propagation constant, see equation (2.109)
$\beta_{\pm}$	β for RCP and LCP waves, respectively
β_0	$= \omega/c_0$, β in absence of curvature, see equation (4.210)
β_1	scc cquations (2.19) and (3.78)
β_2	see equations (2.20) and (3.78)
β_d	$= \sqrt{\varepsilon_f \omega^2/c_0}$, see equation (2.12)
β_i, β_r	real and imaginary parts of β
β_m	see equation (4.133)
β_n	see equation (3.60a)
β_y, β_z	cartesian components of β
$\boldsymbol{\Gamma}$	Christoffel matrix
γ	gyromagnetic ratio
γ_l	orbital gyromagnetic ratio
γ_m	see equation (4.184)
γ_S	spin gyromagnetic ratio
γ'_S	see equation (4.210)
Δ	measure of strength of modulation, see equations (1.89), (2.17) and (2.18)
Δ_l	see equation (5.2)
δ	angle of rotation of polarization, arbitrary angle between direction of wave propagation and dc magnetic field
δ'	angle of axis rotation
δ_0	see equations (2.17) and (2.18)
$\delta_{\pm}$	see equation (3.15)
$\delta_{\parallel}^{\pm}$	see equation (3.79)
$\delta_{\perp}$	see equations (3.81) and (5.2)
δ_{jk}	Kronecker delta function
δ_p	depth of penetration
$\boldsymbol{\varepsilon}$	permittivity tensor for drifting charge carriers
$\boldsymbol{\varepsilon}_f$	dielectric tensor for ferrimagnetic medium
ε_L	permittivity of the background latttice

$\varepsilon, \varepsilon_z$	elements of $\boldsymbol{\varepsilon}$
ε_d	dielectric constant of the dielectric region
ε_{ijk}	Levi–Civita tensor
ε_f	dielectric tensor of ferrimagnetic medium
ε_m	$= \beta_m/\beta$
ζ	normalized coordinate, see equation (4.210)
ζ_1	element of $\boldsymbol{\mu}_i$, see equation (3.27)
ζ_2	see equation (3.28)
ζ_d	normalized thickness
ζ_s	$= -(\eta_s^2 + \mu)/\mu$
$\boldsymbol{\eta}$	viscosity tensor
η	see equation (3.23)
η_{44}	element of $\boldsymbol{\eta}$
η_s	see equation (4.61)
θ, φ	Eulerian angles
θ	angle of incidence, arbitrary angle between direction of wave propagation and applied field, angle between magnetization vector and hexagonal c-axis
θ_0	effective thickness of ferrrimagnetic slab
θ_1, θ_2	angles of incidence and transmission, respectively
$\hat{\theta}$	unit vector in θ-direction
θ_c	see equation (4.195)
$\theta_c(\varphi)$	see equation (2.40)
θ_F	angle of rotation of plane of polarization
θ_x	angle between $\hat{n}$- and x-axis
κ	magnitude of off-diagonal elements of $\boldsymbol{\chi}$, see equation (1.22)
κ_0	magnitude of off-diagonal elements of $\boldsymbol{\mu}$, in a lossless medium
κ^{eff}	element of $\boldsymbol{\mu}^{\text{eff}}$
κ_e	element of $\boldsymbol{\varepsilon}$
κ', κ''	real and imaginary parts of κ, see equation (1.32)
$\boldsymbol{\lambda}$	fourth rank tensor
λ	damping parameter, see equation (4.41)
λ_{ijkl}	elements of $\boldsymbol{\lambda}$
$\boldsymbol{\mu}$	permeability tensor of ferrimagnetic medium
$\boldsymbol{\mu}^{\text{eff}}$	effective $\boldsymbol{\mu}$
$\boldsymbol{\mu}_i$	see equation (3.26)
$\boldsymbol{\mu}_r$	see equation (3.27)
$\boldsymbol{\mu}_l, \boldsymbol{\mu}_s$	magnetic and spin magnetic moments, respectively, of an electron
$\boldsymbol{\mu}^T$	magnetic dipole moment
μ	element of $\boldsymbol{\mu}$, micron ($= 10^{-4}$ cm)

μ_0	magnitude of diagonal elements of $\boldsymbol{\mu}$ in a lossless medium
μ', μ''	real and imaginary parts of μ
$\mu_\pm$	see equation (4.210)
μ_{11}–μ_{33}	components of $\boldsymbol{\mu}''$
μ^{a}_{11}–μ^{a}_{33}	components of $\boldsymbol{\mu}$
μ_{eff}	see equation (2.19)
$\mu^{(0)}_{eff}$	$=(\mu_0^2-\kappa_0^2)/\mu_0$
μ'_{eff}	see equation (2.46)
μ''_{eff}	see equation (2.47)
μ^{y}_{eff}	see equation (3.16)
$\mu_{xx}, \mu_{yy}, \mu_z$	elements of $\boldsymbol{\mu}$
$\mu^{eff}_{xx}, \mu^{eff}_{yy}$	elements of $\boldsymbol{\mu}^{eff}$
ν, ν_c	collision frequency
ξ	normalized coordinate, see equation (4.210)
ξ_1	element of $\boldsymbol{\mu}_i$
ξ_2	see equation (3.28)
ξ_e	see equation (5.20)
ξ_m	see equation (5.2)
ρ	density of solid/elastic medium
ρ_e	charge density
σ	spin–elastic fields coupling constant, see equation (2.93)
σ_e	electrical conductivity of metal
τ_e	delay time
$\tau_d^\pm, \tau_d$	group delay time
τ_f	group delay time at frequency f
Φ	total flux intercepted by pick-up coil
φ, ψ	Eulerian angles
φ	angle between direction of wave propagation and biasing field
φ'	$=\pi/2-\varphi$
$\varphi_\pm$	phase angles of RCP and LCP waves, respectively
φ_0	steering angle
φ_e	phase shift introduced by amplifier, transducers
φ_s	see equation (4.61)
φ_v	see equation (4.101)
$\hat{\varphi}$	unit vector in φ-direction
$\boldsymbol{\chi}$	susceptibility tensor
χ	magnitude of diagonal elements of $\boldsymbol{\chi}$
Ψ_s	see equation (1.99)
Ψ_1, Ψ_2, Ψ_3	see equation (1.100)
ψ	magnetostatic potential
ψ_0	amplitude of ψ
ψ_1	see equation (3.10)

$\psi_2, \psi_3, \psi_4, \psi_5, \psi_6$	see equation (3.14)
ψ^*	complex conjugate of ψ
ψ^{crit}	$= \arctan(\omega_0/\omega_m)^{1/2}$
ψ_i, ψ_f	angles of incidence and reflection, respectively
$\psi_m^{(q)}$	see equation (5.34)
$\psi_p(x, y)$	particular solution to equation (3.71)
ψ_{wb}	angle between wavevector and biasing field
Ω	resonance frequency
Ω_0	$= \omega_0/\omega_m$, see equation (4.180)
$\Omega_1, \Omega_2, \Omega_3$	resonance frequencies of volume, f.m. and f.a. modes, respectively
ω	angular frequency of the wave, precessional frequency of **h**
ω_0	$= \gamma H_0$
$\omega_2, \omega_3, \omega_4, \omega_5, \omega_6, \omega_7, \omega_8$	see equation (3.14)
ω_1', ω_3'	see equations (2.108), (4.159) and (5.2)
ω_2'	see equations (4.163) and (5.2)
ω_4'	$= (\omega_0 + \frac{1}{2}\omega_m)$
ω_5'	see equation (4.19)
ω_8'	see equation (4.20)
$\omega_{2n}', \omega_{3n}'$	see equation (3.60b)
$\omega_\pm$	see equation (5.25)
$\Delta\omega$	see equation (2.111)
$\Delta\omega_h$	$= \gamma \Delta H$
ω_a	$= \gamma H_a$
ω_a^u	$= \gamma H_a^u$
ω_{ex}	see equation (1.48)
ω_c	cyclotron frequency
$\omega_c^\pm$	see equation (4.175)
ω_{cs}	cut-off frequencies for magnetostatic surface waves
ω_l	see equation (2.98)
ω_m	$= 4\pi M_0 \gamma$
ω_p	plasma frequency
ω_r	see equation (1.48)
$\omega_r^\pm$	see equation (4.175)
ω_{res}	resonance frequency
ω_{res}^u	see equation (1.83)
ω_s	see equation (2.96)
ω^t, ω^{td}	see equation (2.98)
ω_σ	see equation (2.108)

1

Introduction

1.1. Ferrimagnetics

All materials interact with externally applied magnetic fields. When the constituents (atoms, molecules, or ions) of a material do not possess a permanent magnetic dipole moment, the interaction of the material with the magnetic field is rather weak and a sample of such a material is repelled from the regions of high magnetic field (diamagnetic behavior). On the other hand, if the constituents of the material do possess a permanent magnetic dipole moment (which may arise from the spin and orbital motion of electrons), it is attracted toward the regions of high magnetic field because the magnetic dipoles have a tendency to align themselves along the direction of the biasing field (paramagnetic behavior). Some of the paramagnetic crystalline solids, when cooled below certain critical temperatures, exhibit magnetic order* even in the absence of an external magnetic field; this leads to spontaneous magnetization. The magnetic ordering takes place on account of exchange interaction, which has a quantum mechanical origin, discussed at length by Anderson (1963a, b). As a consequence of the exchange interaction, the successive magnetic dipoles are generally aligned either parallel (ferromagnetic behavior) or antiparallel. In the latter case, the magnetic lattice can be looked upon as comprising a number of sublattices, each of which consists of identical magnetic dipoles oriented along a specific crystallographic direction. The net magnetization of the crystal is thus the vector sum of the magnetizations of different sublattices; this may be zero (antiferromagnetic behavior) or finite (ferrimagnetic behavior). The critical temperature at which the transition from the paramagnetic state to the magnetically ordered state is observed is called the Curie temperature for ferromagnetic materials and the Néel temperature for antiferromagnetic and ferrimagnetic materials.

* The state in which successive magnetic dipoles are oriented in a specific manner is called a magnetically ordered state.

High-resistivity ferrimagnetic materials exhibit some interesting properties at microwave frequencies; the present book discusses microwave behavior and relatively recent applications of such materials. Ferrimagnetic materials are found to possess a variety of crystallographic structures, such as ferrimagnetic spinels (ferrites), ferrimagnetic garnets, hexagonal magnetoplumbites, etc. The structure and basic properties of some important classes of ferrimagnetics are briefly discussed in Appendix A.

In order to study the microwave response of ferrimagnetics, it is necessary to understand the nature of magnetic dipoles and their interaction with various internal and external fields. This is discussed in the following sections.

1.2. *Magnetic Dipole and Magnetization*

A circulating current gives rise to a magnetic dipole moment. In an atom, electrons move around the nucleus; the associated circulating currents lead to the magnetic dipole moment of the atom. For an atom possessing a single electron moving in a circular orbit of radius a_0, the angular momentum $\mathbf{L}$ and circulating current i are given by*

$$\mathbf{L} = m v_0 a_0 \tag{1.1}$$

and

$$i = -(e v_0 / 2\pi a_0) \tag{1.2}$$

where m, e, and v_0 represent the mass, charge, and orbital speed of the electron. The magnetic dipole moment is given by

$$\begin{aligned} \boldsymbol{\mu}_l &= (\text{current} \times \text{area})/c_0 \\ &= -e v_0 a_0 / 2 c_0 \end{aligned}$$

where $c_0 = 3 \times 10^{10}$ cm/sec is the speed of light in a vacuum. Hence

$$\boldsymbol{\mu}_l = -(e/2mc_0)\mathbf{L} \tag{1.3}$$

Thus, the atom possesses a magnetic dipole moment by virtue of the orbital angular momentum of the electron. Apart from this orbital angular momentum, the electron is also known to have an additional (intrinsic)

* As noted in the Preface, the Gaussian system of units has been adopted in the present book. However, the final appendix describes the method of conversion to S.I. units.

singular momentum called spin.* The magnetic moment corresponding to the spin angular momentum is obtained from quantum mechanical considerations as (Schiff, 1968)

$$\boldsymbol{\mu}_s = -(e/mc_0)\mathbf{S} \tag{1.4}$$

where S is the electron spin angular momentum. In general, the spin and orbital motions of the electron are not independent. In the rest frame of the electron, the nucleus moves around it and constitutes a ring of current which, in turn, leads to a magnetic field at the electron location. The spin magnetic moment of the electron interacts with this magnetic field, giving rise to spin-orbit coupling. When spin and orbital angular momenta are both present, the net magnetic dipole moment $\boldsymbol{\mu}^T$ can be expressed as (Schiff, 1968)

$$\boldsymbol{\mu}^T = -g(e/mc_0)\mathbf{J} \equiv -\gamma\mathbf{J} \tag{1.5}$$

where $\mathbf{J} = \mathbf{L} + \mathbf{S}$ is the total angular momentum, while $\gamma = ge/2mc_0$ is called gyromagnetic ratio.† The so-called g-factor is a measure of the strength of spin-orbit interaction; it is unity for purely orbital motion ($s = 0$) and two for pure spin ($L = 0$). The atoms (or ions) have several electrons in different quantum states. The net magnetic moment of the atom can be obtained by combining the magnetic moments of the various electrons as discussed in Appendix B.

In a magnetic oxide, each magnetic ion is surrounded by several oxygen ions. The outer electron orbits of the magnetic ions are strongly influenced by the electrostatic field of the surrounding ions (crystalline field). Consequently, in most cases,‡ the orbital motion of outer electrons in the ion is not significantly modified when the magnetic field is switched on. Thus, the orbital angular momentum does not contribute appreciably to the magnetic moment of an ion which is embedded in a matrix of anions. This is called "quenching" of orbital angular momentum (Wagner, 1972; White, 1970). When the quenching is complete, only spin contributes to the magnetic moment in which case $g = 2$ and, therefore, $\gamma = 1.76 \times 10^7$ rad/sec Oe.

* The spin of the electron is sometimes visualized by assigning to it a rotational motion about its own axis (analogous to the spin of the earth). This analogy, though quite appealing, is not appropriate since the elctronic spin has a purely quantum mechanical origin.

† It is worth pointing out that some authors include the negative sign appearing in equation (1.5) in the definition of γ itself. The reader should therefore be careful while comparing the subsequent equations and results with those given elsewhere.

‡ In the case of certain ions, e.g., rare earth ions, the magnetic moment arises from inner shell electrons. The orbital motion of these electrons is not significantly affected by the crystalline field, and, therefore, the orbital angular momentum does contribute to the magnetic moment of such ions.

In a piece of normal paramagnetic material, the magnetic dipoles are randomly oriented owing to thermal agitation. An external magnetic field is required to align the dipoles along a specific direction. However, in the case of ferro- or ferrimagnetic materials, the exchange interaction is sufficiently strong to cause the alignment of the dipoles even in the absence of the external magnetic field, provided that the temperauture is less than the critical temperature. For ferromagnetic materials, the magnetization (defined as the net magnetic dipole moment per unit volume) at absolute zero is given by

$$\mathbf{M} = N\boldsymbol{\mu}^{\mathrm{T}} \tag{1.6}$$

where N is the number of magnetic dipoles per unit volume. In the case of ferrimagnetics, the magnetization is given by

$$\mathbf{M} = \sum_i N_i \boldsymbol{\mu}_i^T \tag{1.7}$$

where N_i and $\boldsymbol{\mu}_i^T$ refer, respectively, to the number density and the magnetic moment of the dipoles in the ith sublattice. It is noteworthy that equations (1.6) and (1.7) are valid only for infinite media. In the absence of an external magnetic field, a finite piece of a ferrimagnetic material exhibits a domain structure (Kittel, 1971) which, in turn, suppresses the net magnetic moment of the sample.* However, when a finite sample of appropriate shape† is subjected to a sufficiently strong external dc magnetic field, all magnetic dipoles are aligned parallel to one another and the sample behaves like a single domain. In this state the sample is said to be magnetically saturated and the net magnetization, called saturation magnetization, is given by equation (1.6) or (1.7). It is worth mentioning that only a magnetized ferrimagnet (i.e., a ferrimagnet which is magnetized to saturation) is significantly transparent to microwaves. Therefore, in subsequent development, the medium is regarded as magnetically saturated.

We shall now proceed to discuss the interaction of magnetic dipoles with a magnetic field.

1.3. *A Dipole in a Magnetic Field*

The potential energy of a magnetic dipole in a uniform magnetic field **H** is given by

$$U = -\boldsymbol{\mu}^{\mathrm{T}} \cdot \mathbf{H} = -\mu^{\mathrm{T}} H \cos\theta \tag{1.8}$$

* When domains are present, the magnetization of a single domain is still given by equations (1.6)–(1.7).

† See Section 1.8 for further illustration of this point.

where θ is the angle between the directions of the magnetic dipole moment and the applied magnetic field, as shown in Figure 1.1. A change in the orientation of the dipole causes a variation in the potential energy. The force experienced by the dipole is given by

$$\begin{aligned} \mathbf{F} &= -\nabla U \\ &= -\left(\hat{r}\frac{\partial U}{\partial r} + \hat{\theta}\frac{1}{r}\frac{\partial U}{\partial \theta} + \hat{\varphi}\frac{1}{r\sin\theta}\frac{\partial U}{\partial \varphi} \right) \end{aligned} \tag{1.9}$$

Consequently, the torque $\mathbf{T}_d$ exerted on the dipole by the magnetic field is expressed as

$$\mathbf{T}_d = -\mathbf{r} \times \nabla U \tag{1.10}$$

Since U in equation (1.8) depends only on θ and $\hat{r} \times \hat{\theta} = \hat{\varphi}$, the torque is seen to be

$$\begin{aligned} \mathbf{T}_d &= -\hat{\varphi}(\partial U/\partial \theta) \\ &= \boldsymbol{\mu}^{\mathrm{T}} \times \mathbf{H} \end{aligned} \tag{1.11}$$

This torque is of course equal to the rate of change of angular momentum $\mathbf{J}$. Thus,

$$d\mathbf{J}/dt = \boldsymbol{\mu}^{\mathrm{T}} \times \mathbf{H} \tag{1.12}$$

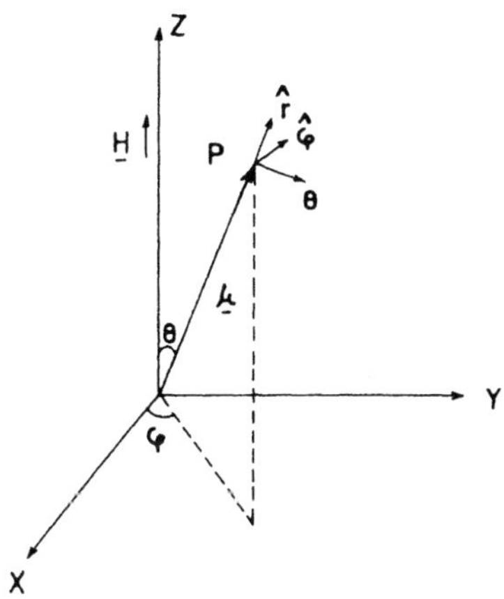

Figure 1.1. A magnetic dipole $\boldsymbol{\mu}$ in an external magnetic field $\mathbf{H}$.

Multiplying both sides by $-\gamma$ and using equation (1.5), we obtain

$$d\boldsymbol{\mu}^{\mathrm{T}}/dt = -\gamma\boldsymbol{\mu}^{\mathrm{T}} \times \mathbf{H} \tag{1.13}$$

We now generalize this equation for application to ferrimagnetics. If N_i is the number of dipoles per unit volume in the ith sublattice and if all the dipoles in a given sublattice are aligned parallel to one another, equation (1.13) is transformed into the following fundamental equation*:

$$d\mathbf{M}/dt = -\gamma\mathbf{M} \times \mathbf{H} \tag{1.14}$$

because $\mathbf{M} = \Sigma_i N_i \boldsymbol{\mu}_i^{\mathrm{T}}$. This equation describes the effect of magnetic field on magnetization. It is seen that the vector $d\mathbf{M}/dt$ is perpendicular to both $\mathbf{M}$ and $\mathbf{H}$. Moreover,

$$d(|\mathbf{M}|^2)/dt = 2\mathbf{M} \cdot d\mathbf{M}/dt = 0 \tag{1.15}$$

This implies that the magnitude of **M** remains constant in time. These considerations lead to the conclusion that **M** behaves like a rigid vector and executes clockwise (right-handed) precession about the direction of the magnetic field, as illustrated in Figure 1.2. The figure shows two states of the magnetization vector, at times t and $t + dt$, respectively. When the magnetic field is constant (H_0), the change in **M** in the interval dt can be written as (see Figure 1.2)

$$|d\mathbf{M}| = |\mathbf{M}| \sin\theta d\varphi$$

Thus

$$|d\mathbf{M}/dt| = |\mathbf{M}| \sin\theta\,(d\varphi/dt) = \gamma|\mathbf{M}|\mathbf{H}_0 \sin\theta$$

from which we have

$$\omega_0 = \gamma H_0 \tag{1.16}$$

where $\omega_0 = d\varphi/dt$ is the angular frequency of the precessional motion and H_0 is the magnitude of the dc magnetic field (biasing field). Evidently, ω_0 as given by equation (1.16) represents the natural precessional frequency of the magnetization vector in a constant magnetic field. This state of motion in which all magnetic dipoles precess in unison about the biasing field is

* Equation (1.14) does not incorporate several interactions present in a real ferrimagnetic material; these interactions will be discussed in Sections 1.5–1.9.

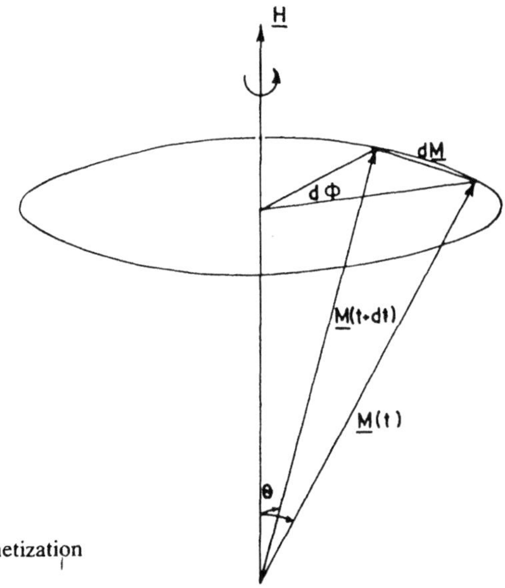

Figure 1.2. The states of the magnetization vector at times t and $t + dt$.

called the "uniform precessional mode." It also follows from equation (1.16) that since $\gamma = 1.76 \times 10^7$ rad/sec Oe, ω_0 is directly obtained in radians/second when H_0 is substituted in oersteds. However, in general, the frequency $f_0 = \omega_0/2\pi$ rather than the angular frequency ω_0 is required; in this case, we have $f_0 = \gamma' H_0$, where $\gamma' = \gamma/2\pi = 2.8 \times 10^6$ Hz/Oe.

Next, consider the situation in which a small time-varying (specifically, sinusoidal) magnetic field **h** is superimposed on $\mathbf{H}_0$. It is assumed that,as time passes, the tip of the vector **h** traces an ellipse about the biasing field direction. The dipoles experience an additional torque on account of **h**. As a consequence, the magnetization vector executes nutation (Goldstein, 1950), which means that an oscillatory motion is superimposed on the uniform precession (Figure 1.3). When the frequency f of the time-varying magnetic field **h** is equal to the precessional frequency f_0 and, in addition, the vector **h** follows the uniform precession (i.e., it executes right-handed circular rotation about the dc magnetic field), the amplitude of precession of the magnetization tends to grow, and efficient transfer of energy from the ac field to the system of dipoles is achieved. This is called the resonance condition. When the frequency and sense of rotation of **h** are not in consonance with the resonance condition, the interaction between the ac field and dipole system may not be very strong and, consequently, the energy transfer is not efficient.

We shall now solve equation (1.14) and obtain the permeability tensor in order to describe, quantitatively, the response of a biased ferrimagnet to a small magnetic excitation.

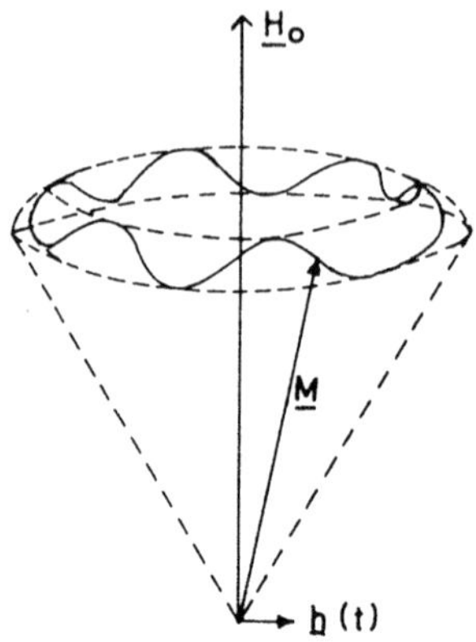

Figure 1.3. Nutation of magnetization **M** in the presence of a magnetic field consisting of a dc and an alternating component.

1.4. *The Permeability Tensor*

Let a ferrimagnetic sample be subjected to a uniform dc magnetic field H_0 acting along the z-axis. The field is assumed to be strong enough to drive the ferrimagnet into magnetic saturation. A small, time-harmonic magnetic field of amplitude **h** and frequency $f = \omega/2\pi$ is superimposed on the dc field. Therefore, the total magnetic field is given by

$$\mathbf{H} = H_0\hat{z} + \mathbf{h}\exp(j\omega t) \qquad (|\mathbf{h}| \ll H_0) \tag{1.17}$$

where $\hat{z}$ is a unit vector along the z-axis. When $\mathbf{h} = 0$, **M** equals the saturation magnetization M_0 and is directed along the z-axis. When **h** is sufficiently small, but not zero, the instantaneous magnetization can be expressed as*

$$\mathbf{M} = M_0\hat{z} + \mathbf{m}\exp(j\omega t) \tag{1.18}$$

$|\mathbf{m}| \ll M_0$ since $|\mathbf{h}| \ll H_0$. Substitution of **H** and **M** from equations (1.17) and (1.18) into equation (1.14) yields

$$j\omega\mathbf{m} = -\gamma M_0\hat{z} \times \mathbf{h} - \gamma H_0\mathbf{m} \times \hat{z} \tag{1.19}$$

In writing this equation, the term involving the cross product of **m** and **h** has been neglected under the linear, small-signal approximation. The

* In general, the expansion of **M** will involve all terms with time dependence $\exp(nj\omega t)$; $n = 0, 1, 2, \ldots$. However, the terms corresponding to $n \geq 2$ can be neglected under the linear, small-signal approximation.

vector equation (1.19) can be rewritten in terms of its components as follows:

$$\left.\begin{aligned} j(\omega/\gamma)m_x + H_0 m_y &= M_0 h_y \\ H_0 m_x - j(\omega/\gamma)m_y &= M_0 h_x \\ m_z &= 0 \end{aligned}\right\} \tag{1.20}$$

These equations can be solved for the components of **m** to obtain

$$\left.\begin{aligned} 4\pi m_x &= \chi h_x + j\kappa h_y \\ 4\pi m_y &= -j\kappa h_x + \chi h_y \\ 4\pi m_z &= 0 \end{aligned}\right\} \tag{1.21}$$

where

$$\left.\begin{aligned} \chi &= \omega_0\omega_m/(\omega_0^2 - \omega^2) \\ \kappa &= \omega\omega_m/(\omega_0^2 - \omega^2) \\ \omega_0 &= \gamma H_0, \qquad \omega_m = 4\pi M_0 \gamma \end{aligned}\right\} \tag{1.22}$$

Equation (1.21) can be expressed in matrix form as

$$\begin{pmatrix} 4\pi m_x \\ 4\pi m_y \\ 4\pi m_z \end{pmatrix} = \begin{pmatrix} \chi & j\kappa & 0 \\ -j\kappa & \chi & 0 \\ 0 & 0 & 0 \end{pmatrix} \begin{pmatrix} h_x \\ h_y \\ h_z \end{pmatrix} \tag{1.23}$$

The permeability tensor is defined through the relation

$$\mathbf{b} = \mathbf{h} + 4\pi\mathbf{m} \equiv \boldsymbol{\mu}\cdot\mathbf{h} \tag{1.24}$$

whence* (Polder, 1949: Lax and Button, 1962)

$$\boldsymbol{\mu} = \begin{pmatrix} \mu & j\kappa & 0 \\ -j\kappa & \mu & 0 \\ 0 & 0 & 1 \end{pmatrix} \tag{1.25}$$

where

$$\mu = 1 + \chi = [\omega_0(\omega_0 + \omega_m) - \omega^2]/(\omega_0^2 - \omega^2) \tag{1.26}$$

* Many authors prefer to define the permeability tensor in such a way that the signs of their off-diagonal elements are opposite to ours. This is due to the difference in sign between our definition of κ [cf. equation (1.22)] and theirs.

It is seen from equations (1.21) and (1.22) that, as expected, the ac magnetization exhibits a singularity when the resonance condition is satisfied, i.e., when $\omega = \omega_0$. Equation (1.21) also indicates that, under the linear approximation, the z-component of the magnetization is constant in time. Moreover, the medium does not respond to an impressed ac field h_z that is directed along the dc magnetic field. These facts have important consequences which will become evident in the subsequent chapters.

It is interesting to examine the behavior of the elements of the permeability tensor under the reversal of the direction of the biasing field. To do so equation (1.14) should be solved again after substituting for **H** and **M** as follows:

$$\left.\begin{aligned}\mathbf{H} &= -H_0\hat{z} + \mathbf{h}\exp(j\omega t)\\ \mathbf{M} &= -M_0\hat{z} + \mathbf{m}\exp(j\omega t)\end{aligned}\right\} \tag{1.27}$$

It is easy to show that the resulting permeability tensor has the same form as equation (1.25) except that the signs of the off-diagonal elements are now reversed.* It will be shown in the following chapters that the concept of biasing field reversal is related to nonreciprocal propagation effects.

In essence, equations (1.24) and (1.25) describe the response of ferrimagnetics to the external ac magnetic field which may be associated with a wave propagating through the medium. In subsequent chapters, we shall discuss the modification of waves (electromagnetic or elastic) propagating in magnetized ferrimagnetics and the application of the associated phenomena to devices. It is natural to expect that significant modification in the wave characteristics caused by the gyromagnetic nature of the medium will occur only when the permeability tensor is significantly different† from the unit tensor. It can be concluded from equations (1.22) and (1.26) that this is so when f is close to the resonance frequency $f_0 = \gamma' H_0$. For biasing fields of reasonable strength, the resonance frequency lies in the microwave range. Therefore, the gyromagnetic effects in ferrimagnetics (and hence their applications) are particularly important in this frequency range.

1.5. *Damping*

The discussion in Sections 1.3 and 1.4 shows that when the frequency of the impressed alternating magnetic field is equal to the gyrofrequency,

* It should be noted that the biasing field reversal can be effected not only by reversing the sign of κ but also by reversing the sign of ω.

† While the elements of the permeability tensor can be calculated directly using equations (1.22) and (1.26), a method for quick evaluation of the same has been developed by Gardiol (1972).

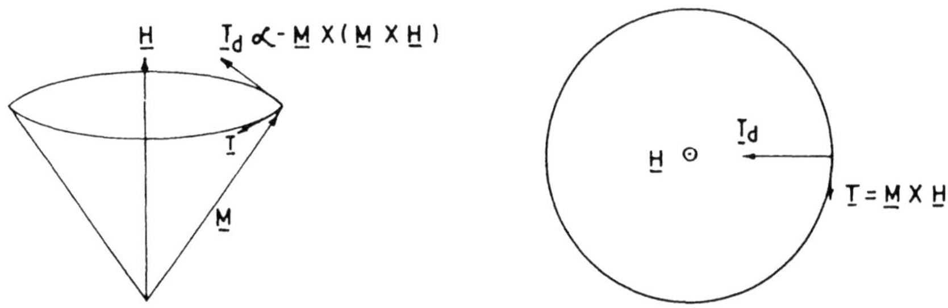

Figure 1.4. Damping torque $\mathbf{T}_d$ acting on the magnetization **M**.

the precessional amplitude grows to infinity. However, in practice, the precession is finite owing to various damping mechanisms, which results in microwave energy loss from the uniform precessional mode. We shall describe these losses phenomenologically because the precise microscopic description of damping is beyond the scope of this book. Several types of phenomenological models have been proposed. Perhaps the simplest and the most suitable one, for our purpose, is that credited to Landau and Lifshitz (1935). The rationale behind this model is as follows:

It was stated in Section 1.4 that under the action of the total magnetic field **H**, the magnetization vector **M** executes precessional motion about **H** on account of the torque $\mathbf{M} \times \mathbf{H}$, which is perpendicular to both **M** and **H**. The effect of damping is to limit the response of the magnetization to the exciting field. In other words, the increase in the precessional amplitude is not as large as in the absence of damping. This can be accounted for by introducing a torque term in equation (1.14), which causes the pulling back of the magnetization toward **H**, as shown in Figure 1.4. Therefore, the rate of change of **M** on account of damping should be perpendicular to $\mathbf{M} \times \mathbf{H}$ as well as to **M*** and be directed "toward" (not "along") **H**:

$$(d\mathbf{M}/dt)_{\text{damping}} = -\frac{\alpha\gamma}{|\mathbf{M}|}\mathbf{M} \times (\mathbf{M} \times \mathbf{H}) \tag{1.28}$$

where α is a dimensionless, positive, phenomenological parameter which describes the magnitude of the loss. When losses are small, the damping term is also small and can be rewritten using equation (1.14) as[†]

$$(d\mathbf{M}/dt)_{\text{damping}} \simeq (\alpha/|\mathbf{M}|)\mathbf{M} \times \partial\mathbf{M}/\partial t \tag{1.29}$$

* The reader interested in microscopic theory of damping is referred to Haas and Callen (1963) and Sparks (1964) among others.

† The Landau–Lifschitz damping term (1.28) was first expressed in this form by Gilbert (1955).

(a)

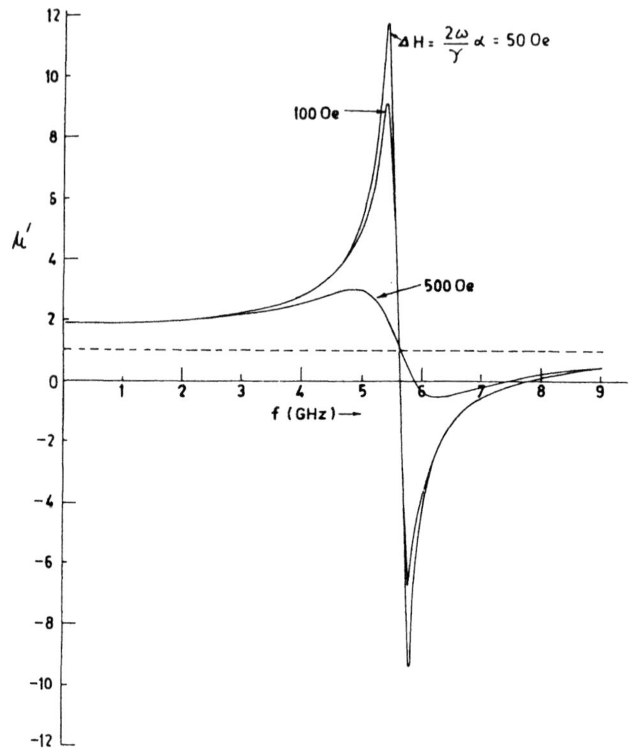

(c)

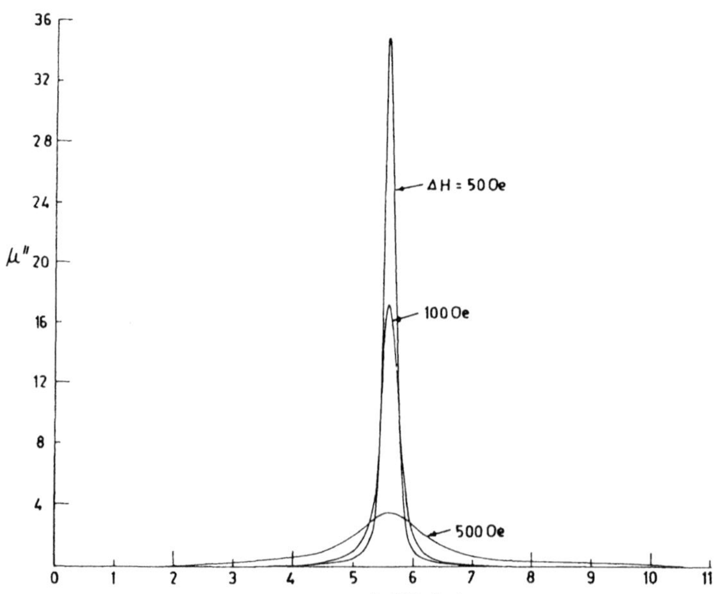

Figure 1.5. Variation of (a) μ', (b) κ', (c) μ'', and (d) κ'' with frequency $f = \omega/2\pi$ for different linewidths ΔH. Other parameters are $4\pi M_0 = 1750\,\text{G}$, $\gamma' = 2.8\,\text{GHz/kOe}$ and $H_0 = 2000\,\text{Oe}$.

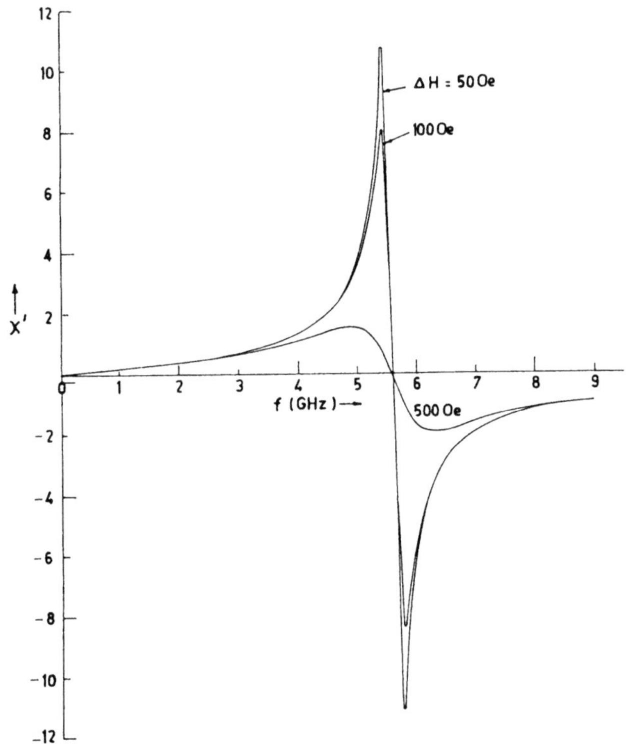

(b)

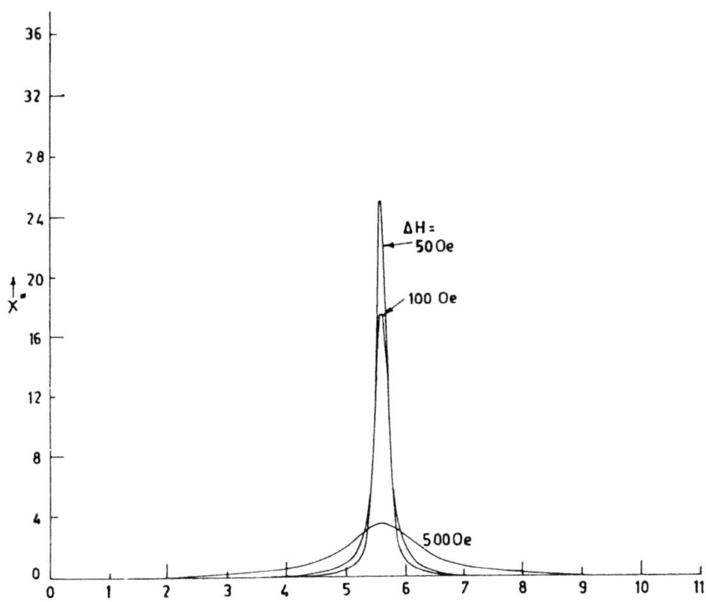

(d)

This term can be introduced on the right-hand side of equation (1.14). Then the development of Section 1.4 can be followed to obtain the modified permeability tensor. Under the linear, small-signal approximation, the damping term reduces to

$$(d\mathbf{M}/dt)_{\text{damping}} = j\omega\alpha\hat{z} \times \mathbf{m} = (j/T)\hat{z} \times \mathbf{m} \tag{1.30}$$

where $T = 1/\omega\alpha$ has the dimensions of time and is called the relaxation time. This term can be included in the right-hand side of equation (1.19), from which it is seen that it is sufficient to replace $\gamma H_0 = \omega_0$ by $\omega_0 + j/T$ in equation (1.22) in order to obtain the susceptibility tensor and hence the permeability tensor. The elements of this tensor are given by

$$\left.\begin{aligned} \mu &= 1 + \frac{(\omega_0 + j/T)\omega_m}{(\omega_0 + j/T)^2 - \omega^2} \equiv \mu' - j\mu'' \\ \kappa &\equiv \frac{\omega\omega_m}{(\omega_0 + j/T)^2 - \omega^2} \equiv \kappa' - j\kappa'' \end{aligned}\right\} \tag{1.31}$$

whence

$$\left.\begin{aligned} \mu' &= 1 + \omega_0\omega_m(\omega_0^2 + 1/T^2 - \omega^2)/D_0 \\ \mu'' &= (\omega_m/T)(\omega_0^2 + \omega^2 + 1/T^2)/D_0 \\ \kappa' &= \omega\omega_m(\omega_0^2 - \omega^2 - 1/T^2)/D_0 \\ \kappa'' &= 2\omega\omega_0\omega_m/TD_0 \\ D_0 &= (\omega_0^2 - \omega^2 - 1/T^2)^2 + 4\omega_0^2/T^2 \end{aligned}\right\} \tag{1.32}$$

The frequency dependence of the real and imaginary parts of μ and κ has been displayed in Figures 1.5 a–d for a typical set of parameters. When $\omega = \omega_0$, $\mu' - 1$ and κ' are very small while μ'' and κ'' exhibit resonance peaks, which is as expected. When α is small, the absorption peaks are stronger and narrower. As α increases, the peak height is reduced and the resonance region broadens out. The dependence of μ', μ'', κ', and κ'' on the dc biasing field, given ω and other parameters, has been displayed in Figures 1.6a–d. The general nature of the curves is in fact the same as in Figure 1.5; note that the behavior for small (large) ω is the same as that for large (small) H_0.

It is evident from equations (1.31) and (1.32) that a knowledge of relaxation time T is necessary to determine the elements of the permeability tensor. In the case of small losses, it is useful to introduce the concept of linewidth ΔH, which is defined as the width of the resonance curves at those points on the dc magnetic field scale (Figures 1.6b–d), where the magnitude of μ'' (or κ'') is half of its peak value.* It follows from equation (1.32) that the peak value of μ'' occurs at $\omega = \omega_0$ and is given by

$$(\mu'')_{\text{peak}} = \tfrac{1}{2}\omega_{\text{m}} T \tag{1.33}$$

where terms involving $1/T^2$ have been neglected. Obviously μ'' is half of its peak value at those values of H_0 which satisfy the condition

$$(\omega_{\text{m}}/T)(\omega_0^2 + \omega^2)/D_0 = \tfrac{1}{2}(\tfrac{1}{2}\omega_{\text{m}} T) \tag{1.34}$$

The approximate solution to this equation is seen to be

$$H_0 = \omega/\gamma \pm 1/\gamma T \tag{1.35}$$

The resonance linewidth is obtained as

$$\Delta H = 2/\gamma T = 2\omega\alpha/\gamma \tag{1.36}$$

Equivalently, the relaxation time T can be written in terms of the observed linewidth as

$$\left.\begin{aligned} T^{-1} &= \tfrac{1}{2}\gamma \Delta H = \tfrac{1}{2}\Delta\omega_{\text{h}} \\ \Delta\omega_{\text{h}} &= \gamma \Delta H \end{aligned}\right\} \tag{1.37}$$

which shows that the relaxation time can be inferred from the resonance absorption curve. Hence the linewidth is a convenient parameter for losswise characterization of the nature of ferrimagnetics. It is seen from equation (1.36) that ΔH is proportional to frequency. The microwave linewidths of ferrimagnetics in the X-band range from about 0.1 Oe in the case of well-prepared single crystals of yttrium-iron-garnet (YIG) to several hundred oersteds for polycrystalline spinels and hexagonal magnetoplumbites (see Appendix A). Thus, in the case when the linewidth is small, one can obtain meaningful results even after neglecting the losses. However, this neglect of losses may lead to conceptual inadequacies in some cases (Gardiol, 1967).

* The position of the peak (on dc magnetic field or frequency scale) in κ'' is the same as that in μ'' provided that the losses are small.

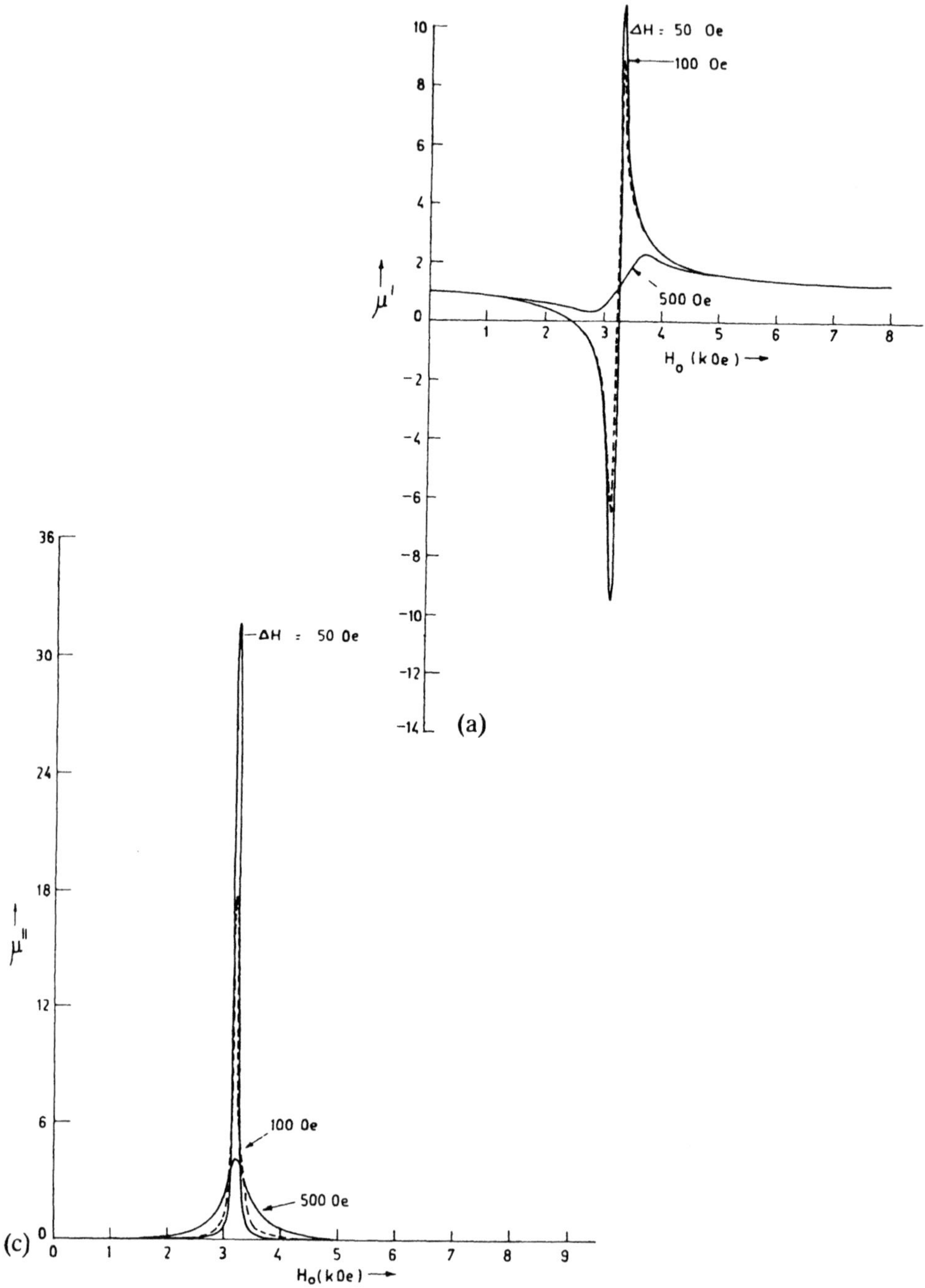

Figure 1.6. Variation of (a) μ', (b) κ', (c) μ'', and (d) κ'' with biasing field H_0 for different linewidths ΔH. Other parameters are the same as in Figure 1.5.

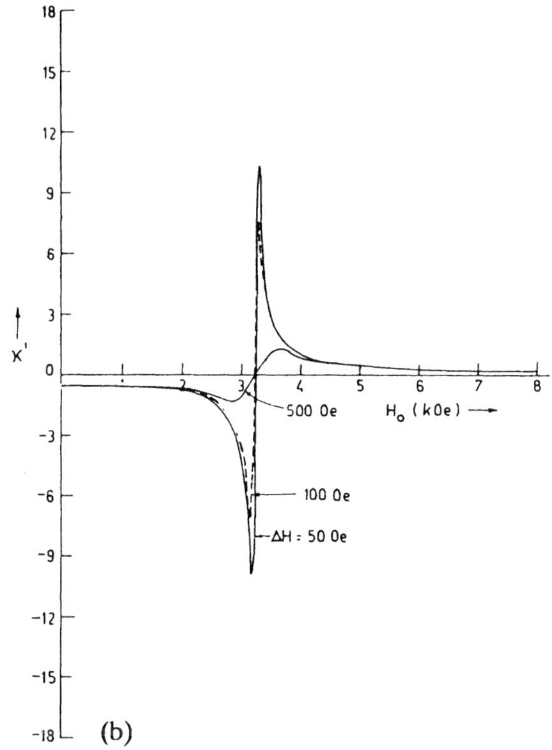

χ'
0
3
6
9
12
15
18
-3
-6
-9
-12
-15
-18
1
2
3
4
5
6
7
8
500 Oe
H_o (kOe)
100 Oe
ΔH = 50 Oe

(b)

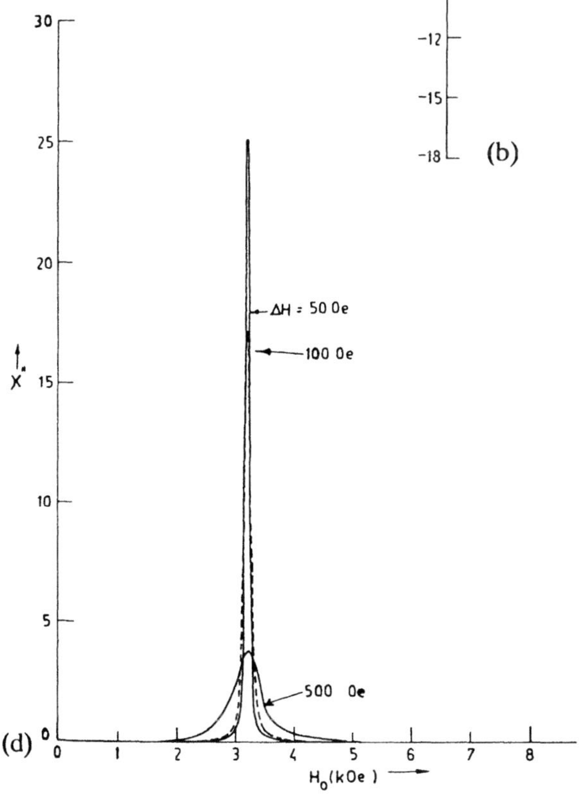

χ''
0
5
10
15
20
25
30
1
2
3
4
5
6
7
8
ΔH = 50 Oe
100 Oe
500 Oe
H_o(kOe)

(d)

1.6. Exchange Interaction

In a real crystal, the electron distribution of neighboring atoms overlap. When the atoms are characterized by nonzero spin, this overlap leads to a quantum mechanical spin-dependent electrostatic interaction, viz., the exchange interaction which, if appreciable, can orient the electron spins of the neighboring atoms either parallel or antiparallel. The exchange energy for two atoms i and j with spins $\mathbf{S}_i$ and $\mathbf{S}_j$, respectively, can be shown from quantum mechanical considerations to be given by*

$$U_{ex} = -2J_{ov}\mathbf{S}_i \cdot \mathbf{S}_j \tag{1.38}$$

where J_{ov} is called the overlap integral (Feynman, 1972) and is a measure of the extent of overlap of the electron distributions of atoms i and j. The overlap integral can be positive or negative, large or small. Our interest lies only in those cases where $|J_{ov}|$ is significantly large.† When J_{ov} is positive, the energy of the system is at a minimum when the spins are parallel; this leads to ferromagnetic order. However, when J_{ov} is negative, antiparallel alignment of successive spins is favored, which leads, in the simplest cases, to two oppositely magnetized interpenetrating sublattices, as shown in Figure 1.7. This situation is also encountered in the case of most ferrimagnetic oxides, where the metal ions (magnetic ions) are separated by negatively charged oxygen ions. The distance between the metal ions is therefore too large for appreciable‡ direct exchange interaction. In these cases, the magnetic order results from various complicated exchange mechanisms (indirect exchange interaction) in which the intermediate ions play a significant role and make effective J_{ov} negative.

When the successive spins are not parallel (antiparallel, for negative J_{ov}), the energy of the system (at absolute zero) will be higher than the minimum when all other interactions are ignored. Consequently, each spin experiences a torque which tends to orient it parallel to the neighboring spins. We will now derive§ an expression for this torque, called exchange torque.

* A simple derivation has been given by Feynman (1972).

† Incidently, even when the overlap integral is small, the magnetic order occurs at sufficiently low temperatures, where the thermal agitation does not dominate over the magnetic order. It is desirable, however, that microwave ferrimagnetics should exhibit magnetic order at room temperature, which is possible only when $|J_{ov}|$ is relatively large.

‡ The exchange interaction is a short-range interaction.

§ The present derivation is by no means rigorous; it has been carried out only to bring out the essential features of the exchange term. A rigorous derivation can be found, for example, in Akhiezer et al. (1968).

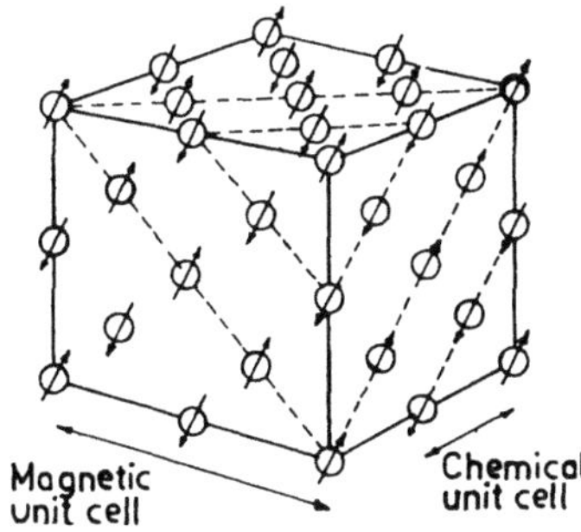

Figure 1.7. Oppositely magnetized interpenetrating sublattices (after Kittel, 1971).

The exchange energy given by equation (1.38) can be expressed in terms of the magnetic moment and magnetization as

$$\begin{aligned} U_{ex} &= -(2/\gamma)J_{ov}\boldsymbol{\mu}_i \cdot \boldsymbol{\mu}_j \\ &= -2J_{ov}(V^2/\gamma^2)\mathbf{M}_i \cdot \mathbf{M}_j \end{aligned} \tag{1.39}$$

where V is the average volume available per magnetic dipole and $\mathbf{M}_i$ $(\cdot \mathbf{M}_j)$ is the magnetization in the neighborhood of the ith (jth) spin. It is often convenient to interpret the spin order as resulting from an effective magnetic field (called the Weiss field) experienced by each dipole. This field may be inferred from equation (1.39) which may be rewritten as

$$U_{ex} = -\mathbf{M}_i \cdot \mathbf{H}_{ji}^{\omega} \tag{1.40}$$

where

$$\left.\begin{aligned} \mathbf{H}_{ji}^{\omega} &= \lambda\, \mathbf{M}_j \\ \lambda &= 2J_{ov}V^2/\gamma^2 \end{aligned}\right\} \tag{1.41}$$

It is seen from equation (1.40) that $\mathbf{H}_{ji}^{\omega}$ represents the component of the Weiss field* at the location of the ith spin due to the jth spin. This

* The net Weiss field experienced by the ith spin is obtained by vector summation of $\mathbf{H}_{ji}^{\omega}$ over all $j \neq i$. When the magnetization is uniform, the Weiss field is independent of the location of i.

magnetic field* exerts a torque $\mathbf{M}_i \times \mathbf{H}_{ji}^{\omega}$ on $\mathbf{M}_i$. Since the exchange interaction is operative only in a short range, it is sufficient to consider the nearest-neighbor interaction. For simplicity, it is assumed that nonuniformity in magnetization occurs only in one dimension, say, along the z-axis; the magnetization is uniform within any plane perpendicular to the z-axis. In this case, the exchange torque experienced by $\mathbf{M}_i$ is given by

$$\begin{aligned} \mathbf{T}^{ex} &= \mathbf{M}_i \times (\mathbf{H}_{i+1,i}^{\omega} + \mathbf{H}_{i-1,i}^{\omega}) \\ &= \lambda \mathbf{M}_i \times (\mathbf{M}_{i+1} + \mathbf{M}_{i-1}) \end{aligned} \tag{1.42}$$

If the magnetic field is directed along the z-axis, we can write

$$\left.\begin{aligned} \mathbf{M}_i &= M_0 \hat{z} + \mathbf{m}_i \\ \mathbf{M}_{i\pm 1} &= M_0 \hat{z} + \mathbf{m}_{i\pm 1} \end{aligned}\right\} \tag{1.43}$$

Moreover,

$$\mathbf{m}_{i\pm 1} = \mathbf{m}_i + \frac{\partial \mathbf{m}_i}{\partial z} a_1 + \frac{1}{2}\frac{\partial^2 \mathbf{m}_i}{\partial z^2} a_1^2 + \cdots \tag{1.44}$$

where a_1 is the interdipole spacing. The substitution from equations (1.43) and (1.44) into equation (1.42) results in

$$\begin{aligned} \mathbf{T}^{ex} &= \lambda \left[(M_0 \hat{z} + \mathbf{m}_i) \times \left(2M_0 \hat{z} + 2\mathbf{m}_i + a_1^2 \frac{\partial^2 \mathbf{m}_i}{\partial z^2} + \cdots \right)\right] \\ &= H_i^{\omega} a_1^2 \hat{z} \times \frac{\partial^2 \mathbf{m}_i}{\partial z^2} + \cdots \end{aligned} \tag{1.45}$$

This suggests that the general exchange term for the three-dimensional case can be written as

$$\mathbf{T}^{ex} = \frac{H^{\omega} a_1^2}{|\mathbf{M}|} \mathbf{M} \times \nabla^2 \mathbf{M} \tag{1.46}$$

It should be recognized that this form of the exchange term is a consequence of the inversion symmetry, owing to which the first-order deriva-

* It should be pointed out that the concept of the Weiss field is introduced only for convenience. The Weiss field is not a real magnetic field and hence it does not directly enter Maxwell's equations. For example, no dc current, as defined by $\nabla \times \mathbf{H}^{\omega} = \mathbf{J}_c$, flows in order to account for the nonuniformity of the Weiss field in those regions of space where the magnetization is nonuniform and steady.

tives cancel out in equation (1.45). Most of the ferrimagnetics satisfy* this condition and, therefore, equation (1.46) is widely applicable. It is also worth noting that the terms involving the fourth and still higher-order derivatives have been neglected in equation (1.45). Consequently, the characteristic distance of variation (i.e., the wavelength) should be much larger than interspin spacing in order that equation (1.42) may be valid.

The exchange torque given by equation (1.46) should be added to the right-hand side of equation (1.14) so as to make the latter applicable in situations where magnetization is spatially nonuniform (spin wave). The explicit effect of exchange interaction on a permeability tensor can be examined by studying the variation of the oscillating part of the magnetization $\mathbf{m}$ as $\exp j(\omega t - \mathbf{k}\cdot\mathbf{r})$, where $\mathbf{k} = \beta\hat{n}$. The linear term in the exchange torque is seen to be

$$\begin{aligned} \mathbf{T}^{ex} &= H^{\omega} a_1^2(-\beta^2)\hat{z} \times \mathbf{m} \\ &= H^{\omega} a_1^2 \beta^2 \mathbf{m} \times \hat{z} \end{aligned} \tag{1.47}$$

This term can be introduced on the right-hand side of equation ((1.19), the solution to which reveals that the permeability tensor has still the same form as in equation (1.25). The only modification on account of the exchange term is that, in the expressions for μ and κ in equations (1.22) and (1.26), ω_0 should be replaced by ω_r, where

$$\left.\begin{aligned} \omega_r &= \omega_0 + \omega_{ex} \\ \omega_{ex} &= \gamma D_{ex} \beta^2 \\ D_{ex} &= H^{\omega} a_1^2 \end{aligned}\right\} \tag{1.48}$$

This shows that the effect of the exchange interaction is significant only when ω_{ex} is comparable to ω_0.

The Weiss field can be estimated from the condition that the magnetic order is disrupted when the exchange energy for a dipole is equal to the thermal energy at the critical temperature, i.e.,

$$\mu_B H^{\omega} \simeq k_B T_C \tag{1.49}$$

where k_B is the Boltzmann constant, equal to 1.38×10^{-16} erg/deg K. Since T_C, the critical temperature, is about 1000 °K for a typical ferrimagnetic material, we have $H^{\omega} \simeq 10^7$ Oe. Since the interatomic spacing a_1 is about 10^{-8} cm, it follows from equation (1.48) that the effect of exchange will be important when $\beta \geqslant 10^5\ \text{cm}^{-1}$.

* There are few exceptions, such as piezomagnetic materials; see for example, Berlincourt, Curran, and Jaffe (1964).

1.7. Magnetocrystalline Anisotropy*

It is observed that in the absence of an external magnetic field, the magnetic dipoles within a typical domain are oriented along certain preferential crystallographic directions or planes. This tendency arises from the interaction of magnetic ions with the crystalline electric field.† The corresponding interaction energy is called magnetocrystalline energy, which is dependent on the orientation of the magnetization with respect to the crystallographic directions. The microscopic origin of the magnetocrystalline anisotropy has been discussed in the past by several authors and most recently by Darby and Isaac (1974). Instead of getting involved in this discussion, we shall adopt a phenomenological (macroscopic) approach because of its simplicity and usefulness in the present context. The starting point of the phenomenological theory is the fact that the expression for the anisotropy energy should have the characteristic symmetry of the crystal structure of the material in question. The microwave ferrimagnetics are mostly cubic (spinels and garnets) and, less commonly, hexagonal (magnetoplumbites). First we take up the case of the cubic ferrimagnetics.

The anisotropy energy U_a depends on the direction cosines α_i ($i = 1, 2, 3$) of the magnetization vector relative to the crystal axes $\langle 100 \rangle$.‡ Thus, α_i are the components of the unit vector $\mathbf{M}/|\mathbf{M}|$. The three orthogonal directions being magnetically equivalent, all terms in the expansion of U_a should be invariant under an interchange among all α_i. Moreover, only even powers of α_i should be involved in the expansion because the opposite ends of any crystal axis are equivalent. Thus, in the case when the coordinate axes are along the cube edges, the anisotropy energy can be expressed as

$$U_a = k_0(\alpha_1^2 + \alpha_2^2 + \alpha_3^2) + k_1(\alpha_1^2\alpha_2^2 + \alpha_2^2\alpha_3^2 + \alpha_3^2\alpha_1^2) + k_2\alpha_1^2\alpha_2^2\alpha_3^2 + \cdots \quad (1.50)$$

Since $\boldsymbol{\alpha}$ is a unit vector, we have $\Sigma_i \alpha_i^2 = 1$. Since the addition or subtraction of a constant to U_a does not alter the physical consequences, equation (1.50) can be written as

$$U_a = k_1(\alpha_1^2\alpha_2^2 + \alpha_2^2\alpha_3^2 + \alpha_3^2\alpha_1^2) + k_2\alpha_1^2\alpha_2^2\alpha_3^2 + \cdots \quad (1.51)$$

* Magnetocrystalline anisotropy should be distinguished from an isotropy associated with wave propagation. A magnetized ferrite (with no magnetocrystalline anisotropy) exhibits anisotropic wave propagation; the wave velocity is different for propagation in different directions relative to the biasing field. However, in this book, the term "anisotropy" invariably refers to anisotropy on acount of the crystalline nature.

† The electric field experienced by a magnetic cation due to the surrounding anions.

‡ See Kittel (1971) for an introduction to the representation of crystallographic axes and planes by Miller indices.

where k_1 and k_2 are phenomenological parameters called the first- and second-order anisotropy constants, respectively. The anisotropy constants can be positive or negative and can be inferred from resonance experiments (von Aulock, 1965).

The case of hexagonal magnetoplumbites can be treated in a similar manner. It is seen that on retaining only the dominant terms, the anisotropy energy can be expressed as*

$$U_a = k_1 \sin^2\theta + k_2 \sin^4\theta \tag{1.52}$$

where θ is the angle between the magnetization vector and the hexagonal unique axis (the c-axis, as it is called). Usually, $k_1 > k_2$ and hence the gross behavior of the hexagonal magnetoplumbite is governed mainly by the value of k_1, which is different for different materials and can be positive or negative. When k_1 is positive, U_a is minimum when $\theta = 0$. Therefore, in equilibrium, the magnetization vector tends to align itself along the c-axis (easy axis†); such materials are called uniaxial. On the other hand, k_1 is negative for certain other materials and, in equilibrium, the energy is minimum when $\theta = 90°$. Hence, the dc magnetization in such materials is confined to the plane (easy plane) which is perpendicular to the c-axis (hard axis); these materials are called planar magnetoplumbites or ferroxplanar.

Since the effect of magnetocrystalline anisotropy is to constrain the magnetization to orient itself along preferential crystallographic axes or planes, the anisotropy energy can be looked upon as a result of the interaction of the magnetization with an equivalent magnetic field called the anisotropy field. Like the exchange field discussed in Section 1.6, and unlike the dipolar field to be discussed in Section 1.8, the anisotropy field is not a real magnetic field. Hence, it is not included in Maxwell's equations. The anisotropy field $\mathbf{H}_A$ is defined through the relation

$$\begin{aligned} U_a &\equiv -\mathbf{M}\cdot\mathbf{H}_A \\ &= -|\mathbf{M}|\,\alpha\cdot\mathbf{H}_A \\ &\simeq M_0\boldsymbol{\alpha}\cdot\mathbf{H}_A \end{aligned} \tag{1.53}$$

* It is not difficult to show that the azimuthal angle φ enters the expression for U_a only in the sixth-order term, on account of hexagonal symmetry. The sixth- and higher-order terms are, in general, very small and, therefore, have been neglected in writing equations (1.52).

† This nomenclature is adopted because a smaller dc magnetic field is required to magnetically saturate the sample along this direction.

where $\boldsymbol{\alpha} = \alpha_1\hat{x} + \alpha_2\hat{y} + \alpha_3\hat{z}$. The anisotropy field $\mathbf{H}_A$ is obtained from this relation as

$$\mathbf{H}_A = -(1/M_0)\nabla_{\alpha}U_a \tag{1.54}$$

where $\nabla_{\alpha}U_a$ represents the gradient of U_a with respect to the components of the vector $\boldsymbol{\alpha}$. We shall now examine the explicit effect of anisotropy on the elements of the permeability tensor in the cases of the cubic and hexagonal materials, respectively.

1.7.1. *Cubic Ferrimagnetics*

The cubic anisotropy and its influence on microwave behavior of ferrimagnetics have been investigated by Kittel (1948), Suhl (1955), Artmann (1956), Schneider (1972), and Vittoria et al. (1973, 1974), among others. The present treatment closely follows Vittoria et al. (1974), but is more general and brings out the physics a little better.

The second- and higher-order anisotropy constants will be neglected since these are known to be very small for most of the cubic ferrimagnetics. As such, the anisotropy energy given by equation (1.51) reduces to

$$U_a = k_1(\alpha_1^2\alpha_2^2 + \alpha_2^2\alpha_3^2 + \alpha_3^2\alpha_1^2) \tag{1.55}$$

The anisotropy field can be deduced from equation (1.54) as

$$\mathbf{H}_A = -\frac{2k_1}{M_0}\left(\sum_i \alpha_i\hat{x}_i - \sum_i \alpha_i^3\hat{x}_i\right) \tag{1.56}$$

where $\hat{x}_1 = \hat{x}$, $\hat{x}_2 = \hat{y}$, and $\hat{x}_3 = \hat{z}$. If the dc field is directed along the z-axis, α_3 is close to unity, while, in linear theory, $\alpha_1^2, \alpha_1^3, \ldots$ are negligible. Equation (1.56) then reduces to

$$\mathbf{H}_A = -(2k_1/M_0)(\alpha_1\hat{x}_1 + \alpha_2\hat{x}_2) \tag{1.57}$$

This term can be added to $\mathbf{H}$ in equation (1.17) and the subsequent procedure described in Section 1.4 can be followed in order to obtain the modified permeability tensor. However, in general, the dc magnetic field may be oriented along an arbitrary direction relative to the cube edges in which case the dc magnetization will also be arbitrarily oriented (though not necessarily coincident with the dc magnetic field). In such a case, the cubic anisotropy can be more conveniently described by making a coordinate transformation to a new (primed) coordinate system, the z'-axis of

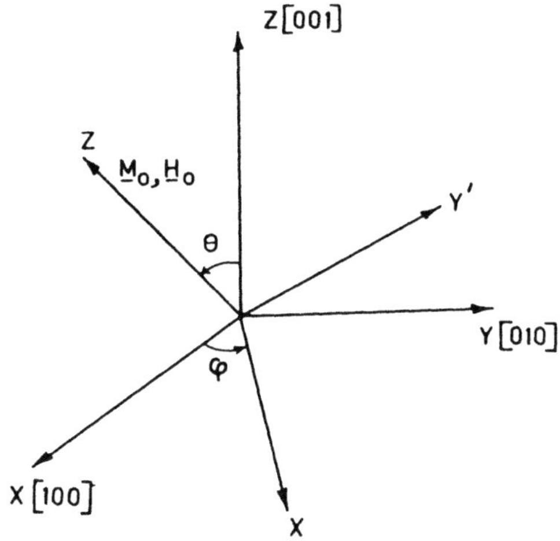

Figure 1.8. Orientation of primed coordinate axes relative to the unprimed ones.

which should be oriented along dc magnetization.* Two independent rotations are required to effect this transformation. It is convenient to consider the rotations through the first two Eulerian angles φ and θ as shown in Figure 1.8; this transformation is similar to the one discussed in detail in Section C.1 of Appendix C. It is further shown in Section C.4 of the same appendix that, under the aforementioned rotations, equation (1.56) transforms to

$$\mathbf{H}_A = -\frac{2k_1}{M_0}\left(\sum_i \alpha_i' \hat{x}_i' - \mathbf{P}\right) \tag{1.58}$$

where $\hat{x}_i'$ and $\hat{\alpha}_i'$ are, respectively, the unit vectors and the components of $\boldsymbol{\alpha}$ in the primed coordinate system. Further, it is shown in Section C.4 that, under the linear approximation, $\mathbf{P}$ is given by

$$\begin{aligned}\mathbf{P} = & -\tfrac{1}{4}\sin^3\theta \sin 4\varphi(\hat{x}_1' + 3\alpha_1'\hat{x}_3') + \tfrac{3}{2}\sin^2\theta \sin^2 2\varphi\, \alpha_1'\hat{x}_1' \\ & + \tfrac{3}{8}\sin\theta \sin 2\theta \sin 4\varphi(\alpha_1'\hat{x}_2' + \alpha_2'\hat{x}_1') + [\cos^3\theta \sin\theta \\ & - \sin^3\theta \cos\theta\,(\sin^4\varphi + \cos^4\varphi)](\hat{x}_2' + 3\alpha_2'\hat{x}_3') \\ & + 3\sin^2\theta \cos^2\theta\,(1 + \sin^4\varphi + \cos^4\varphi)\alpha_2'\hat{x}_2' \\ & + [\cos^4\theta + (\sin^4\varphi + \cos^4\varphi)\sin^4\theta]\hat{x}_3'\end{aligned} \tag{1.59}$$

* The dc magnetization will be parallel to an "effective" dc magnetic field.

The vector **P** can be split into dc and time-varying components as

$$\begin{aligned}\mathbf{P}_{dc} = &-\tfrac{1}{4}\sin^3\theta\sin 4\varphi\hat{x}_1' \\ &+[\cos^3\theta\sin\theta - \sin^3\theta\cos\theta(\sin^4\varphi+\cos^4\varphi)]\hat{x}_2' \\ &+[\cos^4\theta+(\sin^4\varphi+\cos^4\varphi)\sin^4\theta]\hat{x}_3' \end{aligned} \quad (1.60)$$

and

$$\begin{aligned}\mathbf{P}_{ac} = &\tfrac{3}{2}\sin^2\theta(\sin^2 2\varphi\alpha_1' + \tfrac{1}{2}\cos\theta\sin 4\varphi\alpha_2')\hat{x}_1' \\ &+\tfrac{3}{2}\sin^2\theta\cos\theta[\tfrac{1}{4}\sin 4\varphi\alpha_1' + (1+\sin^4\varphi+\cos^4\varphi)\cos\theta\alpha_2']\hat{x}_2' \\ &+[\{3\cos^3\theta\sin\theta - 3\sin^3\theta\cos\theta(\sin^4\varphi+\cos^4\varphi)\}\alpha_2' \\ &-\tfrac{3}{4}\alpha_1'\sin^3\theta\sin 4\varphi]\hat{x}_3' \end{aligned} \quad (1.61)$$

Hence, the anisotropy field can be decomposed into static* and time-varying components as

$$\mathbf{H}_A = \mathbf{H}_{A0} + \mathbf{h}_A \quad (1.62)$$

where

$$\mathbf{H}_{A0} = -\frac{2k_1}{M_0}\hat{x}_3' + \frac{2k_1}{M_0}\mathbf{P}_{dc} \quad (1.63)$$

and

$$\mathbf{h}_A = -\frac{2k_1}{M_0}(\alpha_1'\hat{x}_1' + \alpha_2'\hat{x}_2') + \frac{2k_1}{M_0}\mathbf{P}_{ac} \quad (1.64)$$

If $\mathbf{H}_i$ is the internal dc field† due to external sources, the effective dc field $\mathbf{H}_0$ which causes dc magnetization along the x_3'-axis (z'-axis) is given by

$$\begin{aligned}\mathbf{H}_0 &= H_0\hat{x}_3' = \mathbf{H}_i + \mathbf{H}_{A0} \\ &= \mathbf{H}_i - \frac{2k_1}{M_0}\hat{x}_3' + \frac{2k_1}{M_0}\mathbf{P}_{dc}\end{aligned} \quad (1.65)$$

* It will be seen later that, unlike the exchange field, the dc part of the anisotropy field leads to a nonvanishing torque experienced by magnetization.

† When a finite sample is subjected to a uniform external field the macroscopic field inside the sample is, in general, different from the external field; this will be discussed in detail in Section 1.8.

Thus, in order to cause an effective dc field H_0 along $\hat{x}'_3$, the required internal field, on account of external sources, should be equal to

$$\mathbf{H}_i = (H_0 + 2k_1/M_0)\hat{x}'_3 - (2k_1/M_0)\mathbf{P}_{dc} \tag{1.66}$$

In general, $\mathbf{H}_0$ and $\mathbf{H}_i$ are not parallel for arbitrary orientation of $\hat{x}'_3$. However, $\mathbf{H}_0$ is parallel to $\mathbf{H}_i$ if

$$\sin^3\theta \sin 4\varphi = 0 \tag{1.67}$$

and

$$\cos^3\theta \sin\theta - \sin^3\theta \cos\theta\,(\sin^4\varphi + \cos^4\varphi) = 0 \tag{1.68}$$

It is seen that equations (1.67) and (1.68) are simultaneously satisfied for (i) $\theta = 0$, (ii) $\varphi = 0$, $\theta = \pi/4$ and for (iii) $\varphi = \pi/4$, $\theta = \arccos(1/\sqrt{3})$. We consider these cases one by one.

1. $\theta = 0$; magnetization along the [100]-axis and its equivalents (cube edges).

It follows from equation (1.60) that $\mathbf{P}_{dc} = \hat{x}'_3$. Thus, equation (1.66) implies that

$$H_0 = H_i \tag{1.69}$$

i.e., the cubic anisotropy does not contribute to the dc field if the dc magnetization is oriented along a cube edge.

2. $\varphi = 0$, $\theta = \pi/4$; dc magnetization along the [110]-axis and its equivalents (face diagonals).

It follows from equation (1.60) that $\mathbf{P}_{dc} = \frac{1}{2}\hat{x}'_3$. Hence, it is seen from equation (1.66) that

$$H_0 = H_i - k_1/M_0 \tag{1.70}$$

3. $\varphi = \pi/4$, $\theta = \arccos(1/\sqrt{3})$; dc magnetization along [111]-axis and its equivalents (body diagonals).

It is seen from equation (1.60) that $\mathbf{P}_{dc} = \frac{1}{3}\hat{x}'_3$. It follows from equation (1.66) that

$$H_0 = H_i - 4k_1/3M_0 \tag{1.71}$$

For a material with positive k_1, the effective dc field is evidently smaller than the applied field for any direction of magnetization other than the [100]-axis. As such, a larger dc magnetic field will be required to magnetically saturate the medium along any direction other than the [100]-axis. Hence the [100]-axis is an "easy axis" while the [111]-axis is a "hard axis." However, when k_1 is negative, the effective dc field is larger than the applied field for any axis other than the [100]-axis. In this case, the [111]-axis is an easy axis while the [100]-axis is a hard axis.

Now we shall consider the time-dependent component of the anisotropy field, which can be expressed with the help of equations (1.61) and (1.64) as

$$\mathbf{h}_{\mathrm{A}} = h_{\mathrm{A}x}\hat{x}_1' + h_{\mathrm{A}y}\hat{x}_2' + h_{\mathrm{A}z}\hat{x}_3' \tag{1.72}$$

where

$$\left.\begin{aligned} h_{\mathrm{A}x} &= p_x m_x' + q_x m_y' \\ h_{\mathrm{A}y} &= p_y m_x' + q_y m_y' \\ h_{\mathrm{A}z} &= p_z m_x' + q_z m_y' \end{aligned}\right\} \tag{1.73}$$

and

$$\left.\begin{aligned} M_0 p_x &= \frac{2k_1}{M_0}\left(\tfrac{3}{2}\sin^2\theta\sin^2 2\varphi - 1\right) \\ M_0 p_y &= \frac{2k_1}{M_0}\cdot\tfrac{3}{4}\sin^2\theta\cos\theta\sin 4\varphi \\ M_0 p_z &= -\frac{2k_1}{M_0}\cdot\tfrac{3}{4}\sin^3\theta\sin 4\varphi \\ M_0 q_x &= M_0 p_y \\ M_0 q_y &= \frac{2k_1}{M_0}\left[3\sin^2\theta\cos^2\theta(1+\sin^4\varphi+\cos^4\varphi) - 1\right] \\ M_0 q_z &= \frac{2k_1}{M_0}\left[3\cos^3\theta\sin\theta - 3\sin^3\theta\cos\theta\,(\sin^4\varphi+\cos^4\varphi)\right] \end{aligned}\right\} \tag{1.74}$$

Now we can solve equation (1.14) after substituting for **H** and **M** as follows:

$$\mathbf{H} = H_0\hat{x}_3' + \mathbf{h} + \mathbf{h}_{\mathrm{A}}$$

and

$$\mathbf{M} = M_0\hat{x}_3' + \mathbf{m}$$

Following the procedure described in Section 1.4, the permeability tensor is obtained as*

$$\boldsymbol{\mu} = \begin{pmatrix} \mu_{11} & j\kappa & 0 \\ -j\kappa & \mu_{22} & 0 \\ 0 & 0 & 1 \end{pmatrix} \tag{1.75}$$

where

$$\left.\begin{aligned} \mu_{11} &= 1 + \omega_m(\omega_0 - \gamma M_0 q_y)/D_1 \\ \mu_{22} &= 1 + \omega_m(\omega_0 - \gamma M_0 p_x)/D_1 \\ \kappa &= \omega_m(\omega - j\gamma M_0 q_x)/D_1 \\ D_1 &= (\omega_0 - \gamma M_0 p_x)(\omega_0 - \gamma M_0 q_y) - \gamma^2 M_0^2 q_x^2 - \omega^2 \end{aligned}\right\} \tag{1.76}$$

The resonance frequency is obtained from equation (1.76) as

$$\omega_{res} = [(\omega_0 - \gamma M_0 p_x)(\omega_0 - \gamma M_0 q_y) - \gamma^2 M_0^2 q_x^2]^{1/2} \tag{1.77}$$

which, in general, depends on θ and φ. Specifically, for magnetization within the (100)-planes, we have $\theta = (2n+1)\pi/2$ and the resonance frequency is seen to be

$$\omega_{res} = \gamma\left\{(H_0 + 2k_1/M_0)\left[H_0 + \frac{2k_1}{M_0}(1 - \tfrac{3}{2}\sin^2 2\varphi)\right]\right\}^{1/2} \tag{1.78}$$

It should be noted that H_0 is the effective dc field and is related to the internal dc field H_i as in equation (1.65). In particular, when dc magnetization is along the [100]-axis, we have $\sin 2\varphi = 0$ and $H_0 = H_i$. The resonance frequency thus reduces to

$$\omega_{res} = \gamma(H_i + 2k_1/M_0)$$

When the magnetization is along an arbitrary axis within a (110)-plane, we have $\varphi = (2n+1)\pi/4$ and ω_{res}, given by equation (1.77), reduces to

$$\omega_{res} = \gamma\left\{\left[H_0 + \frac{2k_1}{M_0}(1 - \tfrac{3}{2}\sin^2\theta)\right]\left[H_0 + \frac{2k_1}{M_0}(1 - \tfrac{9}{8}\sin^2 2\theta)\right]\right\}^{1/2} \tag{1.79}$$

* This permeability tensor ignores losses and exchange effects; to take losses into account, ω_0 in equation (1.76) should be replaced by $(\omega_0 + j/T)$ whereas to take exchange effects into consideration ω_0 should be replaced by ω, given by equation (1.48).

In the special case of magnetization along the [110]-axis, we have $\theta = \pi/2$, while H_0 is given by equation (1.70). The resonance frequency (1.79) reduces to

$$\omega_{res} = \gamma[(H_i - 2k_1/M_0)(H_i + k_1/M_0)]^{1/2}$$

The resonance frequency for the case of magnetization along the [111]-axis also follows from equation (1.79) with $\cos^2\theta = \frac{1}{3}$ and H_0 as given by equation (1.71):

$$\omega_{res} = \gamma(H_i - 4k_1/3M_0)$$

1.7.2. *Hexagonal Magnetoplumbites*

The magnetization of hexagonal magnetoplumbites is, in general, very tightly bound to the easy axis (uniaxial case) or to the easy plane (planar case). It follows that extremely large dc magnetic fields may be required to magnetize the sample away from the easy axis or easy plane. Hence, in most cases, a uniaxial material is magnetized* along the c-axis while a planar material is magnetized along an arbitrary direction within the easy plane. The magnetocrystalline anisotropy energy of a hexagonal magnetoplumbite is given by equation (1.52) and can be rewritten as

$$U_a = k_1(1 - \alpha_y^2) + k_2(1 - \alpha_y^2)^2 \tag{1.80}$$

where $\alpha_y = \cos\theta$; the relevant coordinate system and the angle θ are shown in Figure 1.9. The anisotropy field is obtained from equations (1.54) and (1.80) as

$$\mathbf{H}_A = \left[\frac{2k_1}{M_0} + \frac{4k_2}{M_0}(1 - \alpha_y^2)\right]\alpha_y\hat{y} \tag{1.81}$$

In what follows we discuss the permeability tensors of uniaxial and planar magnetoplumbites separately.

1.7.2.1. *Uniaxial Magnetoplumbite*

The dc field is applied along the c-axis which is the y-axis. It follows from Figure 1.9 that $\theta \simeq 0$. Consequently, α_y can be replaced by unity. The

* As discussed in Appendix A, an oriented uniaxial magnetoplumbite is magnetically saturated even in the absence of the external bias while, for magnetic saturation, an oriented planar magnetoplumbite does require a small dc field, acting along an arbitrary direction within the easy plane.

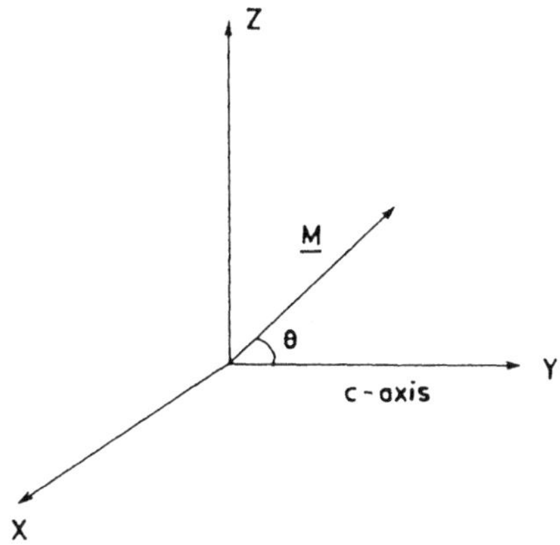

Figure 1.9. The hexagonal c-axis and magnetization.

anisotropy field, given by equation (1.81), reduces to

$$\left.\begin{aligned} \mathbf{H}_A &= H_a^u \hat{y} \\ H_a^u &= 2k_1/M_0 \end{aligned}\right\} \tag{1.82}$$

It is evident that the only effect of anisotropy is to add the term H_a^u to the applied field; H_a^u acts as a built-in magnetic field. Therefore, the permeability tensor for a uniaxial magnetoplumbite magnetized along the c-axis is the same as that for an isotropic ferrimagnet except that H_0 is replaced by $(H_0 + H_a^u)$. The resonance condition for a uniaxial magnetoplumbite is then given by

$$\omega_{res}^u = \gamma(H_0 + H_a^u) \equiv (\omega_0 + \omega_a^u) \tag{1.83}$$

where

$$\omega_a^u = \gamma H_a^u$$

It follows that if H_a^u is sufficiently large (which is indeed so for many uniaxial magnetoplumbites; Appendix A), the resonance can be realized at extremely high frequencies even when the applied field is very small or zero. This fact makes uniaxial magnetoplumbites suitable for devices operated in the millimeter–wavelength region. A drawback with these materials, however, is that their resonance linewidths are generally much broader than those of garnets and spinels.

1.7.2.2. *Planar Magnetoplumbite*

In this case the external dc field is applied along an arbitrary direction (say the z-axis) within the easy (xz) plane. As such, θ is very close to 90°

and α_y is very small. Neglecting the third and higher powers of α_y, the anisotropy field given by equation (1.81) reduces to

$$\mathbf{H}_A = -H_a \alpha_y \hat{y} \tag{1.84}$$

where

$$H_a = -(2k_1 + 4k_2)/M_0 \tag{1.85}$$

and

$$\alpha_y = \cos\theta = m_y/M_0 \tag{1.86}$$

It follows from equations (1.84)–(1.86) that the magnetization experiences an additional torque $M_0\hat{z} \times \mathbf{H}_A = H_a m_y \hat{x}$. Hence, we should introduce the term $-\gamma H_a m_y \hat{x}$ in the right-hand side of equation (1.19) and adopt a procedure analogous to that in Section 1.4. The permeability tensor assumes the form (Bady, 1961)

$$\boldsymbol{\mu} = \begin{pmatrix} \mu_{xx} & j\kappa & 0 \\ -j\kappa & \mu_{yy} & 0 \\ 0 & 0 & 1 \end{pmatrix} \tag{1.87}$$

where

$$\left.\begin{aligned} \mu_{xx} &= [(\omega_0 + \omega_a)(\omega_0 + \omega_m) - \omega^2]/D_2 \\ \mu_{yy} &= [\omega_0(\omega_0 + \omega_a + \omega_m) - \omega^2]/D_2 \\ \kappa &= \omega\omega_m/D_2 \\ D_2 &= \omega_0(\omega_0 + \omega_a) - \omega^2 \\ \omega_a &= \gamma H_a \end{aligned}\right\} \tag{1.88}$$

It is seen that the form of this permeability tensor is the same as that for cubic ferrimagnetics, such as equation (1.75). However, the difference is that for certain planar magnetoplumbites, the magnitudes of μ_{xx} and μ_{yy} can be considerably different, which is not so for any known ferrimagnetic spinel or garnet. This fact has important consequences, which will be discussed later.

The effect of losses can be incorporated by adding the damping term from equation (1.30) to the right-hand side of equation (1.19). The elements of the resulting permeability tensor are obtained as

$$\left.\begin{aligned}
\mu_{xx} &= [(\omega_0+\omega_a)(\omega_0+\omega_m)-\omega^2+j\Delta\omega_h(\omega_0+(\omega_a+\omega_m)/2)]/D_3\\
\mu_{yy} &= [\omega_0(\omega_0+\omega_a+\omega_m)-\omega^2+j\Delta\omega_h(\omega_0+(\omega_a+\omega_m)/2)]/D_3\\
\kappa &= \omega\omega_m/D_3\\
\Delta &= \mu_{xx}\mu_{yy}-\kappa^2\\
&= [(\omega_0+\omega_m)(\omega_0+\omega_a+\omega_m)-\omega^2+j\Delta\omega_h(\omega_0+\tfrac{1}{2}\omega_a+\omega_m)]/D_3\\
D_3 &= \omega_0(\omega_0+\omega_a)-\omega^2+j\Delta\omega_h(\omega_0+\tfrac{1}{2}\omega_a)\\
\Delta\omega_h &= \gamma\Delta H
\end{aligned}\right\} \quad (1.89)$$

It is worth noting that equations (1.88) and (1.89) are applicable to the case when the plane xz is the easy plane. The expressions for μ_{xx} and μ_{yy} should be interchanged in the case when the yz plane is the easy plane. It follows from equation (1.88) that resonance occurs when

$$\omega = [\omega_0(\omega_0+\omega_a)]^{1/2} \quad (1.90)$$

For a planar magnetoplumbite with a significantly large anisotropy field H_a, the resonance frequency is clearly much larger than that for cubic ferrimagnetics with a given applied field H_0. Thus, planar magnetoplumbites, like uniaxial magnetoplumbites, are strong candidates for applications in millimeter-wave devices.

1.8. *Dipole–Dipole Interaction: Demagnetization*

A magnetic dipole is a source of magnetic field and it also interacts with the magnetic field of other sources (dipoles). In an external magnetic field, the dipole tends to align itself along the field direction; the net magnetic field at any point in space is equal to the vector sum of the external field and the field on account of the dipole. When several dipoles are present, the field experienced by any particular dipole is the resultant of the external field and the dipolar field (i.e., the field due to all other dipoles). Thus, when a piece of ferrimagnetic material is placed in an initially uniform magnetic field, the field experienced by dipoles in the medium (internal field) is, in general, nonuniform and different from the external field. Since H_0, appearing in equations (1.17) and (1.22), is the field experienced by the dipoles, which is the resultant of the internal field H_i and any other fields (such as a dc anisotropy field), it is necessary to determine the internal field in order to correctly compute the elements of the permeability tensor.

When the medium is infinite, the contributions (at any point) of the

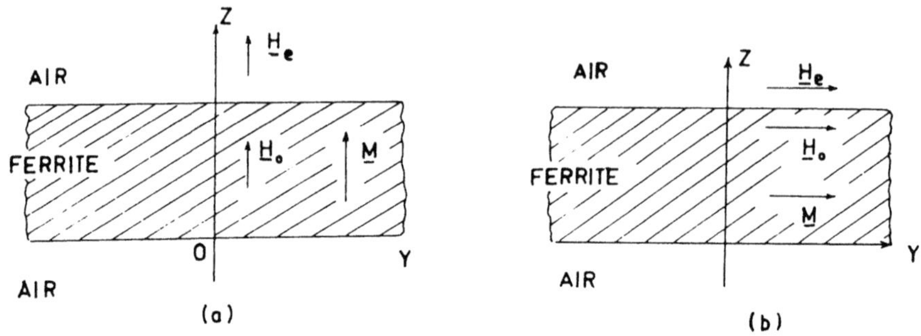

Figure 1.10(a). Magnetization perpendicular to the plane of the sheet. (b) Magnetization parallel to the plane of the sheet.

fields due to the various dipoles cancel out and the internal field is the same as the applied field. However, this is not true for a realistic, finite sample of arbitrary shape. In such cases, the internal field is generally determined by solving appropriate Maxwell's equations* subject to the usual boundary conditions and symmetry restrictions.

As a simple example,† consider a ferrimagnetic sheet of a finite thickness placed in an initially uniform static magnetic field H_e directed perpendicular to its face, as shown in Figure 1.10a. When the sheet extends to infinity in the xy-plane, it is reasonable to assume, on grounds of symmetry, that the magnetization is uniform and is directed along the applied field. If H_i is the internal field, the normal components of the magnetic induction within and outside the sheet are, respectively, given by

$$B_z^{\text{in}} = H_i + 4\pi M_0$$

and

$$B_z^{\text{out}} = H_e$$

The field matching condition yields

$$H_i = H_e - 4\pi M_0 \tag{1.91}$$

Thus, the internal field is equal to the external field minus the saturation magnetization.‡ This is physically understood in terms of the pole forma-

* The dipolar field, unlike the exchange field or anisotropy field, is a real magnetic field and is, therefore, included in Maxwell's equations. Hence it can be obtained by solving Maxwell's equations subject to appropriate boundary conditions.

† In this specific example, only symmetry and boundary conditions are used; no explicit solution of Maxwell's equations is required.

‡ Usually, $4\pi M_0$ (measured in gauss) rather than M_0 is called the saturation magnetization.

tion at the two surfaces of the sheet. The field produced by these poles acts opposite to the direction of the applied or magnetizing field and is, therefore, called a "demagnetizing field." The magnitude of a demagnetizing field turns out to be equal to the saturation magnetization. Thus, the net field (which is equal to the vector sum of the applied field and demagnetizing field) is as given by equation (1.91).

When the applied field is parallel to the face of the sheet, as shown in Figure 1.10b, one can match the tangential components of the magnetic field; thus

$$H_i = H_e \tag{1.92}$$

i.e., the internal field is the same as the applied field. This is expected because no pole formation occurs when the sheet extends to infinity in the direction of magnetization. Even when the sheet is of finite extent in the xy-plane, equations (1.91) and (1.92) will be approximately valid provided that the aspect ratio of the slab, defined as the length in the direction of magnetization divided by slab thickness, is large compared to unity.

The evaluation of the demagnetizing field is much more difficult when the sample has an arbitrary shape. We shall outline the general procedure* to obtain the internal field. In the case of a nonconducting ferrimagnetic sample placed in an initially uniform dc magnetic field, Maxwell's equations are decoupled and the problem is rendered magnetostatic. The internal field is then determined by solving the following set of equations:

$$\nabla \times \mathbf{H} = 0 \tag{1.93a}$$

$$\nabla \cdot \mathbf{B} = 0 \tag{1.93b}$$

The (nonuniform) magnetization at any point inside the medium can be expressed as

$$\mathbf{M}(\mathbf{r}) = M_0 \boldsymbol{\alpha}(\mathbf{r}) \tag{1.94}$$

where $\boldsymbol{\alpha}(\mathbf{r})$ is a unit vector in the direction of magnetization. It is evident that $\boldsymbol{\alpha}$ will be directed along the internal dc field:

$$\boldsymbol{\alpha}(\mathbf{r}) = \mathbf{H}_i(\mathbf{r})/|\mathbf{H}_i(\mathbf{r})| \tag{1.95}$$

The internal field can be expressed as a sum of the (uniform) applied field $\mathbf{H}_e$ and the demagnetizing field $\mathbf{H}_d$; the latter is produced by the dipoles of the medium. Thus,

* The present discussion follows the classic paper by Joseph and Schlömann (1965).

$$\mathbf{H}_i(\mathbf{r}) = \mathbf{H}_e + \mathbf{H}_d(\mathbf{r}) \tag{1.96}$$

Since $\mathbf{H}_e$ is uniform, it follows from equations (1.93a) and (1.96) that

$$\nabla \times \mathbf{H}_d = 0$$

which implies that $\mathbf{H}_d$ can be expressed as the gradient of a scalar field:

$$\mathbf{H}_d(\mathbf{r}) = M_0 \nabla \Psi_s \tag{1.97}$$

Since $\mathbf{B}_i = \mathbf{H}_i(\mathbf{r}) + 4\pi \mathbf{M}(\mathbf{r})$, it follows from equations (1.93) and (1.97) that

$$\nabla^2 \Psi_s = -4\pi M_0 \nabla \cdot \boldsymbol{\alpha}(\mathbf{r}) \tag{1.98}$$

The general solution to this equation can be expressed as

$$\Psi_s(\mathbf{r}) = \int \frac{\boldsymbol{\alpha}(\mathbf{r}') \cdot (\mathbf{r}' - \mathbf{r})}{|\mathbf{r}' - \mathbf{r}|^3}\, d^3 r' \tag{1.99}$$

We construct the solution as a power series in $\varepsilon = M/|\mathbf{H}_e|$:

$$\Psi_s = \Psi_1 + \varepsilon \Psi_2 + \varepsilon^2 \Psi_3 + \cdots \tag{1.100}$$

If $\mathbf{H}_e = H_e \hat{z}$, then Ψ_1 and Ψ_2 can be determined,* following Joseph and Schlömann (1965), as

$$\left.\begin{aligned} \Psi_1 &= \int \frac{(z' - z)\, d^3 r'}{|\mathbf{r}' - \mathbf{r}|^3} \\ \Psi_2 &= \int \frac{(x' - x)\partial \Psi_1/\partial x + (y' - y)\partial \Psi_1 \partial y}{|\mathbf{r}' - \mathbf{r}|^3}\, d^3 r' \end{aligned}\right\} \tag{1.101}$$

The volume integrals are to be evaluated over the volume of the sample. Once the Ψ's are evaluated, it is possible to obtain $\mathbf{H}_d$, and hence $\mathbf{H}_i$, using equations (1.96), (1.97), and (1.100).

The demagnetizing fields for a variety of sample geometries have been evaluated by different workers. Perhaps the most important shape is an ellipsoid—the only case for which the demagnetizing field is uniform† and can be expressed as

* Ψ_3, Ψ_4, etc., are seldom useful in practice.

† Maxwell (1904) was the first to show that the magnetization would be uniform provided that the surface of the sample is a curve of second degree; the only curve of second degree leading to a finite sample is ellipsoid. See for example Stratton (1941) and Sommerfeld (1952) for further discussion.

$$(\mathbf{H}_d)_i = -4\pi \sum_j N_{ij} M_j \tag{1.102}$$

where N_{ij} forms the "demagnetizing tensor." It is evident that, though uniform, $\mathbf{H}_d$ and hence $\mathbf{H}_i$ are not necessarily parallel to $\mathbf{H}_e$. However, if the coordinate axes are oriented along the principal axes of the ellipsoid (so that $\mathbf{H}_e$ is directed along a principal axis), the demagnetizing tensor turns out to be diagonal. Writing $N_x = N_{xx}$, etc., we have

$$(\mathbf{H}_d)_i = -N_i(4\pi M_i) \tag{1.103}$$

where the demagnetizing factors add up to unity; i.e.,

$$N_x + N_y + N_z = 1 \tag{1.104}$$

The demagnetizing factors for a given ellipsoid depend on its shape or, equivalently, on the ratio of the axes. Let $a \geqslant b \geqslant c \geqslant 0$ represent the semiaxes of the ellipsoid. Following Osborn (1945), the demagnetizing factors for different axis ratios have been displayed in Figure 1.11a–c.

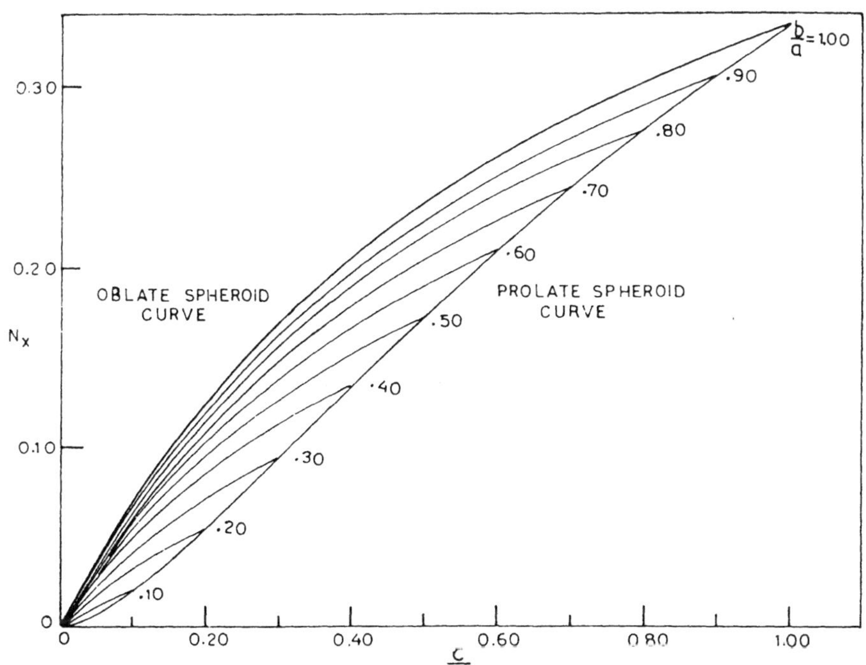

Figure 1.11a. Variation of N_x with c/a for different b/a (after Osborne, 1945).

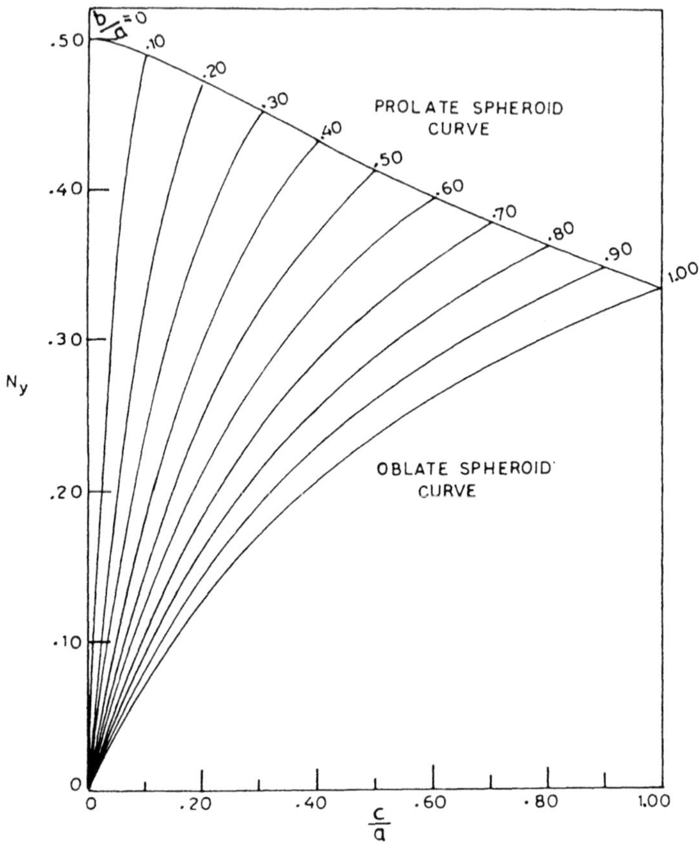

Figure 1.11b. Variation of N_y with c/a for different b/a (after Osborne, 1945).

Hence, the internal field can be computed using equations (1.96) and (1.103).

Many sample shapes of practical interest, such as a sphere, a thin rod, a thin disc, etc., are in fact special or limiting cases of the ellipsoid. The demagnetizing factors for these shapes are as follows:

Sphere

$$N_x = N_y = N_z = \tfrac{1}{3}$$

Thin rod (z-axis along the axis of the rod)

$$N_x = N_y = \tfrac{1}{2}; \qquad N_z = 0$$

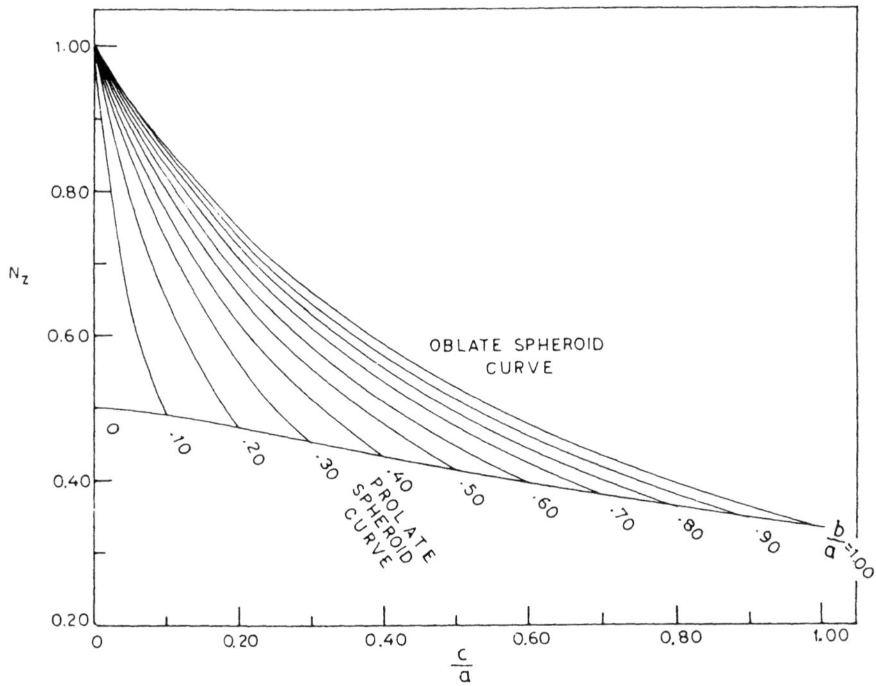

Figure 1.11c. Variation of N_z with c/a for different b/a (after Osborne, 1945).

Thin disk (z-axis perpendicular to the disk)

$$N_x = N_y = 0; \qquad N_z = 1$$

Many other sample shapes of practical interest, such as a rectangular prism, a thick cylinder, etc., are nonellipsoidal for which the demagnetizing field would be nonuniform. The first-order demagnetizing field can be expressed as (Joseph and Schlömann, 1965)

$$[\mathbf{H}_d^{(1)}(\mathbf{r})]_i = -4\pi M_0 \sum_j N_{ij}^{(1)}(\mathbf{r})\alpha_j \tag{1.105}$$

The diagonal elements of $\boldsymbol{N}$ still add up to unity.

Figures 1.12–1.15 show the rectangular prism geometry and the demagnetizing factor in different directions for samples of different sizes.

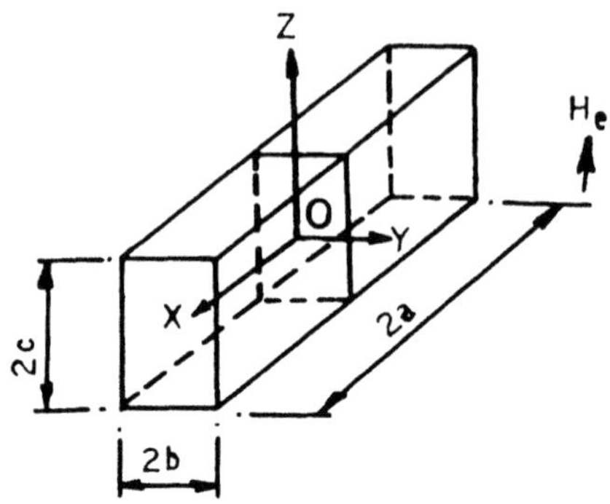

Figure 1.12. Coordinate system as used in calculations pertaining to rectangular prisms. The origin of the coordinate system coincides with the center of the prism. The dc field is applied along the z-direction (after Joseph and Schlömann, 1965).

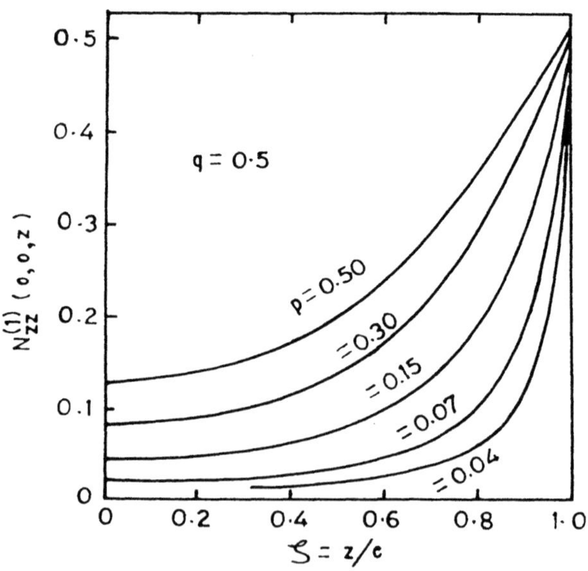

Figure 1.13. The spatial variation of the first-order demagnetizing factor $N_{zz}^{(1)}$ along a coordinate axis of the prism ($x = y = 0$) in terms of a reduced variable $\zeta = z/c$ (z is the distance from the center of the prism) for various sample shapes ($q = a/c = 0.5$, $p = b/c$). Note that $N_{zz}^{(1)}(-z) = N_{zz}^{(1)}(z)$ (after Joseph and Schlömann, 1965).

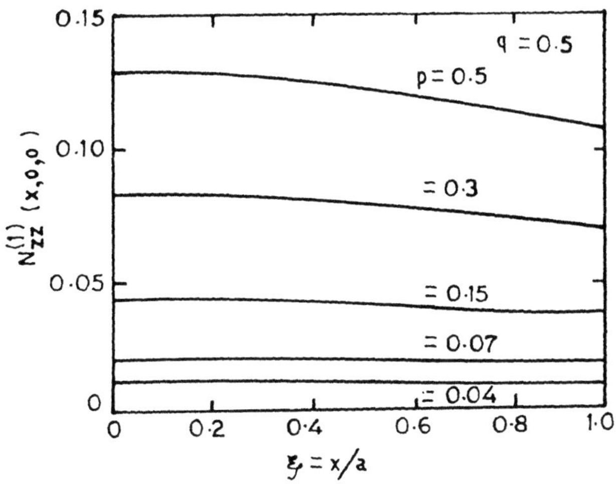

Figure 1.14. The spatial variation of the first-order demagnetizing factor $N_{zz}^{(1)}$ along a coordinate axis of the prism ($y = z = 0$) in terms of a reduced variable $\xi = x/a$ (x is the distance from the center of the prism) for various sample sizes ($q = a/c = 0.5$, $p = b/c$). Note that $N_{zz}^{(1)}(-x) = N_{zz}^{(1)}(x)$ (after Joseph and Schlömann, 1965).

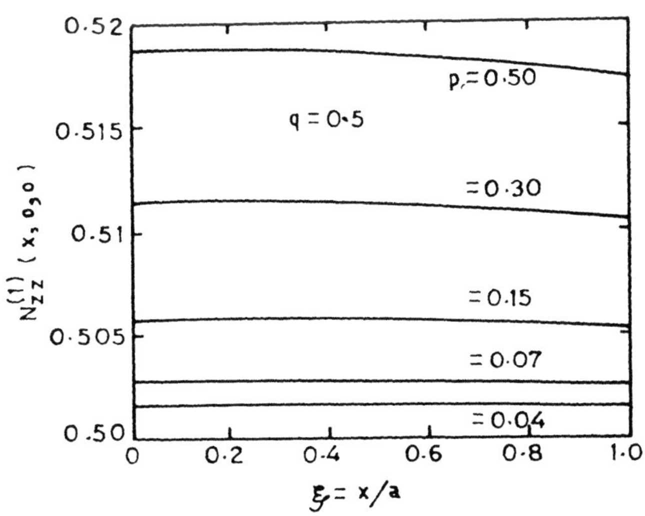

Figure 1.15. The spatial variation of the first-order demagnetizing factor $N_{zz}^{(1)}$ along an axis on the edge of the prism ($y = 0$, $z = c$) in terms of a reduced variable $\xi = x/a$ for various sample sizes $q = a/c = 0.5$, $p = b/c$. Note that $N_{zz}^{(1)}(-x) = N_{zz}^{(1)}(x)$ (after Joseph and Schlömann, 1965).

The corresponding configuration and demagnetizing factor for a circular cylinder have been shown in Figures 1.16–1.19. The second-order demagnetizing factors for semi-infinite samples have been computed by Joseph and Schlömann (1965).

The foregoing discussion has been restricted to first-order dc demagnetization theory. The reader interested in alternative approaches and advanced problems of dc demagnetization theory is referred to Joseph

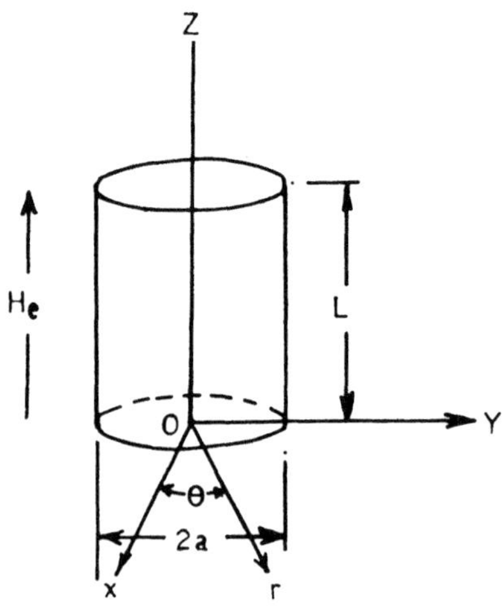

Figure 1.16. Coordinate system used for calculations pertaining to the circular cylinder. The origin of coordinates coincides with the center of the circular endface of the cylinder that is oriented perpendicular to the dc field H_e applied along the z-direction (after Joseph and Schlömann, 1965).

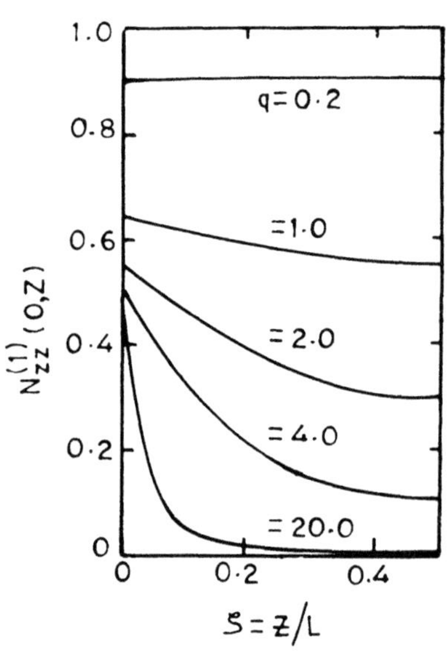

Figure 1.17. The spatial variation of the first-order demagnetizing factor $N_{zz}^{(1)}$ along the central axis of the cylinder in terms of a reduced variable $\zeta = z/L$ for various sample shapes ($q = L/a$). Note that $N_{zz}^{(1)}(z) = N_{zz}^{(1)}(L - z)$ (after Joseph and Schlömann, 1965).

(1966, 1967) and Schlömann (1970). Furthermore, the so-called ac demagnetization theory (which deals with the demagnetization of an alternating field applied to finite samples) has been developed by a number of authors including Kittel (1947, 1948), Berk and Lengyel (1955), von Aulock and Rowan (1957), Waldron (1957, 1959), Lewandowsky (1964), and Helszajn and McStay (1969).

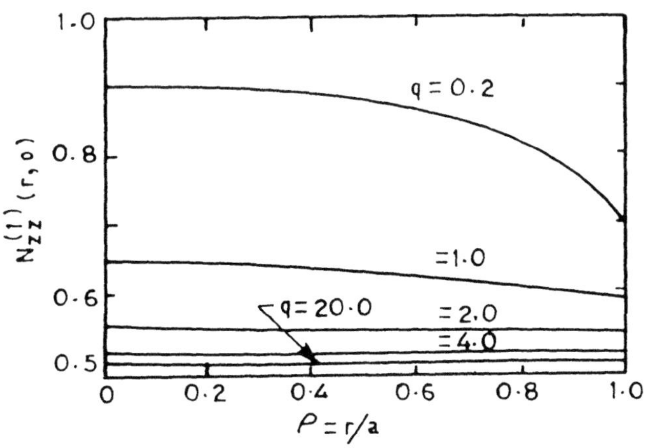

Figure 1.18. The spatial variation of the first-order demagnetizing factor $N_{zz}^{(1)}$ in the endface of a circular cylinder ($z = 0$) as a function of the distance from the central axis (z) in terms of a reduced variable $\rho = r/a$ for various sample shapes ($q = L/a$) (after Joseph and Schlömann, 1965).

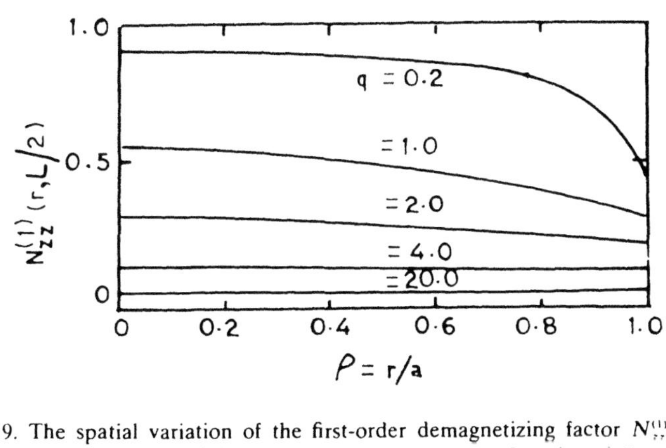

Figure 1.19. The spatial variation of the first-order demagnetizing factor $N_{zz}^{(1)}$ in a plane parallel to the endfaces and equidistant from both ($z = L/2$) as a function of distance from the central (z) axis in terms of a reduced variable $\rho = r/a$ for various sample shapes ($q = L/a$) (after Joseph and Schlömann, 1965).

1.9. Magnetoelastic Interaction

In any ferrimagnetic material, a coupling exists between the lattice and the system of magnetic dipoles. This essentially means that a magnetic disturbance in the spin system (spin wave) can give rise to a mechanical disturbance (elastic wave) in the lattice and vice versa. In a biased ferrimagnetic material, this coupling (called magnetoelastic coupling) can result from various mechanisms,* such as spin–orbit interaction (crystal anisotropy), dipole–dipole interaction, exchange interaction, etc. For example, in the presence of an elastic disturbance, the ion cores in the crystal are displaced from their equilibrium positions. The surrounding electron clouds also follow the motion of the respective ions. This orbital reorientation of electron clouds modifies the states of successive electron spins as a consequence of spin-orbit coupling which, in turn, gives rise to a magnetic disturbance and hence to the magnetoelastic interaction. Similar arguments apply to the cases of dipole–dipole and exchange contributions toward the magnetoelastic interaction.

As in the case of magnetocrystalline anisotropy, we shall adopt a phenomenological (continuum) approach while describing the magnetoelastic interaction. This presumes some knowledge of the continuum elastic wave theory, the relevant results of which are summarized in Appendix D. We shall limit the present discussion to cubic ferrimagnetics.

As described by Tiersten (1964), Brown (1965), and Auld (1968), the expression for magnetoelastic interaction energy can be obtained by the application of symmetry properties and conservation laws. The first terms in the expansion of the interaction energy are linear in strain, but, quadratic in direction cosines of **M** as a result of the required inversion symmetry. Thus, in the limit of small strain and small rf magnetization, the magnetoelastic energy can be expressed as

$$U_{\text{me}} = \lambda_{ijkl}\alpha_i\alpha_j S_{kl} \tag{1.106}$$

where λ_{ijkl} are the components of a fourth-rank tensor $\boldsymbol{\lambda}$. This tensor is symmetric with respect to the interchange of k and l and of i and j. These restrictions reduce the number of independent elements of the $\boldsymbol{\lambda}$-tensor to 36. Further reduction in this number is achieved when the condition $\Sigma_i \alpha_i^2 = 1$ is utilized. Moreover, it is required that the interaction energy should remain invariant under the (cubic) symmetry operations, e.g., under a rotation by $\pi/2$ about a cube edge. A careful treatment leads to the conclusion that only two independent constants are required to character-

* For details, see Smit and Wijn (1959), Kanamori (1963), and Auld (1965).

ize the magnetoelastic interaction energy of the cubic ferrimagnet. As such, the interaction energy assumes the following form:

$$U_{me} = b_1(\alpha_x^2 S_{xx} + \alpha_y^2 S_{yy} + \alpha_z^2 S_{zz}) + 2b_2(\alpha_x\alpha_y S_{xy} + \alpha_y\alpha_z S_{yz} + \alpha_z\alpha_x S_{zx}) \quad (1.107)$$

where b_1 and b_2 are called the first and second magnetoelastic constants, respectively. The factor 2 in the second term is not a universal choice, but it is seen to lead to simpler expressions. In abbreviated subscript notations (see Appendix D), the interaction energy (1.107) can be expressed as

$$U_{me} = b_1(\alpha_1^2 S_1 + \alpha_2^2 S_2 + \alpha_3^2 S_3) + b_2(\alpha_1\alpha_2 S_6 + \alpha_2\alpha_3 S_4 + \alpha_3\alpha_1 S_5) \quad (1.108)$$

This interaction energy leads to a stress and a torque experienced by magnetic dipoles. In analogy with torques due to exchange interaction and anisotropy fields, the magnetoelastic torque also can be looked upon as arising on account of an effective magnetic field which is given by

$$\mathbf{H}_{me} = -(1/M_0)\nabla_\alpha U_{me} \quad (1.109)$$

Moreover, the effective stress on account of magnetoelastic coupling is given in consonance with equation (D.25) as

$$T_I^{me} = \frac{\partial U_{me}}{\partial S_I} \quad (1.110)$$

The substitution of U_{me} from equation (1.108) into (1.109) and (1.110) leads to the desired expressions for $\mathbf{H}_{me}$ and $\boldsymbol{T}^{me}$, respectively. In linearized theory, if the dc magnetization is along the z-axis, α_3 is unity whereas the second- and higher-order product terms involving α_1 and α_2 can be neglected. As such, $\mathbf{H}_{me}$ and $\boldsymbol{T}^{me}$ are, respectively, obtained as

$$\mathbf{H}_{me} = -(b_2 S_5 \hat{x}_1 + b_2 S_4 \hat{x}_2 + 2b_1 S_3 \hat{x}_3)/M_0 = \mathbf{h}_{me} \quad (1.111)$$

and

$$\boldsymbol{T}^{me} = \begin{pmatrix} T_1^{me} \\ T_2^{me} \\ T_3^{me} \\ T_4^{me} \\ T_5^{me} \\ T_6^{me} \end{pmatrix} \quad (1.112)$$

where

$$\left.\begin{aligned} T_1^{\text{me}} &= T_2^{\text{me}} = T_6^{\text{me}} = 0 \\ T_3^{\text{me}} &= b_1 \\ T_4^{\text{me}} &= b_2\alpha_2 \\ T_5^{\text{me}} &= b_2\alpha_1 \end{aligned}\right\} \tag{1.113}$$

In equation (1.111), we have equated $\mathbf{H}_{\text{me}}$ to $\mathbf{h}_{\text{me}}$ since no dc term is involved. On the other hand, the magnetoelastic stress has a dc component $T_3^{\text{me}} = b_1$.

When the direction of magnetization relative to the cube edges is arbitrary, it is more convenient to work in a transformed coordinate system; such a transformation of interaction energy has been discussed in Section C.5 of Appendix C. The magnetoelastic magnetic field and magnetoelastic stress can be obtained using equations (1.109) and (1.110) along with equation (C.28) of Appendix C.

2

Plane Waves in Ferrimagnetics

The study of the propagation of plane waves in ferrimagnetics of infinite extent is necessary if one wishes to understand the propagation phenomena in devices (of finite extent). Further, in many practical situations, the wavelength* is much smaller than the sample size and the waves in regions far (few wavelengths) from the boundaries behave as plane waves. Thus, in this chapter, we discuss the propagation of electromagnetic waves (Section A) and magnetoelastic waves (Section B) in ferrimagnetics of infinite extent.

PART A. *ELECTROMAGNETIC WAVES*

2.1. *Electromagnetic Field Equations*

It is well known that the electromagnetic field quantities inside any material medium are governed by Maxwell's equations†:

$$\left.\begin{aligned} \nabla\cdot\mathbf{B} &= 0 \\ \nabla\cdot\mathbf{D} &= 4\pi\rho_e \\ \nabla\times\mathbf{E} &= -\frac{1}{c_0}\frac{\partial\mathbf{B}}{\partial t} \\ \nabla\times\mathbf{H} &= \frac{4\pi}{c_0}\mathbf{J}_c + \frac{1}{c_0}\frac{\partial\mathbf{D}}{\partial t} \end{aligned}\right\} \tag{2.1}$$

* To be more precise, the wavelength inside the medium.

† Gaussian system of units is used.

In the case of electrical insulator ferrimagnetics, the charge density ρ_e and current density $\mathbf{J}_c$ can both be equated to zero. Moreover, since our interest lies in magnetized ferrimagnetics, the total instantaneous magnetic field $\mathbf{H}$ consists of a dc component $\mathbf{H}_0$ and a time-varying component $\mathbf{h}$, on account of the wave, i.e.,

$$\mathbf{H} = \mathbf{H}_0 + \mathbf{h}(t) \tag{2.2}$$

Consequently, the total instantaneous magnetic induction also consists of a static component $\mathbf{B}_0$ and a time-varying component $\mathbf{b}$, i.e.,

$$\mathbf{B} = \mathbf{B}_0 + \mathbf{b}(t) \tag{2.3}$$

It follows from equations (2.1)–(2.3) that the field vectors $\mathbf{E}$, $\mathbf{D}$, $\mathbf{H}$, and $\mathbf{b}$ associated with electromagnetic waves in a ferrimagnetic insulator necessarily satisfy the following set of equations:

$$\left.\begin{aligned} \nabla \cdot \mathbf{b} &= 0 \\ \nabla \cdot \mathbf{D} &= 0 \\ \nabla \times \mathbf{E} &= -\frac{1}{c_0}\frac{\partial \mathbf{b}}{\partial t} \\ \nabla \times \mathbf{h} &= \frac{1}{c_0}\frac{\partial \mathbf{D}}{\partial t} \end{aligned}\right\} \tag{2.4}$$

Since any physical disturbance can be expressed as an integral over harmonic components,* it is useful to investigate, at first, the behavior of a harmonic wave of angular frequency ω and finally sum over the various frequency components in order to obtain a complete solution. Thus, in complex representation, the field vectors vary as $\exp(j\omega t)$; Maxwell's equations (2.4) reduce to

$$\left.\begin{aligned} \nabla \cdot \mathbf{b} &= 0 \\ \nabla \cdot \mathbf{D} &= 0 \\ \nabla \times \mathbf{E} &= -j\frac{\omega}{c_0}\mathbf{b} \\ \nabla \times \mathbf{h} &= j\frac{\omega}{c_0}\mathbf{D} \end{aligned}\right\} \tag{2.5}$$

Of these eight scalar equations, only six are independent because the first

* See, for example, Lighthill (1964).

two can, in fact, be deduced by taking the divergence of both sides of the remaining equations. Obviously, the resulting six equations are insufficient to determine the 12 unknown quantities **E**, **D**, **h**, and **b**; each vector has three components. Therefore, in order to obtain a complete solution (in principle) the set of equations (2.5) should be supplemented by another six equations involving the field vectors. These are the so-called constitutive relations which characterize the ferrimagnetic medium electromagnetically, i.e., describe the response of the medium to impressed electric and magnetic fields.

It is an experimental fact that in most of the material media, including the majority of ferrimagnetics [and excluding magnetoelectrics (Freeman and Schmid, 1975) and moving media (Kong, 1975)], the electric and magnetic interactions are not coupled; i.e., a pure electric field causes only dielectric polarization while a pure magnetic field causes only magnetization. Hence, the magnetic induction **b** depends only on **h** (and not on **E** or **D**), while the displacement vector **D** depends only on **E**. It is also an experimental fact that for sufficiently small amplitudes of **E** and **h** the relationship between **D** and **E**, as well as that between **b** and **h**, is linear. The latter has already been discussed in Chapter 1 and is expressed as

$$\mathbf{b} = \boldsymbol{\mu} \cdot \mathbf{h} \tag{2.6}$$

For a single crystal, **D** and **E** are related through the dielectric tensor $\boldsymbol{\varepsilon}_f$. However, the dielectric anisotropy is small for most of the ordinary microwave ferrimagnetics. Therefore, it can be ignored,* rendering the dielectric constant a scalar. Thus,

$$\mathbf{D} = \varepsilon_f \mathbf{E} \tag{2.7}$$

Substituting for **b** and **D** from equations (2.6) and (2.7) in (2.5), the relevant field equations for ferrimagnetics can be expressed as

$$\left.\begin{aligned} \nabla \times \mathbf{E} &= -j(\omega/c_0)\boldsymbol{\mu} \cdot \mathbf{h} \\ \nabla \times \mathbf{h} &= j(\omega/c_0)\varepsilon_f \mathbf{E} \end{aligned}\right\} \tag{2.8}$$

These equations are applicable only to ferrimagnetic insulators. However, if the electrical conductivity is nonzero and equal to σ_e, it is easy to show that equations (2.8) continue to hold provided that ε_f is replaced by $(\varepsilon_f - 4j\pi\sigma_e/\omega)$. Electromagnetic wave phenomena in ferrimagnetics can be investigated by solving equations (2.8) subject to the appropriate boundary conditions.

* In the case of polycrystalline ferrimagnetics, dielectric anisotropy averages out to zero.

2.2. Exchange-Free Plane Waves

We consider the propagation of a uniform plane wave in magnetized ferrimagnetics, neglecting the exchange effects. As such, the following description would lead to meaningful conclusions only if the wavenumber is smaller than about $10^5\,\mathrm{cm}^{-1}$ (see Section 1.6); the explicit effect of exchange interaction on plane waves will be considered later in Section (2.4). The permeability tensor may be adopted from equation (1.87) as its form is general enough to apply to hexagonal as well as cubic ferrimagnetics. However, the cubic anisotropy is usually very small and, in general, does not significantly affect the wave propagation characteristics. On the other hand, the hexagonal planar anisotropy is significantly large for certain magnetoplumbites (see Appendix A). Therefore, we shall discuss in detail the case of hexagonal planar magnetoplumbite from which the results for a uniaxial or an isotropic material follow as a special case.

It is convenient to make a coordinate transformation to a new (primed) coordinate system which is oriented with respect to the original (unprimed) one through the first two Eulerian angles φ and θ, as shown in Figure C.1 of Appendix C. The details of the coordinate transformation have been given in Appendix C, where it has also been shown that Maxwell's equations are form-invariant while the permeability tensor is transformed to that given by equations (C.17) and (C.18).

2.2.1. The Propagation Constant

Consider the propagation of a uniform plane wave along the y'-axis (see Figure C.1), which, as discussed in Appendix C, can be oriented along an arbitrary direction relative to the biasing field by an appropriate choice of φ and θ. In this case, the space–time variations of the field vectors are of the form $\exp j(\omega t - \beta y')$. Substituting this dependence in equations (2.8) and dropping the primes (for convenience), we obtain

$$-j\beta\hat{y} \times \mathbf{E} = -j(\omega/c_0)\boldsymbol{\mu} \cdot \mathbf{h} \tag{2.9}$$

$$-j\beta\hat{y} \times \mathbf{h} = j(\omega/c_0)\varepsilon_f\mathbf{E} \tag{2.10}$$

where $\boldsymbol{\mu}$ is given by equations (C.17) and (C.18) of Appendix C. The elimination of $\mathbf{E}$ from equations (2.9) and (2.10) leads to

$$\beta^2\hat{y} \times (\hat{y} \times \mathbf{h}) + \beta_d^2\boldsymbol{\mu} \cdot \mathbf{h} = 0 \tag{2.11}$$

where

$$\beta_d^2 = \varepsilon_f\omega^2/c_0^2 \tag{2.12}$$

The vector equation (2.11) can be written, in extenso, as

$$[\mu'_{11} - (\beta/\beta_d)^2]h_x + \mu'_{12}h_y + \mu'_{13}h_z = 0 \tag{2.13}$$

$$\mu'_{21}h_x + \mu'_{22}h_y + \mu'_{23}h_z = 0 \tag{2.14}$$

$$\mu'_{31}h_x + \mu'_{32}h_y + [\mu'_{33} - (\beta/\beta_d)^2]h_z = 0 \tag{2.15}$$

A nontrivial solution to equations (2.13)–(2.15) exists only if the determinant of the coefficients of **h** vanishes; this condition leads to the following expression* for the propagation constant $\beta_\pm$:

$$\left(\frac{\beta_\pm}{\beta_d}\right)^2 = \frac{\delta_0 + (\Delta - \mu')\cos^2\theta}{2(\sin^2\theta + \mu''\cos^2\theta)}$$
$$\pm \frac{[(\Delta - \mu)^2\cos^4\theta + (\mu' - \mu'')(2\Delta - \delta)\cos^2\theta + (\delta_0^2 - 4\Delta)\sin^2\theta]}{2(\sin^2\theta + \mu''\cos^2\theta)} \tag{2.16}$$

where

$$\begin{aligned} \delta_0 &= \mu' + \mu'' = \mu_{xx} + \mu_{yy} \\ \Delta &= \mu_{xx}\mu_{yy} - \kappa^2 \end{aligned} \tag{2.17}$$

In the case of a hexagonal planar medium, the elements of the permeability tensor are given by equations (1.89) from which the various relevant quantities involved in equation (2.16) can be obtained in terms of the basic parameters as follows:

$$\left.\begin{aligned} D_3 &= [\omega_0(\omega_0 + \omega_a) - \omega^2 + j\Delta\omega_h(\omega_0 + \tfrac{1}{2}\omega_a)] \\ \Delta &= [(\omega_0 + \omega_m)(\omega_0 + \omega_a + \omega_m) - \omega^2 \\ &\quad + j\Delta\omega_h(\omega_0 + \tfrac{1}{2}\omega_a + \omega_m)]/D_3 \\ \delta_0 &= 2[\omega_0(\omega_0 + \omega_a + \omega_m) + \tfrac{1}{2}\omega_a\omega_m - \omega^2 \\ &\quad + j\Delta\omega_h(\omega_0 + \tfrac{1}{2}\omega_a + \omega_m)]/D_3 \\ (\delta_0^2 - 4\Delta) &= \omega_m^2\{\omega_a^2 + 4\omega^2 - (\Delta\omega_h)^2 - 4j\Delta\omega_h[\omega_0 + \tfrac{1}{2}(\omega_a + \omega_m)]\}/D_3^2 \\ \mu' - \mu'' &= \omega_a\omega_m\cos 2\varphi/D_3 \\ (2\Delta - \delta_0) &= 2\omega_m(\omega_0 + \tfrac{1}{2}\omega_a + \omega_m + \tfrac{1}{2}j\Delta\omega_h)/D_3 \\ (\Delta - \mu') &= \omega_m(\omega_0 + \omega_m + \omega_a\sin^2\varphi - \tfrac{1}{2}j\Delta\omega_h)/D_3 \end{aligned}\right\} \tag{2.18}$$

* While comparing this cxpression with the ones available in the literature, it should be recognized that, in the present coordinate system (Figure C.1), θ is complementary to the angle between the direction of propagation and the direction of the dc magnetic field.

It is evident from equation (2.16) that for a given direction (φ, θ) not more than two independent waves with propagation constants β_+ and β_- can propagate. The polarization characteristics for these waves can be inferred from equations (2.13)–(2.16). We shall describe some specific cases before taking up the general case.

2.2.2. *Special Cases*

2.2.2.1. *Lossless Isotropic Ferrimagnetics**

In this case $H_a = 0$ and $\Delta H = 0$; the latter condition leads to the conclusion that β is either purely real or purely imaginary. Real values of β correspond to lossless propagation while imaginary values of β correspond to cut-off. In this section we discuss the propagation along and perpendicular to the dc magnetic field.

(a) *Propagation Perpendicular to* $\mathbf{H}_0$. Since the wave propagates along the y'-axis, we set $\theta = 0$ and $\varphi = 0$ in order to make the y'-axis coincident with the y-axis. Thus, the dispersion relations for the two modes are obtained as

$$(\beta_1/\beta_d)^2 = (\mu^2 - \kappa^2)/\mu \equiv \mu_{eff} \tag{2.19}$$

and

$$\beta_2 = \beta_d \tag{2.20}$$

It follows that the mode with propagation constant β_2 is not affected by the magnetic nature of the medium and propagates as it would in a dielectric. On the other hand, the mode with propagation constant β_1 does interact with, and is therefore influenced by, the magnetic nature of the medium. In analogy with optics, these modes may be called ordinary (O-) and extraordinary (E-) modes, respectively. The polarization characteristics of the modes can be inferred from equations (2.13)–(2.15) after substituting $\theta = 0$, $\varphi = 0$, and the corresponding propagation constant. It is seen that, for the O-mode, h_z is nonzero while h_x and h_y are identically equal to zero. Moreover, equation (2.10) reveals that $E_y = E_z = 0$ while E_x is nonzero. Thus, the O-mode is a transverse electromagnetic (TEM) wave; there is no component of electric or magnetic field along the direction of propagation (y-axis). The propagation characteristics for this mode can be physically explained as follows: When the wave magnetic field is oriented

* The detailed analysis of this case was first presented by Eberhardt, Horvath, and Knerr (1970).

parallel to the dc magnetic field, there is no change in the z-component of the magnetization. Therefore, the behavior of the medium is the same as it would have been had all the spins been clamped together. Obviously, the propagation characteristics in this case should be governed only by the dielectric constant of the medium.

In case of E-mode, it is seen from equations (2.13)–(2.15) and (2.10) that the only nonvanishing field components are E_z, h_x, and h_y. Thus, the magnetic field of the wave has a longitudinal component h_y. However, the electric field is still transverse to the direction of propagation and, therefore, the wave may be called a TE-mode. The explicit dependence of the propagation constant on basic parameters is revealed on substituting for μ and κ in equation (2.19) from equation (2.18); thus

$$(\beta_1/\beta_d)^2 = \frac{(\omega_0 + \omega_m)^2 - \omega^2}{\omega_0(\omega_0 + \omega_m) - \omega^2} \tag{2.21}$$

The dispersion curve (f–β diagram) for a typical set of parameters is shown in Figure 2.1. Since β^2 is negative for $[\omega_0(\omega_0 + \omega_m)]^{1/2} < \omega < (\omega_0 + \omega_m)$, there is no wave propagation in this interval; the upper and lower frequencies define the cut-off limit ($\beta = 0$) and resonance limit ($\beta \to \infty$), respectively. When ω is much less than $[\omega_0(\omega_0 + \omega_m)]^{1/2}$ or much greater than $\omega_0 + \omega_m$, it is seen (Figure 2.1) that the dispersion characteristics are not significantly different from that of a dielectric (broken line). As such, these frequency limits define the approximate range within and around which the ferrimagnetics exhibit interesting microwave characteristics.

The phase velocity is obtained as

$$v_{ph} \equiv \frac{\omega}{\beta} = \frac{c_0}{\sqrt{\varepsilon_f}} \left[\frac{\omega_0(\omega_0 + \omega_m) - \omega^2}{(\omega_0 + \omega_m)^2 - \omega^2} \right]^{1/2} \tag{2.22}$$

The normalized phase velocity (v_{ph}/c_0) has been displayed in Figure 2.2 for the same set of parameters as for Figure 2.1. It is seen that, near the resonance and cut-off limits, the phase velocity is considerably different from its dielectric limit $c_0/\sqrt{\varepsilon_f}$. At very low frequencies ($\omega \to 0$), v_{ph} is close to the dielectric limit except when $\omega_0 \ll \omega_m$, in which case it is much smaller than the dielectric limit. The phase velocity approaches zero as the frequency is increased to its resonance value. The phase velocity is infinite at the cut-off limit and decreases monotonically when the frequency is increased beyond cut-off, finally approaching the dielectric limit when $\omega \gg (\omega_0 + \omega_m)$. Evidently, magnetically tunable phase shift can be achieved in the case of operation around a desired frequency.

The velocity with which a wave packet of central frequency ω

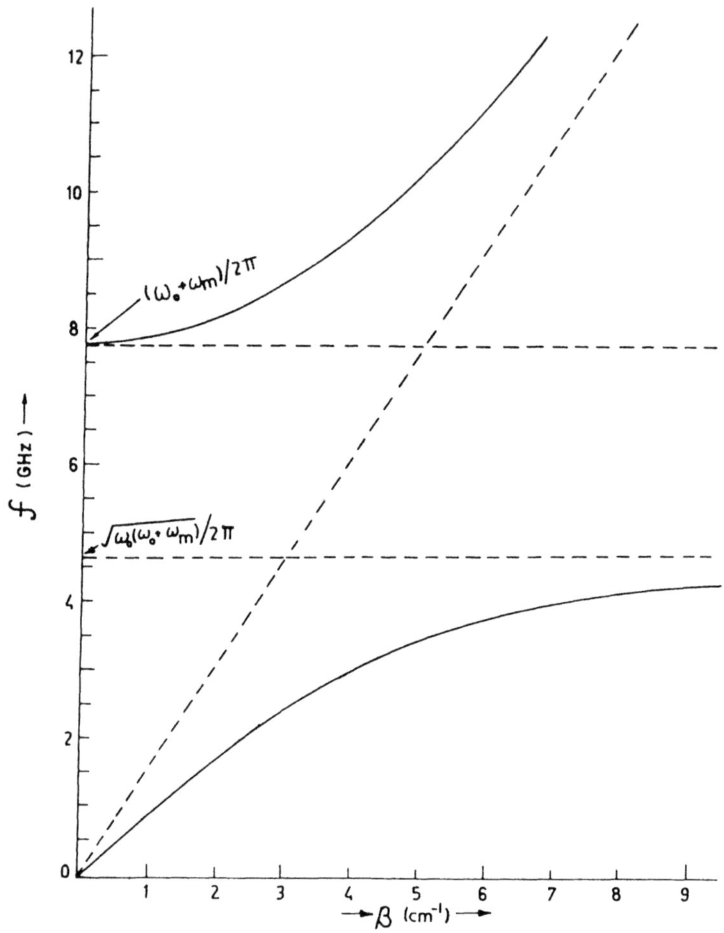

Figure 2.1. Dispersion curve ($f \equiv \omega/2\pi$ vs. β) for plane wave propagation perpendicular to the biasing field. The parameters are $H_0 = 1000$ Oe and $4\pi M_0 = 1750$ G.

propagates is the group velocity v_g which is obtained as

$$v_g \equiv \frac{\partial \omega}{\partial \beta} = \frac{c_0^2}{\varepsilon_f v_{ph}} \frac{\omega_0(\omega_0 + \omega_m) - \omega^2}{c_0^2\beta^2/\varepsilon_f + (\omega_0 + \omega_m) - 2\omega^2} \tag{2.23}$$

The normalized group velocity (v_g/c_0) is plotted as a function of frequency in Figure 2.3. It is seen that the group velocity is very small in the vicinity of resonance* as well as cut-off and it approaches the dielectric limit only

* It is worth mentioning that, in the region of strong dispersion around resonance, the group velocity, signal velocity, and velocity of energy flow are different from one another, as discussed by Brillouin (1960).

considerably far from these limits. This suggests that appreciable magnetically tunable group delay can be achieved in the propagation region near the resonance limit.

(b) *Propagation Parallel to* $\mathbf{H}_0$. In this case we set $\theta = 90°$ along with $H_a = 0$ and $\Delta H = 0$, in equation (2.16) which, in turn, reduces to

$$(\beta_{\pm}/\beta_d)^2 = \mu \pm \kappa \tag{2.24}$$

As in the preceding case, the polarization characteristics of the modes can be inferred from equations (2.13)–(2.15). It is seen that $h_z = 0$ and $E_z = 0$ for each mode. Moreover, $h_x = jh_y$ for the β_+-mode while $h_x =$

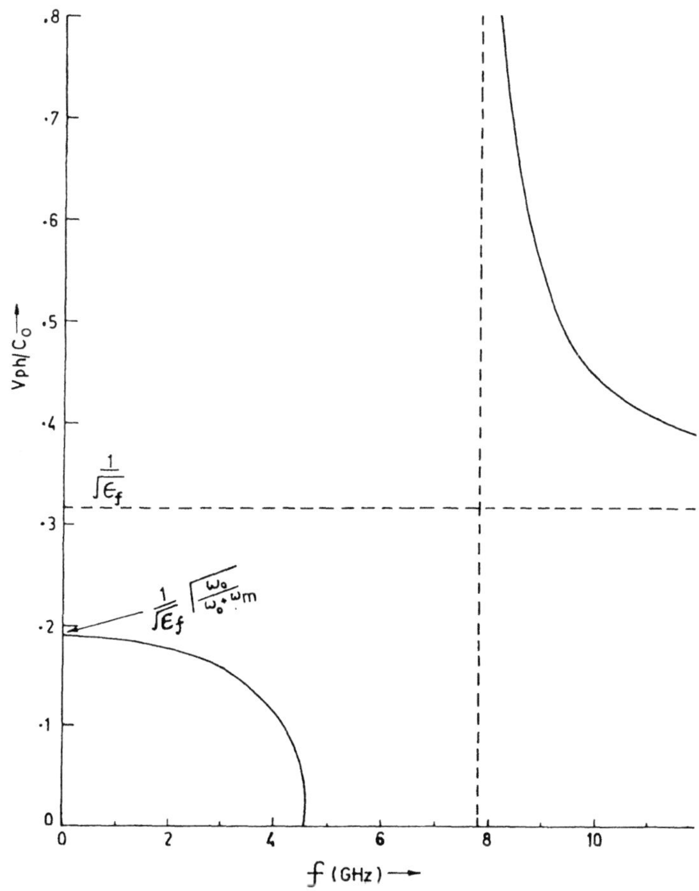

Figure 2.2. Variation of normalized phase velocity (v_{ph}/c_0) with frequency f for propagation perpendicular to the biasing field. The parameters are the same as for Figure 2.1.

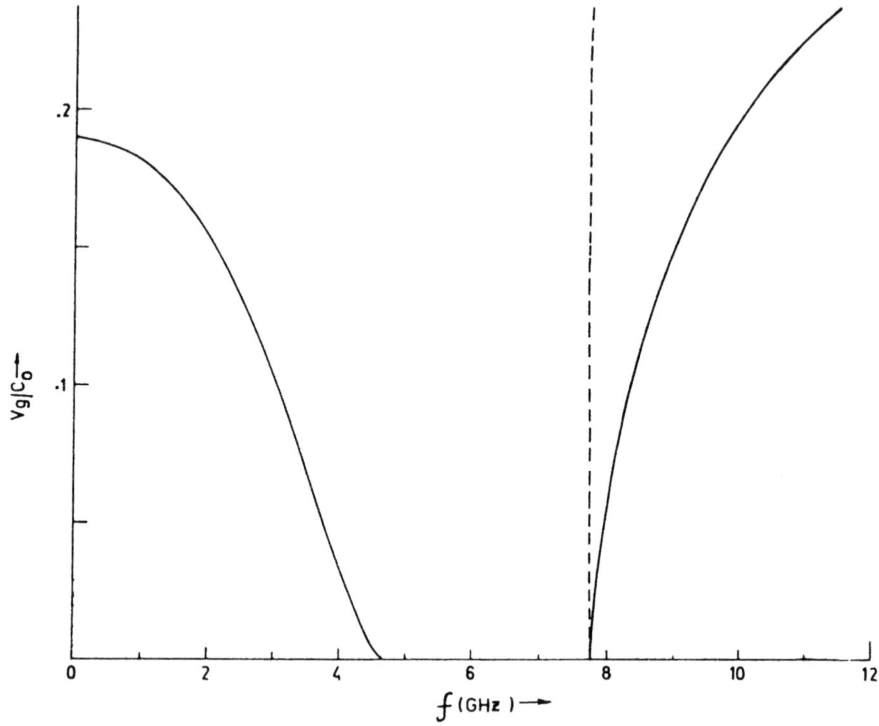

Figure 2.3. Variation of normalized group velocity (v_g/c_0) with frequency for propagation perpendicular to the biasing field. The parameters are the same as for Figure 2.1.

$-jh_y$ for the β_--mode. It follows that the modes with propagation constants β_+ and β_- are right-(R) and left-(L) handed circularly polarized (CP), respectively. The explicit dependence of the propagation constant on basic parameters is obtained on substituting for μ and κ from equation (2.18) in (2.24); this yields

$$\beta_\pm/\beta_d = \left(\frac{\omega_0 + \omega_m \mp \omega}{\omega_0 \mp \omega}\right)^{1/2} \tag{2.25}$$

The dispersion curves for the two modes for a typical set of parameters are shown in Figure 2.4. It is seen that the LCP mode is practically nondispersive and corresponds to the O-mode discussed in the preceding section. However, the RCP mode is strongly dispersive and resembles the E-mode propagating perpendicular to the dc field. It follows from equation (2.25) that the LCP mode propagates at all frequencies while there is no propagation of the RCP mode when $\omega_0 < \omega < (\omega_0 + \omega_m)$; the cut-off range

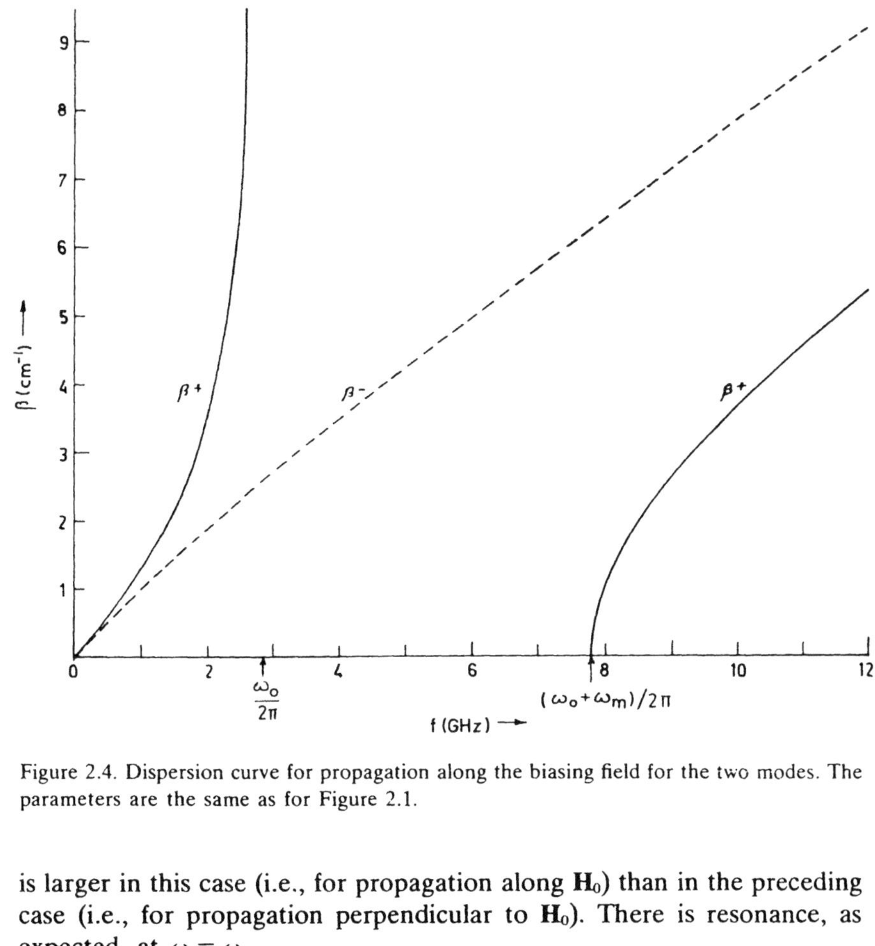

Figure 2.4. Dispersion curve for propagation along the biasing field for the two modes. The parameters are the same as for Figure 2.1.

is larger in this case (i.e., for propagation along $\mathbf{H}_0$) than in the preceding case (i.e., for propagation perpendicular to $\mathbf{H}_0$). There is resonance, as expected, at $\omega = \omega_0$.

The behavior of the RCP and LCP modes can be qualitatively understood as follows. It was discussed in Chapter 1 that, in the presence of a dc magnetic field, the magnetic dipoles in the medium execute a clockwise* precession about the field with the precessional frequency ω_0. The magnetic field of an RCP wave propagation along the dc magnetic field also rotates in the clockwise direction and hence enhances the amplitude of precession of the dipoles. Therefore, the RCP mode interacts strongly with the medium and, specifically, induces resonance when the two frequencies are equal, i.e., when $\omega = \omega_0$. On the other hand, the tendency of the LCP mode is obviously to oppose the precession of dipoles. Hence, this mode interacts rather weakly with the medium.

* This is so when one looks along the dc field direction.

It is worth considering the effect of reversal of the direction of propagation on the modes. Physically, the effect of path reversal is the same as that of the reversal of the direction of the dc magnetic field. The latter can be effected by changing the sign of κ as discussed in Section 1.3. It follows from equations (2.24) and (2.13)–(2.15) that the expressions for propagation constants for RCP and LCP modes are interchanged when the direction of wave propagation is reversed. Thus, the propagation of a mode of specific polarization is, in general, nonreciprocal. Moreover, in the case of propagation opposite to the direction of the dc field, it is the LCP mode which strongly interacts* with the medium and leads to resonance under favorable conditions.

The phase velocity of the two modes is given by

$$v_{\text{ph}}^{\pm} = \frac{c_0}{\sqrt{\varepsilon_f}} \left(\frac{\omega_0 \mp \omega}{\omega_0 + \omega_m \mp \omega} \right)^{1/2} \tag{2.26}$$

The normalized phase velocities for the two modes are displayed in Figure 2.5. It is seen that in a broad frequency range, the phase velocities of the two modes are appreciably different from each other. It will be shown below that this leads to the rotation of the plane of polarization of an electromagnetic wave propagating along the dc field (Faraday rotation).

The group velocity is obtained as

$$v_g^{\pm} = \frac{c_0^2}{\varepsilon_f} \frac{2\beta_{\pm}(\omega \mp \omega_0)}{3\omega^2 \mp 2\omega(\omega_0 + \omega_m) - c_0^2 \beta_{\pm}^2 / \varepsilon_f} \tag{2.27}$$

The normalized group velocity for the two modes is shown in Figure 2.6. The general nature of the group velocity for the RCP mode is similar to that for E-mode discussed in the preceding section. The LCP mode exhibits practically constant group velocity except in the low-frequency range, where it is slightly dispersive.

Faraday Rotation. If a linearly polarized electromagnetic wave is directed along the dc field, its polarization cannot remain unchanged during the propagation, because it is not a characteristic mode of propagation in the medium. A linearly polarized wave can, however, be resolved into two counterrotating circularly polarized waves, each of which does propagate as a mode. The state of polarization after a certain distance of propagation can be determined by recombining the RCP and LCP modes, at that point, in their correct phases. When the amplitudes of the RCP and LCP modes are the same (unequal attenuation is ignored), the wave

* See Bolle and Lewin (1973) for a further discussion.

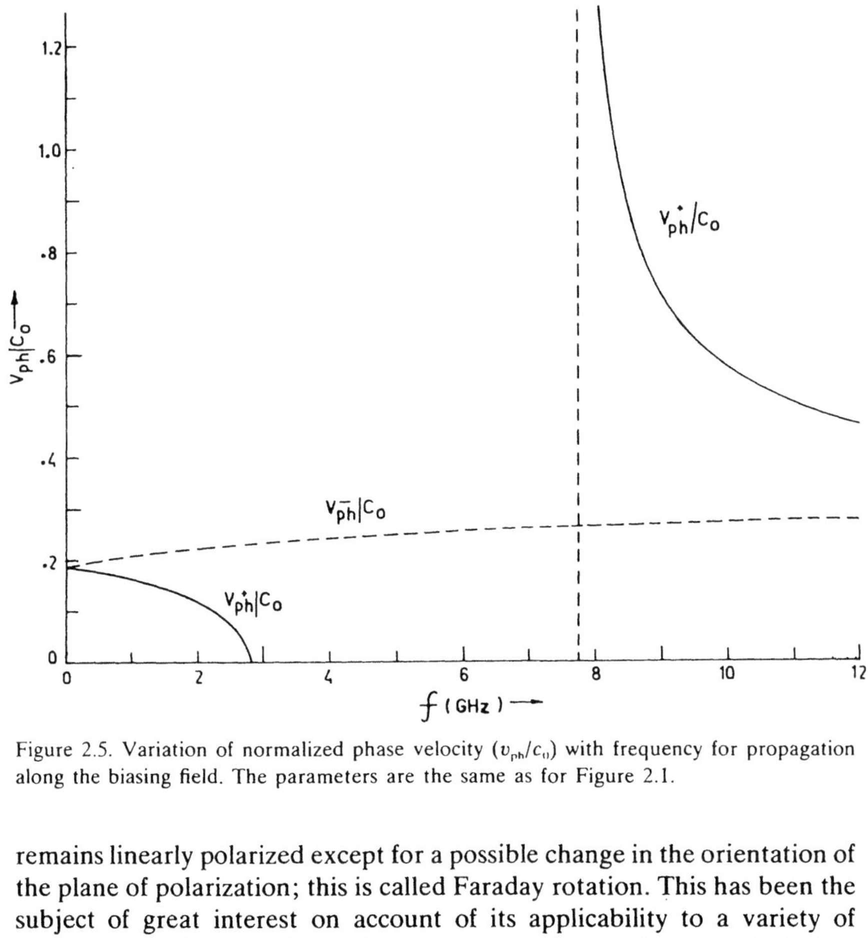

Figure 2.5. Variation of normalized phase velocity (v_{ph}/c_0) with frequency for propagation along the biasing field. The parameters are the same as for Figure 2.1.

remains linearly polarized except for a possible change in the orientation of the plane of polarization; this is called Faraday rotation. This has been the subject of great interest on account of its applicability to a variety of waveguide devices, as discussed in numerous books and reviews including the ones by Hogan (1952, 1953), Lax and Button (1962), Waldron (1961), Gurevich (1965), von Aulock and Fay (1968). In what follows, we shall present an elementary discussion of Faraday rotation in ferrimagnetics in order to bring out the essential features of this effect.

Consider a linearly polarized plane wave at $z = 0$. For polarization along the x-axis, the electric field is given by

$$\boldsymbol{\mathscr{E}}(z = 0) = \mathscr{E}_0 \cos \omega t \hat{x}$$

In complex notations, $\boldsymbol{\mathscr{E}}$ can be represented by $\mathbf{E}$ as

$$\begin{aligned}\mathbf{E} &= \mathscr{E}_0 \exp(j\omega t)\hat{x} \\ &= \tfrac{1}{2}\mathscr{E}_0(\hat{x} + j\hat{y})\exp(j\omega t) + \tfrac{1}{2}\mathscr{E}_0(\hat{x} - j\hat{y})\exp(j\omega t)\end{aligned}$$

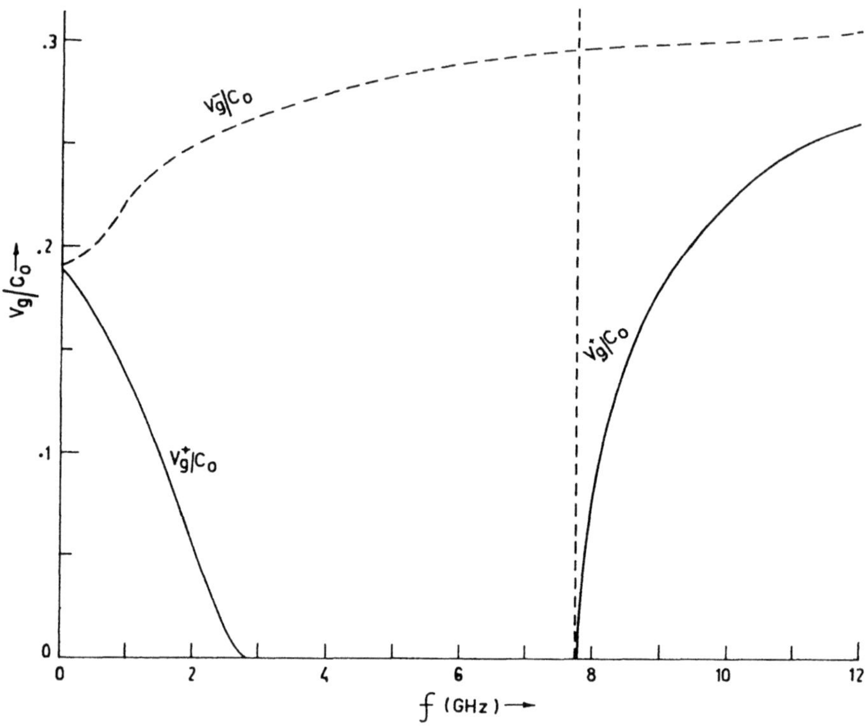

Figure 2.6. Variation of normalized group velocity (v_g/c_0) with frequency for propagation along the biasing field. The parameters are the same as for Figure 2.1.

The two terms represent LCP and RCP modes, respectively. After propagating a distance l, the LCP and RCP modes undergo phaseshifts by $\beta_- l$ and $\beta_+ l$, respectively. Thus, neglecting absorption, the electric field at $z = l$ is given by

$$\mathbf{E}(z = l) = \tfrac{1}{2}\mathscr{E}_0(\hat{x} + j\hat{y})\exp(j\omega t - \beta_- l) + \tfrac{1}{2}\mathscr{E}_0(\hat{x} - j\hat{y})\exp(j\omega t - \beta_+ l)$$

Converting back to real quantities, we have

$$\begin{aligned}\boldsymbol{\mathscr{E}}(z = l) &= \tfrac{1}{2}\hat{x}\mathscr{E}_0[\cos(\omega t - \beta_+ l) + \cos(\omega t - \beta_- l)] \\ &\quad + \tfrac{1}{2}\mathscr{E}_0\hat{y}[\sin(\omega t - \beta_+ l) - \sin(\omega t - \beta_- l)] \\ &= \mathscr{E}_0 \cos\left[\omega t - \frac{\beta_+ + \beta_-}{2} l\right]\left[\hat{x}\cos\left(\frac{\beta_- - \beta_+}{2} l\right)\right. \\ &\quad \left. + \hat{y}\sin\left(\frac{\beta_- - \beta_+}{2} l\right)\right] \qquad (2.28)\end{aligned}$$

Evidently, the electric field is still linearly polarized with polarization along the unit vector $\hat{n}_r$:

$$\hat{n}_r = \hat{x}\cos\left(\frac{\beta_- - \beta_+}{2}l\right) + \hat{y}\sin\left(\frac{\beta_- - \beta_+}{2}l\right) \tag{2.29}$$

Thus, the angle θ_F through which the plane of polarization rotates is given by

$$\theta_F = \arccos(\hat{n}_r \cdot \hat{x}) = \frac{\beta_- - \beta_+}{2}l$$

Hence,

$$\theta_F/l = (\beta_- - \beta_+)/2 \tag{2.30}$$

Substituting for $\beta_\pm$ from equation (2.25), we obtain the rotation per unit length as

$$\theta_F/l = \frac{\omega\sqrt{\varepsilon_f}}{2c_0}[(\mu - \kappa)^{\frac{1}{2}} - (\mu + \kappa)^{\frac{1}{2}}] \tag{2.31}$$

The introduction of (unequal) losses for the two modes leads to "Faraday ellipticity." The detailed treatment of Faraday rotation and ellipticity can be found in the literature cited earlier.

2.2.2.2. *Lossless Hexagonal Planar Magnetoplumbites*

(a) *Propagation Perpendicular to* $\mathbf{H}_0$. Proceeding as in Section 2.2.2.1, it is seen that, as before, one of the modes does not interact with the magnetic nature of the medium. The propagation constant for the other (interacting) mode is obtained as

$$(\beta/\beta_d)^2 = \Delta/(\mu_{xx}\sin^2\varphi + \mu_{yy}\cos^2\varphi) \tag{2.32}$$

The polarization characteristics for the modes are qualitatively the same as in the corresponding case of isotropic ferrimagnetics. It is clear from equation (2.32) that the propagation constant is different for propagation in different directions within the plane perpendicular to the dc field. Substituting for Δ, μ_{xx}, and μ_{yy}, from equation (1.88), the propagation constant transforms to

$$\beta = \beta_d \left[\frac{(\omega_0 + \omega_m)(\omega_0 + \omega_a + \omega_m) - \omega^2}{\omega_0(\omega_0 + \omega_a + \omega_m) + \omega_a\omega_m \sin^2\varphi - \omega^2} \right]^{1/2} \tag{2.33}$$

It follows that there is no propagation when

$$\omega_0(\omega_0 + \omega_a + \omega_m) + \omega_a\omega_m \sin^2\varphi < \omega^2 < (\omega_0 + \omega_m)(\omega_0 + \omega_a + \omega_m) \equiv \omega_4^2$$

and the resonance occurs at the lower limit. In the special case of propagation along the hard axis (y-axis), we have $\varphi = 0$ and resonance occurs when $\omega = [\omega_0(\omega_a + \omega_0 + \omega_m)]^{1/2} \equiv \omega_2$. On the other hand, in the case of propagation in a direction within the easy plane but perpendicular to the dc field, we have $\varphi = 90°$ and the resonance is achieved for

$$\omega = [(\omega_0 + \omega_a)(\omega_0 + \omega_m)]^{1/2} \equiv \omega_3$$

For planar magnetoplumbites with large anisotropy (see Appendix A), ω_3 is appreciably different from ω_2 for $H_0 \ll H_a$ and a strong asymmetry (in the propagation characteristics) about the biasing field is expected. As such, the microwave behavior of transversely biased planar magnetoplumbites is significantly affected by the orientation of the hard axis; some interesting consequences of this effect were theoretically predicted by Sodha and Srivastava (1976) and Srivastava (1976a, b) and are discussed later in the present monograph.

(b) *Propagation Parallel to* $\mathbf{H}_0$. In this case, the propagation constant is obtained as

$$\left(\frac{\beta_\pm}{\beta_d}\right)^2 = \frac{\delta_0 \pm (\delta_0^2 - 4\Delta)^{1/2}}{2} \tag{2.34}$$

Substituting for δ_0 and Δ from equation (2.18), we have

$$\left(\frac{\beta_\pm}{\beta_d}\right)^2 = \frac{\omega_0(\omega_0 + \omega_a + \omega_m) + \frac{1}{2}\omega_a\omega_m - \omega^2 \pm \omega_m(\omega^2 + \frac{1}{4}\omega_a^2)^{1/2}}{\omega_0(\omega_0 + \omega_a) - \omega^2} \tag{2.35}$$

Numerical calculations show that the mode with propagation constant β_- is practically nondispersive, as in the case of an isotropic ferrite. The mode with propagation constant β_+, on the other hand, exhibits strong dispersion. This mode is nonpropagating in the frequency range

$$\omega_1 \equiv [\omega_0(\omega_0 + \omega_a)]^{1/2} < \omega < \omega_4$$

the lower limit represents the resonance condition. It is not difficult to show that the modes are right- and left-handed elliptically polarized in contrast to the circularly polarized modes in the case of isotropic ferrimagnetics.

2.2.3. *Lossless Ferrimagnetics: The General Case*

2.2.3.1. *Mode Classification*

It is evident from equation (2.16) that, at the most, two independent waves with propagation constants β_+ and β_- can propagate in a given direction (φ, θ). In general, these modes are elliptically polarized. Only one of these modes has the phase of the wave magnetic field appropriate for enhancing the clockwise gyration of the magnetization vector about the biasing field, thus leading to the possibility of resonance under suitable conditions. This wave is termed the resonant mode or, in analogy with optics, extraordinary or E-mode. The other mode is the nonresonant or ordinary (O) mode. While identifying these modes, one should be particularly careful in choosing the sign before the square root appearing in the numerator of the right-hand side of equation (2.16) to avoid jumping from one branch of the dispersion curves to the other when the parameters are varied. In order to appreciate this point better, we reconsider the special case of propagation parallel to the dc field. The substitution for δ_0 and Δ in equation (2.34) from (2.18) (without any further simplification) yields

$$\left(\frac{\beta_\pm}{\beta_d}\right)^2 = \frac{\omega_2^2 - \omega^2 + \frac{1}{2}\omega_a\omega_m}{\omega_1^2 - \omega^2} \pm \left[\frac{\omega_m^2(\omega^2 + \frac{1}{4}\omega_a^2)}{(\omega_1^2 - \omega^2)^2}\right]^{1/2} \tag{2.36}$$

For consistency, only a positive quantity will be taken out of the radical sign in this equation. When $\omega < \omega_1$, equation (2.36) reduces to

$$\left(\frac{\beta_\pm}{\beta_d}\right)^2 = \frac{\omega_2^2 - \omega^2 + \frac{1}{2}\omega_a\omega_m \pm \omega_m(\frac{1}{4}\omega_a^2 + \omega^2)^{1/2}}{\omega_1^2 - \omega^2} \tag{2.37}$$

On the other hand when $\omega > \omega_1$, equation (2.36) reduces to

$$\left(\frac{\beta_\pm}{\beta_d}\right)^2 = \frac{\omega_2^2 - \omega^2 + \frac{1}{2}\omega_a\omega_m \mp \omega_m(\frac{1}{4}\omega_a^2 + \omega^2)^{1/2}}{\omega_1^2 - \omega^2} \tag{2.38}$$

The behavior of the RHSs of equations (2.37) and (2.38) can be examined in the limit when $\omega \to \omega_1$. Since $\beta \to \infty$ at resonance, it is seen that the wave with propagation constant $\beta_+(\beta_-)$ is the resonant E-mode if ω is less (greater) than ω_1. The reverse is true for the nonresonant O-mode. In other words, if we decide to take only positive quantities out of the radical sign, the wave branch with propagation constant β_+ (β_-) in the region $\omega < \omega_1$ corresponds to the wave branch with propagation constant β_- (β_+) in the region $\omega > \omega_1$. This conclusion is also valid for propagation in an arbitrary direction relative to the biasing field.

2.2.3.2. *Propagating and Nonpropagating Regions*

Since equation (2.16) is symmetric in θ and φ, it is sufficient to investigate its consequences only for $0 \leqslant \varphi \leqslant \pi$ and $0 \leqslant \theta \leqslant \pi/2$. The behavior of the modes is investigated in this section in the following regimes.

(a) $0 < \omega < \omega_1$. It is seen from equations (C.18) and (2.18) that μ'', δ_0, and $(\Delta - \mu')$ are all positive in this range. Therefore, according to equation (2.16), β_e^2 $(= \beta_+^2)$ is positive. A comparison of terms appearing within square brackets in the numerator of the right-hand side of equation (2.16) shows that β_0^2 $(= \beta_-^2)$ is also positive. Hence both the modes propagate in this frequency range.

(b) $\omega_1 < \omega < \omega_2$. In this range, δ_0 and $(\Delta - \mu')$ are both negative. Consequently, for the E-mode, the numerator in equation (2.16) is negative throughout the range. Thus, the propagation constant for the E-mode is now β_-. The wave propagation would still be possible if the denominator were also negative, the condition for which is obtained as

$$\theta < \theta_c(\varphi) \tag{2.39}$$

where

$$\cos^2 \theta_c(\varphi) = \frac{\omega^2 - \omega_1^2}{\omega_m(\omega_0 + \omega_a \sin^2 \varphi)} \tag{2.40}$$

In the case of the O-mode, the numerator is positive/negative if

$$(\Delta - \mu')^2 \cos^4 \theta + (\mu' - \mu'')(2\Delta - \delta_0)\cos^2 \theta + (\delta_0^2 - 4\Delta)\sin^2 \theta \gtrless [-\delta_0 - (\Delta - \mu')\cos^2 \theta]^2$$

or if

$$\sin^2 \theta + \mu'' \cos^2 \theta \gtrless 0 \tag{2.41}$$

which implies that both the numerator and denominator in the right-hand side of equation (2.16) have the same sign for the frequency range in question. Hence the O-mode propagates in all directions for $\omega_1 < \omega < \omega_2$.

The analysis can be carried out for other frequency ranges, such as $\omega_2 < \omega < \omega_3$, $\omega_3 < \omega < \omega_4$, and $\omega > \omega_4$, in a similar fashion. The results are summarized as follows. The O-mode propagates in all directions for all frequencies and differs little from the case of clamped spins. The E-mode propagates in all directions for $\omega < \omega_1$ and $\omega > \omega_4$. However, when $\omega_1 < \omega < \omega_4$, the E-mode propagates only in the specific directions for which the conditions (2.39) and (2.40) hold. Figure 2.7 shows the propagating and nonpropagating region for the E-mode in θ–φ space. The data for Zn_2Y planar magnetoplumbite (Appendix A) have been used in the numerical calculations. For each ω, there is one curve $\theta = \theta_c(\varphi, \omega)$ in the θ–φ space. The figure shows only the quadrant $0 \leq \theta \leq \pi/2$, $0 \leq \varphi \leq \pi$; the curves in the remaining quadrants can be completed by symmetry about θ- and φ-axes. The propagation of a wave of given frequency ω is allowed along a certain direction characterized by angles θ_0 and φ_0 provided that $\theta_0 < \theta_c(\omega, \varphi_0)$, i.e., when the point (θ_0, φ_0) lies in between the

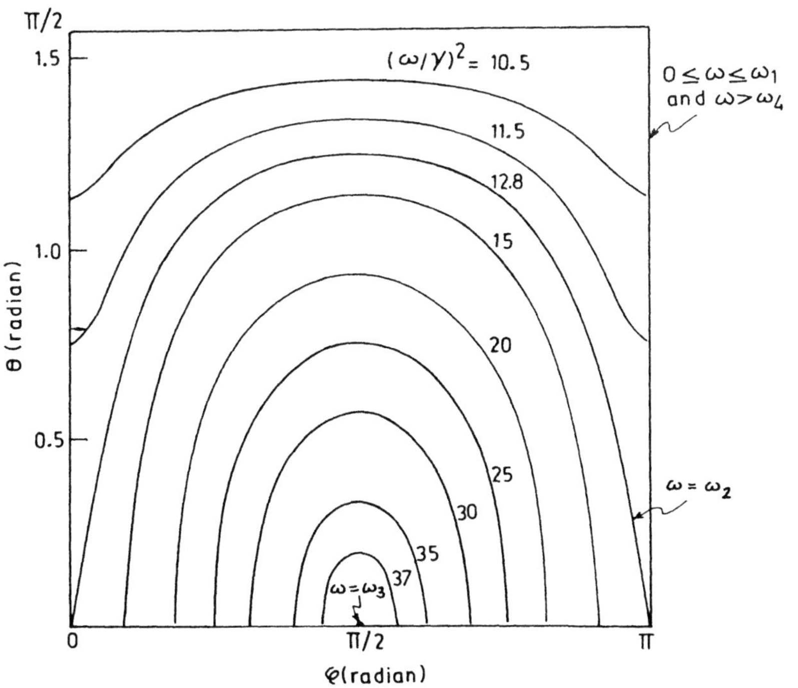

Figure 2.7. The function $\theta = \theta_c(\varphi, \omega)$ in θ–φ space for different $(\omega/\gamma)^2$. The parameters are $H_0 = 1000$ Oe, $4\pi M_0 = 2800$ G, and $M_a = 9000$ Oe (after Srivastava, 1977).

corresponding curve and the φ-axis. The various resonances and cut-off limits are clearly represented in the figure.

The frequency dependence of the effective permeability defined by $(\beta/\beta_d)^2$ is displayed in Figure 2.8, for the two modes propagating along different directions (θ, φ). Only two curves, one each for $\theta = 0°$ and $90°$, have been shown for the O-mode; there is no φ-dependence in either case. The curves for all other combinations of θ and φ lie in between the two. Clearly, the O-mode is practically nondispersive. On the other hand, the E-mode is strongly dispersive except for frequencies above resonance. The effective permeability for propagation within the plane containing the hard axis and biasing field axis (i.e., the case in which $\varphi = 0$) does not vary significantly with θ. But, as φ increases, the effective permeability exhibits stronger dependence on θ. For a given θ ($\neq \pi/2$), the resonance occurs at higher frequencies for larger values of φ.

The explicit effect of planar anisotropy on effective permeability has been shown in Figure 2.9. The curves have been drawn corresponding to the three specific cases: (i) $\theta = 0$, (ii) $\theta = 90°$, $\varphi = 0$, and (iii) $\theta = 90°$, $\varphi = 90°$. It is seen that as the anisotropy field is increased, there is a general shift of the region of strong dispersion toward higher frequencies. The

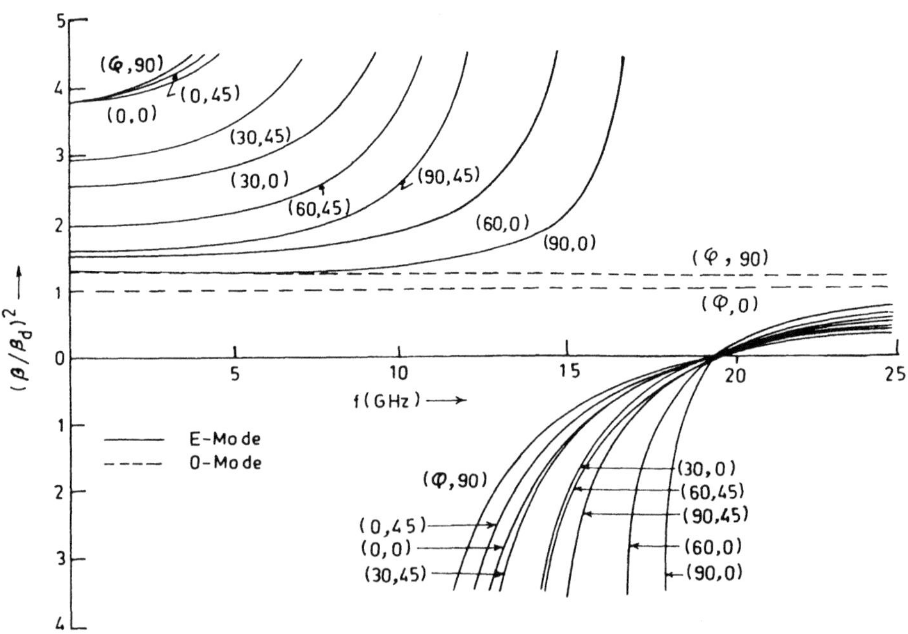

Figure 2.8. Frequency dependence of $(\beta/\beta_d)^2$ for propagation along the direction (φ, θ). The parameters are the same as for Figure 2.7 (after Srivastava, 1977).

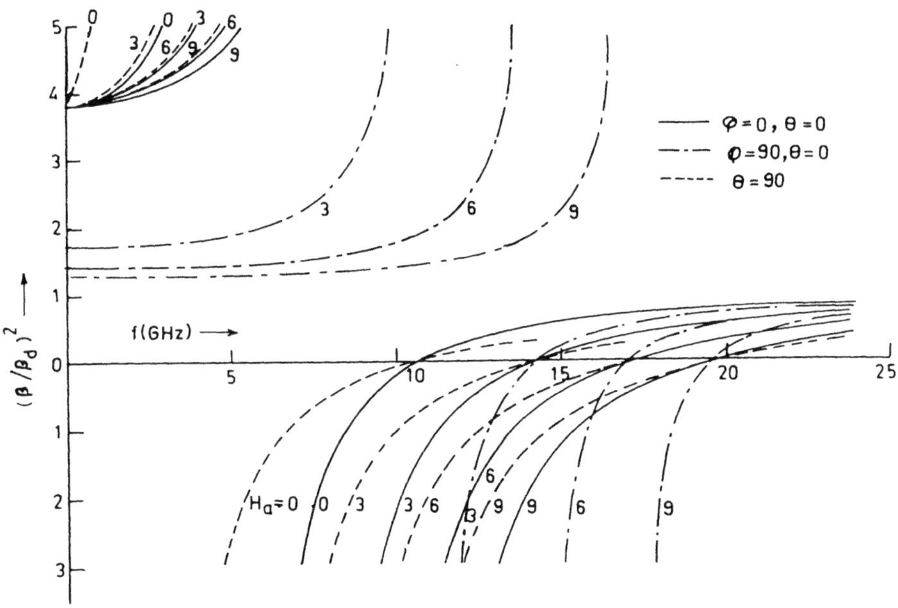

Figure 2.9. Frequency dependence of $(\beta/\beta_d)^2$ for different values of the anisotropy field H_a. The curves for $\theta = 90°$ have not been completed above their respective cut-offs (after Srivastava, 1977).

resonance frequency is largest for propagation along the direction perpendicular to the plane containing the hard axis and the direction of the biasing field. Although the general shape of dispersion curves is not significantly affected by the planar anisotropy,* there is no quantitative equivalence between isotropic and planar ferrimagnetics particularly when the anisotropy field is large compared to the biasing field and saturation magnetization.

2.2.4. *Effect of Damping*†

When the resonance linewidth ΔH is nonzero, the propagation constant given by equation (2.16) is rendered complex; this leads to attenuation of the wave as it propagates down the medium. The loss factor $\alpha_{\pm}$, defined as the imaginary part of the propagation constant, can be obtained from numerical solution of equation (2.16). The results have been

* Because this shape is the characteristic of anomalous dispersion (loss neglected).

† This section follows Lax and Button (1962).

displayed in Figure 2.10 for a typical set of parameters. The O-mode is practically lossless while the E-mode exhibits strong attenuation around the resonance. The effect of increased linewidth on the O-mode is rather insignificant. However, the absorption curve for the E-mode broadens out as ΔH increases.

Sometimes, it is advantageous to have an approximate analytical expression for the absorption coefficient in the limit of small linewidth. This is easily obtained for the special case of propagation in isotropic ferrimagnetics in directions parallel and perpendicular to the dc field. In the case of propagation along the dc field, the loss factor for the two modes can be obtained from equation (2.24). If μ and κ are assumed to be complex and equal to $\mu' - j\mu''$ and $\kappa' - j\kappa''$, respectively, the imaginary part of the propagation constant is obtained as

$$\alpha_{\pm} = \frac{\omega}{c_0}\left(\frac{\varepsilon_f}{2}\right)^{1/2} \{[(\mu' \pm \kappa')^2 + (\mu'' \pm \kappa'')^2]^{1/2} - (\mu' \pm \kappa')\}^{1/2} \tag{2.42}$$

Numerical calculations show that α_- is, in general, quite small. When the dc magnetic field is smaller than its resonance value such that μ' and κ' are large and negative and μ'' and $\kappa'' \simeq 0$ (see Figure 1.6), α_+ in equation (2.42)

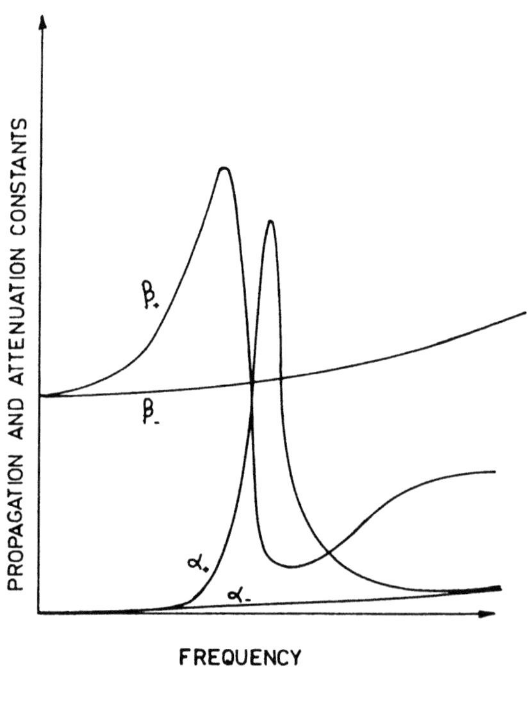

Figure 2.10. Qualitative plot of frequency dependence of the propagation and attenuation constants for the two modes of propagation along the biasing field.

reduces to

$$\alpha_+ \simeq (\omega/c)(\varepsilon_f|\mu' + \kappa'|)^{1/2} \tag{2.43}$$

On the other hand, in the region above resonance, where μ' and κ' are large and positive and μ'', $\kappa'' \simeq 0$, α_+ reduces to

$$\alpha_+ = (\omega\sqrt{\varepsilon_f}/2c_0)(\mu'' + \kappa'')/(\mu' + \kappa')^{1/2} \tag{2.44}$$

If an RCP wave is directed along the dc magnetic field, its attenuation coefficient is α_+, which is large in the vicinity of resonance. However, when an RCP wave is directed against the dc field, its absorption coefficient is α_- (see Section 2.2.2), which is quite small. Hence, the attenuation of a wave of given polarization is nonreciprocal. Moreover, if a plane wave is directed along the dc field, the constituent RCP and LCP components obviously undergo unequal attenuation during propagation and recombine to lead to elliptical rather than linear polarization. The result is Faraday ellipticity.

In the case of propagation perpendicular to the dc field, the attenuation coefficient for the E-mode is obtained from equation (2.19) as

$$\alpha_e = (\omega/c_0)(\varepsilon_f/2)^{1/2}[(\mu'^2_{\text{eff}} + \mu''^2_{\text{eff}})^{1/2} - \mu'_{\text{eff}}]^{1/2} \tag{2.45}$$

where

$$\mu'_{\text{eff}} = (\mu'^2 - \kappa'^2)/\mu' \tag{2.46}$$

$$\mu''_{\text{eff}} = (1 + \kappa'^2/\mu'^2)/\mu'' - 2\kappa''\kappa'/\mu' \tag{2.47}$$

If the linewidth is small, $\mu''_{\text{eff}} \simeq 0$. In this case, the approximate expressions for α_e in the cut-off region ($\mu'_{\text{eff}} < 0$) and in the region above resonance ($\mu'_{\text{eff}} > 1$) are obtained as

$$\alpha_e \simeq (\omega/c_0)(\varepsilon_f|\mu'_{\text{eff}}|)^{1/2} \tag{2.48}$$

and

$$\alpha_e \simeq (\omega\sqrt{\varepsilon_f}/2c_0)\mu''_{\text{eff}}/(\mu'_{\text{eff}})^{1/2} \tag{2.49}$$

respectively. The approximate expressions in equations (2.43), (2.44), (2.48), and (2.49) are often useful in the evaluation of losses. In more complicated problems, e.g., guided wave problems, electromagnetic wave reflection, etc., it is quite difficult to estimate losses without getting involved in rigorous numerical analysis. Some approximate methods for evaluation of losses in a variety of situations have been discussed by Rosenbaum (1964) and Gupta and Srivastava (1979).

2.2.5. Comparison with Waves in Magnetoplasmas

Some important propagation characteristics of electromagnetic plane waves in magnetized ferrimagnetics and magnetoplasma are quite similar (Suhl and Walker, 1954). Consequently, several studies (e.g., Engineer and Nag, 1965) of electromagnetic waves in magnetoplasma have followed analogous studies in ferrimagnetics. Similarly, some investigations of electromagnetic waves in ferrimagnetics (e.g., Srivastava, 1978a) have also been motivated by the corresponding studies in magnetoplasmas. Still, there seems to be ample scope for "cross-talk" between these two apparently different areas. Hence, it is worthwhile to compare the propagation of electromagnetic waves in ferrimagnetics with that in magnetoplasmas.

When a solid state or gaseous electron plasma is subjected to a dc magnetic field acting along the z-axis, in the first approximation the electrons move in circular orbits in the xy-plane. The relation between an impressed harmonic electric field **E** and the resulting displacement vector **D** is given by (Steele and Vural, 1969):

$$\mathbf{D} = \varepsilon_{\mathrm{L}} \boldsymbol{\varepsilon} \cdot \mathbf{E} \tag{2.50}$$

where

$$\boldsymbol{\varepsilon} = \begin{pmatrix} \varepsilon & j\kappa_e & 0 \\ -j\kappa_e & \varepsilon & 0 \\ 0 & 0 & \varepsilon_z \end{pmatrix} \tag{2.51}$$

$$\left.\begin{aligned} \varepsilon &= 1 - \omega_{\mathrm{p}}^2(\omega - j\nu)/\omega[(\omega - j\nu)^2 - \omega_{\mathrm{c}}^2] \\ \kappa_e &= \omega_{\mathrm{p}}^2\omega_{\mathrm{c}}/\omega[(\omega - j\nu)^2 - \omega_{\mathrm{c}}^2] \\ \varepsilon_z &= 1 - \omega_{\mathrm{p}}^2[\omega(\omega - j\nu)] \end{aligned}\right\} \tag{2.52}$$

$$\left.\begin{aligned} \omega_{\mathrm{p}}^2 &= 4\pi n_e e^2/m^*\varepsilon_{\mathrm{L}} \\ \omega_{\mathrm{c}} &= eH_0/m^*c_0 \end{aligned}\right\} \tag{2.53}$$

In these expressions, n_e, e, m^*, H_0, c_0, ε_{L}, ν, and ω represent, respectively, the electron density, electronic charge, effective mass, dc magnetic field, speed of light, permittivity of the background lattice, collision frequency and angular frequency of the impressed harmonic field.

The expressions in equations (2.52) and (2.53) are applicable to simple semiconductor plasmas. In the case of gaseous plasmas, $\varepsilon_{\mathrm{L}} = 1$ and $m^* = m$. It is evident from equations (2.51) and (2.52) that the dielectric tensor is quite similar to the permeability tensor for magnetized ferrimagnetics. As a consequence, several characteristics of electromagnetic wave propagation in magnetoplasma resemble those in the case of ferrimagnetics. For instance, it is easy to show that, when collisions are ignored ($\nu = 0$), the modes propagating along the dc field (z-axis) are circularly polarized, thereby leading to the Faraday rotation of a plane polarized wave. The propagation constant for the two modes is given by

$$\beta_{\pm} = \sqrt{\varepsilon_L}(\omega/c_0)(\varepsilon \pm \kappa_e)^{1/2} \tag{2.54}$$

which can be compared with equation (2.24). The RCP wave (β_+) exhibits resonance at $\omega = \omega_c$, the cyclotron frequency, which, incidently, corresponds to the resonance frequency, $\omega_0 = \gamma H_0$, in the case of ferrimagnetics.

For propagation perpendicular to the propagation field (say along the y-axis), only one of the modes interacts with the gyrotropic nature of the medium. The propagation constant for this mode is seen to be

$$\beta = \sqrt{\varepsilon_L}(\omega/c_0)\left(\frac{\varepsilon^2 - \kappa_e^2}{\varepsilon}\right)^{1/2} \tag{2.55}$$

which can be compared with equation (2.19). This mode is characterized by field quantities h_z, E_x, and E_y and is a TM mode. The other mode does not interact with the gyrotropic nature of the medium. It is a pure TEM mode with field vectors E_z and h_x and the propagation constant is given by

$$\beta = \sqrt{\varepsilon_L}\sqrt{\varepsilon_z}\,\omega/c_0 \tag{2.56}$$

which can be compared with equation (2.20).

The preceding discussion illustrates clearly the analogy between ferrimagnetics and magnetoplasmas as regards plane wave propagation. However, it should be emphasized that the analogy between the two media is not complete in all circumstances; the difference becomes evident particularly in the case of loaded waveguide problems. Although the partial differential equations describing the field quantities are the same in the two cases, the different relevant boundary conditions lead to significantly different results, as discussed by Gabriel and Brodwin (1966).

2.3. *The Magnetostatic Approximation*

It was noted earlier that except for frequencies around a certain range, the ferrimagnetics behave more or less like dielectrics. In that case, the propagation constant is given approximately by $\beta = \varepsilon_f^{1/2}\omega/c_0$, while the wave electric and magnetic fields are of comparable magnitudes, as one would expect for a simple dielectric. In the frequency range in which the phase and group velocities differ appreciably from those in a dielectric, it is worth comparing the magnitudes of the electric and magnetic fields of the wave. We follow the treatments of Auld (1960) and Steele and Vural (1969).

For simplicity, the discussion will be limited to the case of a lossless isotropic ferrite. For propagation within the xz-plane, along a direction making an angle δ with the z-axis, the propagation constant can be obtained from equation (2.16) on substituting $\varphi = -\pi/2$ and $\theta = \pi/2 - \delta$ along with $H_a = 0$ and $\Delta H = 0$. With equations (2.13)–(2.15), the components of $\mathbf{h}$, except for a common scaling factor, can be obtained as (Steele and Vural, 1969)

$$\mathbf{h} = \begin{bmatrix} -j(\mu(\beta_d/\beta)^2 - 1)/\kappa \\ (\beta_d/\beta)^2 \\ -j\{(\mu(\beta_d/\beta)^2 - 1)/\kappa\}\{\sin\delta\cos\delta/(\sin^2\delta - (\beta_d/\beta)^2)\} \end{bmatrix} \exp j(\omega t - \beta\hat{n}\cdot\mathbf{r}) \tag{2.57}$$

where $\hat{n} = \cos\delta\hat{z} + \sin\delta\hat{x}$. The electric field of the wave can be obtained by substituting for $\mathbf{h}$ in equations (2.10), which yields

$$\mathbf{E} = \frac{\beta_d}{\beta\sqrt{\varepsilon_f}} \begin{bmatrix} \cos\delta \\ j\{(\mu(\beta_d/\beta)^2 - 1)/\kappa\}\{\cos\delta/(\sin^2\delta - (\beta_d/\beta)^2)\} \\ -\sin\delta \end{bmatrix} \exp j(\omega t - \beta\hat{n}\cdot\mathbf{r}) \tag{2.58}$$

Now consider propagation perpendicular to the dc field ($\delta = 90°$) in the case when $\beta \gg \beta_d$. It is seen from Figure 2.1 that $\omega \simeq [\omega_0(\omega_0 + \omega_m)]^{1/2}$ in which case $\mu \to 0$ while $\kappa \to -(1 + \omega_m/\omega_0)^{1/2}$. Under these approximations, the wave magnetic field in equation (2.57) reduces to

$$\mathbf{h} = \begin{bmatrix} -j/(1 + \omega_m/\omega_0)^{1/2} \\ (\beta_d/\beta)^2 \\ 0 \end{bmatrix} \exp j(\omega t - \beta x) \tag{2.59}$$

The electric field in equation (2.58) reduces to

$$\mathbf{E} = -\left(\frac{\beta_d}{\beta}\right)\frac{1}{\sqrt{\varepsilon_f}}\begin{pmatrix} 0 \\ 0 \\ -1 \end{pmatrix} \exp j(\omega t - \beta x) \tag{2.60}$$

Evidently, since $\beta \gg \beta_d$, it is possible to neglect h_y and $\mathbf{E}$, and thus we have

$$\mathbf{h} = \begin{bmatrix} -j/(1+\omega_m/\omega_0)^{1/2} \\ 0 \\ 0 \end{bmatrix} \exp j(\omega t - \beta x) \tag{2.61}$$

It follows that $\nabla \times \mathbf{h} = 0$; the waves may be called magnetostatic.

Next we consider propagation along the dc magnetic field ($\delta = 0$) in the limit $\beta \gg \beta_d$. It is evident from Figure 2.4 that $\omega \simeq \omega_0$ in which case $\mu, \kappa \to \infty$. Therefore, $\mathbf{h}$ in equation (2.57) and $\mathbf{E}$ in equation (2.58) reduces to

$$\mathbf{h} = \begin{bmatrix} j(\beta_d/\beta)^2 \\ (\beta_d/\beta)^2 \\ 0 \end{bmatrix} \exp j(\omega t - \beta z) \tag{2.62}$$

and

$$\mathbf{E} = \frac{\beta_d}{\beta} \frac{1}{\sqrt{\varepsilon_f}} \begin{pmatrix} 1 \\ -j \\ 0 \end{pmatrix} \exp j(\omega t - \beta z) \tag{2.63}$$

respectively. It follows that $\mathbf{E}$ and $\mathbf{h}$ are still circularly polarized and that the components of $\mathbf{E}$ and $\mathbf{h}$ are quite large in magnitude. Hence, the magnetostatic condition is not satisfied in this case. For propagation along an arbitrary direction ($\delta \neq 90°$ or $0°$), the magnetostatic condition is approximately satisfied except in the neighborhood of $\delta = 0°$.

The magnetostatic approximation was first suggested by Mercereau and Feynman (1956) in order to resolve the spurious modes in ferromagnetic resonance of a single-crystal sphere, as observed by White and Solt (1956). Walker (1957) solved the following set of magnetostatic equations for a sphere:

$$\left.\begin{aligned} \nabla \times \mathbf{h} &= 0 \\ \nabla \cdot (\boldsymbol{\mu} \cdot \mathbf{h}) &= 0 \end{aligned}\right\} \tag{2.64}$$

Since then, this set of equations has been extensively used to investigate a wide variety of magnetic wave problems. While the magnetostatic approximation greatly simplifies the mathematical analysis, the results are almost as good as the ones obtained from rigorous analysis. Nevertheless, it is desirable that each result based on magnetostatic approximation be checked by rigorous analysis; there are situations (see, e.g., Srivastava 1978b), apart from the $\delta = 0°$ case discussed above, in which the magnetostatic analysis can lead to basically wrong results in certain domains, even when $\beta \gg \beta_d$.

2.4. Exchange-Dominated Plane Waves

When the wavelength is so small that the wavenumber exceeds $10^4\,\text{cm}^{-1}$ or so, the effect of exchange interaction starts to influence the plane wave characteristics. It was noted in Section 1.6 that the exchange interaction can be accounted for by replacing H_0 by $H_0 + D_{ex}\beta^2$; $D_{ex} = H^{\omega} a_1^2$ in the permeability tensor, as is revealed by equation (1.48). Accordingly, the propagation constant β for plane waves in isotropic ferrites, with allowance for exchange effects, can be obtained from equation (2.16) on replacing H_0 by $H_0 + D_{ex}\beta^2$ and substituting $H_a = 0$:

$$\left(\frac{\beta}{\beta_d}\right)^2 = \frac{(\mu^2 - \mu - \kappa^2)\sin^2\delta + 2\mu}{2(\mu\sin^2\delta + \cos^2\delta)} \pm \frac{[(\mu^2 - \mu - \kappa^2)^2\sin^4\delta + 4\kappa^2\cos^4\delta]^{1/2}}{2(\mu\sin^2\delta + \cos^2\delta)} \tag{2.65}$$

where $\delta = \pi/2 - \theta$ is the angle between the dc field (z-axis) and the direction of propagation while μ and κ incorporate the effect of exchange in a manner discussed above. Typical dispersion curves for the two modes in the case of propagation along ($\delta = 0°$) and perpendicular ($\delta = 90°$) to the dc field are shown in Figure 2.11; the curves for intermediate values of δ lie in between the two sets of curves plotted in the figure. Comparison with exchange-free plane waves (see Figures 2.1 and 2.4) shows that the O-mode is unaffected by the exchange interaction. The dispersion curves for the E-mode are also practically unchanged as long as $\beta \leqslant 10^5\,\text{cm}^{-1}$. However, beyond this limit, the frequency increases monotonically with β. The region $\beta \geqslant 10^5\,\text{cm}^{-1}$ is often called the "spin wave region." The intermediate region, defined by $\beta_d \ll \beta < 10^5\,\text{cm}^{-1}$, in which phase and group velocities are both small, is called the "magnetostatic region." In both regions, the propagation effects are negligible while the wave characteristics are governed by dipolar and exchange effects, respectively.

The dispersion relation for magnetostatic and spin waves can be obtained directly from magnetostatic equations (2.64) in a much simpler way. Assuming $\mathbf{h}$ to be of the form $\exp j(\omega t - \mathbf{k}\cdot\mathbf{r})$; $\mathbf{k} = \beta\hat{n}$, the first of equations (2.64) leads to

$$\mathbf{k} \times \mathbf{h} = 0 \tag{2.66}$$

which implies that the dipolar field $\mathbf{h}$ is parallel to $\hat{\mathbf{n}}$. With use of the second of equations (2.64) and $\mathbf{b} = \mathbf{h} + 4\pi\mathbf{m}$, it is easy to show that

$$\mathbf{h} = -4\pi(\mathbf{m}\cdot\mathbf{k})\mathbf{k}/\beta^2 \tag{2.67}$$

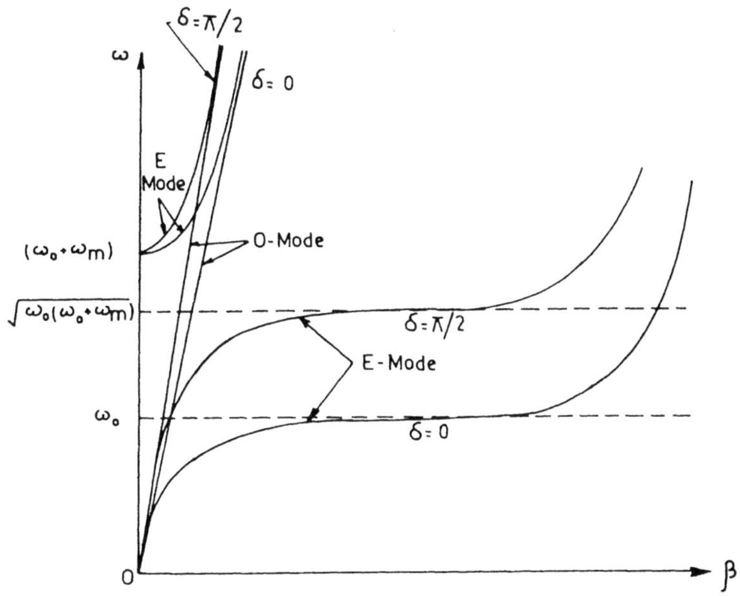

Figure 2.11. Qualitative dispersion diagram for plane waves propagating along ($\delta = 0$) and perpendicular ($\delta = \pi/2$) to the biasing field. The O-mode is unaffected by exchange interaction while the E-mode is modified in the high-wavenumber region (greater than about 10^5 cm^{-1} for a typical set of parameters).

The dipolar field (2.67) and the exchange field as obtained* from equation (1.46) can be included in **H** in equation (1.14); the solution to the latter would, in turn, yield the desired dispersion relation. However, we offer the following simpler procedure.

In consonance with the first of equations (2.64), the magnetostatic potential is defined as

$$\mathbf{h} = \nabla\psi \tag{2.68}$$

In the case of isotropic ferrimagnetics magnetized along the z-axis, the permeability tensor has a form given by equation (1.25). Substitution in the second of equations (2.64) yields

$$[\mu(\partial^2/\partial x^2 + \partial^2\partial y^2) + \partial^2/\partial z^2]\psi = 0 \tag{2.69}$$

* The exchange field $\mathbf{H}_{ex}$ is such that $\mathbf{T}_{ex} = -\gamma\mathbf{M}\times\mathbf{H}_{ex}$; it follows from equation (1.46) that $\mathbf{H}_{ex} = -(D_{ex}/\gamma M_0)\nabla^2\mathbf{M}$.

Now consider the propagation of a plane wave within the xz-plane at an angle δ relative to the z-axis. As such $\partial/\partial y = 0$, and ψ has the form

$$\psi = \psi_0 \exp j(\omega t - \beta \hat{n} \cdot \mathbf{r}) \tag{2.70}$$

where

$$\hat{n} = \cos\delta \hat{z} + \sin\delta \hat{x} \tag{2.71}$$

Substitution of equations (2.70) and (2.71) into equation (2.69) leads to

$$\beta^2(\mu \sin^2\delta + \cos^2\delta)\psi = 0 \tag{2.72}$$

Nontrivial solution to this equation requires that

$$\mu \sin^2\delta + \cos^2\delta = 0 \tag{2.73}$$

Substitution for μ from equation (1.26), with allowance for exchange, leads to the following dispersion relation:

$$\omega = [(\omega_0 + \gamma D_{ex}\beta^2)(\omega_0 + \gamma D_{ex}\beta^2 + \omega_m \sin^2\delta)]^{1/2} \tag{2.74}$$

The dispersion curves for $\delta = 0°$ and $90°$ are shown in Figure 2.12. The uniform precessional mode ($\beta = 0$) for $\theta = 0°$ is the only mode that corresponds to $\omega = \omega_0$. All other modes (different β and θ) have characteristic frequencies above ω_0. Figure 2.12 can be compared with Figure

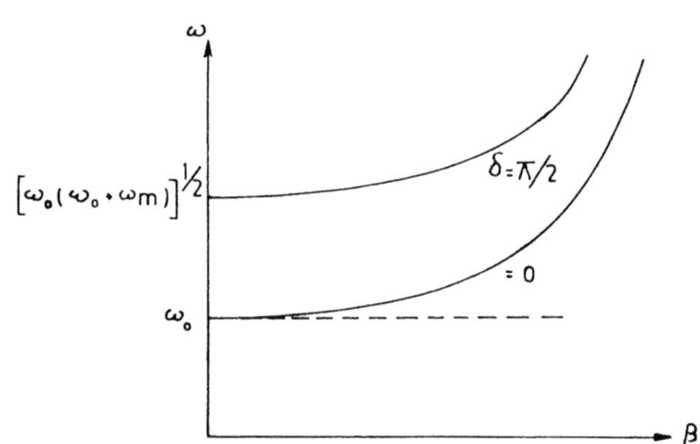

Figure 2.12. Dispersion diagram for plane waves under the magnetostatic approximation. A comparison with Figure 2.11 shows that the magnetostatic approximation leads to correct results in a high-wavenumber region. It should be noted that the uniform precessional mode characterized by frequency $\omega = \omega_0$ is nondegenerate.

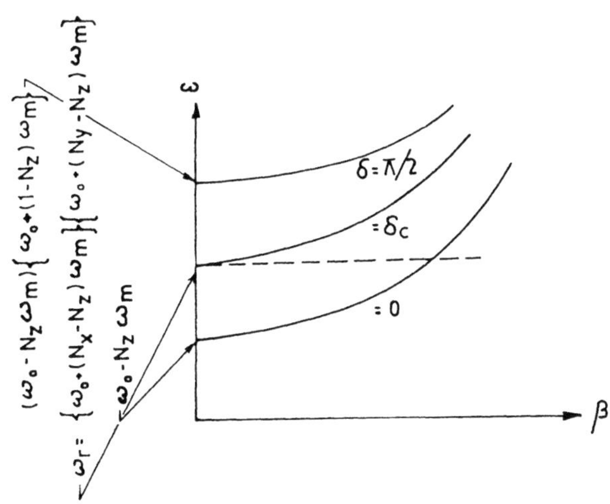

Figure 2.13. Spin wave manifold for a sample of ellipsoidal shape. It should be noted that unlike in the case of infinite medium (Figure 2.12), the uniform precessional mode (broken line) for a finite sample is degenerate with spin waves propagating in different directions $0 < \delta < \delta_c$, for corresponding values of the wavenumber β.

2.11, which was obtained from rigorous analysis. The excellent agreement in the region $\beta \gg \beta_d$ is the justification for application of magnetostatic approximation in this region.

It should be emphasized that equation (2.74) has been obtained for an infinite medium. When a finite sample is considered, it is necessary* to replace ω_0 by $\omega_0 - N_z\omega_m$, and the result is

$$\omega = [(\omega_0 - N_z\omega_m + \gamma D_{ex}\beta^2)(\omega_0 - N_z\omega_m + \gamma D_{ex}\beta^2 + \omega_m \sin^2\delta)]^{1/2} \quad (2.75)$$

Figure 2.13 shows the spin wave dispersion relation for a finite sample for $\delta = 0$ and $\pi/2$. The frequency of the uniform precessional mode (i.e., the resonance frequency ω, the expression for which is derived in Appendix E) for a finite sample is also shown in the figure by a broken line. It is seen that, unlike in the case of an infinite medium (Figure 2.12), the frequency of the uniform precessional mode for a finite sample lies within the "spin

* The wavelength of the spin waves is so small that a finite sample of macroscopic dimensions is big enough to ignore the effect of boundaries. Hence, the dispersion relation is unchanged except that H_0 is to be replaced by the internal dc field which is equal to $H_0 - N_z(4\pi M_0)$.

wave manifold." Thus, in the case of a finite ferrimagnetic sample, there exist spin wave modes (characterized by different combinations of δ and β), having frequencies the same as that of the uniform precessional mode. The existence of these so-called degenerate spin wave modes was first recognized by Clogston et al. (1956), who subsequently explained the anomalously large observed linewidths of ferrimagnetics in terms of the transfer of energy from the uniform precessional mode to the degenerate spin wave modes.

Next we consider the use of hexagonal planar magnetoplumbites. It is easy to show that the magnetostatic potential satisfies the following equation:

$$(\mu_{xx}\partial^2/\partial x^2 + \mu_{yy}\partial^2/\partial y^2 + \partial^2/\partial z^2)\psi = 0 \tag{2.76}$$

where μ_{xx} and μ_{yy} are as given by equation (1.88), assuming the xz-plane to be the easy plane. The effect of spin waves can be included in the expressions for μ_{xx} and μ_{yy} on replacing H_0 by $[H_0 + D_{ex}(k_x^2 + k_z^2) + D'_{ex}k_y^2]$, recognizing the fact that the y-axis (hard axis) is not equivalent to the x- and z-axes, both of which lie within the easy plane. If the direction of the wavevector $\mathbf{k} = \beta\hat{n}$ is defined by polar angles (θ, φ), we have

$$\left.\begin{aligned} k_x &= \beta \sin\theta \cos\varphi \\ k_y &= \beta \sin\theta \sin\varphi \\ k_z &= \beta \cos\theta \end{aligned}\right\} \tag{2.77}$$

Substituting $\psi = \psi_0 \exp(j\omega t - j\beta\hat{n}\cdot\mathbf{r})$ along with equation (2.77) in (2.76), we obtain the dispersion relation as (Schlömann et al., 1963)

$$\begin{aligned} \omega^2 = &\{\omega_0 + \gamma[D_{ex}(k_x^2 + k_z^2) + D'_{ex}k_y^2]\} \\ &\times\{\omega_0 + \gamma[D_{ex}(k_x^2 + k_z^2) + D'_{ex}k_y^2] + \omega_m \sin^2\theta\} \\ &+ \omega_a\{\omega_0 + \omega_m \sin^2\theta\cos^2\varphi + \gamma[D_{ex}(k_x^2 + k_z^2) + D'_{ex}k_y^2]\} \end{aligned} \tag{2.78}$$

In the case of a finite sample, ω_0 should be replaced by $(\omega_0 - N_z\omega_m)$ as noted above. The qualitative dispersion curves for an infinite planar magnetoplumbite are shown in Figure 2.14. The main conclusion is the strong φ-dependence of the spin wave spectrum, particularly when H_a is large, which is true for Zn_2Y or Co_2Y planar magnetoplumbites (see Appendix A).

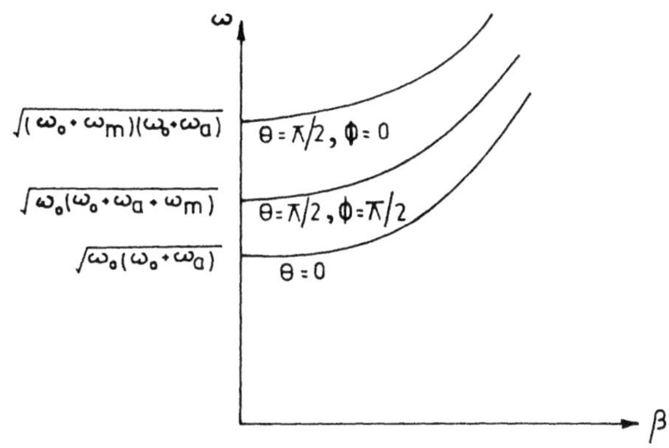

Figure 2.14. Spin wave manifold for infinite hexagonal planar magnetoplumbite. The dispersion curves exhibit a strong φ-dependence particularly when H_a is large and $\theta \neq \pi/2$ (after Schlömann et al., 1963).

PART B. *MAGNETOELASTIC WAVES*

2.5. *Magnetoelastic Field Equations*

As discussed in Section 1.9 of Chapter 1, the magnetoelastic interaction leads to an effective magnetic field (magnetoelastic magnetic field) and an effective stress (magnetoelastic stress). The introduction of this field in equation (1.14) and of the stress in equation (D.27) of Appendix D, leads to the following coupled equations, which govern magnetoelastic wave propagation in a ferrimagnet magnetized along the x_3-axis:

$$d\mathbf{M}/dt = -\gamma M_0 \hat{x}_3 \times \left(\mathbf{h} + \frac{1}{M_0} D_{\mathrm{ex}} \nabla^2 \mathbf{M} + \mathbf{h}_{\mathrm{A}} + \mathbf{h}_{\mathrm{me}}\right) - \gamma \mathbf{m} \times (H_0 \hat{x}_3 + H_{\mathrm{A0}}) \tag{2.79}$$

and

$$\rho \partial^2 u_i / \partial t^2 = \nabla_{iJ} C_{Jk} \nabla_{kl} u_l + \nabla_{iJ} \partial U_{\mathrm{me}} / \partial S_J \tag{2.80}$$

where the body force term has been dropped in equation (2.80). We will now discuss the various terms appearing in these equations. In equation (2.79), $\mathbf{h}$ is the wave magnetic field as given by equation (2.67) in the magnetostatic approximation. As noted in Section (2.3), the magnetostatic approximation should lead to correct results when $\beta \gg \beta_d$; this condition is satisfied for elastic waves. The term $(1/M_0)D_{ex}\nabla^2\mathbf{M}$ represents the exchange field, which depends strongly on the wavelength; as noted in Section 1.6, this term requires consideration only when the wavenumber exceeds $10^5\,\text{cm}^{-1}$. The next term in the right-hand side of equation (2.79) is the time-varying part of the anisotropy field, which is given by equations (1.72)–(1.74) for cubic ferrimagnetics and by equation (1.84) for planar magnetoplumbites. The field $\mathbf{h}_{me}$ refers to the magnetoelastic magnetic field as given by equation (1.111) for a cubic material. The field $\mathbf{H}_{A0}$ represents the dc part of the anisotropy field, which is given by equation (1.63) for cubic ferrimagnetics. In equation (2.80) the first term in the right-hand side is the usual force per unit volume on account of elastic stress, while the second term represents the effect of magnetoelastic interaction, which, in turn, can be inferred from equations (1.110), (1.112) and (1.113).

Equations (2.79) and (2.80) are valid for lossless media. The effect of magnetic damping can be included in equation (2.79), as before (Section 1.5), by adding the Landau–Lifshitz damping term, which amounts to replacing H_0 by $H_0 + j/\gamma T$ in the final result. The elastic damping is included by making C_{IJ} complex as discussed in Appendix D.

Examination of equations (2.79) and (2.80) shows that elastic variables enter the magnetization equation (2.79) through $\mathbf{h}_{me}$, while magnetic variables enter the elastic equation (2.80) through $\boldsymbol{T}^{me}$. The equations are coupled and require simultaneous solution subject to appropriate boundary conditions to allow investigation of magnetoelastic wave phenomena.

In what follows, we discuss the propagation of exchange-free and exchange-coupled magnetoelastic plane waves in magnetically and elastically isotropic and anisotropic ferrimagnetics.

2.6. *Exchange-Free Plane Waves*

In this section we consider the propagation of uniform plane waves ignoring the effect of spin waves, which limits the applicability of the results to the case when $\beta < 10^5\,\text{cm}^{-1}$. As elastic and magnetic anisotropies are also ignored, the treatment is thus restricted to nearly isotropic materials, such as yttrium–iron–garnet (YIG). The effects of exchange and anisotropy will be considered in subsequent sections.

The dispersion relation for magnetoelastic plane waves propagating along an arbitrary direction relative to the dc field was first obtained by Schlömann (1960) and then, in slightly different forms, by Auld (1965, 1971) and Parekh and Bertoni (1973). This section closely follows the excellent paper by Parekh and Bertoni (1973).

2.6.1. The Dispersion Relation

In the absence of exchange and anisotropy terms, the magnetoelastic equations (2.79) and (2.80) reduce to

$$d\mathbf{M}/dt = -\gamma\mathbf{M}_0 \times (\mathbf{h} + \mathbf{h}_{me}) - \gamma\mathbf{m} \times \mathbf{H}_0 \tag{2.81}$$

and

$$\rho\partial^2 u_i/\partial t^2 = \Gamma_{il}u_l + \nabla_{iJ}\partial U_{me}/\partial S_J \tag{2.82}$$

where $\Gamma_{il} = \nabla_{iJ}C_{Jk}\nabla_{kl}$ are the elements of the matrix $\boldsymbol{\Gamma}$ which can be inferred from equation (D.29) in Appendix D. Now consider the propagation of a plane wave in the xz-plane in a direction making an angle θ with the [100] direction (z-axis). The propagation constant is given by

$$\mathbf{k} = k_x\hat{x} + k_z\hat{z} = \beta(\cos\theta\hat{z} + \sin\theta\hat{x}) \tag{2.83}$$

The dipolar field $\mathbf{h}$ is written from equation (2.67) as

$$\begin{aligned}\mathbf{h} &= -4\pi(\mathbf{m}\cdot\mathbf{k})\mathbf{k}/\beta^2 \\ &= -4\pi m_x k_x(k_x\hat{x} + k_z\hat{z})/\beta^2\end{aligned} \tag{2.84}$$

The magnetoelastic magnetic field is written from equation (1.111) as

$$\mathbf{h}_{me} = -(b_2/M_0)(S_5\hat{x} + S_4\hat{y}) - (2b_1/M_0)S_3\hat{z} \tag{2.85}$$

The matrix $\boldsymbol{\Gamma}$ is obtained from equation (D.29) as

$$\boldsymbol{\Gamma} = \begin{bmatrix} c_{11}k_x^2 + c_{44}k_z^2 & 0 & (c_{11} - c_{44})k_xk_z \\ 0 & c_{44}\beta^2 & 0 \\ (c_{11} - c_{44})k_xk_z & 0 & c_{44}k_x^2 + c_{11}k_z^2 \end{bmatrix} \tag{2.86}$$

The force density on account of magnetoelastic stress is given by $F_i^{me} = \nabla_{iJ}\partial U_{me}/\partial S_J$, which can be obtained from equations (D.13), (1.110), and (1.112), in matrix form, as

$$\mathbf{F}^{me} = -\frac{jb_2}{M_0}\begin{pmatrix} k_zm_x \\ k_zm_y \\ k_xm_x \end{pmatrix} \tag{2.87}$$

The various quantities obtained in equations (2.83)–(2.87) can be substituted in equations (2.81) and (2.82) which, in turn, lead to the following set of coupled equations:

$$\omega m_x - j\omega_0 m_y - \gamma b_2 k_z u_y = 0 \tag{2.88}$$

$$(\omega_0 + \omega_m k_x^2/\beta^2)m_x - j\omega m_y - j\gamma b_2 k_z u_x - j\gamma b_2 k_x u_z = 0 \tag{2.89}$$

$$\gamma b_2(B_r k_x^2 + k_z^2 - k_t^2)u_x + \gamma b_2(B_r - 1)k_x k_z u_z + j\sigma k_z m_x = 0 \tag{2.90}$$

$$\gamma b_2(\beta^2 - k_t^2)u_y + j\sigma k_z m_y = 0 \tag{2.91}$$

$$\gamma b_2(B_r - 1)k_x k_z u_x + \gamma b_2(k_x^2 + B_r k_z^2 - k_t^2)u_z + j\sigma k_x m_x = 0 \tag{2.92}$$

where

$$\left.\begin{aligned} c_l^2 &= c_{11}/\rho, \qquad c_t^2 = c_{44}/\rho \\ B_r &= c_l^2/c_t^2 = c_{11}/c_{44} \\ \sigma &= \frac{\gamma b_2^2}{\rho M_0 c_t^2} = \gamma b_2^2/M_0 c_{44} \\ k_t &= \omega/c_t, \qquad k_l = \omega/c_l \end{aligned}\right\} \tag{2.93}$$

It may be noted that c_l and c_t are the speeds of uncoupled longitudinal and transverse elastic waves, respectively. Equations (2.88)–(2.92) can be rewritten in a convenient matrix form as

$$\begin{bmatrix} \omega & -j\omega_0 & 0 & -k_z & 0 \\ \omega_0 + \omega_m k_x^2/\beta^2 & -j\omega & -jk_z & 0 & -jk_z \\ j\sigma k_z & 0 & B_r k_x^2 + k_z^2 - k_t^2 & 0 & (B_r - 1)k_x k_z \\ 0 & j\sigma k_z & 0 & \beta^2 - k_t^2 & 0 \\ j\sigma k_x & 0 & (B_r - 1)k_x k_z & 0 & k_x^2 + B_r k_z^2 - k_t^2 \end{bmatrix} \begin{bmatrix} m_x/\gamma b_2 \\ m_y/\gamma b_2 \\ u_x \\ u_y \\ u_z \end{bmatrix} = 0 \tag{2.94}$$

The existence of a nontrivial solution requires that the determinant of the square matrix in the left-hand side of equation (2.94) be equal to zero, the numerical solution to which leads to the refractive index diagrams (RID).* It is, however, more convenient to investigate the dispersion characteristics for plane waves propagating in a fixed direction from the polar rather than the cartesian form of the dispersion relation. The substitutions $k_x = \beta \sin\theta$ and $k_z = \beta\cos\theta$ in equations (2.94) lead to the polar form of the dispersion relation as (Schlömann, 1960)

$$\begin{aligned} &(k_t^2 - B_r\beta^2)\{(k_t^2 - \beta^2)^2(\omega^2 - \omega_S^2) - \sigma\beta^2(k_t^2 - \beta^2) \\ &\quad \times [(\omega_S^2/\omega_0)\cos^2\theta + \omega_0\cos^2 2\theta] - \sigma^2\beta^4\cos^2\theta\cos^2 2\theta\} \\ &\quad - \sigma\beta^2\sin^2 2\theta(k_t^2 - \beta^2)[\omega_0(k_t^2 - \beta^2) + \sigma\beta^2\cos^2\theta] = 0 \end{aligned} \tag{2.95}$$

* Loci of points (k_x, k_z) for a fixed frequency ω.

where

$$\omega_S^2 = \omega_0(\omega_0 + \omega_m \sin^2\theta) \tag{2.96}$$

After some algebra, the dispersion relation (2.95) can be expressed in a more compact form, as given by Auld (1965, 1971), as

$$(\omega_S^2 - \omega^2) - \frac{b_2\beta^2}{\rho M_0 H_0}\left(\frac{\omega^2\cos^2\theta}{\omega_{td}^2 - \omega^2} + \frac{\omega^2\cos^2 2\theta}{\omega_t^2 - \omega^2} + \frac{\omega_0^2\sin^2 2\theta}{\omega_l^2 - \omega^2}\right) = 0 \tag{2.97}$$

where

$$\left.\begin{aligned} &\omega_l^2 = c_l^2\beta^2, \qquad \omega_t^2 = c_t^2\beta^2 \\ &\omega_{td}^2 = \omega_t^2 - [(b_2\beta)^2/\rho M_0 H_0]\cos^2\theta \end{aligned}\right\} \tag{2.98}$$

The general nature of the dispersion of exchange-free plane waves in isotropic ferrimagnetics can be understood in terms of the interaction between spin and elastic fields. In the absence of this coupling ($\sigma = 0$), the general dispersion relation yields a magnetostatic wave with dispersion relation $\omega = \omega_S$ and three elastic waves: one longitudinal wave with the dispersion relation $\omega = c_l\beta$ and two degenerate shear waves with orthogonal polarization and common dispersion relation $\omega = c_t\beta$. Even in the presence of magnetoelastic interaction, the coupling between magnetostatic and elastic waves is not appreciable unless the frequency and wavevector of the uncoupled waves are comparable. This happens around the frequency at which the phase velocities of magnetostatic and elastic waves are of the same order. Evidently, the magnetoelastic coupling is appreciable only near the crossover region, i.e., the region in which the dispersion curves of the individual uncoupled waves cross each other.

First we shall consider the special cases of propagation along and perpendicular to the dc field and then take up the general case.

2.6.2. *Special Cases*

(a) *Propagation Along* $\mathbf{H}_0$. When $\theta = 0°$, the dispersion relation (2.95) reduces to

$$(\beta^2 - k_l^2)(\beta^2 - k_t^2/\delta_\parallel^+)(\beta^2 - k_t^2/\delta_\parallel^-) = 0 \tag{2.99}$$

where

$$\delta_\parallel^\pm = (\omega_0 - \sigma \mp \omega)/(\omega_0 \mp \omega) \tag{2.100}$$

The solution $\beta = k_l$ represents an uncoupled longitudinal elastic wave propagating along the dc field. However, the two shear waves are coupled to the magnetostatic waves and get modified. The dispersion relation for the coupled quasi-shear waves is given by

$$\beta_{\pm}^2 = k_t^2/\delta_{\parallel}^{\pm} \tag{2.101}$$

For the wave with propagation constant β_+, it is seen from equations (2.88)–(2.91) that

$$m_x/m_y = u_x/u_y = +j \tag{2.102}$$

which implies that the wave is an RCP wave. Similarly, the other wave can be shown to be an LCP wave. The coupling parameter σ being small (about 1 MHz), $\delta_{\parallel}^{-} \sim 1$ and, therefore, the LCP mode is only weakly coupled to the spin system. On the other hand, $\delta_{\parallel}^{+}$ is a strong function of ω, particularly around the resonance: $\delta_{\parallel}^{+} \to \infty$ as $\omega \to \omega_0$ and $\delta_{\parallel}^{+} \to 0$ at $\omega = \omega_0 - \sigma$. It is also seen that the wave propagation is forbidden in the range $(\omega_0 - \sigma) \leqslant \omega \leqslant \omega_0$. The existence of this forbidden gap was first recognized by Parekh and Bertoni (1973).

The circular polarization of the quasi-shear modes suggests the possibility of Faraday rotation of a linearly polarized shear elastic wave propagating along the biasing field, and this is discussed in the following section.

Faraday Rotation. Faraday rotation of magnetoelastic plane waves was predicted by Kittel (1958) and Vlasov and Ishmukhametov (1960). The experimental observations by Bömmel and Dransfeld (1960) and by Matthews and LeCraw (1962) verified the prediction. Following equation (2.30), we can express the rotation per unit length as

$$\begin{aligned} \theta_F/l &= \tfrac{1}{2}(\beta_- - \beta_+) \\ &= \tfrac{1}{2}k_t(1/\sqrt{\delta_{\parallel}^{-}} - 1/\sqrt{\delta_{\parallel}^{+}}) \end{aligned} \tag{2.103}$$

Bömmel and Dransfeld (1960) observed magnetoelastic Faraday rotation in a thin YIG disk near the crossover region, where unequal absorptions for the two modes leads to rather large ellipticity. Matthews and LeCraw (1962) employed a longitudinally magnetized YIG rod and carried out the experiment in a region away from the crossover point. In this case, $|\omega_0 - \omega| \gg \sigma$ and the expression for rotation per unit length can be approximated to

$$\theta_F/l \simeq \tfrac{1}{2}k_t\sigma\omega/(\omega^2 - \omega_0^2 + \omega_0\sigma) \tag{2.104}$$

where it is assumed that the demagnetizing field correction has been applied to H_0. Sufficiently far away from the crossover region, even the term $\omega_0\sigma$ can be ignored when compared to the other term. Thus,

$$\theta_F/l \simeq \tfrac{1}{2}k_t\sigma\omega/[\omega^2 - (H - 4\pi N_z M_0)^2] \tag{2.105}$$

where H is the applied field. Substituting for σ and k_t from equation (2.93) in (2.105), we have

$$|\theta_F/l| = \frac{\omega^2 b_2^2}{2\rho\gamma M_0 c_t^3}[(H - 4\pi N_z M_0)^2 - (\omega/\gamma)^2]^{-1} \tag{2.106}$$

It follows that the variation of rotation per unit length with dc field can lead to a numerical estimate for b_2. Figure 2.15 shows the experimental curve of Matthews and LeCraw (1962). The slope of the curve yields the value of b_2 as 6×10^6 ergs/cc, which compares well with the independently obtained (Clark et al., 1963) value 6.4×10^6 ergs/cc.

(b) *Propagation Perpendicular to* $\mathbf{H}_0$. When $\theta = 90°$, the dispersion relation (2.95) leads to

$$(\beta^2 - k_l^2)(\beta^2 - k_t^2)(\beta^2 - k_t^2/\delta_\perp) = 0 \tag{2.107}$$

where

$$\left.\begin{aligned} \delta_\perp &= (\omega^2 - \omega_\sigma^2)/(\omega^2 - \omega_3'^2) \\ \omega_3'^2 &= \omega_0(\omega_0 + \omega_m) \\ \omega_\sigma^2 &= \omega_0(\omega_0 + \omega_m - \sigma) = \omega_3'^2 - \omega_0\sigma \end{aligned}\right\} \tag{2.108}$$

It is seen that the longitudinal wave does not couple to the spin system as in the case of propagation along $\mathbf{H}_0$. Moreover, one shear wave is also

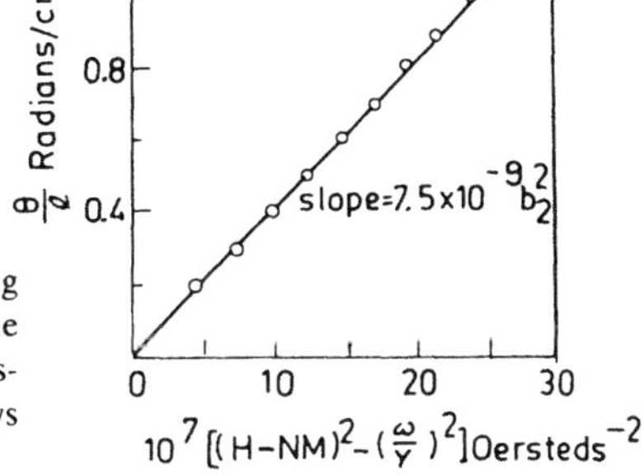

Figure 2.15. Experimental curves for the biasing field dependence of rotation per unit length. The slope of the curve is a measure of the magnetoelastic coupling. Note that $N = 4\pi N_z$ (after Matthews and LeCraw, 1962).

uncoupled; the particle displacement of this wave is along the y-axis, i.e., out of the plane containing the biasing field and the direction of propagation. However, the other shear wave exhibits strong coupling to the spin system. The particle displacement for this wave is along the biasing field, and the dispersion relation is given by

$$\beta = k_t/\sqrt{\delta_\perp} \tag{2.109}$$

It follows from equation (2.109) that there is no propagation in the frequency range $\omega_\sigma < \omega < \omega_3'$. Substitution for β from equation (2.109) in equations (2.88)–(2.92) shows that the magnetization **m** is, in general, elliptically polarized.

2.6.3. *General Case*

When the direction of propagation is arbitrary, each elastic wave is coupled to the magnetostatic wave and vice versa. We shall describe the following methods to examine magnetoelastic plane wave propagation: (i) dispersion curves: the plot of frequency vs. wavenumber for a fixed direction, and (ii) refractive index diagram (RID): the loci points (k_x, k_z) for a fixed frequency.

2.6.3.1. *Dispersion Curves*

When θ is arbitrary, it is seen from the numerical solution of the dispersion relation that, in general, all three elastic waves couple to the spin system. The specific values of θ for which one or more elastic waves do not interact with the spin system are (i) $\theta = 0°$, (ii) $\theta = 90°$, and (iii) $\theta = 45°$. In the last case, the shear wave with particle displacement in the plane containing $\mathbf{H}_0$ and $\mathbf{k}$ is uncoupled. Figure 2.16 shows the dispersion diagram for an arbitrary $\theta \neq 0°, 45°$, or $90°$. The dispersion curves consist of four

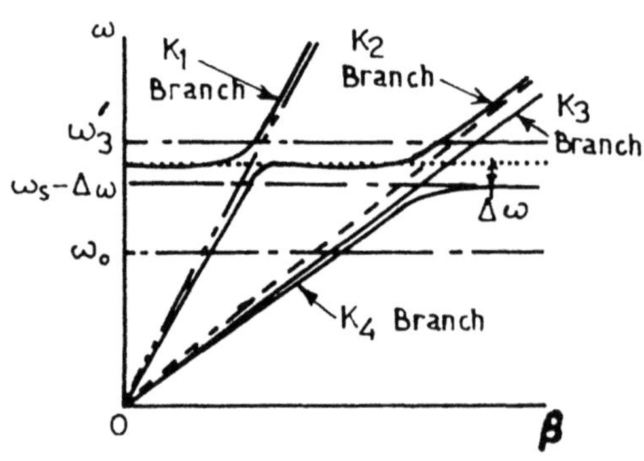

Figure 2.16. Dispersion diagram for exchange free magnetoelastic plane wave corresponding to $\theta \neq 0°, 45°$, or $90°$. The splitting $\Delta\omega$ is shown exaggerated only to emphasize that the coupling is strong not only at the cross over point but also for all larger values of the wavenumber (after Parekh and Bertoni, 1973).

branches which are labeled as K_1, K_2, K_3, and K_4. The K_3 branch is only slightly perturbed as compared to the corresponding shear wave branch in the absence of magnetoelastic coupling. Therefore, the quasi-shear wave represented by the K_3 branch is only weakly coupled to the spin system. Far from the crossover points, the remaining branches also correspond to the waves which are more or less uncoupled. However, the interaction is strong in the vicinity of the crossover points, and the character of a plane wave changes from quasi-magnetostatic on one side of the crossover point to quasi-longitudinal or quasi-shear on the other.

It is deduced from equation (2.95) that $\beta \to \infty$ when

$$\begin{aligned} \omega &= \{\omega_S^2 - \sigma[\omega_0(\cos^2\theta + \cos^2 2\theta + \sin^2 2\theta/B_r) + \tfrac{1}{4}\omega_m \sin^2 2\theta \\ &\qquad - \sigma\cos^2\theta\,(\cos^2 2\theta + \sin^2 2\theta/B_r)]\}^{1/2} \\ &\equiv \omega_S - \Delta\omega \end{aligned} \tag{2.110}$$

As σ is very small, an approximate expression for $\Delta\omega$ is obtained from equation (2.110) as

$$\Delta\omega = (\sigma/2\omega_S)[\omega_0(\cos^2\theta + \cos^2 2\theta + \sin^2 2\theta/B_r) + \tfrac{1}{4}\omega_m \sin^2 2\theta] \tag{2.111}$$

The range $(\omega_S - \Delta\omega) \leqslant \omega \leqslant \omega_S$ represents the frequency interval in which no wave propagation is possible. As θ increases from 0° to 90°, the "forbidden band" $\Delta\omega$ decreases monotonically from $(\omega_0 - \omega_\sigma)$ to $(\omega_3' - \omega_\sigma)$. The variation of the band gap with θ is shown in Figure 2.17.

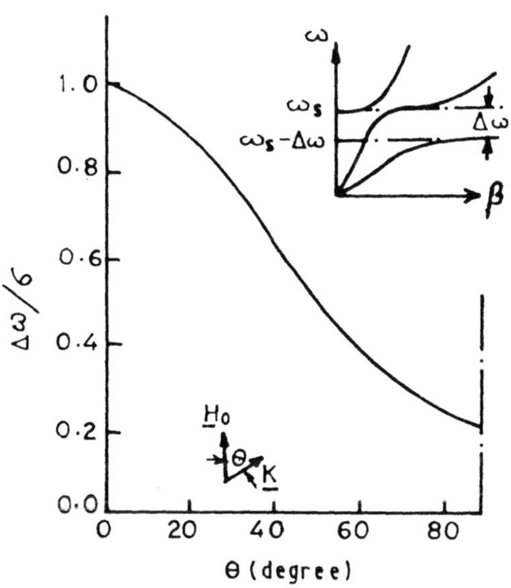

Figure 2.17. Variation of normalized cut-off band $\Delta\omega/\sigma$ with θ for Ga-YIG. The parameters are $H_0 = 50$ Oe, $4\pi M_0 - 300$ G, $\sigma/2\pi - 1.272$ MHz, and $B_r = 4$ (after Parekh and Bertoni, 1973).

As θ increases from 0° to 90°, the resonance frequency ω_S of purely magnetostatic plane waves (which is also the frequency at which $\beta \to 0$ for the quasi-magnetostatic part of the K_1 branch) increases from ω_0 to ω_3'. Correspondingly, the frequency $(\omega_S - \Delta\omega)$, at which $\beta \to \infty$ for the quasi-magnetostatic part of the K_4 branch, increases from $(\omega_0 - \sigma)$ to ω_σ. The general nature of the dispersion curves is, however, the same as shown in Figure 2.16.

2.6.3.2. *Refractive Index Diagrams*

The dispersion relation represented by equation (2.95) is an even function of k_x and k_z. Thus, it is sufficient to perform calculations only for the principal quadrant in the k_x–k_z plane; the curves in the remaining quadrants can be completed by symmetry. The RID for a typical set of data is plotted in Figures 2.18–2.20.

It is worth considering the RID corresponding to uncoupled waves before taking up the case of coupled magnetoelastic plane waves. The longitudinal elastic wave is represented by a circle of radius $\beta = \omega/c_l$. The two degenerate shear elastic waves are represented by a single circle of radius $\beta = \omega/c_t$. However, the magnetostatic wave branch exists only in the frequency range $\omega_0 < \omega < \omega_3'$ and is represented by a radial straight line starting from the origin and extending to infinity.* As ω increases from ω_0 to ω_3', this line rotates from the k_z-axis to the k_x-axis. Since the magnetoelastic coupling is important only in the crossover region, the RIDs differ appreciably from those of the uncoupled waves only in the frequency range $\omega_0 < \omega < \omega_3'$. It follows that the nature of the RID is basically governed by the variation of ω_S [and hence of $(\omega_S - \Delta\omega)$] with θ.

When $\omega < (\omega_0 - \sigma)$, the RID has three branches corresponding to the K_2, K_3, and K_4 branches of the dispersion curves displayed in Figure 2.16. When ω is smaller than $\omega_0 - \sigma$, the wavenumbers vary only slightly with θ. Consequently, the K_2 branch is approximately a circle of radius ω/c_l, while the K_3 and K_4 branches are almost coincident circles of radius ω/c_t. When $\omega \leqslant (\omega_0 - \sigma)$, the K_4 branch remains a circle for $\theta \simeq 90°$, but deviates from a circle as $\theta \to 0°$. The character of the RID for ω slightly below $(\omega_0 - \sigma)$ is illustrated in the computed diagram given by the solid curves in Figure 2.18. Although not resolvable in the figure, the K_2 branch lies slightly outside the circle of radius ω/c_l except for $\theta = 0°$ and 90°, where it touches this circle. Similarly, the K_3 branch also lies slightly outside the circle of radius ω/c_t touching it at $\theta = 45°$ and 90°. As ω approaches $(\omega_0 - \sigma)$ from

* This open branch occurs on account of the magnetostatic approximation; the inclusion of exchange effects makes it closed.

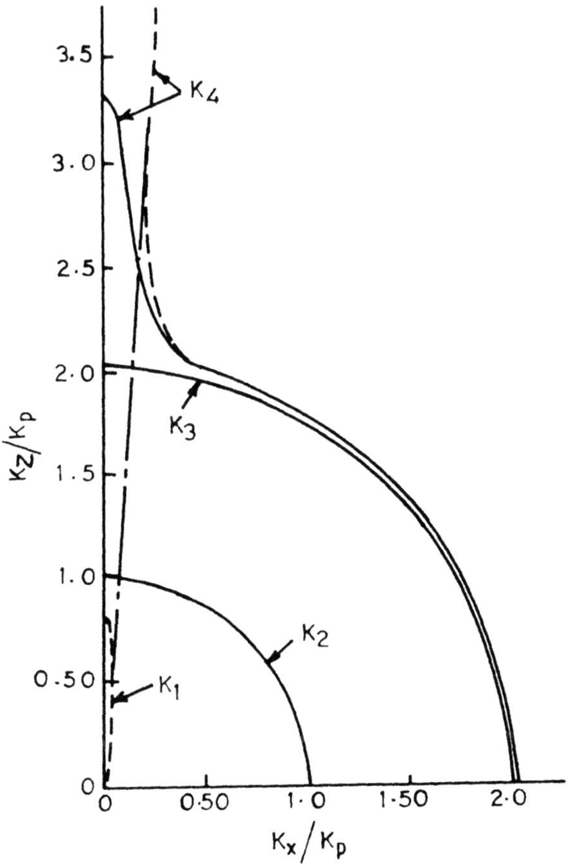

Figure 2.18. Refractive index diagram for exchange-free plane magnetoelastic waves corresponding to frequencies (i) $f = 0.138\,\text{GHz}$ (ω slightly below $(\omega_0 - \sigma)$: solid curves) and (ii) $f = 0.140\,\text{GHz}$ (ω slightly above $(\omega_0 - \sigma)$: broken curves). The parameters are the same as for Figure 2.17. The curves corresponding to the two frequencies are almost coincident for large k_x (after Parekh and Bertoni, 1973).

below, the bulge in the K_4 branch intersects the k_z-axis at a value of k_z that increases monotonically and approaches infinity as $\omega \rightarrow (\omega_0 - \sigma)$.

When $(\omega_0 - \sigma) < \omega < \omega_0$, the resonance frequency ω_S lies above ω for all θ, but $(\omega_S - \Delta\omega)$ moves up past ω as θ increases from 0° to 90°. Thus, the K_1 branch is again absent from the RID. However, the K_4 branch is an open one, which approaches infinity along a radial line whose angle with the k_z-axis is found from the condition $\omega = \omega_S(\theta) - \Delta\omega(\theta)$. This open branch corresponds to a quasi-magnetostatic plane wave and exists in the RID in the range $(\omega_0 - \sigma) < \omega < \omega_\sigma$. In Figure 2.18, this open branch is represented by a dashed curve which has been plotted for $\omega \geqslant \omega_0$. The K_1 branch appears in the RID in the range $\omega_0 < \omega < \omega_3'$, and is represented by a dashed K_1 curve in Figure 2.18 for the same frequency as was used to compute the dashed portion of the K_4 branch.

As ω increases from $(\omega_0 - \sigma)$ to ω_σ, the angle between the k_z-axis and the open branch increases, as does the angle at which K_1 approaches the origin. This behavior is shown in Figure 2.19 for $f = 0.25$ GHz $(\omega_0 < 2\pi f < \omega_\sigma)$. The figure also shows clearly the change in K_2 from a value slightly below k_t to a value approaching k_l. The wavevector **k** and the direction for the group velocity $\mathbf{v}_g$ are indicated at various points. The group velocity is normal to the curves. It is seen that **k** and $\mathbf{v}_g$ are not parallel on account of anisotropy in propagation. This anisotropy is strongest for a quasi-magnetostatic wave.

Since ω_σ exceeds $(\omega_S - \Delta\omega)$ for all θ, the open branch, K_4, does not appear in the RID when $\omega > \omega_\sigma$. In the range $\omega_\sigma < \omega < \omega_3'$ the K_2 branch moves from slightly below k_t to k_l as θ increases from 0° to 90°; the transition occurs for θ close to 90°. The K_1 branch also goes to the origin at an angle approaching 90°. As ω increases slightly above ω_3', the endpoint of the K_1 branch moves out from the origin to a point along the k_x-axis, as

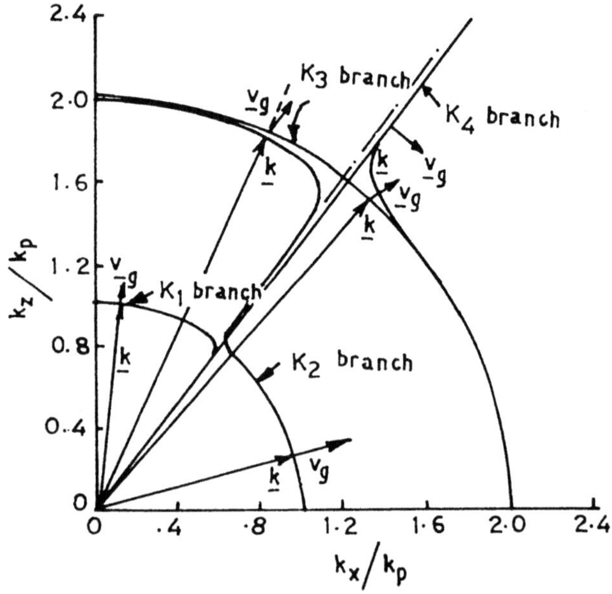

Figure 2.19. Refractive index diagram for magnetoelastic plane waves corresponding to frequency $f = 0.25$ GHz $(\omega_0 < \omega < \omega_\sigma)$. The parameters are the same as for Figure 2.17 (after Parekh and Bertoni, 1973).

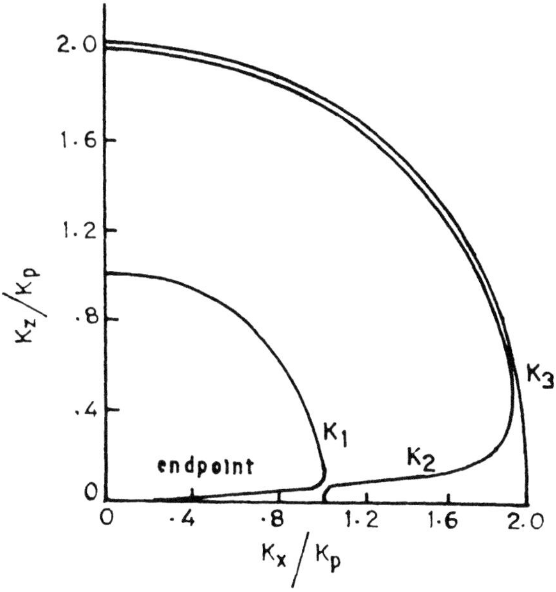

Figure 2.20. Refractive index diagram for magnetoelastic plane waves corresponding to frequency $f = 0.37041$ GHz (ω slightly above ω_3'). The parameters are the same as for Figure 2.17 (after Parekh and Bertoni, 1973).

shown in Figure 2.20. When $\omega > \omega_3'$, the endpoint of the K_1 branch tends to $k_x = k_l$, while the corresponding endpoint on the k_x-axis of the K_2 branch tends toward k_t. This means essentially that the effect of magnetoelastic interaction becomes insignificant when $\omega \gg \omega_3'$.

2.6.3.3. *Polarization of Modes*

The polarization characteristics of magnetoelastic plane waves propagating along an arbitrary direction relative to the biasing field can be inferred from equations (2.88)–(2.92). It is convenient to resolve particle displacement in longitudinal and shear rather than cartesian components. If $\hat{s} = \hat{y} \times \hat{n}$ represents a unit vector in the xz-plane but perpendicular to the direction of propagation, the particle displacement $\mathbf{u}$ can be expressed as

$$\mathbf{u} = u_l\hat{n} + u_y\hat{y} + u_S\hat{s}$$

The substitution for u_x and u_z in terms of u_l and u_S in equations (2.88)–(2.92) yields

$$\left.\begin{aligned} \frac{u_l}{u_S} &= \frac{k_t^2-\beta^2}{k_l^2-B_r\beta^2}\tan 2\theta \\ \frac{u_y}{u_S} &= -j\frac{\cos\theta}{\cos 2\theta}\frac{\omega(k_t^2-\beta^2)}{\omega_0(k_t^2-\beta^2)+\sigma\beta^2\cos^2\theta} \\ \frac{m_y}{m_x} &= -j\frac{\omega(k_t^2-\beta^2)}{\omega_0(k_t^2-\beta^2)+\sigma\beta^2\cos^2\theta} \end{aligned}\right\} \tag{2.112}$$

The variations of u_y/u_l and u_y/u_S with θ have been shown in Figures 2.21a and b, respectively, for the same data as for Figure 2.19. The four solutions

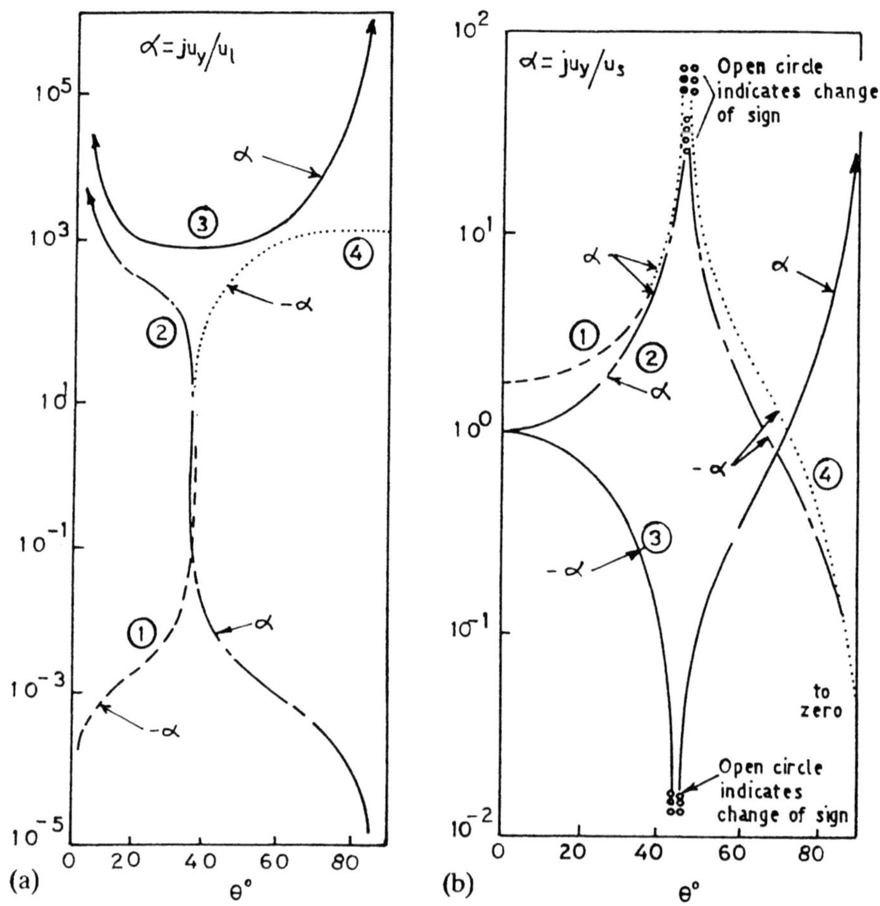

Figure 2.21. Variation of (a) u_y/u_l and (b) u_y/u_s, with θ for magnetoelastic plane waves of frequency $f = 0.25$ GHz, the parameters being the same as for Figure 2.17 (after Parekh and Bertoni, 1973).

shown in each of the curves correspond to the four branches in the RID. Unless the wavenumber lies close to the crossover region, the polarizations of the modes are very nearly those of longitudinal waves and elliptically polarized shear waves. The shear component of a quasi-longitudinal wave or the longitudinal component of a quasi-shear wave are significant only for predominantly magnetostatic waves.

At $\theta = 0°$, the K_1 branch corresponds to a pure longitudinal elastic wave, the K_2 branch to a strongly coupled quasi-shear wave which is right-handed circularly polarized with respect to $\mathbf{H}_0$, and the K_3 branch to a weakly coupled quasi-shear wave with left-handed circular polarization. At $\theta = 90°$, the K_2 branch corresponds to a decoupled longitudinal wave, the K_3 branch to a decoupled shear wave with linear polarization along y-axis, and the K_4 branch to a coupled quasi-shear wave with linear polarization along $\mathbf{H}_0$. For the K_3 branch, the motion is essentially confined to the ys plane for all values of θ, changing from LCP at $\theta = 0$ to linear polarizaton along $\hat{s}$ for $\theta = 45°$ to RCP at $\theta \simeq 70°$, finally assuming linear polarization along the y-axis at $\theta = 90°$.

The polarization of the magnetization vector $\mathbf{m}$ for the corresponding conditions is shown in Figure 2.22. The vector $\mathbf{m}$ is right-handed elliptically polarized (REP) for $0 < \theta < \pi/2$ (and LEP for $\pi/2 < \theta < \pi$; see Part A) for the K_1, K_2, and K_4 branches. The K_3 branch, which represents a weak coupling between the elastic and spin systems, corresponds to LEP for $0 < \theta < \pi/2$ (and REP for $\pi/2 < \theta < \pi$), with the ellipticity $|m_y/m_x|$ decreasing from 1 at $\theta = 0°$ to under 10^{-3} at $\theta = 45°$ and then increasing to infinity at $\theta = 90°$. The ellipticity $|m_y/m_x|$ of the magnetoelastic plane wave corresponding to the K_1 branch has a constant value of sightly below 2 for the entire range of θ for this branch, i.e., $0 < \theta < 37.3°$. For the K_2 branch, the ellipticity increases monotonically from 1 at $\theta = 0°$ to 2 at $\theta = 37.3°$ and then remains constant at this value for large θ. For the K_4 branch, the ellipticity has a maximum value of about 3 at $\theta \simeq 60°$, which decreases to zero on either side.

2.7. *Effect of Exchange*

In the last section, we discussed exchange-free magnetoelastic plane waves in ferrimagnetics. The effect of exchange can be taken into account, as before, by replacing ω_0 by ω_r in equations (2.95)–(2.107) and ω_r is given by

$$\omega_r = \omega_0 + \omega_{ex} a_1^2 \beta^2$$

where ω_{ex} can be inferred from equation (1.48) of Chapter 1. Since the

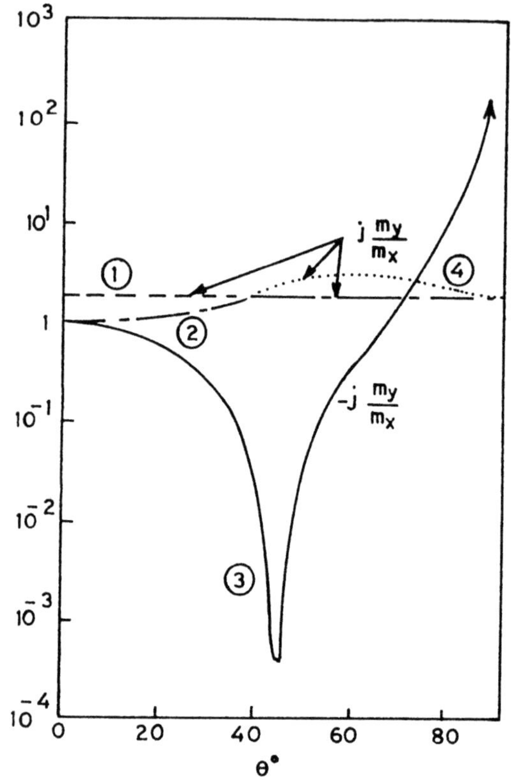

Figure 2.22. Variation of m with θ for magnetoelastic plane waves of frequency $f = 0.25$ GHz, the parameters being the same as for Figure 2.17 (after Parekh and Bertoni, 1973).

frequencies involved in the case of magnetoelastic waves are, in general, below about 1 GHz, relatively smaller values of β ($\sim 10^4\,\text{cm}^{-1}$) define the approximate limit beyond which the effect of exchange needs to be considered. For example, in the case of YIG, $c_{44} = 7.64 \times 10^{11}$ dyn/cm^2 and $\rho = 5.17$ g/cc; the speed of the uncoupled transverse elastic waves [i.e., $v_t = (c_{44}/\rho)^{1/2}$] comes out to be about 3×10^5 cm/sec. For $f \simeq 1$ GHz, the wavenumber $\beta \sim 10^4\,\text{cm}^{-1}$. This clearly shows that the coupled waves undergo modification, on account of exchange, around and beyond the crossover region between spin wave and transverse elastic wave branches. A typical dispersion curve is shown in Figure 2.23. The comparison with exchange-free plane waves shows that, as expected, only the K_4 branch and quasi-magnetic part of the K_2 branch are appreciably modified by exchange effects. It is also seen that the band gap $\Delta\omega$ no longer exists. The concept of band gap is, nevertheless, useful in comparing the position in terms of frequencies of the various branches of dispersion curves.

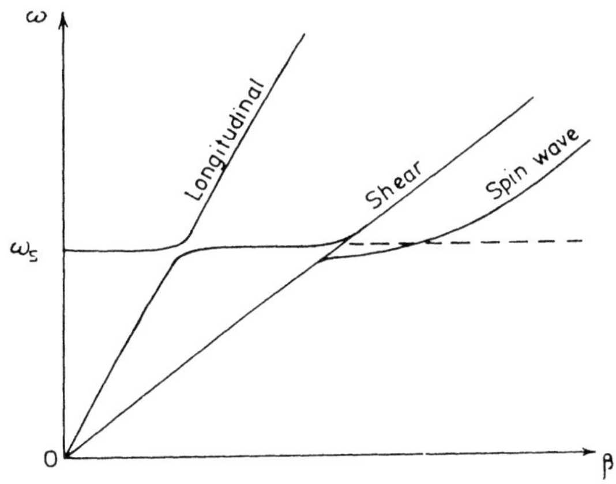

Figure 2.23. Qualitative dispersion curve for exchange-coupled plane magnetoelastic waves, showing that the effect of exchange is important when the wavenumber is large.

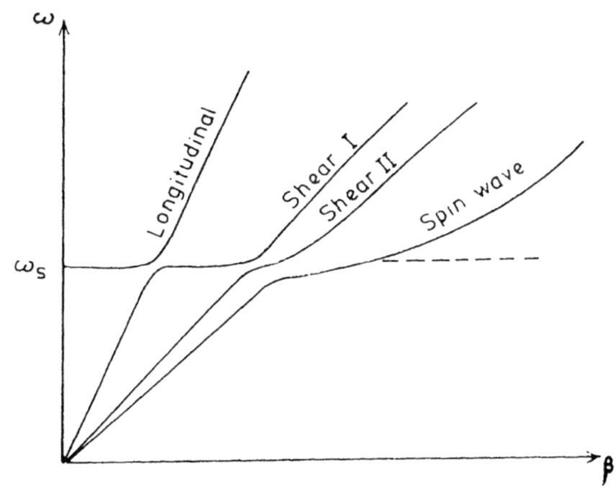

Figure 2.24. Qualitative dispersion curve for magnetoelastic plane waves in elastically anisotropic ferrimagnetic material.

2.8. *Effect of Anisotropy*

The inclusion of magnetocrystalline anisotropy modifies the dispersion curve for the spin wave branch. In the case of cubic ferrimagnetics, the effect is small owing to their small magnetocrystalline anisotropy. However, the introduction of hexagonal planar anisotropy causes appreciable change in the spin wave branch and hence in the coupled magnetoelastic wave branch. The magnetoelastic crossover frequency and wavenumbers are accordingly affected. The general nature of the curves for a specific direction of propagation, however, remains more or less the same.

When elastic anisotropy is included, the three elastic wave branches undergo modification. The degeneracy between the two transverse elastic waves is removed for propagation in an arbitrary direction relative to the crystallographic axes, and the dispersion curve assumes the form shown in Figure 2.24.

The most general case of magnetoelastic plane waves in a conducting and anisotropic ferromagnetic metal has been treated by Vittoria et al. (1974).

3

Wave Propagation Across Interfaces

The propagation of electromagnetic and magnetoelastic plane waves in ferrimagnetics was discussed in the last chapter; however, the analysis did not take into account the effect of interfaces (of ferrimagnetics with other media) which are significant in realistic situations. This chapter presents a discussion of the propagation (reflection and transmission) of electromagnetic and magnetoelastic plane waves in structures having plane interfaces between dielectrics and ferrimagnetics. Such a study is useful in understanding more complicated wave propagation phenomena and also has direct applications for some devices. The reflection and transmission of electromagnetic waves are discussed in Part A while that of magnetoelastic waves is considered in Part B.

PART A. *ELECTROMAGNETIC WAVE PROPAGATION*

3.1. *Ferrimagnetic Half Space*

The reflection of a uniform plane wave from a magnetized ferrimagnetic half space has been studied by several authors. The special case of normal incidence was first investigated by Seavey and Tannenwald (1957). Mueller (1971) considered the general case of reflection and transmission of an obliquely incident wave at the interface between two arbitrarily magnetized ferrimagnetic half spaces. It was, however, recognized by Srivastava (1978) that, in general, the reflection of electromagnetic microwaves from magnetized ferrimagnetics is phase and amplitudewise nonreciprocal; this investigation also explains qualitatively the broad banding effect of dielectric loading in ferrimagnetic devices. Reflection from magnetized ferrimagnetics from the viewpoint of ray propagation and energy flow was discussed by Gupta and Srivastava (1979), who also gave a ray model of attenuated total reflection. The following discussion is based on the investigations of Srivastava (1978) and Gupta and Srivastava (1979).

3.1.1. *Reflection and Transmission Coefficients*

Consider the reflection of a plane wave of angular frequency ω, initially propagating in a lossless dielectric, from a hexagonal planar magnetoplumbite occupying the region $y \geqslant 0$ (Figure 3.1). The electric vector of the incident wave is assumed to be parallel to the dc magnetization, which, in turn, is oriented perpendicular to the plane of incidence. The wavevectors for the incident, reflected, and transmitted waves are given by

$$\left.\begin{aligned} \mathbf{k}^{i} &= \beta_0(-\hat{x}\sin\psi_i + \hat{y}\cos\psi_i) \\ \mathbf{k}^{r} &= \beta_0(-\hat{x}\sin\psi_i - \hat{y}\cos\psi_i) \\ \mathbf{k}^{t} &= \beta(-\hat{x}\sin\psi_t + \hat{y}\cos\psi_t) \end{aligned}\right\} \tag{3.1}$$

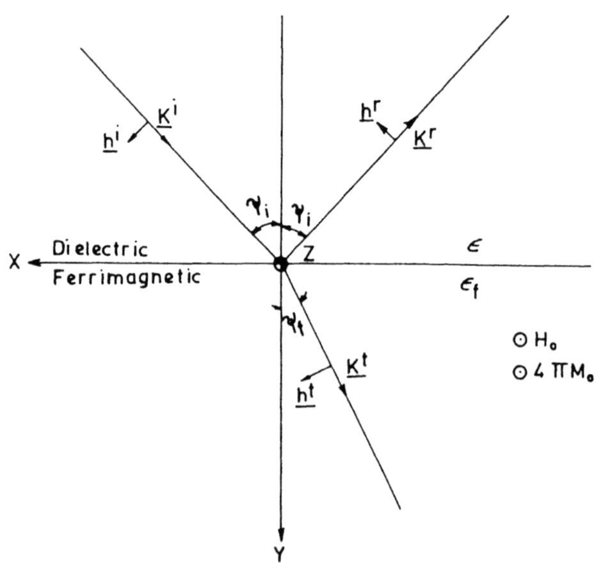

Figure 3.1. Configuration showing the reflection of an electromagnetic plane wave at a dielectric–ferrimagnetic interface. The wave electric vector, dc magnetic field, and dc magnetization are parallel to one another and perpendicular to the plane of the paper.

where ψ_i and ψ_f represent the angles of incidence and refraction,* respectively. The propagation constant β_0 for incident and reflected waves (in the dielectric) is given by

$$\beta_0^2 = \varepsilon\omega^2/c_0^2 \tag{3.2}$$

where ε is the dielectric constant of the dielectric. The wavenumber β for the transmitted wave is given in accordance with equation (2.32) as

$$(\beta/\beta_d)^2 = (\mu_{xx}\mu_{yy} - \kappa^2)/(\mu_{xx}\sin^2\psi_f + \mu_{yy}\cos^2\psi_f) \tag{3.3}$$

where $\beta_d^2 = \varepsilon_f\omega^2/c_0^2$; ε_f is the dielectric constant of the ferrimagnetic medium, while μ_{xx}, μ_{yy}, and κ are given by equations (1.89) of Chapter 1 for the case when the easy plane is oriented parallel to the interface (case 1). The expressions for μ_{xx} and μ_{yy} should be interchanged when the easy plane is perpendicular to the interface. The electric field vectors for the incident, reflected, and transmitted waves may be expressed as

$$\left.\begin{aligned} \mathbf{E}^i &= \hat{z}\exp j(\omega t - \mathbf{k}^i\cdot\mathbf{r}) \\ \mathbf{E}^r &= \hat{z}R_+\exp j(\omega t - \mathbf{k}^r\cdot\mathbf{r}) \\ \mathbf{E}^f &= \hat{z}T_+\exp j(\omega t - \mathbf{k}^f\cdot\mathbf{r}) \end{aligned}\right\} \tag{3.4}$$

where R_+ and T_+ are the reflection and transmission coefficients, respectively. The corresponding tangential components of magnetic field vectors are obtained from Maxwell's equations as follows:

$$\left.\begin{aligned} h_x^i &= \frac{c_0\beta_0\cos\psi_i}{\omega}E_z^i \\ h_x^r &= -\frac{c_0\beta_0\cos\psi_i}{\omega}E_z^r \\ h_x^f &= \frac{c_0\beta_0}{\omega\Delta}(\mu_{yy}\cos\psi_f - j\kappa\sin\psi_f)E_z^f \end{aligned}\right\} \tag{3.5}$$

The matching of the phases of $\mathbf{E}$ and $\mathbf{h}$ at the interface $y = 0$ leads to the generalized Snell's law:

$$\beta_0\sin\psi_i = \beta(\psi_f)\sin\psi_f \tag{3.6}$$

An explicit expression for ψ_f may be obtained by eliminating from equations (3.3) and (3.6); thus

* It is easy to show that the angle of reflection is the same as the angle of incidence.

$$\sin^2 \psi_f = 1/(1 + \Delta/q\mu_{yy} - \mu_{xx}/\mu_{yy}) \equiv 1/P_0 \tag{3.7}$$

where

$$q = \varepsilon \sin^2 \psi_i / \varepsilon_f \tag{3.8}$$

Substitution for μ_{xx}, μ_{yy}, and Δ from equation (1.89) in equation (3.7) leads to the following expression for P_0:

$$P_0 = \frac{(\omega_0 + \omega_m)(\omega_0 + \omega_a + \omega_m) - \omega^2 - q\omega_a\omega_m + j\Delta\omega_h(\omega_0 + \frac{1}{2}\omega_a + \omega_m)}{q\{\omega_0(\omega_0 + \omega_a + \omega_m) - \omega^2 + j\Delta\omega_h[\omega_0 + \frac{1}{2}(\omega_a + \omega_m)]\}} \tag{3.9}$$

It is worth noting that q is always positive; it is smaller than unity for all ψ_i, when $\varepsilon < \varepsilon_f$. But, when $\varepsilon > \varepsilon_f$, it is seen that $q \gtrless 1$ if $\psi_i \gtrless \psi_1$, where

$$\sin^2 \psi_1 = \varepsilon_f / \varepsilon \tag{3.10}$$

The continuity of tangential components of **E** and **h** leads to the following expressions for the reflection and transmission coefficients:

$$R_\pm = \frac{\Delta \cos \psi_i - (\beta/\beta_0)(\mu_{yy} \cos \psi_f \mp j\kappa \sin \psi_f)}{\Delta \cos \psi_i + (\beta/\beta_0)(\mu_{yy} \cos \psi_f \mp j\kappa \sin \psi_f)} \tag{3.11}$$

and

$$T_\pm = \frac{2\Delta \cos \psi_i}{\Delta \cos \psi_i + (\beta/\beta_0)(\mu_{yy} \cos \psi_f \mp j\kappa \sin \psi_f)} \tag{3.12}$$

where R_- and T_- represent the reflection and transmission coefficients when the incident and reflected waves are interchanged, i.e., when the path of rays is reversed.

3.1.2. *Regions of Ordinary and Total Reflection*

In this section, the ferrimagnetic medium is assumed to be lossless. Thus, ΔH in the expressions for μ_{yy}, κ, and Δ, namely, equations (1.89), is equated to zero. The region of ordinary reflection (OR) is defined by $0 < \sin^2 \psi_f < 1$. When $\sin^2 \psi_f$ exceeds unity, it can be shown from equations (3.4) and (3.5) along with the principle of causality that the field inside the ferrimagnetic region would be evanescent; $\cos \psi_f = -j(\sin^2 \psi_f - 1)^{1/2}$ and $|R| = 1$. This implies total reflection (TR) of the incident wave. When $\sin^2 \psi_f$ is negative, the generalized Snell's law, such as equation (3.6), along with the causality condition, implies that $\beta = -j|\beta^2|^{1/2}$ and $\sin \psi_f = j|\sin^2 \psi_f|^{1/2}$. Consequently, $|R|$ is unity and again we have total reflection.

The condition for the occurrence of OR/TR can be written from equation (3.7) as

$$\Delta / q\mu_{yy} - \mu_{xx}/\mu_{yy} \gtrless 0 \tag{3.13}$$

Following the standard techniques of inequality theory, this condition has been analyzed (Srivastava, 1977, 1978) in different frequency ranges in order to classify the regions of OR and TR. The results for cases 1 and 2 are

Table 3.1. Regions of OR/TR for the Case When the Easy Plane Is Parallel to the Interface

Ratio of dielectric constants	Angle of incidence	Effect	Frequency range
$\varepsilon/\varepsilon_t < 1$	$\psi_i < \psi_4$	OR	$0 < \omega < \omega_2$
		TR	$\omega_2 < \omega < \omega_4$
		OR	$\omega > \omega_4$
	$\psi_i > \psi_4$	OR	$0 < \omega < \omega_2$
		TR	$\omega_2 < \omega < \omega_6$
		OR	$\omega > \omega_6$
$1 < \varepsilon/\varepsilon_t < F_1$	$\psi_i < \psi_4$	Same as for $\varepsilon/\varepsilon_t < 1$	
	$\psi_4 < \psi_i < \psi_1$	OR	$0 < \omega < \omega_2$
		TR	$\omega_2 < \omega < \omega_6$
		OR	$\omega > \omega_6$
	$\psi_i > \psi_1$	OR	$0 < \omega < \omega_2$
		TR	$\omega > \omega_2$
$F_1 < \varepsilon/\varepsilon_t < F_2$	$\psi_i < \psi_4$	Same as for $\varepsilon/\varepsilon_t < 1$	
	$\psi_4 < \psi_i < \psi_1$	Same as for $1 < \varepsilon/\varepsilon_t < F_1$	
	$\psi_1 < \psi_i < \psi_2$	OR	$0 < \omega < \omega_2$
		TR	$\omega > \omega_2$
	$\psi_i > \psi_2$	TR	$0 < \omega < \omega_5$
		OR	$\omega_5 < \omega < \omega_2$
		TR	$\omega > \omega_2$
$\varepsilon/\varepsilon_t > F_2$	$\psi_i < \psi_4$	Same as for $\varepsilon/\varepsilon_t < 1$	
	$\psi_4 < \psi_i < \psi_1$	Same as for $1 < \varepsilon/\varepsilon_t < F_1$	
	$\psi_1 < \psi_i < \psi_2$	Same as for $F_1 < \varepsilon/\varepsilon_t < F_2$	
	$\psi_2 < \psi_i < \psi_3$	TR	$0 < \omega < \omega_5$
		OR	$\omega_5 < \omega < \omega_2$
		TR	$\omega > \omega_2$
	$\psi_i > \psi_3$	TR	$0 < \omega < \omega_2$
		OR	$\omega_2 < \omega < \omega_5$
		TR	$\omega > \omega_5$

summarized in Tables 3.1 and 3.2, respectively, for all physically realizable combinations of the relevant parameters. The various quantities appearing in these tables are defined below:

$$\left.\begin{aligned}
F_1 &= 1 + \omega_m/(\omega_0 + \omega_a) \\
F_2 &= 1 + (\omega_0 + \omega_m)/\omega_a \\
F_3 &= 1 + \omega_m/\omega_0 \\
\sin^2 \psi_2 &= \varepsilon_f F_1/\varepsilon \\
\sin^2 \psi_3 &= \varepsilon_f F_2/\varepsilon \\
\sin^2 \psi_4 &= \frac{\varepsilon_f(\omega^2 - \omega_4^2)}{\varepsilon(\omega^2 - \omega_3^2)} \\
\sin^2 \psi_5 &= \varepsilon_f F_3/\varepsilon \\
\sin^2 \psi_6 &= \frac{\varepsilon_f(\omega^2 - \omega_4^2)}{\varepsilon(\omega^2 - \omega_2^2)} \\
\omega_2 &= [\omega_0(\omega_0 + \omega_a + \omega_m)]^{1/2} \\
\omega_3 &= [(\omega_0 + \omega_a)(\omega_0 + \omega_m)]^{1/2} \\
\omega_4 &= [(\omega_0 + \omega_m)(\omega_0 + \omega_a + \omega_m)]^{1/2} \\
\omega_5 &= \left[(\omega_0 + \omega_m)\left(\omega_0 + \omega_a - \frac{\omega_m}{q-1}\right)\right]^{1/2} \\
\omega_6 &= \left[(\omega_0 + \omega_m)\left(\omega_0 + \omega_a + \frac{\omega_m \varepsilon_f}{\varepsilon_f - \varepsilon}\right)\right]^{1/2} \\
\omega_7 &= \left[(\omega_0 + \omega_a + \omega_m)\left(\omega_0 + \frac{\omega_m}{1-q}\right)\right]^{1/2} \\
\omega_8 &= \left[(\omega_0 + \omega_m)\left(\omega_0 + \frac{\omega_m \varepsilon_f}{\varepsilon_f - \iota}\right)\right]^{1/2}
\end{aligned}\right\} \qquad (3.14)$$

The dependence of ψ_f on ψ_i (more precisely, of $S_f \equiv \sin^2 \psi_f$ on $S_i \equiv \sin^2 \psi_i$) for various ω has been numerically investigated using equation (3.7). The results for $\varepsilon < \varepsilon_f$ and for $\varepsilon/\varepsilon_f > F_2$ in case 1 have been displayed in Figures 3.2 and 3.3, respectively. The region for the occurrence of OR is defined by $0 < S_f < 1$, while $S_f < 0$ and $S_f > 1$ correspond to TR. The shaded region in each figure represents those combinations of ψ_i and ψ_f which cannot be realized for any ω whatsoever. When $\varepsilon < \varepsilon_f$, the curves (Figure 3.2) are similar to the ones which would be obtained for an isotropic ferrimagnetic half space subjected to a large dc field. However, the case when $\varepsilon/\varepsilon_f > F_2$ (Figure 3.3) has additional interesting features. Firstly, for each ω, the curve $S_f = S_f(S_i)$ passes through the region $S_f > 1$ in some range of the angle of incidence. Hence, there is TR for every ω in some range of ψ_i. Another interesting feature is the existence of a frequency range $\omega_2 < \omega < \omega_6$ in which TR occurs when the angle of incidence is *smaller* than a certain critical angle; this is in contrast to most of the situations involving reflection of waves, wherein TR occurs only when the angle of incidence *exceeds* the critical angle. This effect can result only in case 1 when $\varepsilon/\varepsilon_f > F_2$, which is possible only for a hexagonal planar magnetoplumbite. In Figure 3.3, this region of "anomalous reflection" is obtained for frequencies between about 9.8 and 14.4 GHz.

In the case of isotropic ferrimagnetics, we have $H_a = 0$ implying $\omega_6 = \omega_8$. It is seen from equations (3.7) and (3.9) that $\psi_f = \psi_i$ for all ψ_i when $\omega = \omega_6 = \omega_8$; the incident ray does not deviate from its path as it propagates across the interface. This is so because the wavenumbers in the two media are the same when $\omega = \omega_6$. Nevertheless, $|R| > 0$, because the

Table 3.2. Regions of OR/TR for the Case When the Easy Plane Is Perpendicular to the Interface

Ratio of dielectric constants	Angle of incidence	Effect	Frequency range
$\varepsilon/\varepsilon_f < 1$	$\psi_i < \psi_6$	OR	$0 < \omega < \omega_3$
		TR	$\omega_3 < \omega < \omega_4$
		OR	$\omega > \omega_4$
	$\psi_i > \psi_6$	OR	$0 < \omega < \omega_3$
		TR	$\omega_3 < \omega < \omega_8$
		OR	$\omega > \omega_8$
$1 < \varepsilon/\varepsilon_f < F_3$	$\psi_i < \psi_6$	Same as for $\varepsilon/\varepsilon_f < 1$	
	$\psi_6 < \psi_i < \psi_1$	OR	$0 < \omega < \omega_3$
		TR	$\omega_3 < \omega < \omega_8$
		OR	$\omega > \omega_8$
	$\psi_i > \psi_1$	OR	$\omega < \omega_3$
		TR	$\omega > \omega_3$
$\varepsilon/\varepsilon_f > F_3$	$\psi_i < \psi_6$	Same as for $\varepsilon/\varepsilon_f < 1$	
	$\psi_6 < \psi_i < \psi_1$	Same as for $1 < \varepsilon/\varepsilon_f < F_3$	
	$\psi_1 < \psi_i < \psi_5$	OR	$\omega < \omega_3$
		TR	$\omega > \omega_3$
	$\psi_i > \psi_5$	TR	$0 < \omega < \omega_7$
		OR	$\omega_7 < \omega < \omega_3$
		TR	$\omega > \omega_3$

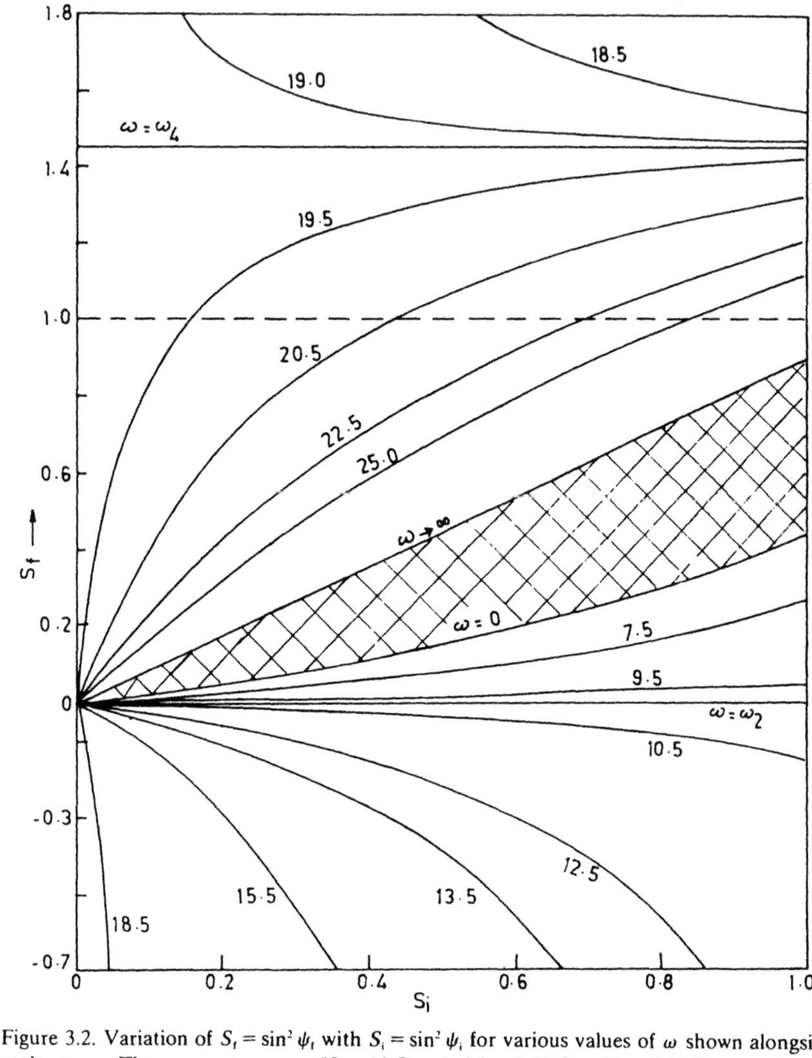

Figure 3.2. Variation of $S_f = \sin^2 \psi_f$ with $S_i = \sin^2 \psi_i$ for various values of ω shown alongside each curve. The parameters are $H_0 = 1$ kOe, $4\pi M_0 = 2.8$ kOe, $H_a = 8.5$ kOe, $\varepsilon = 9$, and $\varepsilon_f = 10$. The region of OR is defined by $0 < S_f < 1$ while $S_f < 0$ and $S_f > 1$ correspond to TR. The shaded region represents those combinations of S_i and S_f which cannot be realized by variation of ω (after Srivastava, 1978).

impedances offered to the wave by the two media are different, even though the wavenumbers are the same.

When the easy plane is oriented parallel to the interface (case 1), it follows from equations (3.6), (3.7), and (3.9) that resonance ($\beta \to \infty$) occurs for all ψ_i when $\omega \to \omega_2$, in which case $\psi_f \to 0$ for all ψ_i. On the other hand, when the easy plane is oriented perpendicular to the interface (case 2), the resonance occurs for $\omega \to \omega_3$.

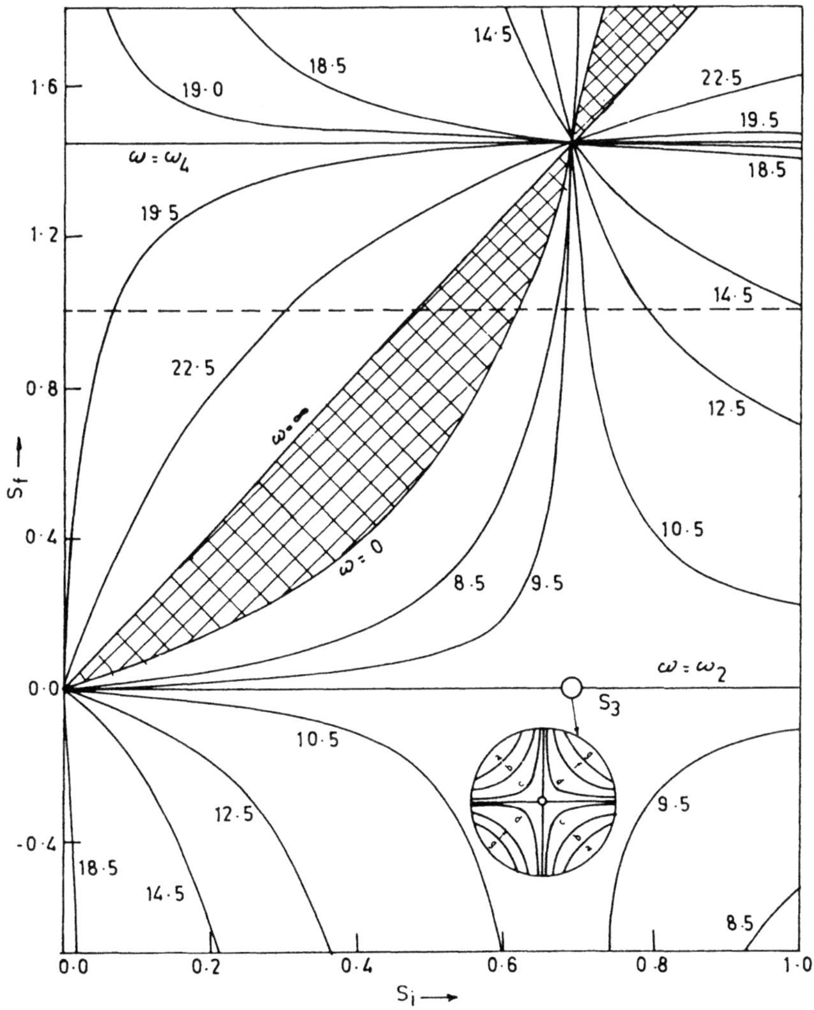

Figure 3.3. Variation of S_f with S_i for different magnitudes of ω. The parameters are the same as for Figure 3.2 except that $\varepsilon = 21$, corresponding to $\varepsilon/\varepsilon_f > F_2$. In the frequency range defined by $9.8 < f < 14.4$ GHz, TR occurs when the angle of incidence is *smaller* than a certain critical angle; this is in contrast with most other situations wherein TR requires the angle of incidence to exceed a certain critical angle (after Srivastava, 1978).

3.1.3. *Phase Shift in Total Reflection*

As discussed earlier, TR occurs in the case of a lossless medium when S_f is negative or when it exceeds unity. In each case, the reflection coefficient given by equation (3.11) can be expressed as

$$R_{\pm} = \frac{R_1 + jR_2^{\pm}}{R_1 - jR_2^{\mp}} \equiv \exp(j\delta_{\pm}) \tag{3.15}$$

where

$$\left.\begin{aligned} R_1 &= \mu_{\text{eff}}^{y} \cos\psi_i; \qquad \mu_{\text{eff}}^{y} = \Delta/\mu_{yy} \\ R_2^{\pm} &= [\sin^2\psi_i - (\beta/\beta_0)^2]^{1/2} \pm \kappa \sin\psi_i/\mu_{yy} \\ (\beta/\beta_0)^2 &= \varepsilon_f \mu_{\text{eff}}^{y}/\varepsilon + (1 - \mu_{xx}/\mu_{yy})\sin^2\psi_i \end{aligned}\right\} \tag{3.16}$$

Since R_1 and $R_2^{\pm}$ depend on the biasing field, the phase shift δ is magnetically tunable. Moreover, δ_+ and δ_- differ from each other when $\kappa \neq 0$ and $\psi_i \neq 0$. Thus, the phase shift is nonreciprocal, which suggests the possibility of obtaining magnetically tunable differential phase shift. The expression for differential phase shift is obtained from equations (3.15) and

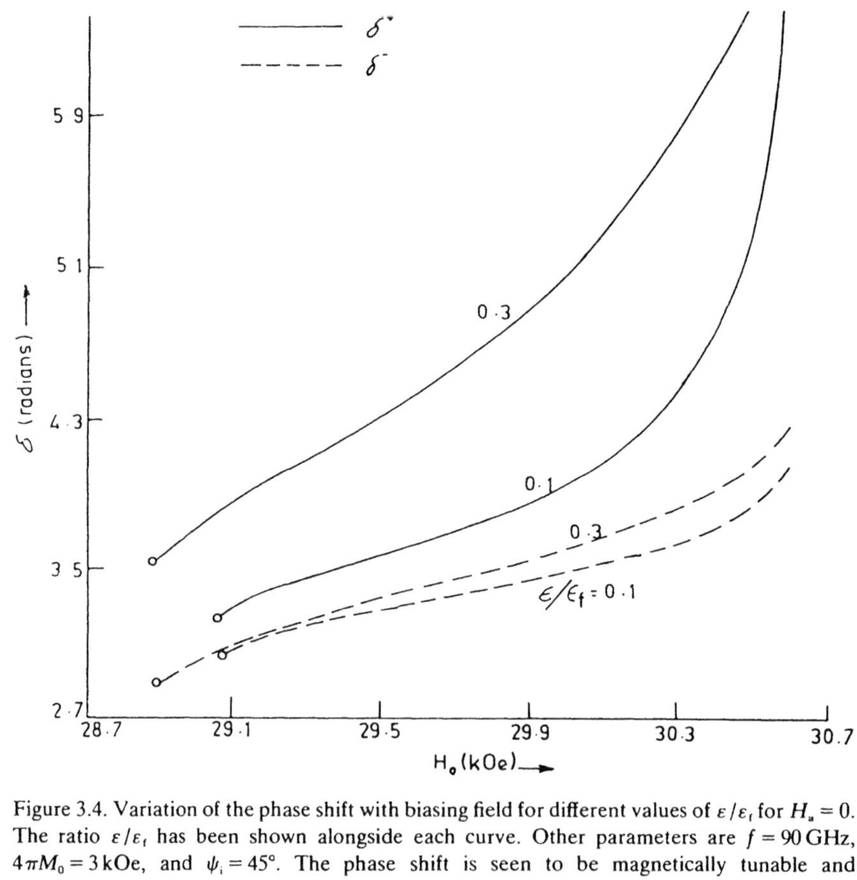

Figure 3.4. Variation of the phase shift with biasing field for different values of $\varepsilon/\varepsilon_f$ for $H_a = 0$. The ratio $\varepsilon/\varepsilon_f$ has been shown alongside each curve. Other parameters are $f = 90$ GHz, $4\pi M_0 = 3$ kOe, and $\psi_i = 45°$. The phase shift is seen to be magnetically tunable and nonreciprocal (after Srivastava, 1978).

(3.16) as

$$(\delta_+ - \delta_-) = 2 \arctan\left(\frac{\kappa \sin 2\psi_i}{\Delta \cos^2 \psi_i + \sin^2 \psi_i - \varepsilon_f \mu_{yy}/\varepsilon}\right) \tag{3.17}$$

The results of numerical calculations based on equation (3.17) have been displayed in Figures 3.4–3.8. Figure 3.4 shows the variation of $\delta_\pm$ with the biasing field for an isotropic ferrimagnetic medium ($H_a = 0$) when $\psi_i = 45°$ and $\varepsilon/\varepsilon_f = 0.1$ and 0.3. Of the two, δ_+ is larger and more strongly

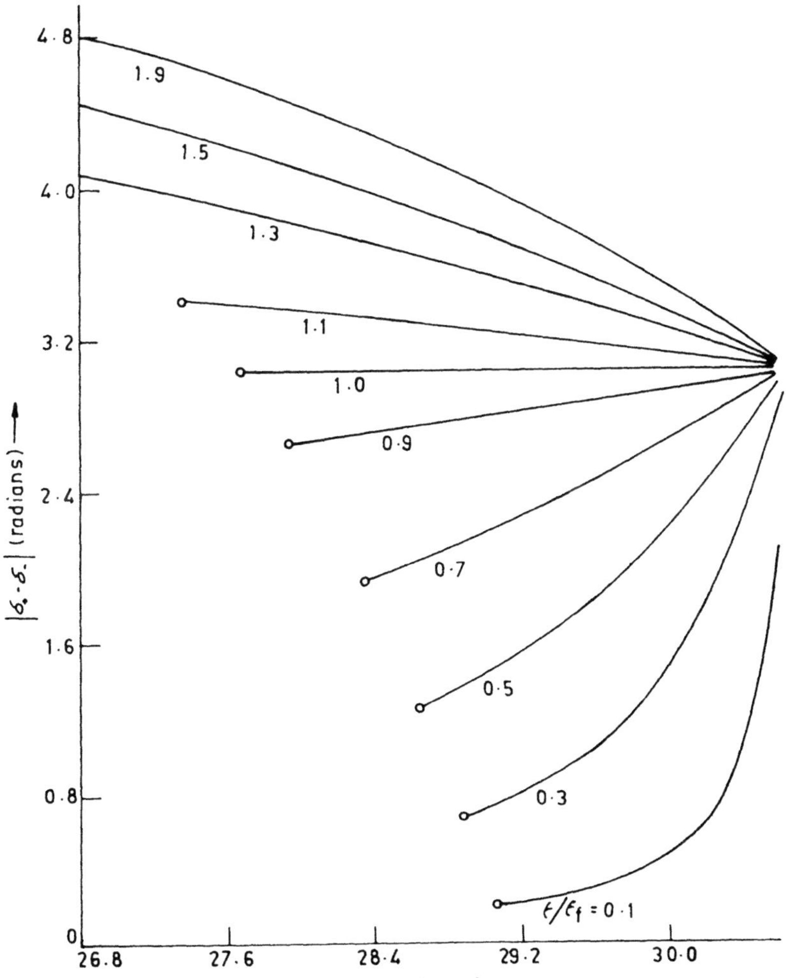

Figure 3.5. Variation of differential phase shift with biasing field for different values of $\varepsilon/\varepsilon_f$ for $H_a = 0$. The parameters are the same as those for Figure 3.4. The differential phase shift at the upper biasing field limit for TR approaches a constant value which is independent of the dielectric constant ratio (after Srivastava, 1978).

dependent on the biasing field. The differential phase shift $(\delta_+ - \delta_-)$ is appreciable and appears to vary significantly with the ratio of the dielectric constants, being larger for the higher value of $\varepsilon/\varepsilon_f$. The differential phase shift due to reflection from an isotropic ferrimagnetic half space has been plotted in Figure 3.5 as a function of the biasing field for various $\varepsilon/\varepsilon_f$, the angle of incidence being 45°. As $\varepsilon/\varepsilon_f$ increases, initially there is a strong increase in differential phase shift throughout the range of the biasing field, apart from a reduction in lower limit of the biasing field required for TR. When $\varepsilon = \varepsilon_f$, the differential phase shift is independent of the biasing field and when $\varepsilon > \varepsilon_f$, it decreases with H_0.

It is evident that the increase in ε is followed by an increase in the range of H_0, which leads to differential phase shift. Equivalently, for a fixed H_0, the frequency range of occurrence of TR increases with increasing ε. This result is of general character in the sense that the presence of a dielectric in proximity to a ferrimagnetic medium increases the frequency range of interest of the various propagation effects; this dielectric loading, in general, increases the bandwidth of the ferrimagnetic devices. The dependence of the differential phase shift on the angle of incidence is shown in Figure 3.8 for $\varepsilon/\varepsilon_f = 0.1$ for different H_0. The differential phase shift vanishes when $\psi_i = 0$ or $\pi/2$. For lower biasing field strengths, maximum differential phase shift is obtained at $\psi_i \simeq 45°$. However, as the biasing field increases, this maximum shifts to higher angles. In the intermediate biasing field region, which is well separated from the resonance and cut-off limits, the optimum differential phase shift occurs when ψ_i is in the range 45° to 60°. This range is quite convenient from the viewpoint of experiments. Figures 3.6 and 3.7 are analogous to Figure 3.5, in which the differential phase shift obtainable from a Co_2Y planar ferrite has been plotted for cases 1 and 2, respectively. Owing to large anisotropy ($H_a = 40$ kOe; see von Aulock, 1965), the required biasing field is much smaller for planar magnetoplumbites than for isotropic ferrimagnetics for the, otherwise, same parameters. For a given $\varepsilon/\varepsilon_f$, the differential phase shift has a tendency to be small in case 1 and large in case 2. When $\varepsilon = \varepsilon_f$ and $\psi_i = 45°$, it can be shown from equation (3.17) that the differential phase shift is independent of the biasing field:

$$|\delta_+ - \delta_-| = 2 \arctan [2\omega/(\omega_m \mp \omega_a)] \tag{3.18}$$

The upper and lower signs in the right-hand side correspond to cases 1 and 2, respectively. Since the differential phase shift at the upper biasing field limit attains this value (i.e., $|\delta_+ - \delta_-|_{\varepsilon=\varepsilon_f}$) for all values of $\varepsilon/\varepsilon_f$, the nature of the curves in Figures 3.5–3.7 can be understood on the basis of equation (3.18). When H_a is large, $|\delta_+ - \delta_-|_u$ at the upper biasing field limit is small in case 1 (it approaches zero as $H_a \to \infty$). However, when $H_a \to 0$ (isotropic

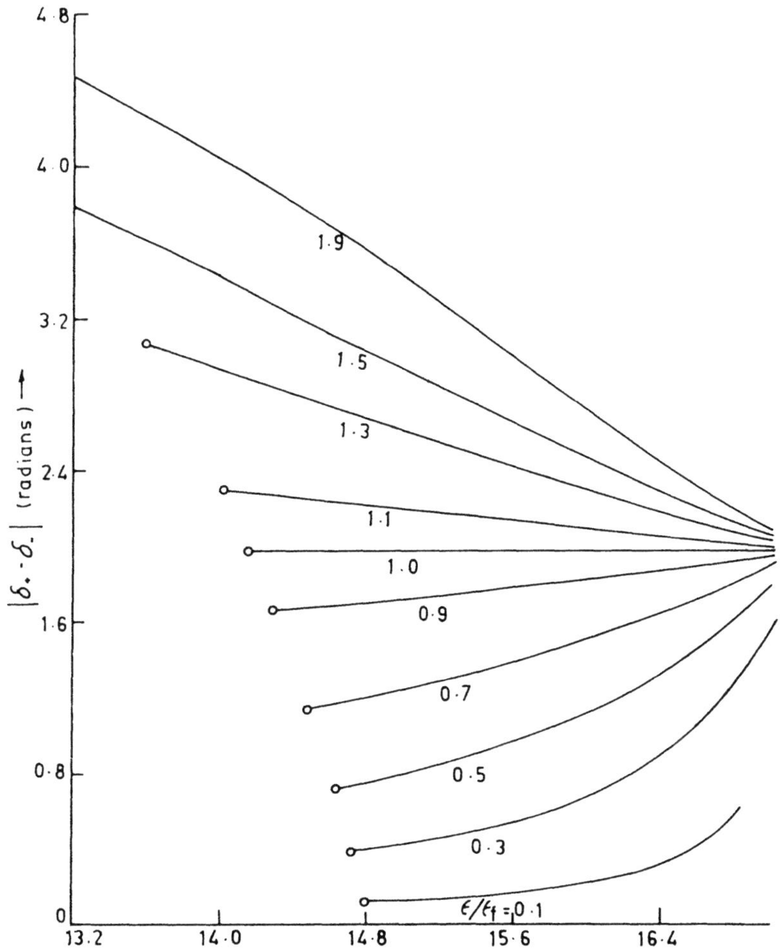

Figure 3.6. Variation of the differential phase shift with biasing field for different values of $\varepsilon/\varepsilon_f$ for a Co_2Y planar magnetoplumbite in case 1. The anisotropy field is $H_a = 40$ kOe while the other parameters are the same as those for Figure 3.4. One effect of planar anisotropy is to modify the range of TR. The magnitude of the differential phase shift is affected by a change (decrease) in the differential phase shift at the upper biasing field limit, when compared with the case of isotropic ferrimagnetics, e.g., Figure 3.5 (after Srivastava, 1978).

material, Figure 3.5), $|\delta_+ - \delta_-|_u$ is relatively large and, in case 2, it is the largest; $|\delta_+ - \delta_-|_u \to \pi$ as $H_0 \to \infty$. Consequently, for a given $\varepsilon/\varepsilon_f$, $|\delta_+ - \delta_-|$ is, in general, small in case 1 and large in case 2. Since the range of the biasing field in which TR occurs is the largest in case 2, the slope of the differential phase shift vs. biasing field curve is largest in case 2 and smallest

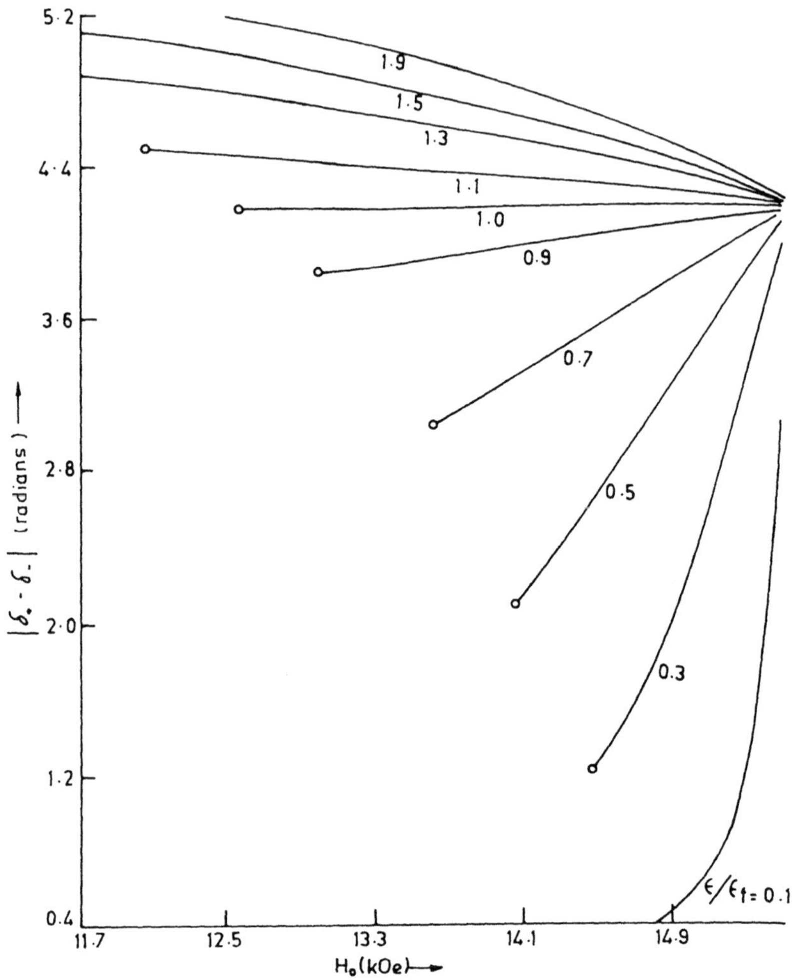

Figure 3.7. Variation of differential phase shift with biasing field for different values of $\varepsilon/\varepsilon_f$ for a Co_2Y planar magnetoplumbite in case 2. The parameters are the same as those for Figure 3.6 (after Srivastava, 1978).

in case 1 (Figures 3.5–3.7). The planar anisotropy is thus seen to control the magnitude and gradient of the differential phase shift. It may be noted that when the angle of incidence is greater (smaller) than 45°, the ratio $\varepsilon/\varepsilon_f$ for which the differential phase shift is independent of the biasing field is slightly greater (smaller) than unity (results not shown). It is concluded that, in general, the presence of dielectrics in the proximity of ferrimagnetics strongly influences the wave phenomena for which, apparently, only

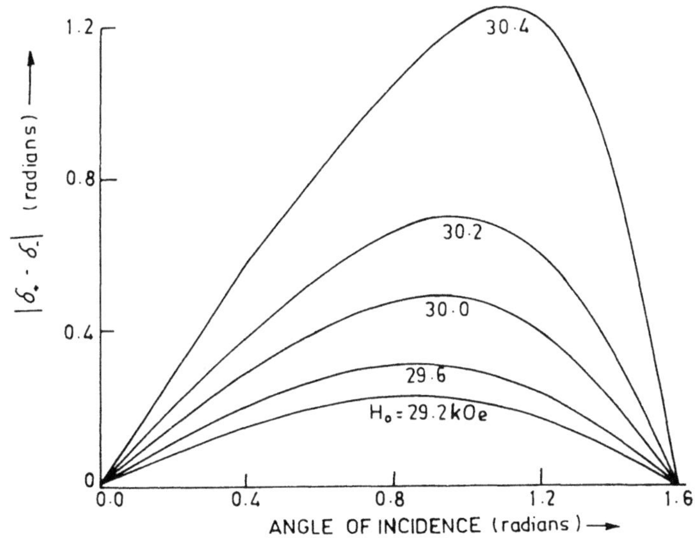

Figure 3.8. Variation of the differential phase shift with angle of incidence for different biasing fields. The parameters are $f = 90\,\mathrm{GHz}$, $4\pi M_0 = 3\,\mathrm{kOe}$, $H_a = 0$, and $\varepsilon/\varepsilon_f = 0.1$ (after Srivastava, 1978).

the latter is responsible. In more complicated configurations, a consequence of this "dielectric loading" is the increase of the bandwidth of phase shift devices (Ince and Temme, 1969; von Aulock and Fay, 1968). The studies also suggest the possibility of application of ferrimagnetics in tunable reflection beam phase shifters (Culshaw, 1961) for operation in the millimeter-wave region.

3.1.4. *Reflection from Lossy Ferrimagnetics*

When the ferrimagnetic half space is considered to be lossless, it is easy to show from equation (3.11) that the power reflected in the cases of OR or TR is reciprocal; the interchange of the incident and reflected beams has no effect* on $|R|$. However, when losses are considered, κ, μ_{xx}, and μ_{yy} are all complex. It then follows from equation (3.11) that $|R_+|$ differs from $|R_-|$. The results of numerical calculations of $|R_\pm|$ have been shown in Figures 3.9–3.13.

* It is worth mentioning that, under certain conditions, $|R_+|$ may differ from $|R_-|$ even when losses are neglected. An example is the case when the hard axis of the planar magnetoplumbite is oriented at an arbitrary angle (different from $\pi/2$ in case 1 or 0 in case 2) to the interface.

Figure 3.9 shows the variation of reflection loss with biasing field in the case when $\psi_i = 45°$, $\varepsilon/\varepsilon_f = 0.1$, and $H_a = 0$. It is seen from Figure 3.9 that for a given linewidth ΔH, both the forward and reverse reflection losses initially increase with the biasing field, attain maxima, and then decrease on further increasing the biasing field (the portions of the curves where the reverse reflection loss $|R_-|$ decreases occur at relatively high biasing field

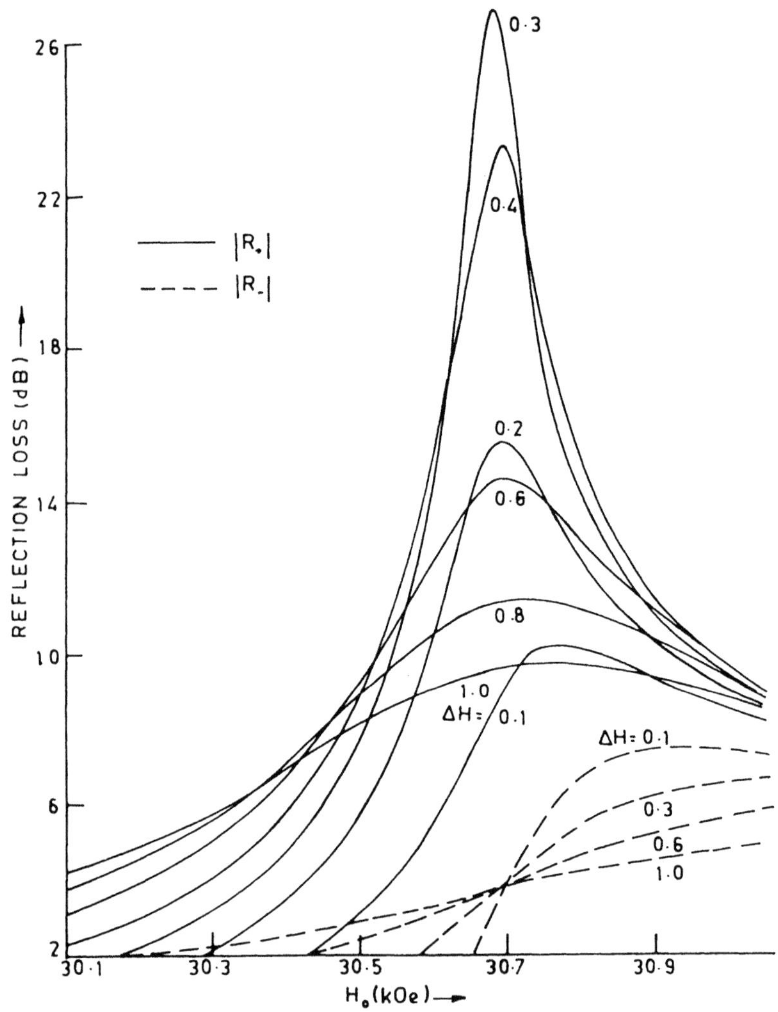

Figure 3.9. Variation of forward and reverse reflection losses with biasing field for different linewidths ΔH. The parameters are $f = 90\,\text{GHz}$, $4\pi M_0 = 3\,\text{kOe}$, $\varepsilon/\varepsilon_f = 0.1$, $\psi_i = 45°$, and $H_a = 0$ (after Srivastava, 1978).

strengths and are not observable in the figure). In general, the peaks in forward reflection loss $|R_+|$ are much stronger and narrower than those in reverse reflection loss. Moreover, the peaks in the forward and reverse losses are rather well separated in the biasing field scales. Specifically, the peak in $|R_+|$ occurs at a smaller biasing field. As such, the reverse loss is small at the biasing field strength at which the forward reflection loss is large. This suggests the possibility of application of such a configuration to a millimeter-wave reflection beam isolator, analogous to the solid state magnetoplasma isolator (Kanda and May, 1975). As discussed earlier, the reflected amplitudes $|R_+|$ and $|R_-|$ are equal when the linewidth is zero. As the linewidth increases, the isolation, defined as forward reflection loss (or just reflection loss) minus reverse reflection loss (or insertion loss), also increases and attains a value more than 20 dB when $\Delta H \simeq 300$ Oe. Further increase in linewidth results in reduced isolation. The variation of $|R_\pm|$ with angle of incidence ψ_i for different $\varepsilon/\varepsilon_f$ has been plotted in Figures 3.10

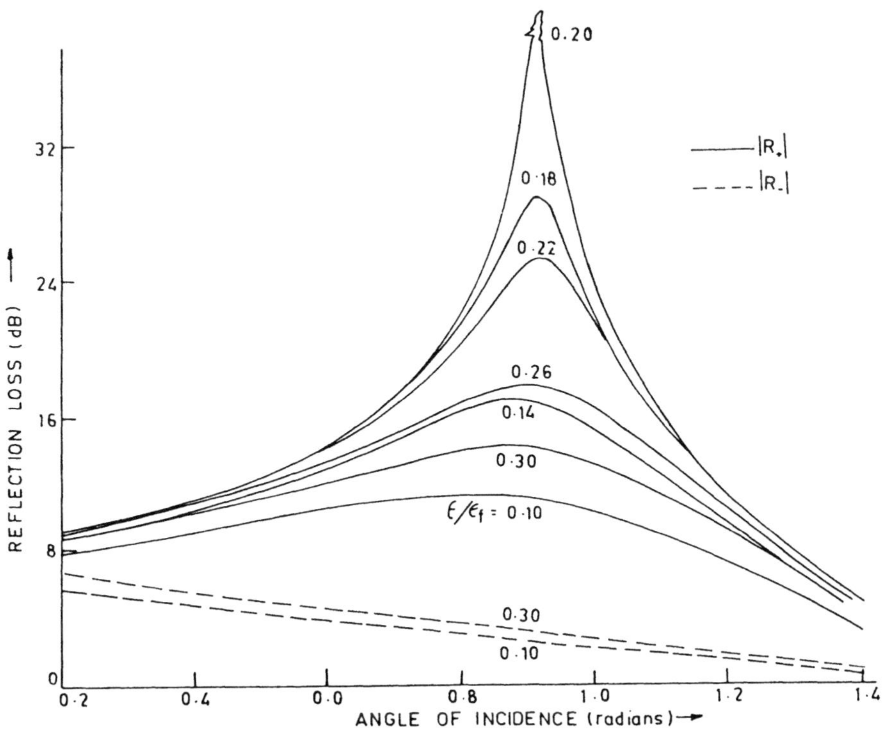

Figure 3.10. Variation of the reflection loss with angle of incidence for different values of $\varepsilon/\varepsilon_f$. The parameters are $f = 90$ GHz, $H_a = 0$, $H_0 = 30.6$ kOe, $4\pi M_0 = 3.0$ kOe, $H_a = 0$, and $\Delta H = 700$ Oe (after Srivastava, 1978).

and 3.11 for $\Delta H = 700$ and 1100 Oe, respectively. The variation of $\varepsilon/\varepsilon_f$ does not significantly alter the angle of incidence at which maximum isolation occurs. However, the magnitude of peak isolation is a sensitive function of $\varepsilon/\varepsilon_f$; initially it increases with $\varepsilon/\varepsilon_f$ and then starts decreasing after attaining a maximum. This maximum occurs for $\Delta H = 700$ Oe when $\varepsilon/\varepsilon_f \simeq 0.2$ and for $\Delta H = 1100$ Oe when $\varepsilon/\varepsilon_f \simeq 0.27$; the larger the linewidth, the larger is the ratio $\varepsilon/\varepsilon_f$ required for maximum isolation.

Figures 3.12 and 3.13 represent $|R_{\pm}|$ for a hexagonal planar magnetoplumbite in cases 1 and 2, respectively. The general behavior as regards the ratio $\varepsilon/\varepsilon_f$ and the linewidth is the same as that for the isotropic ferrimagnetics. However, there are certain differences too. The angle of incidence at which maximum isolation occurs is much larger in case 1 (and smaller in case 2) than that for the corresponding isotropic case. In fact, the optimum isolation requires the incident ray to be directed away from the hard axis.

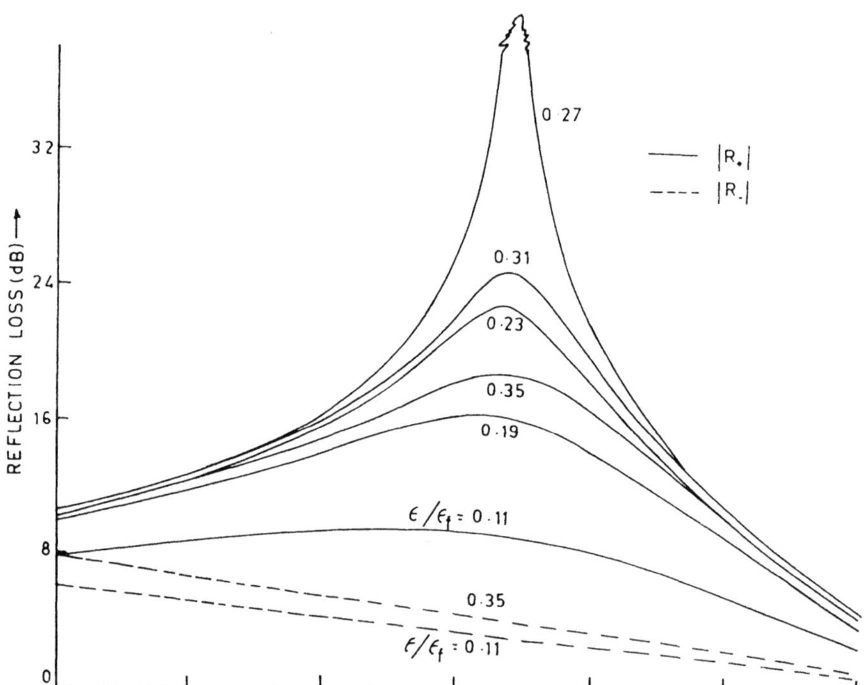

Figure 3.11. Variation of the reflection loss with angle of incidence for different values of $\varepsilon/\varepsilon_f$. The parameters are the same as those for Figure 3.10 except that $\Delta H = 1100$ Oe. The higher the linewidth, the larger the ratio $\varepsilon/\varepsilon_f$ required for maximum isolation (after Srivastava, 1978).

The orientation of the hard axis can thus control the angle of incidence at which optimum isolation is obtained. Moeover, unlike the case of isotropic materials, the angle of incidence at which maximum isolation is achieved is different for different $\varepsilon/\varepsilon_f$. The possible application of these results to a reflection beam isolator has been discussed by Srivastava (1978).

3.1.5. *Energy Flow in Total Reflection: Lateral Shift*

In earlier sections, the reflection from a ferrimagnetic half space was treated by solving the wave equation in different regions and applying appropriate boundary conditions to obtain the desired results. This mathematically precise procedure does not provide much physical insight into the phenomena occurring at the interface. For instance, it does not

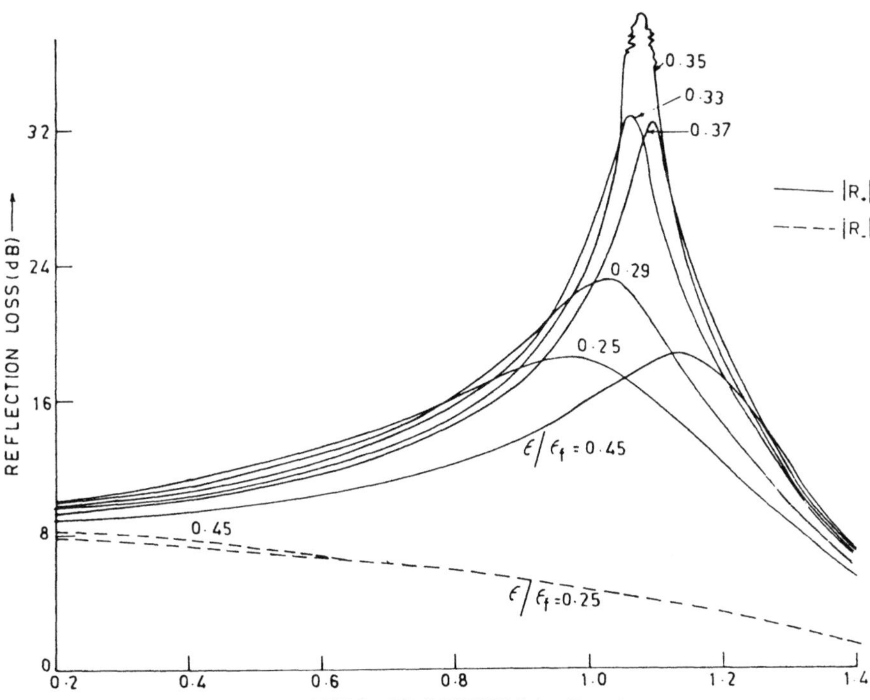

Figure 3.12. Variation of reflection loss with angle of incidence for different values of $\varepsilon/\varepsilon_f$ for a planar magnetoplumbite in case 1 when $H_0 = 17.17\,\text{kOe}$, $H_a = 40\,\text{kOe}$, and $\Delta H = 1.1\,\text{kOe}$. Other parameters are the same as those for Figure 3.10 (after Srivastava, 1978).

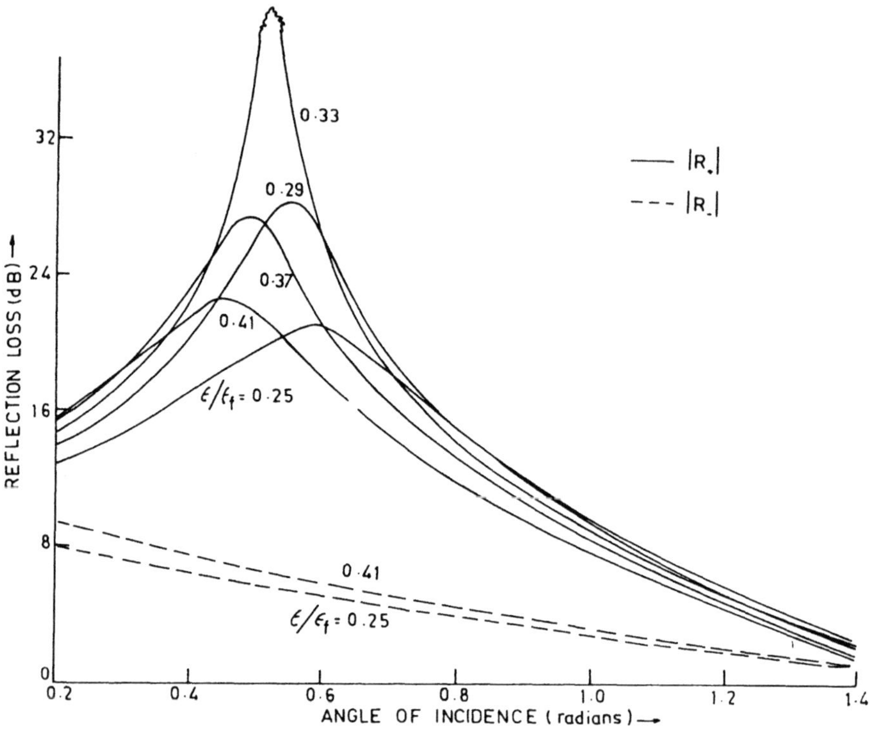

Figure 3.13. Variation of reflection loss with angle of incidence for different values $\varepsilon/\varepsilon_f$ for a planar magnetoplumbite in case 2. The parameters are the same as those for Figure 3.12 (after Srivastava, 1978).

clearly bring out penetration, propagation, and absorption of energy in the reflecting medium. Owing to this lack of physical insight, it is difficult to make valid approximations and develop simple models for the analysis of rather complicated problems, e.g., guided wave propagation in ferrimagnetics.* In what follows, we have considered the reflection of electromagnetic waves at a dielectric–ferrite interface from the viewpoint of ray propagation and energy flow and also described, in the subsequent section, an approximate ray model for attenuated total reflection from a weakly absorbing ferrimagnetic material.

In the case when losses are neglected ($\Delta H = 0$), the time-averaged power flow in the ferrimagnetic region is obtained from equations (3.4) and

* In recent years, guided wave propagation in dielectric optical waveguides has been treated by zig-zag ray models; see for example Kogelnik (1975) and Tien (1977).

Figure 3.14. The path of a ray in lossless total reflection: (a) geometrical reflection and (b) actual reflection. The ray propagation parallel to the interface in the reflection medium is a logical necessity in order to explain finite (nonzero) energy flux parallel to the interface in the reflecting medium.

(3.5) as

$$\mathbf{P}^{\mathrm{t}}_{\pm} = \frac{c_0}{8\pi}\,\mathrm{Re}(\mathbf{E}^{\mathrm{t}} \times \mathbf{h}^{\mathrm{t}*})$$

$$= -\hat{x}\,\frac{c_0^2\beta_0}{8\pi\omega\mu_{\mathrm{eff}}}\left\{\sin\psi_{\mathrm{i}} \pm \frac{\kappa}{\mu}\,[\sin^2\psi_{\mathrm{i}} - (\beta/\beta_0)^2]^{1/2}\right\} |T_{\pm}|$$

$$\times \exp\{-2[\sin^2\psi_{\mathrm{i}} - (\beta/\beta_0)^2]^{1/2} y\} \tag{3.19}$$

Thus, in the event of total reflection, there is a finite time-averaged power flow in the evanescent field set up in the ferrimagnetic region.* This may appear to be surprising since, the reflected and incident powers being equal in the case of lossless TR, it is difficult to conceive of a power flow in the second medium as it would imply a violation of the law of conservation of energy. The logical solution to this apparent paradox has been discussed by Renard (1964) and is explained in Figure 3.14. The electromagnetic energy enters the second (reflecting) medium, where it propagates a certain distance parallel to the interface, and then returns to the first medium. Evidently, there is no way to detect this "lateral shift" of rays in the case of reflection of a uniform plane wave of infinite extent. However, the lateral shift of a well-collimated beam (of finite transverse extent) should be observable; several experimental and theoretical investigations of this effect in dielectrics have been reported and are summarized by Lotsch (1970/71).

Now we proceed to derive an expression for the lateral drift of an electromagnetic beam totally reflected at a dielectric–ferrimagnetic interface. Figure 3.15 shows the lateral shift of a beam (bundle of rays). It is

* It is worth mentioning that the existence of finite energy flux parallel to the interface in the event of TR is not merely a characteristic of ferrites, but is common to reflection phenomena involving any medium and any kind of waves, e.g., electromagnetic, elastic, quantum mechanical, etc.; see for example Lotsch (1970/71).

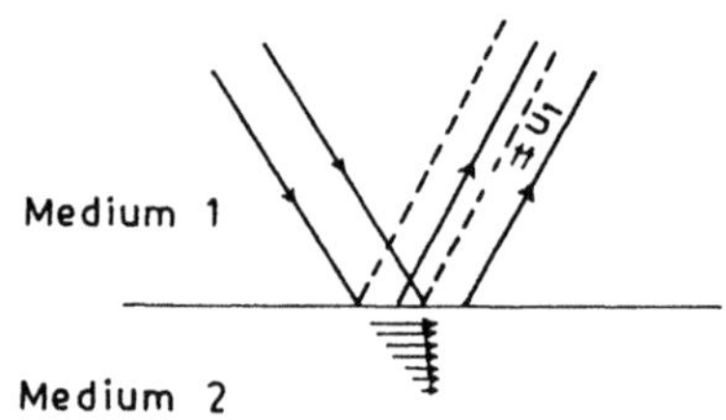

Figure 3.15. The lateral shift of a bounded beam as a consequence of total reflection.

evident that the time-averaged flux of energy of the incident wave across the strip whose width is equal to the lateral shift must be equal to the time-averaged energy flux parallel to the interface in the entire second (ferrimagnetic) medium. Thus, the lateral shift is given by

$$D \cos \psi_i = \frac{\int_0^\infty P^t_{-x}\, dy}{|\mathbf{P}^r|}$$

Equivalently,

$$D = \frac{\int_0^\infty P^t_{-x}\, dy}{|\mathbf{P}^r| \cos \psi_i} = \int_0^\infty P^t_x\, dy / P^r_y \tag{3.20}$$

With equations (3.4) and (3.5), P^r_y is evaluated as

$$P^r_{-y} = -P^r_y = [c_0^2 \beta_0^2 \cos \psi_i / (8\pi\omega)] |R_\pm|^2 \tag{3.21}$$

Substitution of P^r_{-y} from equation (3.21) and of P^t_{-x} from equation (3.19) into equation (3.20) leads to (Gupta and Srivastava, 1979)

$$D_\pm = \frac{2\mu_{\text{eff}} \cos \psi_i (\sin \psi_i \pm \kappa\eta/\mu)}{\beta_0 \eta \{\mu_{\text{eff}}^2 \cos^2 \psi_i + [\eta \pm (\kappa/\mu) \sin \psi_i]^2\}} \tag{3.22}$$

where

$$\eta = [\sin^2 \psi_i - (\beta/\beta_0)^2]^{1/2} \tag{3.23}$$

It follows from equation (3.22) that the lateral shift is nonreciprocal; its magnitude changes when the path of rays is reversed. Specifically, in the case of total reflection of a normally incident wave ($\psi_i = 0$; $\mu_{\text{eff}} < 0$) equation (3.22) reduces to

$$D_{\pm} = \pm \frac{2|\kappa/\mu|}{\beta_0(|\mu_{\mathrm{eff}}| + \varepsilon_f/\varepsilon)} \tag{3.24}$$

which implies that D_+ and D_- have the same magnitude but opposite signs. This might give the impression that there are two distinct shifts for the same incident ray. However, a closer examination reveals that there is no such inconsistency; the meaning of negative shift is explained in Figure 3.16. Figure 3.17 shows paths of a normally incident ray and of two rays which are incident from the opposite sides of the normal. The shift is positive for ray 1 and negative for ray 2. In the limit when $\psi_i \to 0$, it is, therefore, natural that D_+ and D_- should approach same magnitude with opposite signs. The occurrence of a finite shift for $\psi_i = 0$ can be understood in terms of excitation (on account of the incident wave) of elliptical precession of the magnetization vector about the biasing field; this precession leads to the emergence of an energy packet from a point on the interface different from the point at which the electromagnetic energy enters the ferrimagnetic medium. This is a purely magnetic effect and has

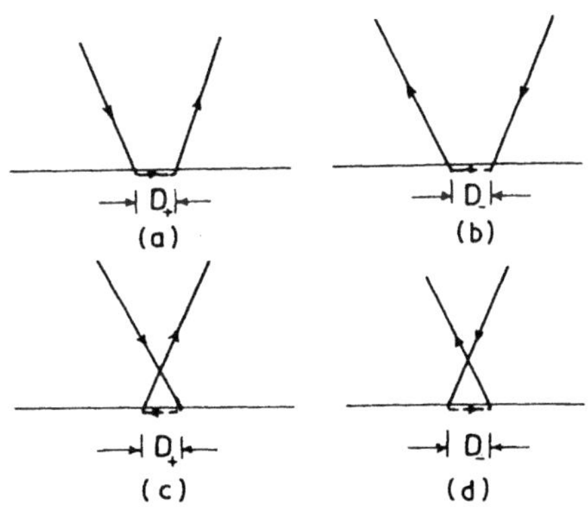

Figure 3.16. Illustration of positive and negative shifts in cases of forward (+) and reverse (−) incidence; (a) and (b) correspond to positive shift while (c) and (d) correspond to negative shift (after Gupta and Srivastava, 1979).

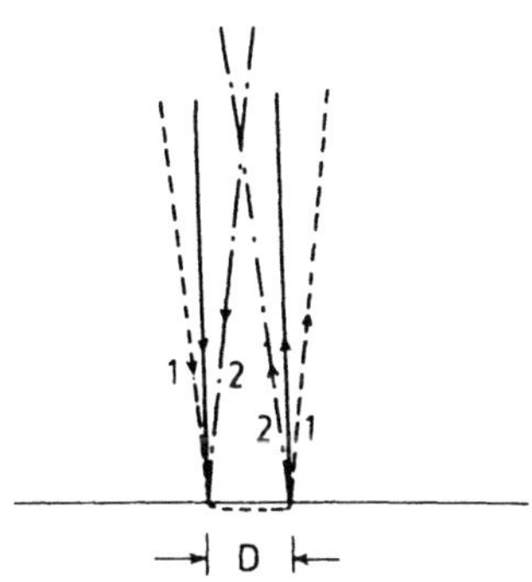

Figure 3.17. Illustration of total reflection of a normally incident ray and of two rays incident at small angles from the opposite sides of the normal. Ray 1 undergoes a positive lateral shift while ray 2 undergoes a negative shift. Obviously, in the limit of normal incidence, the shifts of the two rays approach the same magnitude but opposite signs (after Gupta and Srivastava, 1979).

no analogue in the case of TR from a lossless dielectric. When the angle of incidence is different from zero, the resulting lateral shift consists of magnetic and dielectric contributions; the dielectric contribution has a tendency to cause reciprocal positive shift while the magnetic effect leads to nonreciprocal contribution of a sign which depends on the direction of gyration of magnetization. Figures 3.18 and 3.19 show the dependence of

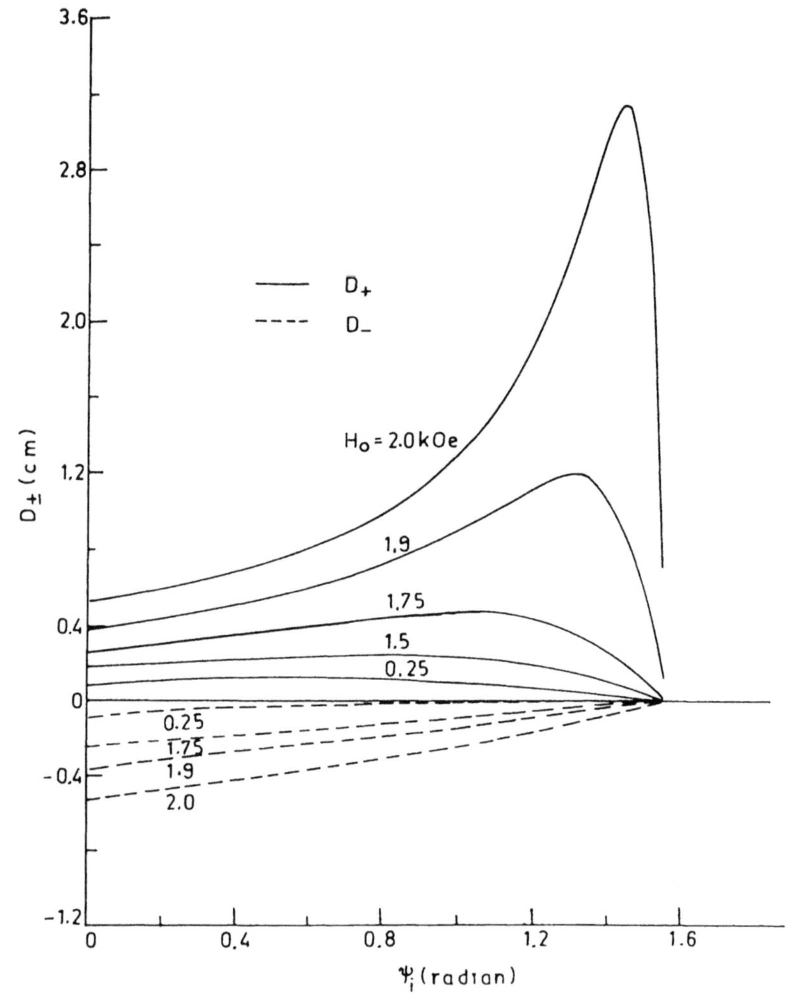

Figure 3.18. Variation of the lateral shift D_+ (solid lines) and D_- (broken lines) with the angle of incidence for different biasing field strengths shown alongside each curve. Other parameters are $\varepsilon/\varepsilon_f = 1/13$, $f = 9\,\mathrm{GHz}$, and $4\pi M_0 = 3.0\,\mathrm{kOe}$. It is interesting that the shift D_+ increases significantly when the biasing field is increased to its resonance limit. This is plausible since the lateral shift is a measure of the strength of the interaction of the electromagnetic wave with the reflecting medium (after Gupta and Srivastava, 1979).

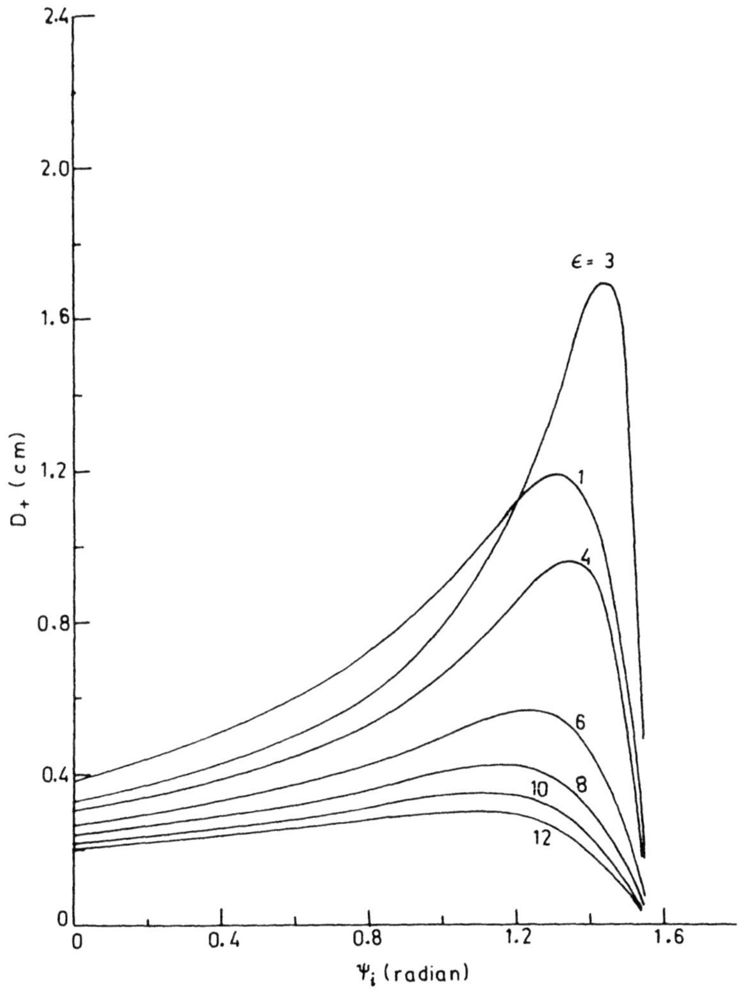

Figure 3.19. Variation of D_+ with angle of incidence for different values of the dielectric constant ratio $\varepsilon/\varepsilon_r$. The parameters are $f = 9\,\text{GHz}$, $\varepsilon_f = 13$, and $H_0 = 1.9\,\text{kOe}$; the value of ε is shown alongside each curve (after Gupta and Srivastava, 1979).

$D_\pm$ on various parameters. It is noteworthy that D_+ exhibits a peak when plotted as a function of the angle of incidence. The occurrence of this peak is a purely magnetic effect since, in the case of TR from a dielectric, the shift decreases monotonically with increasing angle of incidence and approaches zero as $\psi_i \to \pi/2$ (Renard, 1964).

When μ_{eff} is positive, it is not difficult to show that D_+ and D_- are both positive. The shifts $D_\pm$ are both large in the vicinity of resonance (i.e., when $\mu_{\text{eff}} \to \infty$) while $D_{\pm} = 0$ when $\mu_{\text{eff}} = 0$. Numerical calculations (results not shown) indicate that the difference between D_+ and D_- is not significant in this case because the positive μ_{eff} makes TR a predominantly dielectric effect. It is concluded that, in general, the nonreciprocity in lateral shift is large (small) when μ_{eff} is negative (positive). This conclusion continues to hold even if an electromagnetic field different from the plane wave field interacts with the ferrimagnetic surface.* Since the lateral shift is a measure of the strength of interaction between microwaves and ferrimagnetics, the nonreciprocal effects produced by ferrimagnetics in a variety of situations are particularly significant when μ_{eff} is negative.

3.1.6. *Ray Model for Attenuated Total Reflection*

When the magnetic loss is considered, the reflected wave is attenuated on account of absorption in the ferrimagnetic region. In this case, the ray model of reflection, as discussed in the last chapter, should be modified so as to include the effect of absorption. When the linewidth is small (in the first approximation), the effective distance traversed by rays in the ferrimagnetic region would be the same as given by equation (3.22). As such, it is only required to derive an expression for the effective absorption coefficient in order to obtain the approximate reflection coefficient. Assuming ΔH to be small, the permeability tensor can be expanded as

$$\boldsymbol{\mu} = \boldsymbol{\mu}_r - j\Delta H \boldsymbol{\mu}_i \tag{3.25}$$

where $\boldsymbol{\mu}_r$ is the same as the permeability tensor for a lossless medium while $\boldsymbol{\mu}_i$ is given by

$$\boldsymbol{\mu}_i = \begin{pmatrix} \xi_1 & j\zeta_1 & 0 \\ -j\zeta_1 & \xi_1 & 0 \\ 0 & 0 & 0 \end{pmatrix} \tag{3.26}$$

The parameters ξ_1 and ζ_1 are defined as

$$\left.\begin{aligned} \xi_1 &= \frac{\mu_0}{\mu_{\text{eff}}^{(0)}}(\zeta_2 + 2\kappa_0\xi_2) \\ \zeta_1 &= \frac{1}{\mu_{\text{eff}}^{(0)}}\left[(\mu_0^2 + \kappa_0^2)\xi_2 + \frac{\kappa_0}{\mu_{\text{eff}}^{(0)}}\zeta_2\right] \end{aligned}\right\} \tag{3.27}$$

* Because an arbitrary, physically realizable field, incident at an interface, can be expressed in terms of the angular spectrum of plane waves.

μ_0 and κ_0 represent the diagonal and off-diagonal elements of the permeability tensor with loss neglected and $\mu_{\text{eff}}^{(0)} = (\mu_0^2 - \kappa_0^2)/\mu_0$. Furthermore,

$$\left.\begin{aligned} \kappa/\mu &= \kappa_0/\mu_0 - j\Delta H\xi_2 \\ \mu_{\text{eff}} &= \mu_{\text{eff}}^{(0)} - j\Delta H\zeta_2 \end{aligned}\right\} \tag{3.28}$$

The approximate expression for the effective absorption coefficient for rays propagating parallel to the interface in the ferrimagnetic region is obtained as follows. The power lost in a strip of width dx around x is equal to*

$$dP_x^{(t)} = -\frac{\omega}{8\pi}\,\text{Re}[(\boldsymbol{\mu}_i''\cdot\mathbf{h})\cdot\mathbf{h}^*]\,dx\Delta H \tag{3.29}$$

With equations (3.26)–(3.28), equation (3.29) reduces to

$$\frac{dP_x^{(t)}}{dx} = -2\alpha_{\text{eff}}^{\pm} P_x^{(t)} \tag{3.30}$$

where

$$\alpha_{\text{eff}}^{\pm} = \frac{\beta_0\zeta_2\Delta H\{\sin^2\psi_i \pm \eta[\kappa_0/\mu_0 - \mu_{\text{eff}}^{(0)}\xi_2/\zeta_2]\sin\psi_i - \frac{1}{2}\mu_{\text{eff}}^{(0)}\varepsilon_t/\varepsilon\}}{\mu_{\text{eff}}^{(0)}(\sin\psi_i \pm \eta\kappa_0/\mu_0)} \tag{3.31}$$

The integration of equation (3.30) yields

$$P_x^{(t)} = P_0\exp(-2\alpha_{\text{eff}}^{\pm} x) \tag{3.32}$$

which implies that $2\alpha_{\text{eff}}^{\pm}$ is the power absorption coefficient for the rays propagating in the ferrimagnetic region. It follows from the ray model of Figure 3.14 that the fractional power lost by a ray in traversing a distance $D_{\pm}$ in the ferrimagnetic region is given by

$$P_r/P_i \equiv |R_{\pm}|^2 = \exp(-2\alpha_{\text{eff}}^{\pm} D_{\pm}) \tag{3.33}$$

Equivalently,

$$|R_{\pm}| = \exp(-\alpha_{\text{eff}}^{\pm} D_{\pm}) \tag{3.34}$$

* It should be noted that equation (3.29) is valid only when the dispersion and dissipation are both small. (See, for example, Landau and Lifshitz, 1960.) Thus, the subsequent analysis is not applicable in the region close to the resonance.

Numerical calculations have been performed to compare the above approximate expression for $|R|$ with the rigorous one, i.e., equation (3.11). The results have been displayed in Figures 3.20 and 3.21. It is seen that the ray model leads to excellent results except when ΔH is large or when the biasing field is adjusted to its resonance value, which is slightly above 2 kOe for the parameters used in the calculations. The general applicability of the

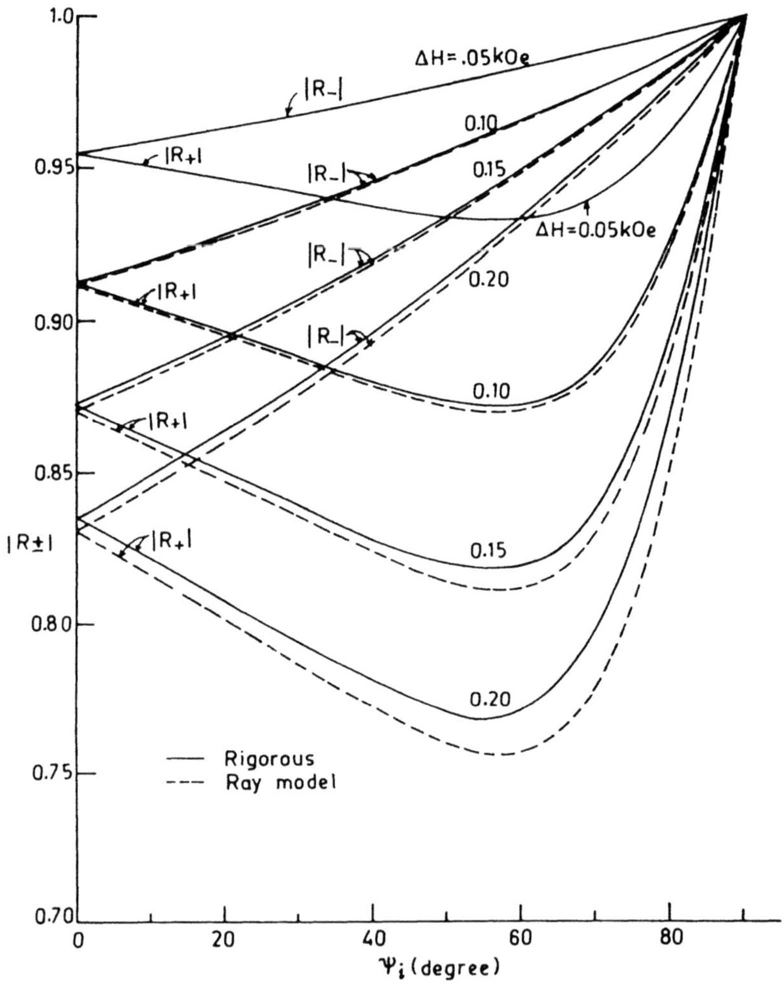

Figure 3.20. The variation of $|R_\pm|$ with ψ_i as obtained from the rigorous formula [equation (3.11), solid lines] and from the ray model [equation (3.34), broken line] with linewidth ΔH as the parameter. Other parameters are $H_0 = 1.75$ kOe, $\varepsilon = 1.0$, $\varepsilon_f = 13.0$, $f = 9$ GHz, and $4\pi M_0 = 3$ kOe. The agreement between the two is excellent except when ΔH is large (after Gupta and Srivastava, 1979).

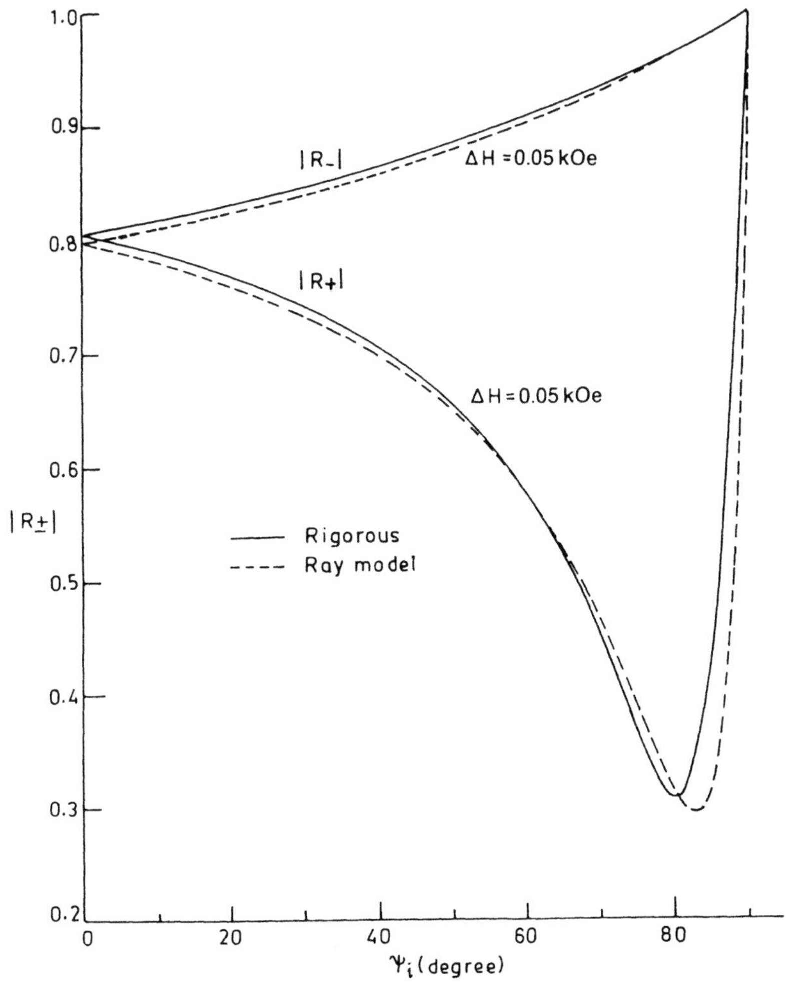

Figure 3.21. Variation of $|R_\pm|$ with angle of incidence for $H_0 = 2.0$ kOe and $\Delta H = 0.05$ kOe, other parameters being the same as those for Figure 3.20. Comparison of Figures 3.20 and 3.21 shows that for $\Delta H = 0.05$ kOe, the agreement is better for the former for which H_0 is farther away from its resonance value (after Gupta and Srivastava, 1979).

ray model makes it possible to obtain useful expressions for $|R|$ in the case of normal incidence. It follows from equations (3.22) and (3.23) that

$$\alpha^{\pm}_{\text{eff}} D_{\pm} = \frac{2\zeta_2 \Delta H \cos\psi_i [\sin^2\psi_i \pm \eta(\kappa_0/\mu_0 - \xi_2 \mu^{(0)}_{\text{eff}}/\zeta_2) - \frac{1}{2}\mu^{(0)}_{\text{eff}} \varepsilon_f/\varepsilon]}{\eta[\mu^{(0)2}_{\text{eff}} \cos^2\psi_i + (\eta \pm \kappa_0/\mu_0 \sin\psi_i)^2]} \quad (3.35)$$

When $\psi_i = 0$, equation (3.55) reduces to

$$\alpha_{\rm eff}^{\pm} D_{\pm} = \frac{(2\varepsilon_f/\varepsilon)^{1/2}\zeta_2 \Delta H}{\sqrt{-\mu_{\rm eff}^{(0)}(\varepsilon_f/\varepsilon - \mu_{\rm eff}^{(0)})}} \tag{3.36}$$

When the biasing field is appreciably different from its resonance limit, it is possible to have $|\mu_{\rm eff}^{(0)}| \ll \varepsilon_f/\varepsilon$, in which case it is seen that

$$|R| \simeq \exp\left(-\sqrt{\frac{2\varepsilon}{\varepsilon_f|\mu_{\rm eff}^{(0)}|}}\,\zeta_2 \Delta H\right) \tag{3.37}$$

Figure 3.22 presents a comparison of the results obtained from equation (3.37) with rigorous results. The agreement is good in intermediate biasing field regions and for small ΔH. When $\varepsilon \simeq \varepsilon_f$ and $\mu_{\rm eff} \gg \varepsilon_f/\varepsilon$, we have

$$|R| \simeq \exp\left(-\frac{\sqrt{2}\,\zeta_2 \Delta H}{|\mu_{\rm eff}^{(0)}|^{3/2}}\right) \tag{3.38}$$

It follows from the preceding discussion that the linewidth can be determined if $|R|$ is known, and this may be useful in the millimeter-wave region.

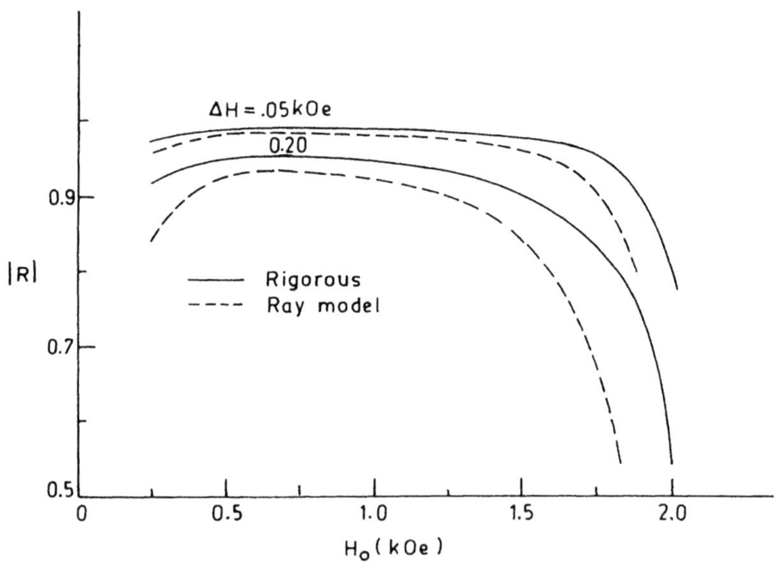

Figure 3.22. Variation with biasing field of $|R_{\pm}|$ obtained from the exact formula [equation (3.11) with $\psi_i = 0$, solid lines] and from equation (3.37), (broken lines) for different ΔH shown in the figure. Other parameters are the same as those for Figure 3.20 (after Gupta and Srivastava, 1979).

It may be possible to generalize the ray formulation given above to analyze electromagnetic and magnetostatic bulk waves in loaded waveguides.*

3.2. *Ferrimagnetic Slab*

Reflection and transmission of electromagnetic waves from ferro- and ferrimagnetic slabs have been investigated by several authors including Suhl and Walker (1954), Seavey and Tannenwald (1957), Heinrich and Meshcheryakov (1969), Phillips (1970), Vittoria et al. (1973, 1975), Sodha and Srivastava (1975), Mueller (1975, 1976), and Lieu et al. (1977). We shall discuss in detail only the case of a normally magnetized slab.

Consider a uniform plane wave of angular frequency ω to be normally incident on a ferrimagnetic slab, $0 \leq z \leq d$, which is magnetized in a direction perpendicular to its face, i.e., along the direction of propagation (z) of the incident wave (see Figure 3.23). The electric field of the incident wave is assumed to be directed along the x-axis. Phenomenologically speaking, a part of the incident wave is transmitted at the dielectric–ferrimagnetic interface $z = 0$. Since the modes of propagation in a ferrimagnetic medium, magnetized along the direction of propagation, are circularly rather than linearly polarized, the transmitted wave (at $z = 0$) will contain a y-component as well as an x-component of the electric field as a consequence of the coupling. The reflected field, which arises on account of radiation from distributed sources in the slab, also contains a y-component. The component of the incident wave transmitted at $z = 0$ propagates as a mixture of RCP and LCP modes; the unequal amplitude

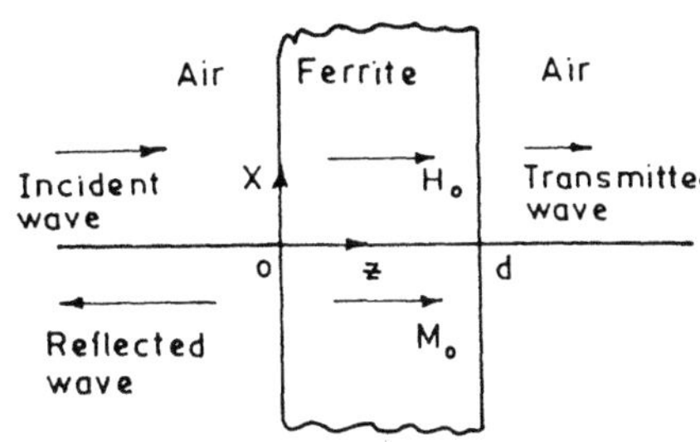

Figure 3.23. Normally magnetized slab ($0 \leq z \leq d$); the incident electromagnetic wave has its electric vector directed along the x-axis.

* In the zig-zag ray formulation of the propagation of guided waves in ferrimagnetics, it should be recognized that the group velocity, signal velocity, and energy velocity are, in general, different from one another since ferrimagnetics are strongly dispersive (Brillouin, 1960).

transmission for the two modes at the second interface ($z = d$) leads to ellipticity (Faraday rotation) in the polarization of the transmitted wave. In this case, the angle of Faraday rotation is defined as the angle between the major axis of the ellipse and the plane of polarization of the incident wave. Moreover, we will see in what follows that the multiple reflections within the ferrite slab cause interference of the various components, which leads to enhancement of the angle of rotation under favorable conditions.

Referring to Figure 3.23, we write the incident, reflected, and transmitted components of the electric field as

$$\left.\begin{aligned} \mathbf{E}^{i} &= \hat{x}\exp j(\omega t - \beta_0 z) \\ \mathbf{E}^{r} &= (R_x\hat{x} + R_y\hat{y})\exp j(\omega t + \beta_0 z) \\ \mathbf{E}^{t} &= (T_x\hat{x} + T_y\hat{y})\exp j[\omega t - \beta_0(z - d)] \end{aligned}\right\} \tag{3.39}$$

where

$$\beta_0 = \omega / c_0 \tag{3.40}$$

The electric field inside the slab consists of right- and left-handed circularly polarized partial waves, propagating along positive and negative z-directions, and is expressed as

$$\begin{aligned} \mathbf{E}^{f} = {} & [A_{+}\exp(j\beta_{+}z) + B_{+}\exp(-j\beta_{+}z)](\hat{x} - j\hat{y})\exp(j\omega t) \\ & + [A_{-}\exp(j\beta_{-}z) + B_{-}(-j\beta_{-}z)](\hat{x} + j\hat{y})\exp(j\omega t) \end{aligned} \tag{3.41}$$

where β_{+} and β_{-} are given by equation (2.24) of Chapter 2. It is more convenient to write the above field expressions in terms of RCP and LCP components, and following Suhl and Walker (1954), the electric field components can be expressed as:

$$\left.\begin{aligned} \mathscr{E}^{i}_{\pm} &= E^{i}_{x} \pm jE^{i}_{y} = \exp j(\omega t - \beta_0 z) \\ \mathscr{E}^{r}_{\pm} &= E^{r}_{x} \pm jE^{r}_{y} = R_{\pm}\exp j(\omega t + \beta_0 z) \\ \mathscr{E}^{t}_{\pm} &= E^{t}_{x} \pm jE^{t}_{y} = T_{\pm}\exp j[\omega t - \beta_0(z - d)] \\ \mathscr{E}^{f}_{\pm} &= E^{f}_{x} \pm jE^{f}_{y} = A_{\pm}\exp j(\omega t + \beta_{\pm}z) + B_{\pm}\exp j(\omega t - \beta_{\pm}z) \end{aligned}\right\} \tag{3.42}$$

where

$$\left.\begin{aligned} R_{\pm} &= R_x \pm jR_y \\ T_{\pm} &= T_x \pm jT_y \end{aligned}\right\} \tag{3.43}$$

The corresponding magnetic field components can be obtained from Maxwell's equations as:

$$\left.\begin{aligned} h_{\pm}^{i} &= \pm j\frac{c_0\beta_0}{\omega}\mathscr{E}_{\pm}^{i} \\ h_{\pm}^{r} &= \mp j\frac{c_0\beta_0}{\omega}\mathscr{E}_{\pm}^{r} \\ h_{\pm}^{t} &= \pm j\frac{c_0\beta_0}{\omega}\mathscr{E}_{\pm}^{t} \\ h_{\pm}^{f} &= \mp \frac{jc_0\beta_{\pm}}{\omega(\mu\pm\kappa)}[A_{\pm}\exp(j\beta_{\pm}z)-B_{\pm}\exp(-j\beta_{\pm}z)]\exp(j\omega t) \end{aligned}\right\} \quad (3.44)$$

It is now necessary to match the tangential components of electric and magnetic fields at each interface. It is evident that matching E_x, E_y, h_x, and h_y is equivalent to matching $\mathscr{E}_{\pm}$ and $h_{\pm}$; this yields the following set of coupled equations:

$$1+R_{\pm}=A_{\pm}+B_{\pm} \quad (3.45)$$

$$\pm j\frac{c_0\beta_0}{\omega}(1-R_{\pm}) = \mp j\frac{c_0\beta_{\pm}}{\omega(\mu\pm\kappa)}(A_{\pm}-B_{\pm}) \quad (3.46)$$

$$A_{\pm}\exp(j\beta_{\pm}d)+B_{\pm}\exp(-j\beta_{\pm}d)=T_{\pm} \quad (3.47)$$

$$\mp\frac{jc_0\beta_{\pm}}{\omega(\mu\pm\kappa)}[A_{\pm}\exp(j\beta_{\pm}d)-B_{\pm}\exp(-j\beta_{\pm}d)] = \pm j\frac{c_0\beta_0}{\omega}T_{\pm} \quad (3.48)$$

Elimination of $A_{\pm}$, $B_{\pm}$, and $R_{\pm}$ from equations (3.45)–(3.48) yields $T_{\pm}$ as follows:

$$T_{\pm}^{-1} = (1+x_{\pm}/a_f)(1+a_f/x_{\pm})\exp(j\beta_{\pm}d)+(1-x_{\pm}/a_f)(1-a_f/x_{\pm})\exp(-j\beta_{\pm}d) \quad (3.49)$$

where

$$x_{\pm}=(\mu\pm\kappa)^{1/2}$$

$$a_f=\sqrt{\varepsilon_f}$$

If $T_{\pm}=|T_{\pm}|\exp(-j\varphi_{\pm})$, it follows from equation (3.49) that

$$\left.\begin{aligned} |T_{\pm}|\cos\varphi_{\pm} &= \tfrac{1}{4}\cos\beta_{\pm}d/\mathscr{D} \\ |T_{\pm}|\sin\varphi_{\pm} &= \tfrac{1}{8}(x_{\pm}/a_f+a_f/x_{\pm})\sin\beta_{\pm}d/\mathscr{D} \end{aligned}\right\} \quad (3.50)$$

where

$$\mathscr{D} = \cos^2 \beta_\pm d + \tfrac{1}{4}(x_\pm/a_f + a_f/x_\pm)^2 \sin^2 \beta_\pm d \tag{3.51}$$

Equation (3.50) implies that the phase angle $\varphi_\pm$ is given by

$$\tan \varphi_\pm = \tfrac{1}{2}(x_\pm/a_f + a_f/x_\pm) \tan \beta_\pm d \tag{3.52}$$

The transmitted electric field can be written as

$$\mathscr{E}'_\pm = |T_\pm| \exp j(\omega t - \varphi_\pm) \tag{3.53}$$

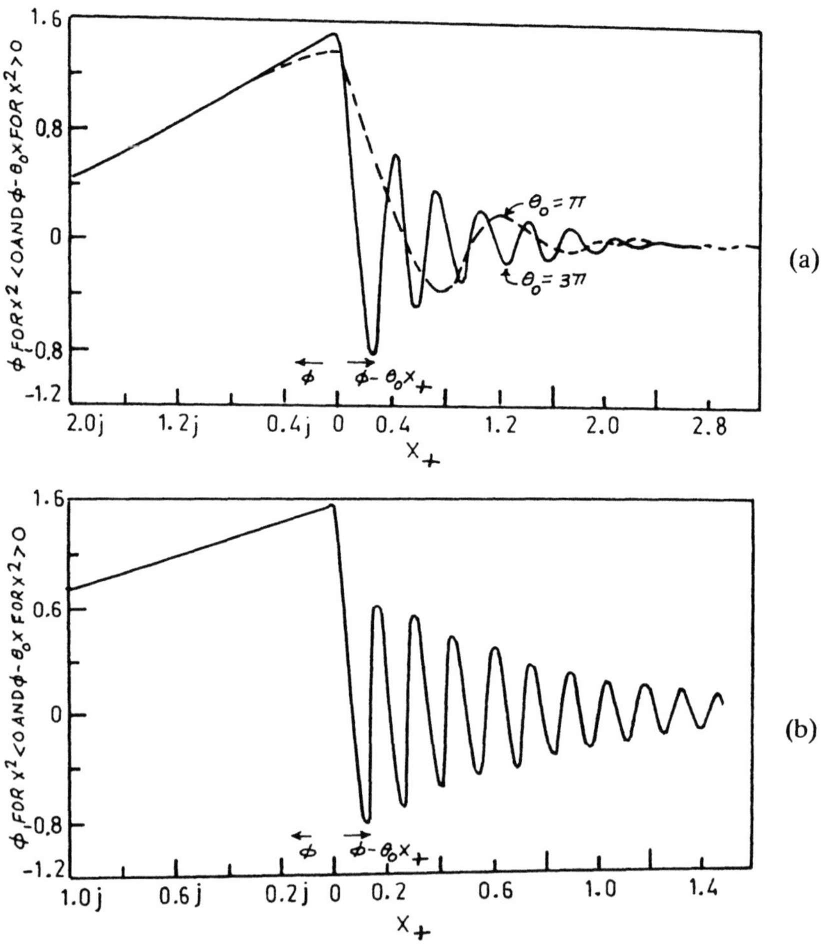

Fig. 3.24. Variation of phase and amplitude of the transmitted RCP wave with x_+ for different effective thicknesses $\theta_0 = \omega a_f d$ when $\varepsilon_f = 10$: (a) phase for $\theta_0 = \pi$ and 3π, (b) phase for $\theta_0 = 7\pi$, (c) $|T_+|$ for $\theta = \pi$ and 3π, and (d) $|T_+|$ for $\theta = 7\pi$ (after Suhl and Walker, 1954).

Conversion into cartesian components leads to

$$\left.\begin{aligned} E_x^t &= \tfrac{1}{2}[|T_+|\cos(\omega t - \varphi_+) + |T_-|\cos(\omega t - \varphi_-)] \\ E_y^t &= \tfrac{1}{2}[|T_+|\sin(\omega t - \varphi_+) - |T_-|\sin(\omega t - \varphi_-)] \end{aligned}\right\} \tag{3.54}$$

Evidently, the inequality of $|T_+|$ and $|T_-|$ makes the transmitted field elliptically polarized. The magnitude of the transmitted field is obtained as

$$\begin{aligned} E^t &= [(E_x^{(t)})^2 + (E_y^{(t)})^2]^{1/2} \\ &= \tfrac{1}{2}[|T_+|^2 + |T_-|^2 + 2|T_+T_-|\cos(2\omega t - \varphi_+ - \varphi_-)]^{1/2} \end{aligned} \tag{3.55}$$

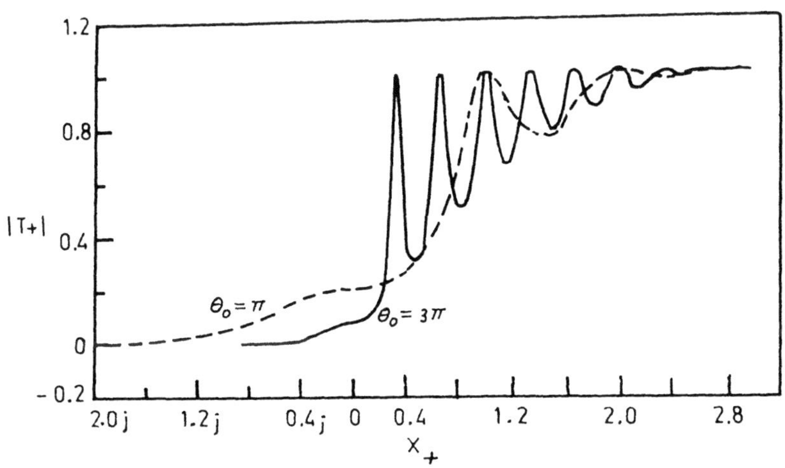

(c)

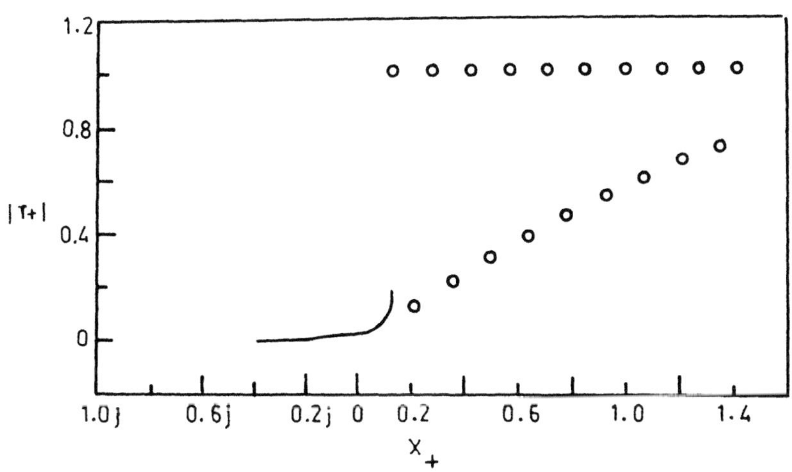

(d)

This oscillates between the limits $\frac{1}{2}(|T_+|+|T_-|)$ and $\frac{1}{2}(|T_+|-|T_-|)$. Moreover, the resultant electric field is maximum when $\omega t = (\varphi_+ + \varphi_-)/2$; this condition defines the major axis of the ellipse. The ratio of minor to major axis is given by

$$\text{Ratio} = \frac{|T_+|-|T_-|}{|T_+|+|T_-|} \tag{3.56}$$

The components of the electric field at $\omega t = (\varphi_+ + \varphi_-)/2$ are obtained from equation (3.54) as

$$\left.\begin{aligned} E_{xt} &= \tfrac{1}{2}[|T_+|+|T_-|]\cos(\varphi_- - \varphi_+)/2 \\ E_{yt} &= \tfrac{1}{2}[|T_+|+|T_-|]\sin(\varphi_- - \varphi_+)/2 \end{aligned}\right\} \tag{3.57}$$

Thus, the orientation of the major axis of the ellipse relative to the direction of polarization of the incident wave field, which is, in fact, the rotation angle, is given by

$$\delta = (\varphi_- - \varphi_+)/2 \tag{3.58}$$

where $\varphi_\pm$ can be inferred from equation (3.52).

If we compare the rotation given by (3.58) with the case of propagation in an infinite medium through a distance equal to the slab thickness, we see that the latter rotation is equal to $\frac{1}{2}(x_- - x_+)d$, as discussed in Chapter 2. The rotation gain is defined as

$$g_1 = \frac{\frac{1}{2}(\varphi_- - \varphi_+)}{\frac{1}{2}(x_- - x_+)d} = \frac{(\varphi_- - \varphi_+)}{(x_- - x_+)d} \tag{3.59}$$

Figure 3.24 shows the phase and amplitude of the transmitted RCP wave as a function of x_+ for different effective thicknesses $\theta_0 = \omega a_f d$. The rotation gain is plotted in Figure 3.25 as a function of θ_0 for different values of a_f. The gain is at a maximum when $\theta_0 = (n+\frac{1}{2})\pi$, and at a minimum when $\theta_0 = n\pi$; the variability of gain is a consequence of interference of different transmitted and reflected partial waves.

Reflection and transmission of electromagnetic waves from a slab magnetized parallel to its face has been investigated by Seavey and Tannenwald (1957). In this case, the reflected and transmitted waves exhibit body resonances as a function of the biasing field.

The oblique incidence of a collimated microwave beam upon a ferrite slab has been investigated by Mueller (1976), who has also discussed the applications to phase plates, polarization interferometer, Babinet compensator, Wollaston prisms, etc. The characteristics of ferrimagnetics that play an important role in these proposed devices are birefringence and magnetic tunability. Mueller (1974) has analyzed more general structures as well, such as ferrimagnetic prisms and lenses. In the latter case there are, in general, two focal lengths corresponding to double refraction.

Electromagnetic wave propagation in a periodic structure consisting of alternate dielectric and ferrimagnetic layers was also investigated by Mueller (1975). In this case, the interference can lead to considerable enhancement of the rotation gain if the ferrimagnetic layers are appropriately magnetized. Reflection and transmission of an electromagnetic wave from a ferrimagnetic slab subjected to a time-dependent (sinusoidally modulated) biasing field has been investigated by Sodha and Srivastava (1975); it is seen that the transmitted and reflected waves are strongly phase and amplitude modulated.

PART B. *MAGNETOELASTIC WAVES*

3.3. *Half Space Problem*

Consider the geometry shown in Figure 3.26, where the half spaces $y > 0$ and $y < 0$ are occupied by cubic magnetoelastic media MEI and

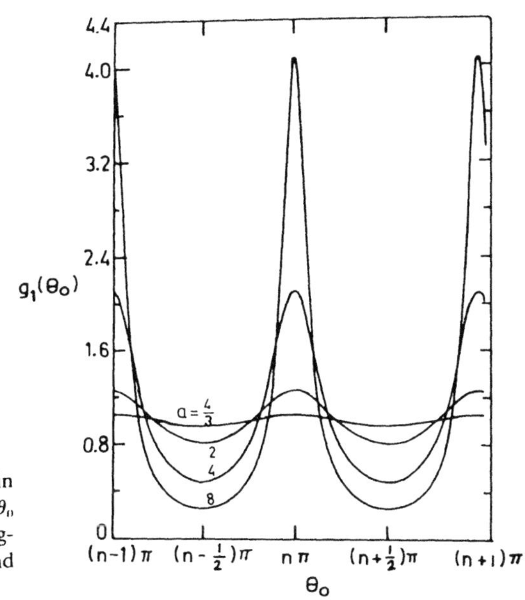

Figure 3.25. Variation of rotation gain $g_1(\theta_0)$ with effective slab thickness θ_0 for different $a_t = \sqrt{\varepsilon_t}$, assuming magnetization to be small (after Suhl and Walker, 1954).

MEII, respectively. The dc magnetic field H_0 is directed along the z-axis, which is perpendicular (outward) to the plane of the paper. A plane magnetoelastic wave, initially propagating in medium I, is incident at the interface $y=0$ at an angle θ_1 with respect to the normal. The elastic displacement of the wave is along the z-axis. The transmitted wave makes an angle θ_2 with the normal, whereas the angle of reflection can be shown to be the same as the angle of incidence θ_1. Since the elastic polarization is along the dc magnetic field, the wavenumbers for plane waves in media I and II can be inferred from equation (2.109) of Chapter 2 as follows:

$$\beta_n = k_t^{(n)}/\sqrt{\delta_\perp^{(n)}} \tag{3.60a}$$

where

$$\left.\begin{aligned} k_t^{(n)} &= \omega/c_t^{(n)} \\ c_t^{(n)} &= \sqrt{c_{44}^{(n)}/\rho_n} \\ \delta_\perp^{(n)} &= \frac{\omega^2 - \omega_{2n}'^2}{\omega^2 - \omega_{3n}'^2} \\ \omega_{3n}'^2 &= \omega_0(\omega_0 + \omega_m^{(n)}) \\ \omega_{2n}'^2 &= \omega_{3n}'^2 - \omega_0\sigma_n \end{aligned}\right\} \tag{3.60b}$$

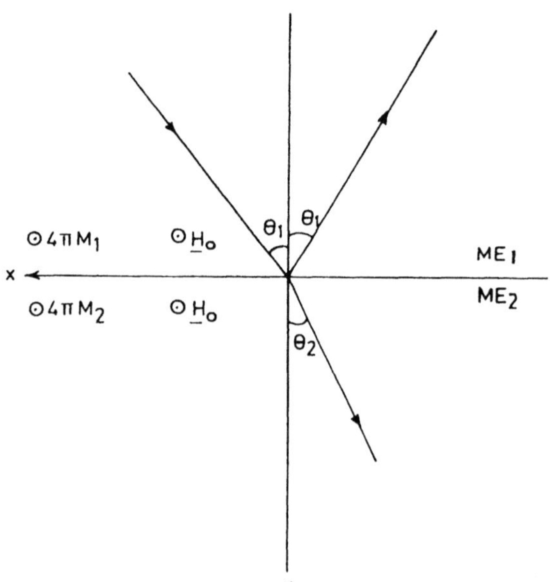

Figure 3.26. Reflection of a plane magnetoelastic wave at the interface between two magnetoelastic ferrimagnetic half spaces.

The index $n = 1$ and 2 refers to medium I and II, respectively. The wavevectors for the incident, reflected, and transmitted waves are expressed as:

$$\left.\begin{aligned} \mathbf{k}^{i} &= \beta_1(-\hat{x}\sin\theta_1 + \hat{y}\cos\theta_1) \\ \mathbf{k}^{r} &= \beta_1(-\hat{x}\sin\theta_1 - \hat{y}\cos\theta_1) \\ \mathbf{k}^{t} &= \beta_2(-\hat{x}\sin\theta_2 + \hat{y}\cos\theta_2) \end{aligned}\right\} \tag{3.61}$$

Since the elastic displacement and magnetic field associated with the wave are uniform along the z-axis, we have $\partial/\partial z = 0$. Furthermore, $u_x = u_y = 0$ since the elastic polarization is along the z-axis. As such, equations (2.82) reduce to the following single equation:

$$-\omega^2 \rho u_z = c_{44}\left(\frac{\partial^2}{\partial x^2} + \frac{\partial^2}{\partial y^2}\right)u_z + \frac{b_2}{M_0}\left(\frac{\partial m_x}{\partial x} + \frac{\partial m_y}{\partial y}\right) \tag{3.62}$$

This equation is applicable to both of the media provided that appropriate values for ρ, c_{44}, b_2, and M_0 are substituted.

With the magnetostatic condition $\mathbf{h} = \nabla\psi$, the equation $\nabla\cdot\mathbf{h} = -4\pi\nabla\cdot\mathbf{m}$ leads to

$$\left(\frac{\partial^2}{\partial x^2} + \frac{\partial^2}{\partial y^2}\right)\psi = -4\pi\left(\frac{\partial m_x}{\partial x} + \frac{\partial m_y}{\partial y}\right) \tag{3.63}$$

The magnetization equation (3.62) transforms into*

$$j\omega\mathbf{m} = -\gamma M_0\hat{z}\times\mathbf{h} - \gamma H_0\mathbf{m}\times\hat{z} - \gamma M_0\hat{z}\times\mathbf{h}_{me} \tag{3.64}$$

With u_x, $u_y = 0$ and $\partial/\partial z = 0$, $\mathbf{h}_{me}$, as given by equation (2.85), reduces to

$$\mathbf{h}_{me} = -\frac{b_2}{M_0}\left(\frac{\partial u_z}{\partial x}\hat{x} + \frac{\partial u_z}{\partial y}\hat{y}\right)$$

and thus equation (3.64) can be expressed in component form as

$$\left.\begin{aligned} j\omega m_x + \omega_0 m_y &= \frac{\omega_m}{4\pi}\frac{\partial\psi}{\partial y} - \gamma b_2\frac{\partial u_z}{\partial y} \\ -\omega_0 m_x + j\omega m_y &= -\frac{\omega_m}{4\pi}\frac{\partial\psi}{\partial x} + \gamma b_2\frac{\partial u_z}{\partial x} \end{aligned}\right\} \tag{3.65}$$

* Incidentally, it is evident from this equation that the effect of reversing the direction of dc magnetization is equivalent to the reversal of the sign of ω. This is consistent with the reversal of the sign of κ to effect the biasing field reversal which was shown in Chapter 1.

The explicit expressions for m_x and m_y are obtained as

$$\left.\begin{aligned}(\omega_0^2-\omega^2)m_x &= \frac{j\omega\omega_m}{4\pi}\frac{\partial\psi}{\partial y}+\frac{\omega_0\omega_m}{4\pi}\frac{\partial\psi}{\partial x}\\ &\quad -j\omega\gamma b_2\frac{\partial u_z}{\partial y}-\gamma b_2\omega_0\frac{\partial u_z}{\partial x}\\ (\omega_0^2-\omega^2)m_y &= \frac{\omega_0\omega_m}{4\pi}\frac{\partial\psi}{\partial y}-j\frac{\omega\omega_m}{4\pi}\frac{\partial\psi}{\partial x}\\ &\quad -\gamma b_2\omega_0\frac{\partial u_z}{\partial y}+j\omega\gamma b_2\frac{\partial u_z}{\partial x}\end{aligned}\right\} \tag{3.66}$$

These equations can be combined to obtain

$$(\omega_0^2-\omega^2)\left(\frac{\partial m_x}{\partial x}+\frac{\partial m_y}{\partial y}\right)=\frac{\omega_0\omega_m}{4\pi}\left(\frac{\partial^2}{\partial x^2}+\frac{\partial^2}{\partial y^2}\right)\psi-\gamma b_2\omega_0\left(\frac{\partial^2}{\partial x^2}+\frac{\partial^2}{\partial y^2}\right)u_z \tag{3.67}$$

Substitution from equation (3.63) into equation (3.67) yields

$$\frac{\partial m_x}{\partial x}+\frac{\partial m_y}{\partial y}=\frac{\gamma b_2\omega_0}{\omega^2-\omega_0(\omega_0+\omega_m)}\left(\frac{\partial^2}{\partial x^2}+\frac{\partial^2}{\partial y^2}\right)u_z \tag{3.68}$$

and equation (3.62) thus transforms into

$$\left(\frac{\partial^2}{\partial x^2}+\frac{\partial^2}{\partial y^2}\right)u_z+\frac{\omega^2\rho}{c_{44}\delta_\perp}u_z=0 \tag{3.69}$$

If the plane wave in question propagates in a direction making an angle θ with the x-axis and has a wavenumber β, the elastic displacement

$$u_z \sim \exp j[\omega t-\beta(x\cos\theta+y\sin\theta)]$$

and this substitution in equation (3.69) yields [in agreement with equation (3.60b)] the following:

$$\beta = k_t/\sqrt{\delta_\perp}$$

Thus, the wavenumbers for plane wave propagation, in an arbitrary direction within the xy-plane, in media I and II are given by equation (3.60b), while the elastic displacement for the incident, reflected, and transmitted waves (see Figure 3.26) can be expressed as

$$\left.\begin{aligned} u_z^{(i)} &= u_0 \exp j(\omega t - \mathbf{k}^i \cdot \mathbf{r}) \\ u_z^{(r)} &= R u_0 \exp j(\omega t - \mathbf{k}^{(r)} \cdot \mathbf{r}) \\ u_z^{(t)} &= T u_0 \exp j(\omega t - \mathbf{k}^{(t)} \cdot \mathbf{r}) \end{aligned}\right\} \tag{3.70}$$

where R and T are the reflection and transmission coefficients, respectively.

In order to obtain ψ, it is seen that the substitution from equation (3.68) into (3.63) leads to

$$\left(\frac{\partial^2}{\partial x^2} + \frac{\partial^2}{\partial y^2}\right)\psi = \frac{4\pi\gamma b_2 \omega_0}{\omega_0(\omega_0 + \omega_m) - \omega^2}\left(\frac{\partial^2}{\partial x^2} + \frac{\partial^2}{\partial y^2}\right)u_z \tag{3.71}$$

In order to obtain the solution for this equation, the appropriate form of u_z can be substituted from equation (3.70) in it. If a function $\psi_p(x, y)$ is such that it satisfies equation (3.71) and is zero when $u_z = 0$, it will be a particular solution to equation (3.71). It is evident that if any solution ψ' of the two-dimensional Laplace equation

$$\left(\frac{\partial^2}{\partial x^2} + \frac{\partial^2}{\partial y^2}\right)\psi' = 0$$

is added to ψ_p, the resulting function $\psi = \psi_p + \psi'$ will also satisfy equation (3.71) and hence will be a general solution of that equation. The general form of ψ' can be expressed as

$$\psi' = D_r \exp(\beta' y)\exp j(\omega t - \beta' x) + D_t \exp(-\beta' y)\exp j(\omega t - \beta' x) \tag{3.72}$$

where D_r and D_t are arbitrary constants and β' is regarded as real in order to ensure that the wave field does not grow or decay along the x-axis because of the inherent uniformity along this axis. Since ψ' and hence ψ cannot grow indefinitely as $|y| \to \infty$, it follows that, for the reflected wave, we have $D_t = 0$, whereas, for the transmitted wave, we have $D_r = 0$. Both D_r and D_t should be zero for the incident wave because the characteristics of the incident wave are independent of the existence or otherwise of the interface $y = 0$. Thus, the magnetostatic potential for the incident, reflected, and transmitted waves is expressed as

$$\left.\begin{aligned} \psi^{(i)} &= P_{me}^{(1)} u_0 \exp j(\omega t - \mathbf{k}^i \cdot \mathbf{r}) \\ \psi^{(r)} &= P_{me}^{(1)} R u_0 \exp j(\omega t - \mathbf{k}^r \cdot \mathbf{r}) + D_r \exp(\beta' y)\exp j(\omega t - \beta' x) \\ \psi^{(t)} &= P_{me}^{(2)} T u_0 \exp j(\omega t - \mathbf{k}^t \cdot \mathbf{r}) + D_t \exp(-\beta' y)\exp j(\omega t - \beta' x) \end{aligned}\right\} \tag{3.73a}$$

where

$$P_{\text{me}}^{(n)} = \frac{4\pi\gamma\omega_0 b_2^{(n)}}{\omega_0(\omega_0 + \omega_{\text{m}}^{(n)}) - \omega^2} \tag{3.73b}$$

The boundary conditions to be applied at the interface $y = 0$ are as follows:

(i) $$u_z^{(i)} + u_z^{(r)} = u_z^{(t)} \tag{3.74}$$

(ii) $$\psi^{(i)} + \psi^{(r)} = \psi^{(t)} \tag{3.75}$$

(iii) $$b_y^{(i)} + b_y^{(r)} = b_y^{(t)} \tag{3.76}$$

(iv) $$c_{44}^{(1)} \frac{\partial u_z^{(i)}}{\partial y} + \frac{b_2^{(1)}}{M_1} m_y^{(i)} + c_{44}^{(1)} \frac{\partial u_z^{(r)}}{\partial y} + \frac{b_2^{(1)}}{M_1} m_y^{(r)} = c_{44}^{(2)} \frac{\partial u_z^{(t)}}{\partial y} + \frac{b_2^{(2)}}{M_2} m_y^{(t)} \tag{3.77}$$

Whereas conditions (ii) and (iii) are the usual boundary conditions on magnetic potential and induction, (i) can be inferred from the first of equations (D.28) of Appendix D. Finally, (iv) can be obtained from the second of equations (D.28) as follows: Since the normal to the interface is along the y-axis, we have

$$\boldsymbol{T} \cdot \hat{n} = \mathbf{T}_y = T_{xy}\hat{x} + T_{yy}\hat{y} + T_{zy}\hat{z} = T_6\hat{x} + T_2\hat{y} + T_4\hat{z}$$

Furthermore, according to equations (D.15) and (1.110), we have

$$T_I = C_{IJ}S_J + \partial U_{\text{me}}/\partial S_I$$

for a magnetoelastic medium. With equations (1.112), (1.113), (D.4), (D.5), and (D.20), it is seen that $T_2 = T_6 = 0$, whereas

$$T_4 = c_{44}(\partial u_z/\partial y) + b_2 m_y/M_0$$

Substitution from equation (3.70) into equation (3.74) yields (at $y = 0$)

$$u_0 \exp(j\beta_1 \sin\theta_1 x) + Ru_0 \exp(j\beta_1 \sin\theta_1 x) = Tu_0 \exp(j\beta_2 \sin\theta_2 x)$$

This condition must be true for all values of x, which is possible only when the phases and amplitudes are separately matched, i.e.,

$$\beta_1 \sin\theta_1 = \beta_2 \sin\theta_2 \tag{3.78}$$

and

$$1 + R = T \tag{3.79}$$

Substitution from equation (3.73) into (3.75), followed by phase and amplitude matching at $y = 0$, yields

$$\beta' = \beta_1 \sin\theta_1 = \beta_2 \sin\theta_2 \tag{3.80}$$

and

$$(1 + R)P_{\text{me}}^{(1)} + R' = TP_{\text{me}}^{(2)} + T' \tag{3.81}$$

where

$$\left.\begin{aligned} R' &= D_r/u_0 \\ T' &= D_t/u_0 \end{aligned}\right\} \tag{3.82}$$

In order to apply the remaining boundary conditions, it is first necessary to determine m_y. Using equations (3.66), (3.70), and (3.73) and dropping the phase variation, we have (at $y = 0$)

$$\left.\begin{aligned} m_y^{(i)} &= -\frac{\gamma b_2^{(1)}\beta_1(\omega\sin\theta_1 - j\omega_0\cos\theta_1)}{\omega_{31}'^2 - \omega^2} u_0 \\ m_y^{(r)} &= -\frac{\gamma b_2^{(1)}\beta_1(j\omega_0\cos\theta_1 + \omega\sin\theta_1)}{\omega_{31}'^2 - \omega^2} Ru_0 + \frac{\omega_m\beta_1\sin\theta_1}{4\pi(\omega_0 - \omega)} R'u_0 \\ m_y^{(t)} &= -\frac{\gamma b_2^{(2)}\beta_2(\omega\sin\theta_2 - j\omega_0\cos\theta_2)}{\omega_{32}'^2 - \omega^2} Tu_0 - \frac{\beta_2\sin\theta_2\omega_m}{4\pi(\omega_0 + \omega)} T'u_0 \end{aligned}\right\} \tag{3.83}$$

Using $b_y = \partial\psi/\partial y + 4\pi m_y$, we have

$$\left.\begin{aligned} b_y^{(i)} &= -\frac{P_{\text{me}}^{(1)}\beta_1\sin\theta_1\omega}{\omega_0} u_0 \\ b_y^{(r)} &= \left(-P_{\text{me}}^{(1)}\frac{\omega}{\omega_0} R + \frac{\omega_0 + \omega_m^{(1)} - \omega}{\omega_0 - \omega} R'\right)\beta_1\sin\theta_1 u_0 \\ b_y^{(t)} &= \left(-P_{\text{me}}^{(2)}\frac{\omega}{\omega_0} T - \frac{\omega_0 + \omega_m^{(2)} + \omega}{\omega_0 + \omega} T'\right)\beta_2\sin\theta_2 u_0 \end{aligned}\right\} \tag{3.84}$$

Using the boundary condition (3.76) and simplifying with the help of equation (3.79), we have

$$(1 + R)(D_{\text{me}}^{(1)} - D_{\text{me}}^{(2)}) + \frac{\omega_0 + \omega_m^{(1)} - \omega}{\omega_0 - \omega} R' + \frac{\omega_0 + \omega_m^{(2)} + \omega}{\omega_0 + \omega} T' = 0 \tag{3.85}$$

where

$$D_{\text{me}} = (\omega/\omega_0)P_{\text{me}} \tag{3.86}$$

Using the boundary condition (3.77) and simplifying with the help of equation (3.79), we have

$$j\beta_1 \cos\theta_1 c_{44}^{(1)}(R-1)\delta_{\perp}^{(1)} + (1+R)(-c_{44}^{(1)}F_{\text{me}}^{(1)} + c_{44}^{(2)}F_{\text{me}}^{(2)} + j\beta_2\cos\theta_2\delta_{\perp}^{(2)}c_{44}^{(2)})$$
$$+\frac{\gamma b_2^{(1)}\beta_1\sin\theta_1}{\omega_0-\omega}R' + \frac{\gamma b_2^{(2)}\beta_2\sin\theta_2}{\omega_0+\omega}T' = 0 \tag{3.87}$$

where

$$F_{\text{me}}^{(n)} = \frac{\sigma_n\omega\beta_1\sin\theta_1}{\omega_0(\omega_0+\omega_{\text{m}}^{(n)})-\omega^2} \tag{3.88}$$

The expression for T' as obtained from equation (3.81) can be substituted into equations (3.85), and this yields

$$R' = \frac{1}{S_1}(1+R)\left[D_{\text{me}}^{(2)} - D_{\text{me}}^{(1)} + (P_{\text{me}}^{(1)} - P_{\text{me}}^{(2)})\frac{\omega_0+\omega_{\text{m}}^{(2)}+\omega}{\omega_0+\omega}\right] \tag{3.89}$$

where

$$S_1 = 2 + \frac{\omega_{\text{m}}^{(1)}}{\omega_0-\omega} + \frac{\omega_{\text{m}}^{(2)}}{\omega_0+\omega} \tag{3.90}$$

Finally, substitution for T' and R' from equations (3.81) and (3.89), respectively, into equation (3.87) leads to the following expression for R:

$$R = \frac{(\beta_1\cos\theta_1 c_{44}^{(1)}S_1\delta_{\perp}^{(1)} - S_1V_{\text{i}}) + j(US_2 + S_1V_{\text{r}})}{(\beta_1\cos\theta_1 c_{44}^{(1)}S_1\delta_{\perp}^{(1)} + S_1V_{\text{i}}) - j(US_2 + S_1V_{\text{r}})} \tag{3.91}$$

where

$$\left.\begin{aligned}
U &= D_{\text{me}}^{(2)} - D_{\text{me}}^{(1)} + \frac{\omega_0+\omega_{\text{m}}^{(2)}+\omega}{\omega_0+\omega}(P_{\text{me}}^{(1)} - P_{\text{me}}^{(2)}) \\
S_2 &= \gamma\beta_1\sin\theta_1\left(\frac{b_2^{(1)}}{\omega_0-\omega} + \frac{b_2^{(2)}}{\omega_0+\omega}\right) \\
V &= V_{\text{r}} + jV_{\text{i}} \\
V_{\text{r}} &= -c_{44}^{(1)}F_{\text{me}}^{(1)} + c_{44}^{(2)}F_{\text{me}}^{(2)} + \gamma b_2^{(2)}(P_{\text{me}}^{(1)} - P_{\text{me}}^{(2)})\frac{\beta_2\sin\theta_2}{\omega_0+\omega} \\
V_{\text{i}} &= \beta_2\cos\theta_2 c_{44}^{(2)}\delta_{\perp}^{(2)}
\end{aligned}\right\} \tag{3.92}$$

Once R is known, it is easy to obtain T, T', and R' using equations (3.79), (3.89), and (3.81).

In the expression for R as given by equation (3.91), the sign of ω should be reversed in order to obtain the reflection coefficient for the case when the incident and reflected waves are interchanged.

The preceding derivation of the reflection coefficient assumes the medium to be magnetically and elastically lossless. The effects of magnetic and elastic losses can be incorporated by replacing ω_0 and c_{44} by $(\omega_0 + j/T)$ and $(c_{44} + j\eta_{44})$, respectively.

In accordance with the earlier discussion (Section 3.1.2), it is seen that when losses are neglected, $\beta_2 \cos \theta_2$ would be purely imaginary when either $\sin^2 \theta_2 < 0$ or $\sin^2 \theta_2 > 1$. In this case, it follows from equation (3.92) that V is real; $V_i = 0$. Finally, it follows from equation (3.91) that $|R| = 1$ implying total reflection. Thus, the conditions for total reflection can be inferred from equations (3.78) and (3.60) by requiring that $\sin^2 \theta_2$ be either negative or exceed unity.

A comparison with the general results of Section 3.1 shows that the magnetoelastic wave reflection is also phase and amplitudewise nonreciprocal. In fact, the entire analysis presented in Section 3.1 can be extended to the present case to ensure that the general conclusions pertaining to reflection of electromagnetic waves are also valid for magnetoelastic waves.

4

Magnetostatic Waves in Layered Planar Structures

The magnetostatic modes in ferrimagnetics were first observed by White and Solt (1956) as spurious peaks in a ferromagnetic resonance experiment. Mercereau and Feynman (1956) described the physical conditions for the occurrence of resonances in the presence of a nonuniform alternating magnetic field. Walker (1957) analyzed the magnetostatic modes of a ferrimagnetic spheroid. Subsequently, the magnetostatic modes in various sample geometries (e.g., sphere, ellipsoid, disc, rod, etc.) were theoretically and experimentally investigated (Walker, 1963). Auld (1960) considered plane wave propagation in an infinite ferrimagnetic medium and showed that, in those regions of the dispersion curves where the wavenumbers are relatively large, the plane wave field satisfies the magnetostatic conditions; thus, the magnetostatic modes are significant even for unbounded media. Damon and Eshbach (1961) investigated the magnetostatic modes in a planar structure, i.e., a semi-infinite ferromagnetic slab. The main purpose of their investigation was to clarify the relationship between the large wavenumber spin wave modes and the magnetostatic modes of a finite sample. As such, they analyzed, in detail, the surface and bulk modes* of a semi-infinite ferrimagnetic slab which is magnetized parallel to its face. It was recognized later (Olson and Yaeger, 1965; Brundle and Freedman, 1968a, b) that appreciable time delays can be obtained at microwave

* It is worth distinguishing between surface and bulk modes before proceeding further. When the energy transported by a guided wave is concentrated at one or more interfaces of the guiding structure, we speak of surface modes. On the other hand, the bulk (or volume) modes are characterized by the distribution of wave energy throughout the transverse cross section of the guiding structure. As such, for instance, the "surface waves" in dielectric optical waveguides, as discussed by several authors (e.g., Tamir, 1973), are, in fact, bulk waves according to the nomenclature followed in the present book.

frequencies from guided surface and bulk magnetostatic waves. The subsequent theoretical and experimental studies of magnetostatic wave propagation in a variety of layered structures have led to the development of several devices. The initial thrust in this area was on surface waves propagating on rather thick YIG plates. In this case the nonuniform internal dc magnetic field makes the theoretical analysis as well as the interpretation of experimental data somewhat difficult. In recent years with the advent of epitaxial growth of high-quality ferrimagnetic films, the magnetostatic wave propagation in layered planar structures has become significant.

In this chapter, we discuss the propagation of surface and bulk magnetic waves in layered planar structures consisting of ferrimagnetics and other media such as dielectrics, metals, and semiconductors. Most of these investigations were carried out with the aim of obtaining appreciable time delays. Since large time delays are associated with slow waves, the emphasis in this chapter is on magnetostatic rather than on fast electromagnetic waves. As such, the analysis has been carried out using the magnetostatic approximation. However, as discussed in Chapters 1 and 2, the magnetostatic approximation does not necessarily yield correct results and requires justification by rigorous analysis for each configuration. Hence, for important configurations, we also present rigorous electromagnetic analyses in addition to the simplified magnetostatic analyses.

The material presented in this chapter can be broadly classified into five different parts. Sections 4.1–4.7 discuss the passive propagation of magnetic waves in the various layered structures involving isotropic ferrimagnetics. The explicit effect of magnetocrystalline anisotropy on magnetic waves is considered in Section 4.8. The traveling wave amplification of magnetostatic waves through the interaction with drifting charges is discussed in Section 4.9, while a general discussion of a wave oscillator has been presented in Section 4.10. Section 4.11 considers transduction of magnetic waves.

4.1. *Ferrimagnetic Half Space*

The simplest configuration for guided magnetic wave propagation is a plane interface between a dielectric (or vacuum) and a ferrimagnetic half space. The study of wave propagation in this configuration is important. Firstly, it helps to understand the characteristics of magnetic wave propagation in more complicated layered structures. Moreover, the half space geometry serves as a good model for the experimental investigations of magnetostatic surface wave propagation on thick ferrimagnetic substrates. The theoretical investigations of surface wave propagation on a ferrimagnetic half space have been reported by several workers including Bresler (1959), Eshbach and Damon (1960), Morgenthaler (1970, 1977), Courtois et al. (1971), Parekh and Ponamgi (1973), Vaslow (1974), Bini et al. (1976), Bardati and Lampariello (1979), and Gupta and Srivastava (1980). The experimental investigations of surface wave propagation on YIG substrates have been reported by a number of workers including Brundle and Freedman (1968a, b), Sparks (1969), Adam (1970), Adam et al. (1970), Sethares and Merry (1973, 1974), and Sethares and Stiglitz (1974). We theoretically analyzed the cases of transverse (Section 4.1.1) and arbitrary (Section 4.1.2) magnetizations. In Section 4.1.3, some experimental results are also discussed.

4.1.1. *Transverse Magnetization*

Consider the configuration shown in Figure 4.1 in which the ferrimagnetic half space is magnetized along the z-axis while the wave propagates along the y-axis. For a guided wave, the amplitude of the wave field decays on both sides of the interface $x = 0$.

As noted in Section 2.4, the magnetostatic potential ψ^{f} in a ferrimagnetic region satisfies the following equation:

$$\left[\mu\left(\frac{\partial^2}{\partial x^2}+\frac{\partial^2}{\partial y^2}\right)+\frac{\partial^2}{\partial z^2}\right]\psi^{f}=0 \tag{4.1}$$

Since μ is unity for nonmagnetic materials, the magnetostatic potential ψ^{d} in a dielectric region satisfies Laplace's equation:

$$\left(\frac{\partial^2}{\partial x^2}+\frac{\partial^2}{\partial y^2}+\frac{\partial^2}{\partial z^2}\right)\psi^{d}=0 \tag{4.2}$$

If the wave field is assumed to be uniform along the dc magnetization, $\partial\psi/\partial z = 0$, and it follows from equations (4.1) and (4.2) that the magnetostatic potential now satisfies the two-dimensional Laplace equation everywhere in space. Thus, the magnetostatic potential in the two regions can be expressed as

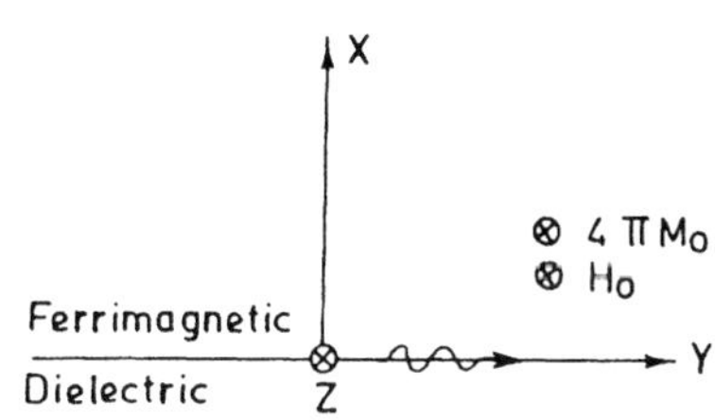

Figure 4.1. Surface wave propagation along the interface between the dielectric and ferrimagnetic half spaces. The ferrimagnetic medium is magnetized along the z-axis (into the plane of the paper) whereas the wave propagates along the y-axis.

$$\left.\begin{aligned} \psi^{\mathrm{d}} &= A \exp(\beta x)\exp j(\omega t - \beta y) \quad \text{for} \quad x \leqslant 0 \\ \psi^{\mathrm{f}} &= B \exp(-\beta x)\exp j(\omega t - \beta y) \quad \text{for} \quad x \geqslant 0 \end{aligned}\right\} \tag{4.3}$$

where, for guided waves, the real part of β should be positive in order that the wave field can decay to zero as $|x| \to \infty$. It is necessary that the tangential component of the magnetic field and the normal component of magnetic induction match at the interface. The first condition is equivalent to matching the ψ's at the interface. The normal component of magnetic induction is given by

$$b_x^{\mathrm{d}} = \partial\psi^{\mathrm{d}}/\partial x = \beta A \exp(\beta x)\exp j(\omega t - \beta y) \tag{4.4}$$

$$\begin{aligned} b_x^{\mathrm{f}} &= \mu \frac{\partial \psi^{\mathrm{f}}}{\partial x} + j\kappa \frac{\partial \psi^{\mathrm{f}}}{\partial y} \\ &= -(\mu - \kappa)\beta B \exp(-\beta x)\exp j(\omega t - \beta y) \end{aligned} \tag{4.5}$$

Matching of the ψ's implies $A = B$, while matching of the b_x's leads to (assuming $\beta \neq 0$)

$$\mu - \kappa + 1 = 0 \tag{4.6}$$

This dispersion relation is applicable to the case of propagation along the positive y-axis. In order to obtain the dispersion relation for propagation along the negative y-axis, it may be noted from Figure 4.1 that the effect of reversal of the direction of propagation is physically equivalent to the effect of reversal of the direction of the biasing field. In Section 1.4, we saw that reversal of the biasing field can be effected by changing the sign of the off-diagonal elements of the permeability tensor. Thus, the dispersion relation for propagation along the negative y-axis is obtained from equation (4.6) as

$$\mu + \kappa + 1 = 0 \tag{4.7}$$

If losses are ignored, substitution for μ and κ from equations (1.22) and (1.26) into the left-hand side of equations (4.6) and (4.7) shows that

$$\mu \mp \kappa + 1 = 2(\omega_0 + \tfrac{1}{2}\omega_{\mathrm{m}} \pm \omega)/(\omega_0 \pm \omega) \tag{4.8}$$

It follows that equation (4.6) cannot be satisfied for any set of parameters while equation (4.7) is satisfied (for arbitrary β) when

$$\omega = \omega_0 + \tfrac{1}{2}\omega_{\mathrm{m}} \tag{4.9}$$

Evidently, no guided magnetostatic propagation is possible along the positive y-axis. However, a wave of frequency given by equation (4.9) (with an arbitrary wavenumber) can propagate along the negative y-axis, i.e., along the $\mathbf{H}_0 \times \hat{n}$ direction, where $\hat{n}$ is the outward normal to the ferrimagnetic surface. (We have $\hat{n} = -\hat{x}$ in the present case.) It is natural to expect that equation (4.9) would not represent the true dispersion in a small-wavenumber region since it has been obtained from magnetostatic analysis. In order to obtain the correct dispersion relation for the (modified) magnetostatic wave and other possible fast modes, we carry out rigorous electromagnetic analysis, following Courtois et al. (1971) and Parekh and Ponamgi (1973).

The electromagnetic equations (2.8) of Chapter 2 can be written, in extenso, as

$$\frac{\partial E_z}{\partial y} - \frac{\partial E_y}{\partial z} = -j\frac{\omega}{c_0}(\mu h_x + j\kappa h_y) \tag{4.10a}$$

$$\frac{\partial E_x}{\partial z} - \frac{\partial E_z}{\partial x} = -j\frac{\omega}{c_0}(-j\kappa h_x + \mu h_y) \tag{4.10b}$$

$$\frac{\partial E_y}{\partial x} - \frac{\partial E_x}{\partial y} = -j\frac{\omega}{c_0} h_z \tag{4.10c}$$

$$\frac{\partial}{\partial y} h_z - \frac{\partial}{\partial z} h_y = j\frac{\omega \varepsilon_f}{c_0} E_x \tag{4.10d}$$

$$\frac{\partial}{\partial z} h_x - \frac{\partial}{\partial x} h_z = j\frac{\omega \varepsilon_f}{c_0} E_y \tag{4.10e}$$

$$\frac{\partial}{\partial x} h_y - \frac{\partial}{\partial y} h_x = j\frac{\omega \varepsilon_f}{c_0} E_z \tag{4.10f}$$

where ε_f represents the dielectric constant of the ferrimagnetic medium. When the variation of the wave field along dc magnetization is ignored, equation (4.10) decouples into two independent sets, one of which contains E_z, h_x, and h_y, while the other contains E_x, E_y, and h_z. As noted in Chapter 2, the second set corresponds to a wave which does not interact with the magnetic nature of the medium and, hence, is of no interest to us. It follows from equations (4.10a, b, and f) that, inside the ferrimagnetic region, the electric field component E_z satisfies the following equation:

$$\left(\frac{\partial^2}{\partial x^2} + \frac{\partial^2}{\partial y^2} + \varepsilon_f \mu_{\text{eff}} \omega^2 / c_0^2\right) E_z^f = 0 \tag{4.11}$$

where $\mu_{\text{eff}} = (\mu^2 - \kappa^2)/\mu$. The wave equation for the electric field in the dielectric region is obtained from equation (4.11) by substituting $\mu = 1$,

$\kappa = 0$ and replacing ε_f by ε_d (the dielectric constant of the dielectric); this yields

$$\left(\frac{\partial^2}{\partial x^2} + \frac{\partial^2}{\partial y^2} + \varepsilon_d \frac{\omega^2}{c_0^2}\right) E_z^d = 0 \tag{4.12}$$

If a variation $\exp j(\omega t - \beta y)$ for E_z^f and E_z^d is assumed in equations (4.11) and (4.12), the electric field in the two regions is obtained as

$$\left.\begin{aligned} E_z^f &= A \exp(-k_f x) \exp j(\omega t - \beta y) \\ E_z^d &= B \exp(k_d x) \exp j(\omega t - \beta y) \end{aligned}\right\} \tag{4.13}$$

where

$$\left.\begin{aligned} k_f^2 &= \beta^2 - \beta_f^2, & k_d^2 &= \beta^2 - \beta_d^2 \\ \beta_f^2 &= \varepsilon_f \mu_{\text{eff}} \frac{\omega^2}{c_0^2}, & \beta_d^2 &= \varepsilon_d \omega^2 / c_0^2 \end{aligned}\right\} \tag{4.14}$$

For guided modes, the field must decay as $|x| \to \infty$; this is possible only when k_f and k_d are real and positive* which, in turn, requires that k_f^2 and k_d^2 be positive. The tangential components of magnetic field can be obtained from equations (4.10a) and (4.10b) as

$$\begin{aligned} h_y^f &= \frac{jc_0}{\omega \mu_{\text{eff}}} \left(\frac{j\kappa}{\mu} \frac{\partial E_z^f}{\partial y} - \frac{\partial E_z^f}{\partial x} \right) \\ &= \frac{jc_0}{\omega \mu_{\text{eff}}} \left(k_f + \frac{\beta \kappa}{\mu} \right) E_z^f \end{aligned} \tag{4.15a}$$

$$h_y^d = -\frac{jc_0 k_d}{\omega} E_z^d \tag{4.15b}$$

Matching of E_z and h_y at the interface $x = 0$ yields the following dispersion relation for surface wave propagation along the $\pm y$ directions:

$$\mu_{\text{eff}} k_d + k_f \pm \beta \kappa / \mu = 0 \tag{4.16}$$

Dividing this by β, we obtain the resonance condition ($\beta \to \infty$) as

* When the media are lossy, both k_d and k_f will be complex, and the amplitude of the wave will decay as the wave propagates. When losses are small the wave can still be described as a guided wave provided that each of k_d and k_f has a positive real part so that, in any transverse plane, the field amplitude may decay to zero as $|x| \to \infty$.

$$\mu_{\text{eff}} + 1 \pm \kappa/\mu = 0 \tag{4.17}$$

Substitution for μ, κ, and μ_{eff} from equations (1.22), (1.26), and (2.19) leads to

$$\frac{2(\omega_0 + \omega_m \mp \omega)(\omega_0 + \frac{1}{2}\omega_m \pm \omega)}{\omega_0(\omega_0 + \omega_m) - \omega^2} = 0 \tag{4.18}$$

For the upper sign (which corresponds to propagation along the $+y$-axis), this equation is apparently satisfied when $\omega = \omega_0 + \omega_m$. However, at this frequency, we have $\mu_{\text{eff}} = 0$, in which case equations (4.15)–(4.17) are invalidated. A careful analysis implies that β is finite at $\omega = \omega_0 + \omega_m$.

Considering the lower sign in equation (4.17) (which corresponds to propagation along the $-y$-axis), we obtain $\omega = \omega_0 + \frac{1}{2}\omega_m$ as the resonance condition, which can be compared with the magnetostatic result (4.9).

Figure 4.2 displays typical dispersion curves for propagation in the $\pm y$ direction as obtained from the numerical solution of equation (4.16). The

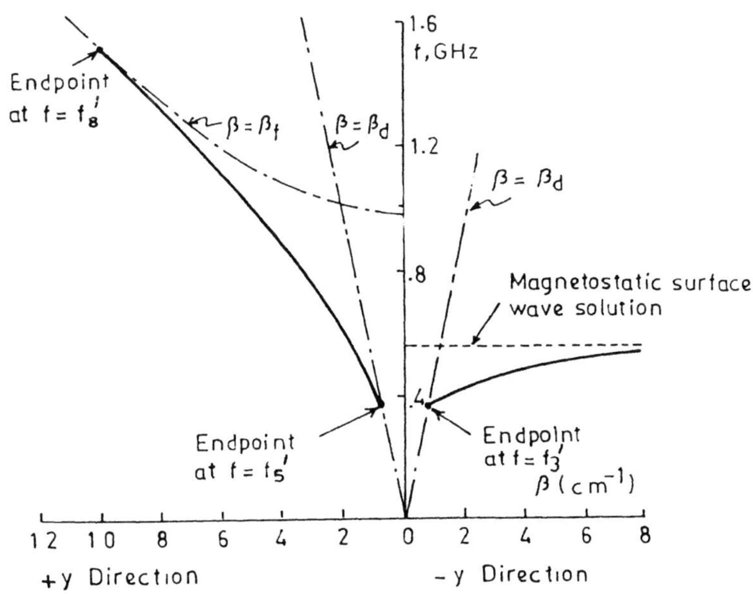

Figure 4.2. Dispersion diagram (wavenumber β vs. wave frequency $f = \omega/2\pi$) for surface wave propagation in the configuration shown in Figure 4.1. The modified magnetostatic mode, propagating along the y-axis in Figure 4.1, has small phase and group velocities while the dynamic mode propagating along the positive y-axis is characterized by large phase and group velocities. For comparison, the magnetostatic wave solution is also shown in the diagram, by a broken line (after Parekh and Ponamgi, 1973).

modified magnetostatic mode propagates in the $\mathbf{H}_0 \times \hat{n}$ direction in the frequency interval $\omega_3' = \omega_0(\omega + \omega_m) < \omega < \omega_0 + \frac{1}{2}\omega_m$. For comparison, the magnetostatic solution has been shown by broken lines. The other mode, corresponding to the upper sign in equation (4.16), propagates in the $-\mathbf{H}_0 \times \hat{n}$ direction. The lower and upper cut-off frequencies for this mode are given by (Parekh and Ponamgi, 1973):

$$\omega_5' = \left\{(\omega_0 + \omega_m)^2 - \omega_m\left[\omega_m + \left(\frac{\varepsilon_f - 2}{\varepsilon_f - 1}\right)\omega_0\right]\right\}^{1/2} \tag{4.19}$$

and

$$\omega_8' = [(\omega_0 + \omega_m)^2 + p_h \omega_m^2]^{1/2} \tag{4.20}$$

where

$$p_h^2 = \frac{\varepsilon_f + (1 + \omega_0/\omega_m)}{2\varepsilon_f - 1} + \frac{[(\varepsilon_f + 1 + \omega_0/\omega_m)^2 + 4\varepsilon_f(\varepsilon_f - 1)(1 + \omega_0/\omega_m)^2]^{1/2}}{2\varepsilon_f - 1} \tag{4.21}$$

This mode is not obtained in magnetostatic analysis and is characterized by comparatively large phase and group velocities. Accordingly, this mode is termed the dynamic mode. As the dispersion of the dynamic mode is small, it can be matched over a wide bandwidth. Hence, this mode is useful in broad-band planar devices (Courtois et al., 1971).

It is worth mentioning that if the ferrimagnetic half space is metallized (i.e., when the dielectric half space is replaced by a perfect conductor), a magnetostatic analysis indicates the existence of a surface wave. However, as noted by Parekh (1975), the rigorous electromagnetic analysis shows that a metallized ferrimagnetic half space cannot support a surface wave.

4.1.2. *Arbitrary Magnetization*

Let us now consider the propagation of a magnetostatic surface wave on a ferrimagnetic half space when the magnetization (as well as the internal dc field) is inclined at an angle α' to the interface; the resulting configuration (Figure 4.3) was first analyzed by Eshbach and Damon (1960). The direction of magnetization is taken as the z-axis of an unprimed coordinate system. A clockwise rotation through angle α' about the y-axis (which is tangent to the interface) leads to the primed coordinate system, the x'-axis of which is perpendicular to the interface. The wave is assumed to propagate in the $y'z'$-plane along a direction making an angle

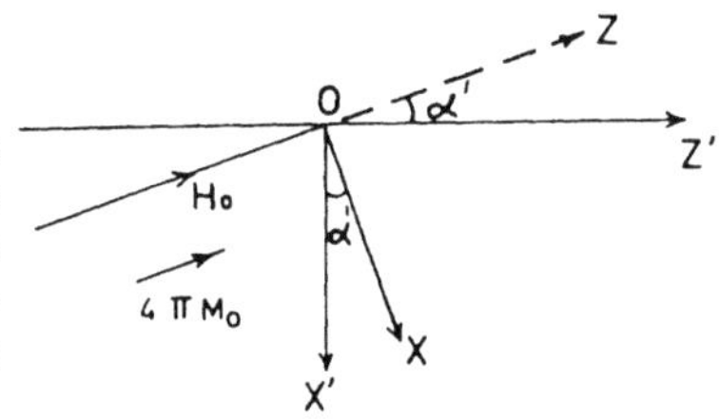

Figure 4.3. Arbitrarily magnetized ferrimagnetic half space; the direction of the internal dc magnetic field H_0 (and hence of dc magnetization) makes an angle α' with the interface. The surface wave propagates in the $y'z'$-plane in a direction making an angle δ' with the y'-axis.

δ' with the y'-axis. The permeability tensor in an unprimed coordinate system is the same as that given by equation (1.25). Following the procedure described in Appendix C, we obtain the transformed permeability tensor (in a primed coordinate system) as:

$$\boldsymbol{\mu}' = \begin{pmatrix} \cos\alpha' & 0 & -\sin\alpha' \\ 0 & 1 & 0 \\ \sin\alpha' & 0 & \cos\alpha' \end{pmatrix} \begin{pmatrix} \mu & j\kappa & 0 \\ -j\kappa & \mu & 0 \\ 0 & 0 & 1 \end{pmatrix} \begin{pmatrix} \cos\alpha' & 0 & \sin\alpha' \\ 0 & 1 & 0 \\ -\sin\alpha' & 0 & \cos\alpha' \end{pmatrix}$$

$$= \begin{pmatrix} \mu\cos^2\alpha' + \sin^2\alpha' & j\kappa\cos\alpha' & (\mu-1)\sin\alpha'\cos\alpha' \\ -j\kappa\cos\alpha' & \mu & -j\kappa\sin\alpha' \\ (\mu-1)\sin\alpha'\cos\alpha' & j\kappa\sin\alpha' & \mu\sin^2\alpha' + \cos^2\alpha' \end{pmatrix} \quad (4.22)$$

It is easy to show from equations (2.64) that, in the primed coordinates, the magnetostatic potential inside the ferrimagnetic medium satisfies the following equation:

$$\left(\mu_{11}\frac{\partial^2}{\partial x'^2} + \mu_{22}\frac{\partial^2}{\partial y'^2} + 2\mu_{13}\frac{\partial^2}{\partial x'\partial z'} + \mu_{33}\frac{\partial^2}{\partial z'^2}\right)\psi^{f} = 0 \quad (4.23)$$

where the μ's are the elements of $\boldsymbol{\mu}'$ in equation (4.22). If the wavevector is expressed as $\beta\hat{s} = \beta_y\hat{y}' + \beta_z\hat{z}'$, the magnetostatic potential outside and inside the magnetic medium (dropping the time dependence) is obtained as:

$$\left.\begin{aligned} \psi^{d} &= A\exp[(\beta_y^2 + \beta_z^2)^{1/2}x']\exp(-j\beta_y y')\exp(-j\beta_z z') \\ \psi^{f} &= B\exp[(\tfrac{1}{2}q_\alpha j - p_\alpha)x']\exp(-j\beta_y y')\exp(-j\beta_z \hat{z}') \end{aligned}\right\} \quad (4.24)$$

where

$$\left.\begin{aligned} q_\alpha &= 2\beta_z\mu_{13}/\mu_{11} \\ p_\alpha &= [\mu(\beta_z^2 + \beta_y^2\mu_{11})]^{1/2}/\mu_{11} \end{aligned}\right\} \quad (4.25)$$

It is required that p_α be real and positive for guided modes. The normal components of the magnetic induction are obtained as

$$b_{x'}^{\mathrm{d}} = (\beta_y^2 + \beta_z^2)^{1/2}\psi^{\mathrm{d}} \tag{4.26}$$

and

$$\begin{aligned} b_{x'}^{\mathrm{f}} &= [(\tfrac{1}{2}jq_\alpha - p_\alpha)\mu_{11} - j\beta_y\mu_{12} - j\beta_z\mu_{13}]^{1/2}\psi^{\mathrm{f}} \\ &= (-\mu_{11}p_\alpha + \beta_y\kappa\cos\alpha')\psi^{\mathrm{f}} \end{aligned} \tag{4.27}$$

Matching of the ψ's and b_x's at $x' = 0$ yields the following dispersion relation for propagation along the $\pm s$-directions (Eshbach and Damon, 1960):

$$\mu_{11}p_\alpha \mp \beta_y\kappa\cos\alpha' + (\beta_y^2 + \beta_z^2)^{1/2} = 0 \tag{4.28}$$

When $\alpha' = 0$ and $\beta_z = 0$, equation (4.28) reduces, as expected, to equations (4.6) and (4.7). Even in the most general case, it is possible to transform equation (4.28) to a simpler form. If δ' represents the angle between the direction of propagation and the y'-axis, we have

$$\tan\delta' = \beta_z/\beta_y \tag{4.29}$$

The substitution for δ', μ, and κ in equation (4.28), followed by algebraic manipulations, leads to

$$\mp 2\omega = \omega_0 \sec\alpha' \sec\delta' + (\omega_0 + \omega_{\mathrm{m}})\cos\alpha'\cos\delta' \tag{4.30}$$

The right-hand side of this equation is positive for any combination of parameters. (α' and δ' are assumed to be in the first quadrant.) It follows that a magnetostatic wave can propagate only in the negative s-direction along which $\mathbf{H}_0 \times \hat{n}$ has a positive projection. Investigating the region of negative μ_{eff}, i.e., $[\omega_0(\omega_0 + \omega_{\mathrm{m}})]^{1/2} < \omega < (\omega_0 + \omega_{\mathrm{m}})$, it is seen that a solution to equation (4.30) is possible only when

$$\omega_0/(\omega_0 + \omega_{\mathrm{m}}) < \cos\alpha'\cos\delta' < 1 \tag{4.31}$$

The relations (4.30) and (4.31) are helpful in understanding the magnetostatic wave propagation characteristics in more complicated layered structures.

Following Adam (1970), we obtain from equation (4.30) the group delay of magnetostatic surface waves propagating in a realistic configura-

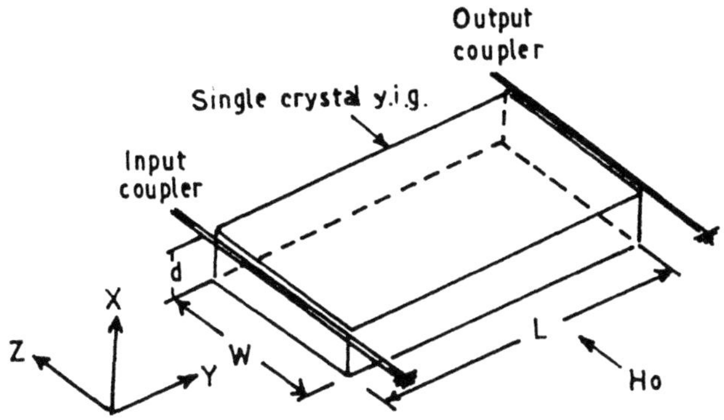

Figure 4.4. A realistic configuration for guiding magnetostatic surface waves. The wave energy is fed into and extracted from the YIG slab by means of fine wire couplers. The biasing field H_0 is parallel to the face of the slab and perpendicular to the direction of propagation (after Adam, 1970).

tion, such as a thick, transversely magnetized ferrimagnetic slab (Figure 4.4). This configuration corresponds* to that in Figure 4.3 with $\alpha' = 0$, in which case equation (4.30) reduces to†

$$\omega = \tfrac{1}{2}\omega_0 \sec\delta' + \tfrac{1}{2}(\omega_0 + \omega_m)\cos\delta' \tag{4.32}$$

Since the extent of the slab is finite in the transverse direction (z-axis), β_z will assume a discrete set of values given by

$$\beta_z = n\pi/w \qquad n = 1, 2, 3, \ldots \tag{4.33}$$

For lower-order modes (which are the ones predominantly excited in an experiment), we have $\beta_y \gg \beta_z$. With equations (4.29) and (4.33), (4.32) can be transformed as

$$\beta_y^2 \simeq \beta_z^2(\omega_0 + \omega_m - \omega)/[\omega_m - 2(\omega - \omega_0)] \tag{4.34}$$

The group velocity is obtained from equation (4.32) when it is differentiated with respect to β_y. This yields

* This is true provided that the transverse wavenumber exceeds the inverse of the thickness of the slab.

† Incidently, this result has also been derived by Benson and Mills (1969) from variational analysis.

$$v_g = \frac{2\beta_y}{\omega_m \beta_z^2}[\omega_m - 2(\omega - \omega_0)]^2$$

$$= \frac{2}{\omega_m \beta_z}(\omega_0 + \omega_m - \omega)^{1/2}[\omega_m - 2(\omega - \omega_0)]^{3/2}$$

Since ω is close to $(\omega_0 + \frac{1}{2}\omega_m)$, the group velocity can be approximated to

$$v_g = \frac{4}{\sqrt{\omega_m}\,\beta_z}(\tfrac{1}{2}\omega_m - \omega + \omega_0)^{3/2} \tag{4.35}$$

The inverse of v_g yields the group delay time for a unit distance of propagation. Since the slab is finite in size with a shape different from ellipsoidal, the internal dc field is not uniform, which renders $\omega_0 = \gamma H_0$ a function of position along the direction of propagation. The internal field can be calculated from the first-order demagnetization theory discussed in Chapter 1. Following Adam (1970), we assume a parabolic internal field variation:

$$H_0(y) = H_c + \delta H y^2/L^2 \tag{4.36}$$

where $H_c = H_e + H_d + H_{AO}$ is the dc field at the center of the slab, H_e is the externally applied field, H_d is the demagnetizing field at the center of the slab, and H_{AO} is the dc anisotropy field. The delay time is obtained as

$$\tau_d = \frac{(4\pi M_0)^{1/2}}{4\gamma}\beta_z \int_{-L/2}^{+L/2} [2\pi M_0 + H_0(y) - \omega/\gamma]^{-3/2} dy \tag{4.37}$$

Equivalently,

$$\tau_d = \frac{(4\pi M_0)^{1/2}\beta_z L}{4\gamma(H_c + 2\pi M_0 - \omega/\gamma)(H_c + 2\pi M_0 - \omega/\gamma + \delta H)^{1/2}} \tag{4.38}$$

This relation was derived by Adam (1970) and is applicable to the propagation of a magnetostatic surface wave on a thick, transversely magnetized ferrimagnetic slab. Results from this expression will be compared with experimental data in Section 4.1.3.

While equation (4.38) takes into account the effect of finite sample width, it has been recently pointed out by Gupta and Srivastava (1980) that this relation, based on magnetostatic dispersion relation (4.28), would lead to approximately correct results up to s-band frequencies and only rigorous electromagnetic analysis correctly predicts the dispersion characteristics at higher frequencies.

4.1.3. *Experimental Results*

While several experimental investigations of magnetostatic modes in ferrite spheres, ellipsoids, rods, and discs were carried out immediately after the observation of magnetostatic modes of a ferrite sample by White and Solt (1956), the first experimental evidence for magnetostatic surface wave propagation on a YIG slab was reported by Brundle and Freedman (1968b). The measurement of group delay of magnetostatic surface waves involves standard pulse techniques. Following Sethares and Merry (1974), the block diagram of the basic experiment is shown in Figure 4.5a, while a typical sample mount is shown in Figure 4.5b. The ground plane (which

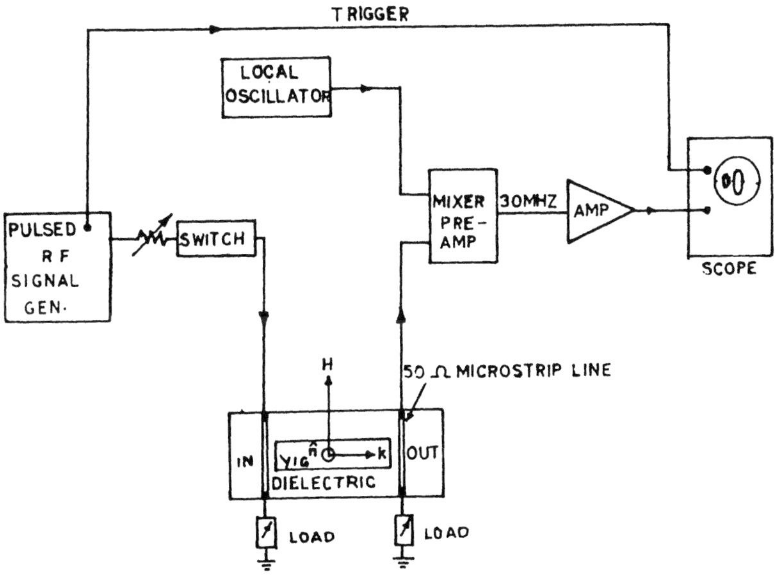

Figure 4.5a. Block diagram of the basic experimental arrangement for the generation and detection of magnetostatic surface waves (after Sethares and Merry, 1974).

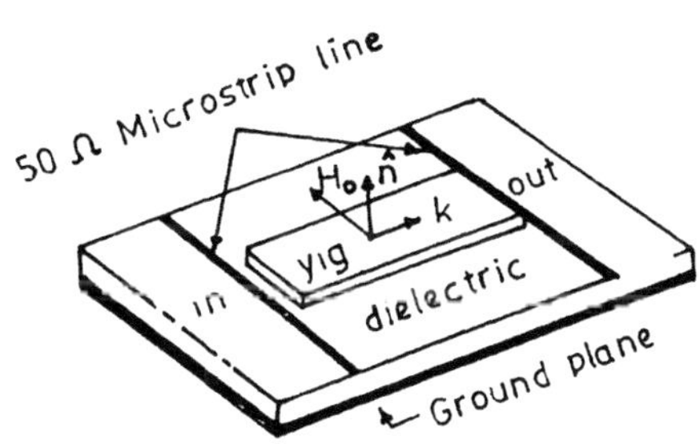

Figure 4.5b. An yttrium iron garnet (YIG) rectangular slab mounted on a microstrip board. A microwave electromagnetic pulse at the input end converts into a magnetostatic surface wave which propagates along the top surface of the sample. The magnetostatic surface wave is reconverted to an electromagnetic pulse at the output end (after Sethares and Merry, 1974).

may be removed when desired) and the dielectric substrate can be cut from commercially available microstrip boards. It is also possible to use an adjustable mount, wherein fine-wire couplers are employed. This arrangement makes it possible to vary the distance between the coupler and sample, thereby providing means to control the strength of the coupling between electromagnetic and magnetostatic waves. The dielectric substrate can be removed or replaced when the ground plane is present (see Section 4.3).

We now present some experimental results. Figure 4.6a shows the experimental configuration of Brundle and Freedman (1968b). The results of a pulse delay test at a carrier frequency of 3 GHz are displayed in Figure 4.6b. It is seen that the group delay time decreases with increasing biasing field. The insertion loss also decreases proportionately; it is about 300 dB/μsec over most of the frequency range. To reduce this large attenuation, Adam (1970) employed chemical polishing of the sample surface using hot orthophosphoric acid as suggested by Basterfield (1969). The minimum power attenuation rate obtained after the chemical polishing was about 54 dB/μsec at 4 GHz.

Figure 4.7 shows the variation of group delay time with biasing field for a typical sample configuration (Figure 4.4). The figure also shows the theoretical group delay time obtained for $n = 1$ from equations (4.38) and (4.33). The agreement between theory and experiment is excellent. Further reduction in the attenuation rate was obtained by Merry and Sethares (1973) through polishing the sample surface using silica gel (Syton). Figure

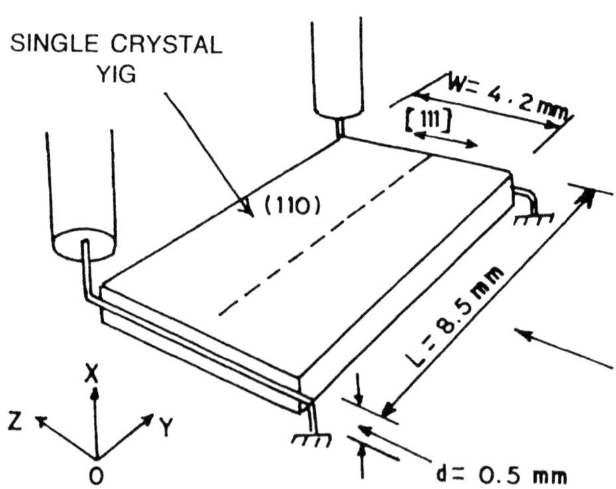

Figure 4.6a. The experimental configuration of Brundle and Freedman (1968b).

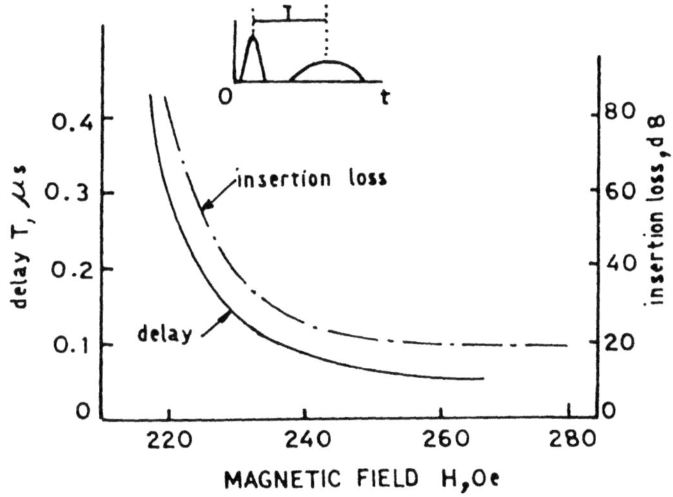

Figure 4.6b. The experimental results of the pulse delay test; variation of pulse delay T and insertion loss (i.e., the resultant of propagation loss and coupling loss) with biasing field (after Brundle and Freedman (1968b).

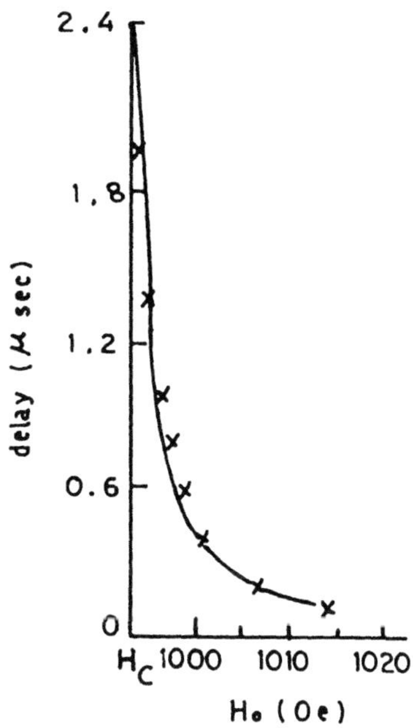

Figure 4.7. Comparison of experimental (crosses) and theoretical (solid line) delay times for a typical sample in the configuration shown in Figure 4.4. The slab dimensions are $L = 8.75$ mm, $W = 2.67$ mm, and $d = 1.2$ mm. The $[\bar{1}00]$, $[00\bar{1}]$, and $[0\bar{1}0]$ crystallographic directions of the YIG single crystal are oriented along the x-, y- and z-axes, respectively. The theoretical calculations have been performed using equation (4.38) with $n = 1$. The quantity $(H_D + H_A)$ has been adjusted to get the best fit with experimental data (after Adam, 1970).

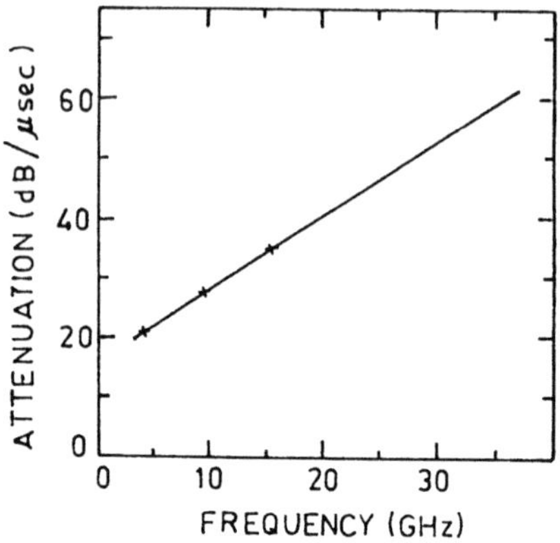

Figure 4.8. Variation of experimentally observed power attenuation rate of magnetostatic surface waves with frequency (after Merry and Sethares, 1973).

4.8 shows the frequency dependence of thc attenuation rate. The linear variation can be understood on the basis of ferromagnetic relaxation theory.* At relatively high frequencies, the attenuation rate L_p can be related to the linewidth as (Merry and Sethares, 1973; Sethares and Stiglitz, 1974):

$$L_p = 76.4\Delta H \qquad [\text{dB}/\mu\,\text{sec}] \tag{4.39}$$

This clearly shows that single-crystal ferrimagnetic slabs (or films) of extremely narrow linewidth are required to obtain a high figure of merit in a magnetostatic wave device. Figure 4.9 is a photograph of such a delay line, described by Sethares and Stiglitz (1974).

In order to assess the utility of magnetostatic waves, it is worth comparing their attenuation with that of surface elastic waves. Such a plot is shown in Figure 4.10. It is seen that above about a few gigahertz, the magnetostatic waves lead to much lower loss than surface elastic waves; this makes magnetostatic waves suitable for microwave signal processing in the gigahertz range.

When the ferrimagnetic slab employed in a delay device is rather thick, the internal dc field is highly nonuniform on account of demagnetization.† In recent years, this difficulty has been resolved by growing high-quality films‡ of ferrimagnetic garnets, specifically, yttrium iron garnet

* Ferromagnetic relaxation theory is beyond the scope of the present book. The interested reader is referred to Sparks (1964).

† One way to avoid this nonuniformity is to surround the principal sample by dummy, metal-coated ferrite samples; see for example Adam et al. (1970).

‡ The various aspects of magnetostatic wave propagation in magnetic films and its applications have been reviewed by Collins et al. (1972), Collins and Pizzarello (1973), Grant et al. (1974), Bongianni (1974), Adam et al. (1975), Adam and Collins (1976), Collins et al. (1977).

Figure 4.9. A photograph of a microwave delay line using YIG. The overall dimensions are 2.25″ × 1.75″ × 0.75″ (after Sethares and Stiglitz, 1974).

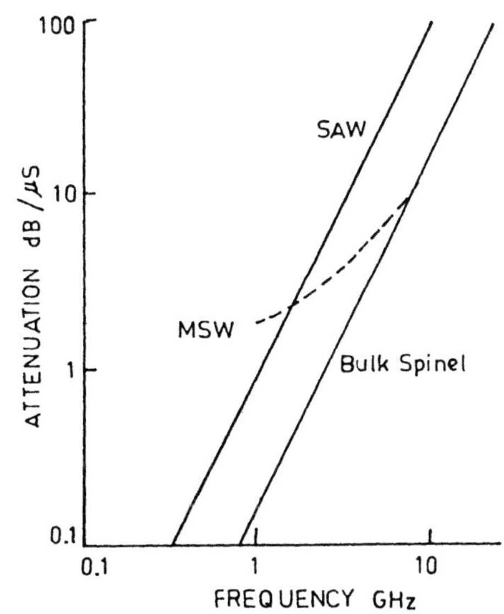

Figure 4.10. Calculated minimum attenuation vs. frequency for magnetostatic waves on the YIG film with $\Delta H = 0.15\,\text{Oe}$ compared with surface elastic waves and bulk elastic waves in spinels (after Adam and Collins, 1976).

(YIG). Early YIG films were grown on highly polished substrates of gadolinium gallium garnet (GGG) by vapor phase epitaxy (Mee et al., 1969). Today, vapor phase epitaxy (VPE) has been replaced by liquid phase epitaxy (LPE), which is a development of the bulk crystal flux growth technique. LPE is more flexible than VPE and leads to reproducible, low-loss single-crystal films (Levinstein et al., 1971; Glass and Elliot, 1975). Garnet films grown up to 5 cm in diameter on a good-quality GGG substrate have less than five defect densities per square centimeter (Adam and Collins, 1976). The uniformity in film thickness can be retained within a range of 10% across 90% of the substrate area with hardly any composition variation or second phases. The rate of growth of the film can range from 0.1 to about 3 μm/min, while the total thickness can be as large as 100 μm. Glass and Elliot (1975) have observed linewidths of about 0.15 Oe at 9.5 GHz at room temperature. This linewidth is comparable to the values reported for the best flux-grown single-crystal YIG and the value is very close to the intrinsic losses of pure defect-free material. As noted by Owens et al. (1975), the selective etching of garnets with phosphoric acid at 270 °C through an SiO_2 mask will make the delay devices more flexible for several applications, including signal routing.

In subsequent sections, we shall consider magnetostatic wave propagation in ferrimagnetic films (which, from an analytical point of view, are slabs of infinite extent in the plane perpendicular to the thickness) surrounded by other media. Most of the experimental results and numerical calculations will be presented for YIG. The interesting microwave characteristics of YIG and the well-developed technology for its growth have led to its almost exclusive use in magnetostatic wave devices.

4.2. *Isotropic Ferrimagnetic Slab*

Before considering magnetostatic propagation in complicated layered structures, it is worth discussing the case of a magnetically isotropic ferrimagnetic slab that is subjected to a dc magnetic field. We have analyzed two specific configurations: the magnetization in the plane of the slab (Section 4.2.1) and the magnetization perpendicular to the plane of the slab (Section 4.2.2).

4.2.1. *Magnetization in the Slab Plane*

When a thin slab or film of an isotropic ferrimagnetic material is magnetized parallel to its face, the internal dc field is approximately the same as the applied field since the effect of demagnetization is negligible (see Section 1.8). This configuration supports surface as well as bulk waves.

We first carry out (Section 4.2.1.1) a two-dimensional magnetostatic analysis of surface wave propagation transverse to the dc magnetization where the wave field has been assumed to be uniform in the direction of dc magnetization. Next, in Section 4.2.1.2, we present a rigorous electromagnetic analysis of transverse propagation. In Section 4.2.1.3, we show the general three-dimensional magnetostatic analysis of surface and bulk modes, considering the possible variation of a wave field along the dc magnetization. Some experimental results and applications are discussed in Section 4.2.1.4.

4.2.1.1. *Transverse Propagation: Magnetostatic Analysis*

The configuration to be analyzed is shown in Figure 4.11. The slab occupies the region $0 \leqslant x \leqslant d$ and extends to infinity in the yz-plane. The wave propagates along the y-axis while the biasing field is along the z-axis. The wave field is assumed to be uniform along the dc magnetization. Following the procedure outlined in Section 4.1, we can express the magnetostatic potential in the three regions as

$$\left.\begin{aligned} \psi^{(1)} &= A\exp(\beta x)\exp j(\omega t - \beta y) && (x \leqslant 0) \\ \psi^{f} &= [B\exp(\beta x) + C\exp(-\beta x)]\exp j(\omega t - \beta y) && (0 \leqslant x \leqslant d) \\ \psi^{(2)} &= D\exp(-\beta x)\exp j(\omega t - \beta y) && (x \geqslant d) \end{aligned}\right\} \tag{4.39}$$

Using equations (4.39), (4.4), and (4.5), we obtain the normal components of magnetic induction in the three regions:

$$\left.\begin{aligned} b_x^{(1)} &= \beta\psi^{(1)} \\ b_x^{(2)} &= -\beta\psi^{(2)} \\ b_x^{f} &= \beta[(\mu + \kappa)B\exp(\beta x) - (\mu - \kappa)C\exp(-\beta x)]\exp j(\omega t - \beta y) \end{aligned}\right\} \tag{4.40}$$

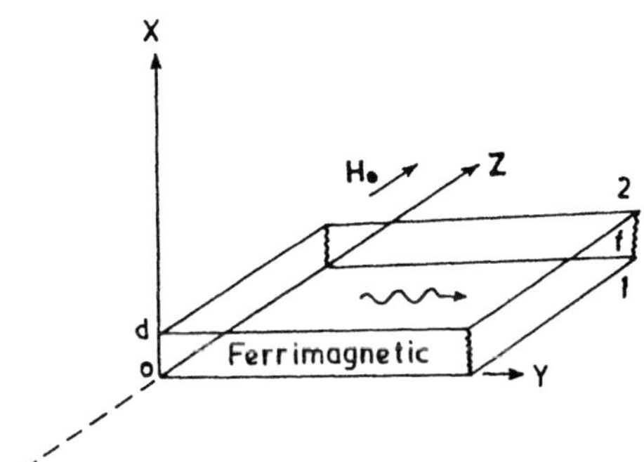

Figure 4.11. The configuration to be analyzed: the ferrimagnetic slab occupies the region $0 \leqslant x \leqslant d$ and extends to infinity in the yz-plane. The wave propagates along the y-axis while the dc magnetization is along the z-axis.

Application of the appropriate boundary conditions at $x = 0$ yields

$$\left.\begin{aligned} B + C &= A \\ (\mu + \kappa)B - (\mu - \kappa)C &= A \end{aligned}\right\} \tag{4.41}$$

while at $x = d$, we have

$$\left.\begin{aligned} B\exp(\beta d) + C\exp(-\beta d) &= D\exp(-\beta d) \\ (\mu + \kappa)B\exp(\beta d) - (\mu - \kappa)\exp(-\beta d) &= -D\exp(-\beta d) \end{aligned}\right\} \tag{4.42}$$

Elimination of A and D from equations (4.41) and (4.42) leads to

$$\left.\begin{aligned} (\mu + \kappa - 1)B &= (\mu - \kappa + 1)C \\ (\mu + \kappa + 1)B\exp(\beta d) &= (\mu - \kappa - 1)C\exp(-\beta d) \end{aligned}\right\} \tag{4.43}$$

Elimination of B and C from equation (4.43) yields the dispersion relation

$$\exp(2\beta d) = \frac{(\mu - \kappa - 1)(\mu + \kappa - 1)}{(\mu - \kappa + 1)(\mu + \kappa + 1)} \tag{4.44}$$

and it can be seen that the reversal of the sign of κ leaves the dispersion relation unchanged. This implies that the waves are reciprocal and the propagation constant and phase and group velocities are the same for propagation along the $\pm y$-directions. However, the expressions for the ψ's and $\mathbf{b}$'s change when the sign of κ is reversed. This means that for a given frequency the wave field and energy distribution for propagation along the $\pm y$-axis are different even though the wavenumber is the same; this is called degeneracy. Figure 4.12 shows a qualitative plot of the field distribution (across the slab) for magnetostatic waves propagating along the $\pm y$-directions. The waves are clearly surface waves since the wave field decays exponentially as one moves away from the interface (on both sides). It is also evident from the figure that the magnetostatic surface wave associated with a ferrimagnetic surface propagates along the $\mathbf{H}_0 \times \hat{n}$ direction.

Substitution for μ and κ from equations (1.22) and (1.26) in equation (4.44) leads to the following dispersion relation for surface wave propagation in a slab:

$$R_g = \exp(2\beta d) = \frac{(\frac{1}{2}\omega_m)^2}{(\omega_0 + \frac{1}{2}\omega_m + \omega)(\omega_0 + \frac{1}{2}\omega_m - \omega)} \tag{4.45}$$

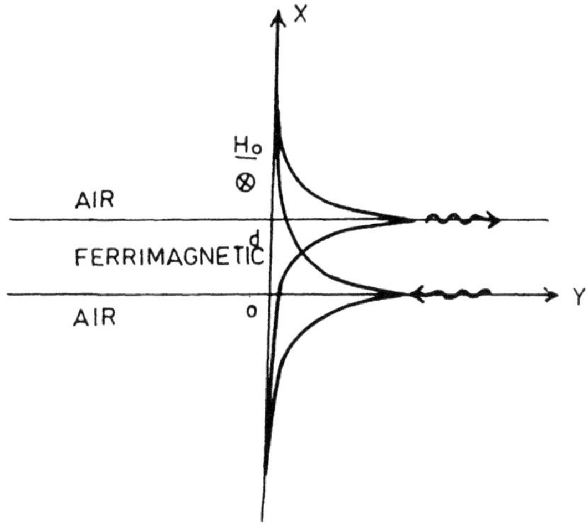

Figure 4.12. Qualitative plot of field distribution across the ferrimagnetic slab; for convenience, the field has been plotted along the y-axis. When the biasing field is along the z-axis (i.e., into the plane of the paper), the mode with energy concentration at the top (bottom) surface propagates along the positive (negative) y-axis, i.e., along the $\mathbf{H}_0 \times \hat{n}$ direction, where $\hat{n}$ is the outward normal at the ferrimagnetic surface.

Since β is positive for guided modes, R_g must exceed unity; it is easy to show that this condition is satisfied when $[\omega_0(\omega_0+\omega_m)]^{1/2} < \omega < (\omega_0 + \frac{1}{2}\omega_m)$. It is seen that $\beta \to \infty$ (resonance) when ω approaches the upper limit whereas $\beta \to 0$ (cut-off limit) when ω approaches the lower limit. Evidently, the frequency range of allowed magnetostatic surface modes depends on the relative magnitudes of ω_0 and ω_m. When $\omega_0 \ll \omega_m$, this frequency range is relatively large. In the limit when $H_0 \to 0$, the range of allowed modes is $0 < \omega < \frac{1}{2}\omega_m$. The frequency range of allowed modes decreases when the biasing field is increased (in order to effect surface wave propagation at higher frequencies). For example, in the case of YIG ($4\pi M_0 = 1750\,\text{G}$), the frequency range of allowed surface modes extends from 3 to 3.85 GHz at $H_0 = 500\,\text{Oe}$, while for $H_0 = 2.5\,\text{kOe}$ the range is only from 9.12 to 9.45 GHz. Thus, the applicability of magnetostatic surface wave devices is usually restricted to frequencies well below the X-band.

The explicit expression for the propagation constant is obtained from equation (4.45) as

$$\beta_d = \frac{1}{2}\ln\left[\frac{(\frac{1}{2}\omega_m)^2}{(\omega_0+\frac{1}{2}\omega_m-\omega)(\omega_0+\frac{1}{2}\omega_m+\omega)}\right] \tag{4.46}$$

It follows that for given ω, ω_0, and ω_m, the propagation constant is inversely proportional to the slab thickness. Figure 4.13a shows the frequency dependence of β_d as obtained from equation (4.46). It is seen that for a given thickness d, the wavenumber β increases monotonically with ω, thereby rendering the slope of the ω vs. β curve positive. Thus, the

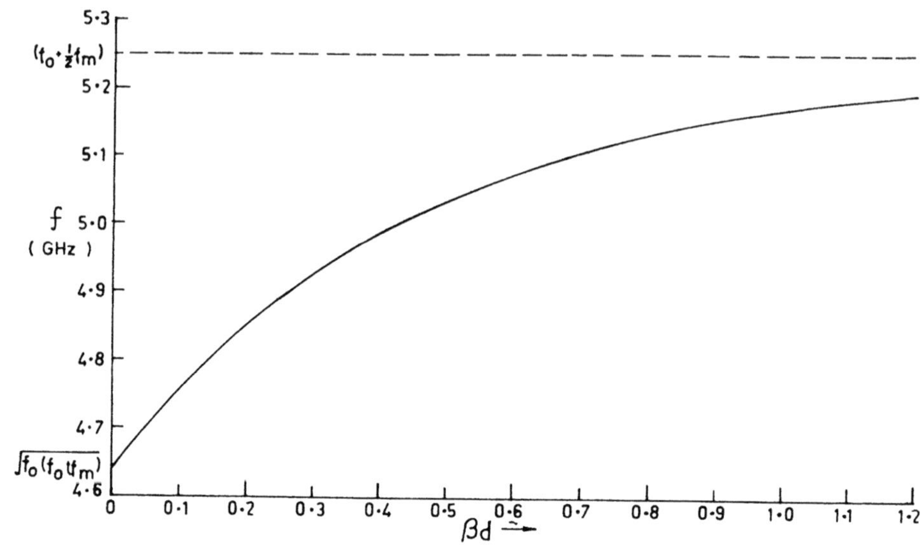

Figure 4.13a. Variation of normalized wavenumber with the wave frequency corresponding to magnetostatic surface wave propagation in the configuration shown in Figure 4.11.

direction of propagation of a wave group or wave packet is the same as the direction of phase propagation; the waves are forward surface waves. The dispersion curves for different slab thicknesses have been plotted on a semilog scale in Figure 4.13b. The curve for $d = 1$ cm can be compared directly with the curve in Figure 4.13a.

The group velocity can be obtained from equation (4.45) as

$$v_g \equiv \partial\omega/\partial\beta$$
$$= (d/\omega)(\tfrac{1}{2}\omega_m)^2 \exp(-2\beta d)$$

Using equation (4.46) yields the time delay per unit distance of propagation as

$$\tau_d = 1/v_g = (\omega/d)/[(\omega_0 + \tfrac{1}{2}\omega_m)^2 - \omega^2] \tag{4.47}$$

It is easy to show that $d\tau_d/d\omega$ is positive throughout the frequency range of the allowed modes. Thus, the group delay time monotonically increases with frequency and is inversely proportional to the slab thickness. Figure 4.13c shows the variation of $d\tau_d$ with wave frequency for a typical set of parameters; the delay time increases monotonically with frequency and approaches infinity at resonance.

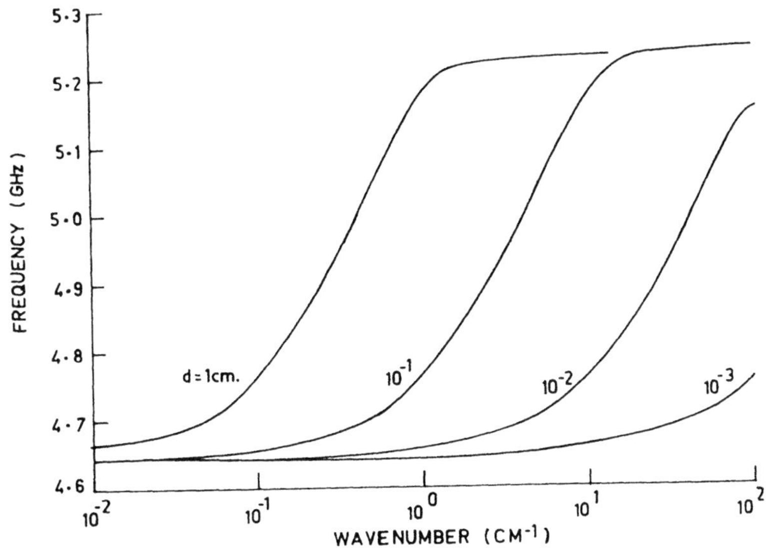

Figure 4.13b. Magnetostatic surface wave dispersion curves for different thicknesses of the ferrimagnetic slab. The curve for $d = 1$ cm can be compared directly with that in Figure 4.13a.

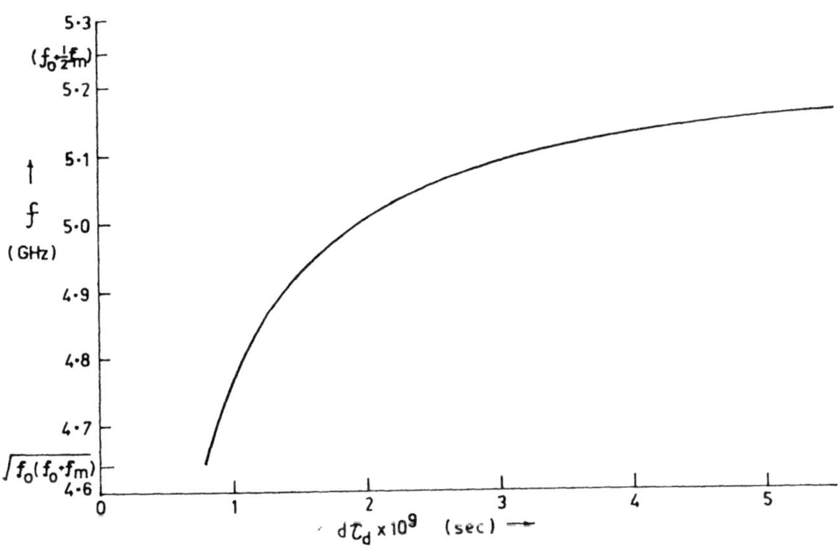

Figure 4.13c. Variation of normalized group delay with frequency of magnetostatic surface waves, corresponding to the dispersion curve shown in Figure 4.13a.

It is worth considering the effect of a small curvature of the slab/film about the direction of the biasing field.*

If the curvature is cylindrical, it is more convenient to work in the cylindrical coordinate system shown in Figure 4.14a. The following transformations are made:

$$\left.\begin{aligned} x &= r\cos\theta \\ y &= r\sin\theta \\ \zeta &= R\ln r/R \\ \xi &= R\theta \end{aligned}\right\} \tag{4.48}$$

If it is assumed that the magnetostatic potential is independent of z, it can be shown from equations (4.1) and (4.48) that the magnetostatic potential satisfies the following equation everywhere in space:

$$\left(\frac{\partial^2}{\partial\zeta^2}+\frac{\partial^2}{\partial\xi^2}\right)\psi=0 \tag{4.49}$$

Now consider the curved film shown in Figure 4.14b. The ferrimagnetic surfaces are at $\zeta=0$ and $\zeta=\zeta_d\equiv R\ln(1+d/R)$. Using the standard procedure, we obtain the dispersion relation for propagation along the ξ-axis as (Srivastava, 1978b):

$$\exp(2\beta\zeta_d)=\frac{(\mu-\kappa-1)(\mu+\kappa-1)}{(\mu-\kappa+1)(\mu+\kappa+1)}\equiv f(\omega) \tag{4.50}$$

which is the same as that for a flat slab [see equation (4.44)] except for the presence of ζ_d instead of d on the left-hand side. If β_0 denotes the wavenumber in the absence of curvature (other parameters remaining the same), we have

$$\exp(2\beta_0 d)=f(\omega)=\exp(2\beta\zeta_d)$$

from which it follows that

$$\left.\begin{aligned} \beta/\beta_0 &= \frac{d/R}{\ln(1+d/R)}>1 \\ \frac{d\omega/d\beta}{d\omega/d\beta_0} &= \frac{\ln(1+d/R)}{d/R}<1 \end{aligned}\right\} \tag{4.51}$$

* Magnetostatic wave propagation in cylindrically curved films is important from the viewpoint of ring interferometers for rotation rate sensing (Newburgh et al., 1974).

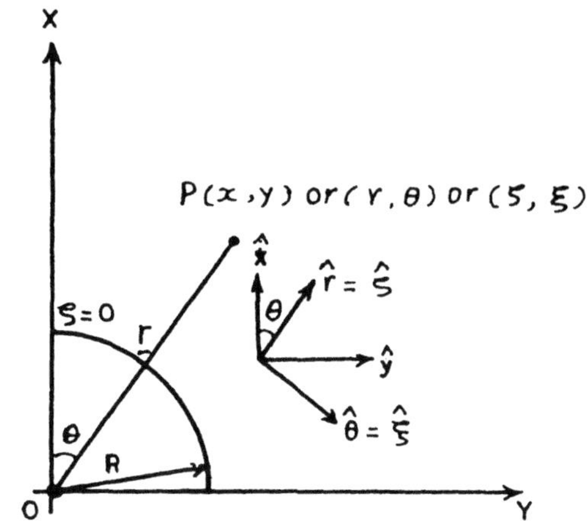

Figure 4.14a. The coordinate system to study propagation along curved surfaces (after Srivastava, 1978b).

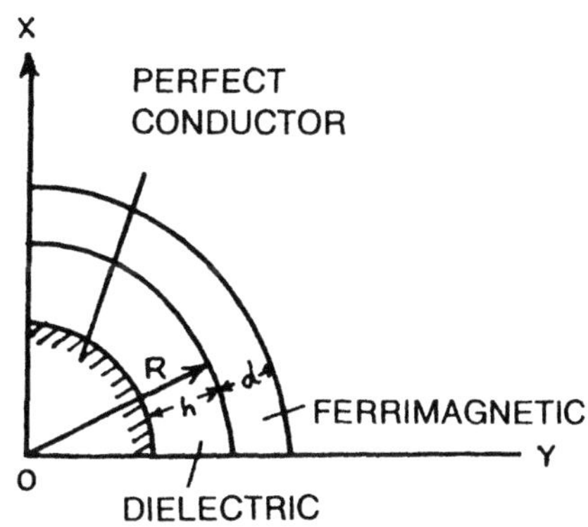

Figure 4.14b. The configuration: curved ferrimagnetic slab backed by grounded dielectric (after Srivastava, 1978b).

Clearly, the effect of curvature is to reduce phase velocity as well as group velocity by the same factor. When d is small, we have

$$(\beta - \beta_0)/\beta_0 \simeq d/2R$$

which would be quite small. Thus, the effect of curvature would be appreciable only when $d/R \gtrsim 0.1$.

4.2.1.2. *Transverse Propagation: Electromagnetic Analysis*

It was seen in Section 4.1 that magnetostatic analysis of surface wave propagation in a transversely magnetized half space yields only the resonance frequency. Rigorous electromagnetic analysis is required to obtain the exact dispersion curve. In addition, electromagnetic analysis also revealed the propagation of a fast, dynamic mode in a direction opposite to the direction of propagation of the magnetostatic wave. Therefore, it is worth doing a rigorous electromagnetic analysis in the present case of the slab and comparing the results with those of the preceding section. The analysis follows Gerson and Nadan (1974).

Referring back to Figure 4.11, we can write the electric field in the three regions from equations (4.11) and (4.12) as

$$\left.\begin{aligned} E_z^{(1)} &= A\exp(k_0x)\exp j(\omega t-\beta y)\\ E_z^{\mathrm{f}} &= [B\exp(k_\mathrm{f}x)+C\exp(-k_\mathrm{f}x)]\exp j(\omega t-\beta y)\\ E_z^{(2)} &= D\exp(-k_0x)\exp j(\omega t-\beta y)\end{aligned}\right\} \qquad (4.52)$$

where

$$\left.\begin{aligned} k_\mathrm{f}^2 &= \beta^2-\varepsilon_\mathrm{f}\frac{\omega^2}{c_0^2}\mu_\mathrm{eff}\\ k_0^2 &= \beta^2-\frac{\omega^2}{c_0^2}\end{aligned}\right\} \qquad (4.53)$$

The relevant tangential component of the magnetic field in the three regions can be obtained from equations (4.15) and (4.52) as

$$\left.\begin{aligned} h_y^{(1)} &= -j\frac{c_0k_0}{\omega}E_z^{(1)}\\ h_y^{\mathrm{f}} &= \frac{jc_0}{\omega\mu_\mathrm{eff}}[(\beta\kappa/\mu-k_\mathrm{f})B\exp(k_\mathrm{f}x)\\ &\qquad +(\beta\kappa/\mu+k_\mathrm{f})C\exp(-k_\mathrm{f}x)]\exp j(\omega t-\beta y)\\ h_y^{(2)} &= \frac{jc_0k_0}{\omega}E_z^{(2)}\end{aligned}\right\} \qquad (4.54)$$

The application of appropriate boundary conditions at $x=0$ and $x=d$, followed by the elimination of constants, leads to the following dispersion relation for propagation along the positive y-axis:

$$\exp(2k_\mathrm{f}d)=\frac{(\mu_\mathrm{eff}k_0-\beta\kappa/\mu-k_\mathrm{f})(\mu_\mathrm{eff}k_0+\beta\kappa/\mu-k_\mathrm{f})}{(\mu_\mathrm{eff}k_0-\beta\kappa/\mu+k_\mathrm{f})(\mu_\mathrm{eff}k_0+\beta\kappa/\mu+k_\mathrm{f})} \qquad (4.55)$$

The invariance of this relation with respect to the reversal of the sign of κ implies reciprocal propagation. Before discussing the numerical implications of this relation, it is worth considering the limiting cases. When μ_{eff} is negative, the large wavenumber approximation of equation (4.55) can be obtained by substituting $k_f \simeq k_0 \simeq \beta$ [which follows from equation (4.53)]; this yields

$$\exp(2\beta d) = \frac{(\mu_{\text{eff}} - \kappa/\mu - 1)(\mu_{\text{eff}} + \kappa/\mu - 1)}{(\mu_{\text{eff}} - \kappa/\mu + 1)(\mu_{\text{eff}} + \kappa/\mu + 1)} \tag{4.56}$$

which can be shown to be equivalent to equation (4.44), i.e., the magnetostatic dispersion relation. It is concluded that the magnetostatic analysis leads to satisfactory results for a transversely magnetized slab when the wavenumbers are large.

Next, consider the case when the slab is very thick; in this case the left-hand side of equation (4.55) becomes very large. As such, the limiting dispersion relation can be obtained by equating the denominator of the right-hand side to zero, i.e.,

$$(\mu_{\text{eff}} k_0 - \beta\kappa/\mu + k_f)(\mu_{\text{eff}} k_0 + \beta\kappa/\mu + k_f) = 0 \tag{4.57}$$

A comparison of this dispersion relation with equation (4.16) shows that it represents the combined dispersion relation for the guided surface waves, provided that the ferrimagnetic medium fills the lower and upper half spaces.

A numerical solution to equation (4.55) for a typical set of parameters yields the dispersion curves displayed in Figure 4.15. It is seen that the modified magnetostatic mode differs from the one obtained from magnetostatic analysis only in the small-wavenumber region. The cut-off wavenumber is finite instead of zero, which is predicted by the magnetostatic analysis. Nevertheless, the frequency range of allowed surface modes remains unaffected, i.e., it extends from $\omega = [\omega_0(\omega_0 + \omega_m)]^{1/2}$ to $\omega = \omega_0 + \frac{1}{2}\omega_m$. The dynamic mode is guided in the small-wavenumber region and is shown in the same figure. The lower cut-off frequency for the dynamic mode is approximately the same as that for the magnetostatic mode while the upper cut-off frequency is finite and thickness-dependent. It exceeds $\omega = \omega_0 + \omega_m$ in all cases and increases with increasing thickness, finally approaching the limit given by equation (4.20) for a half space. The dynamic mode is practically nondispersive in contrast to the magnetostatic mode, which is highly dispersive throughout its frequency range.

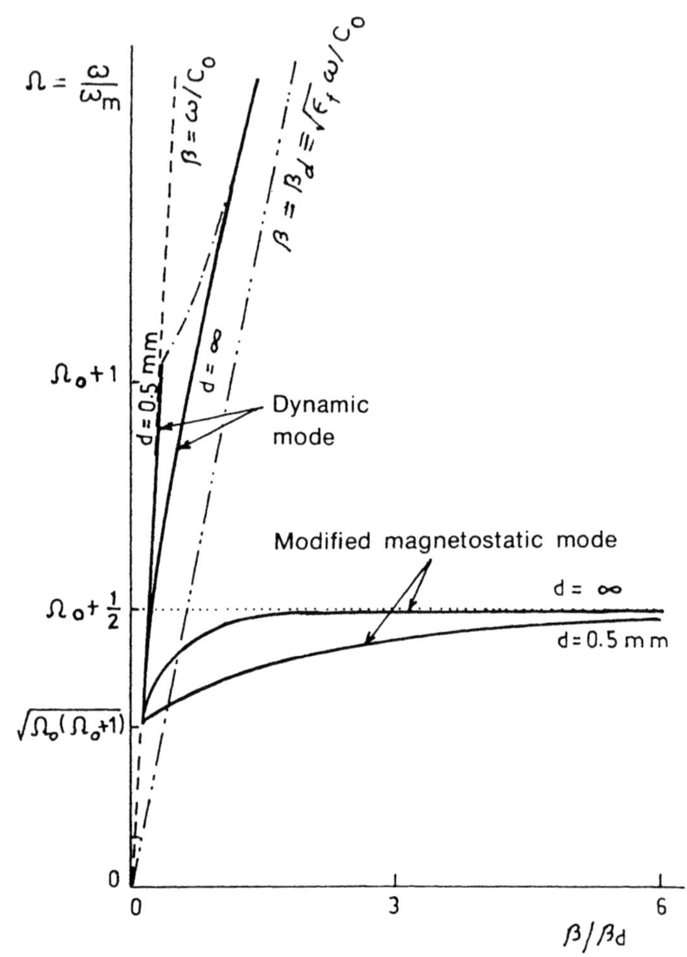

Figure 4.15. Dispersion curves for modified magnetostatic and dynamic modes for $d = 0.05$ cm and $d = \infty$; the latter corresponds to the half space (after Gerson and Nadan, 1974).

4.2.1.3. *General Magnetostatic Analysis*

In the last two sections, the wave field was regarded as uniform in the direction of the dc magnetization. The case when the wave field is allowed to vary harmonically in the direction of dc magnetization has been analyzed by Damon and Eshbach (1961), and a characteristic feature is the occurrence, in the appropriate frequency range, of guided bulk modes apart from the surface modes.

Referring to Figure 4.11 and assuming a variation $\cos k_z z$ along the dc magnetization, we can express the magnetostatic potential in the three regions as

$$\left.\begin{aligned} \psi^{(1)} &= \exp(k_x^e x)\cos k_z z \exp j(\omega t - \beta y) \\ \psi^f &= [B\exp(k_x^i x) + C\exp(-k_x^i x)]\cos k_z z \exp j(\omega t - \beta y) \\ \psi^{(2)} &= D\exp(-k_x^e x)\cos k_z z \exp j(\omega t - \beta y) \end{aligned}\right\} \quad (4.58)$$

where

$$\left.\begin{aligned} k_x^i &= (\beta^2 + k_z^2/\mu)^{1/2} \\ k_x^e &= (\beta^2 + k_z^2)^{1/2} \end{aligned}\right\} \quad (4.59)$$

The normal component of **b** can be obtained from equation (4.58). The standard procedure of application of appropriate boundary conditions and elimination of constants yields the following dispersion relation:

$$\exp(2k_x^i d) = \frac{(\mu k_x^i - \beta\kappa - k_x^e)(\mu k_x^i + \beta\kappa - k_x^e)}{(\mu k_x^i - \beta\kappa + k_x^e)(\mu k_x^i + \beta\kappa + k_x^e)} \quad (4.60)$$

For $k_z = 0$, this dispersion relation reduces, as expected, to equation (4.44). When μ is positive, it follows from equation (4.59) that k_x^i is real (losses neglected) and hence the modes, if they exist, are surface modes. It is not very difficult to show from equation (4.60) that the condition for occurrence of guided surface modes turns out to be (Damon and Eshbach, 1961):

$$\eta_s \equiv k_z/\beta \leqslant (\omega_m/\omega_0)^{1/2} \equiv \cot\varphi_s \quad (4.61)$$

The lower frequency limit of allowed surface modes is $\omega = [\omega_0(\omega_0 + \omega_m)]^{1/2}$, while the upper limit decreases from $\omega = \omega_0 + \frac{1}{2}\omega_m$ to $\omega = [\omega_0(\omega_0 + \omega_m)]^{1/2}$ as η_s varies from zero to the maximum given by equation (4.61).

When μ is negative, it follows from equation (4.59) that k_x^i may be purely imaginary for a certain range of η_s. We write $k_x^i = j\beta\zeta_s$, where $\zeta_s = -(\eta_s^2 + \mu)/\mu$, and substitution in equation (4.60) followed by algebraic manipulation leads to the following dispersion relation:

$$(1 + \eta_s^2) - \kappa^2 - \mu^2\zeta_s + 2\mu\zeta_s^{1/2}(1 + \eta_s^2)^{1/2}\cot\beta\zeta_s^{1/2}d = 0 \quad (4.62)$$

A solution to this equation leads to bulk modes, since k_x^i is imaginary. When $\beta = 0$, i.e., when the phase propagation occurs along the biasing field, the dispersion relation reduces to*

$$2\cot\alpha_s k_z d = \alpha_s - 1/\alpha_s \quad (4.63)$$

* It is more convenient to substitute ζ_s and η_s in terms of k_z and β in equation (4.62) and then take the limit as $\beta \to 0$.

where

$$\alpha_s = (-1/\mu_s)^{1/2} \tag{4.64}$$

Figure 4.16 shows the bulk wave dispersion curves computed from equation (4.63) for a typical set of data. It is seen that the slope of the k_z vs. ω curve is negative. Thus k_z decreases as ω increases. Consequently, the propagation of phase and wave group occurs in opposite directions, and the allowed modes are thus backward waves. The range of backward modes is defined by the condition that μ is negative, which is so when $\omega_0 < \omega < [\omega_0(\omega_0 + \omega_m)]^{1/2}$. The general bulk wave dispersion curves (with arbitrary β) are shown in Figure 4.17 along with the surface modes. It is seen that as k_z decreases to zero, the frequency interval of the bulk modes also decreases to zero. On the other hand, as β approaches zero, the range of surface modes vanishes. The surface and bulk modes can coexist only when the direction of phase propagation makes an arbitrary angle ($\neq 0$ or $\pi/2$) with the direction of the biasing field.

When the direction of phase propagation with respect to the direction of the biasing field is arbitrary, the wavevector is, in general, noncollinear with the group velocity vector or the direction of energy flow. The angle between the wavevector and the group velocity vector is magnetically tunable. The associated "steering" of magnetostatic surface waves was investigated by Pizzarello et al. (1970), and they obtained the angle between the wavevector and group velocity as follows:

If $\mathbf{k} = k_y\hat{y} + k_z\hat{z} = \beta(\sec\theta\hat{y} + \operatorname{cosec}\theta\hat{z})$ represents the wavevector (see

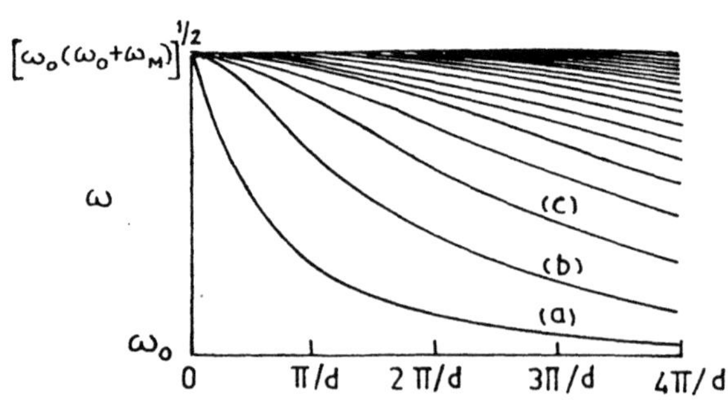

Figure 4.16. Dispersion curves for the case of magnetostatic wave propagation along the biasing field. The curves designated by (a), (b), (c), etc., correspond to $n = 0, 1, 2$, etc. The slope of ω vs. k_z curves is negative throughout the frequency range for all mode orders. Thus the modes are backward waves, for which the directions of energy flow and phase propagation are mutually opposite (after Damon and Eshbach, 1961).

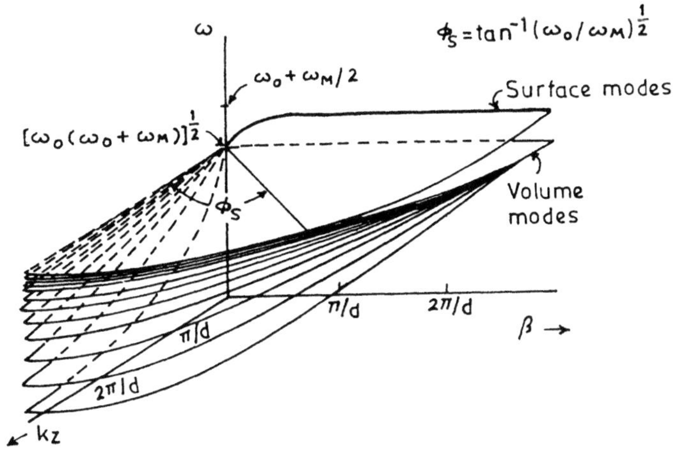

Figure 4.17. General magnetostatic mode spectrum for a ferrimagnetic slab magnetized parallel to its faces. The surfaces and bulk modes are simultaneously allowed only when the angle between the directions of propagation and biasing field exceeds $\varphi_s = \cot^{-1}(\omega_m/\omega_0)^{1/2} = \tan^{-1}(\omega_0/\omega_m)^{1/2}$ (after Damon and Eshbach, 1961).

Figure 4.19), the group velocity can be expressed as

$$\left.\begin{aligned} \mathbf{v}_g &= \nabla_{\mathbf{k}}\omega \\ &= \hat{k}\frac{\partial\omega}{\partial\beta} + \frac{1}{\beta}\frac{\partial\omega}{\partial\theta}(k_y\hat{z} - k_z\hat{y}) \end{aligned}\right\} \tag{4.65}$$

where $\mathbf{k} = \beta\hat{k}$. It follows from this equation that

$$\hat{k}\cdot\mathbf{v}_g = \frac{\partial\omega}{\partial\beta} = |\mathbf{v}_g|\cos\varphi_0 \tag{4.66}$$

and

$$\hat{k}\times\mathbf{v}_g = \frac{1}{\beta}\frac{\partial\omega}{\partial\theta} = |\mathbf{v}_g|\sin\varphi_0 \tag{4.67}$$

Thus

$$\tan\varphi_0 = \frac{\partial\omega/\partial\theta}{\beta\partial\omega/\partial\beta} \tag{4.68}$$

With equation (4.60), the steering angle φ_0 is obtained as

$$\tan\varphi_0 = (G_1 + G_2 + G_3 + G_4 + G_5 + G_6)/G_7 \tag{4.69}$$

where

$$\left.\begin{aligned} G_1 &= \sec^2\theta\tan\theta \\ S_0 &= [(\sec^2\theta + \mu - 1)/\mu]^{1/2} \\ G_2 &= \mu S_0^2 G_3 \cos\theta \\ G_3 &= \sec^2\theta\tan\theta\coth(\beta d S_0 \cos\theta)/S_0 \\ G_4 &= S_0^2\mu\beta d\tan\theta\,\mathrm{cosech}^2(\beta d S_0\cos\theta) \\ G_5 &= -G_4\sec^2\theta/S_0^2\mu \\ G_6 &= \mu G_1 \\ G_7 &= G_4\cot\theta \end{aligned}\right\} \qquad (4.70)$$

Some of the experimental results obtained by Pizzarello et al. (1970) regarding steering of magnetostatic surface waves will be discussed in the following section.

4.2.1.4. *Experimental Results and Applications*

It was seen in Section 4.1 that the measured group delay of magnetostatic surface waves on YIG substrate agrees with the theoretical one, provided that the demagnetizing field is taken properly into account. Since the growth of high-quality epitaxial YIG films of large aspect ratio is now possible, the measured time delays (Bongianni et al., 1969; Sethares, 1975) compare very well with those calculated from equation (4.47) without the need to include the demagnetizing field.

Adam et al. (1973, 1974) have suggested the application of magnetostatic surface waves in YIG film to obtain group delay equalization. The group delay equalizers have been used in long-haul millimeter waveguide systems to equalize the dispersion inherent in the system which, in turn, varies from channel to channel and hop to hop. The conventional group delay equalizers based on lumped and distributed propagation structures have delay characteristics that are fixed during manufacture and cannot be varied in situ. Adam et al. (1974) have demonstrated the applicability of magnetically tunable, magnetostatic surface wave equalizers which have the added advantage of compactness.

The application of magnetostatic surface waves to rotation rate sensing has been suggested by Newburgh et al. (1974). The conventional rotation rate sensors are based on the principle that the circulation times of two light rays traversing a closed optical path on a disc in opposite directions are unequal provided that the disc is rotating. The magnitude of this effect turns out to be inversely proportional to the product of phase velocity and wavelength. Since, for a given wavelength, the phase velocity of magnetostatic surface waves is orders of magnitude smaller than that of light, a rotation rate sensor based on magnetostatic surface waves should have much higher sensitivity than a conventional optical interferometer. The experimental arrangement to guide a clockwise and a counterclockwise wave is shown in Figure 4.18a. It consists of two YIG films epitaxially grown on the surface of a gadolinium–gallium–garnet cylinder. The dc magnetic fields in the two films are in opposite directions. Since a magnetostatic surface wave propagates in the $\mathbf{H}_0 \times \hat{n}$ direction, a clockwise propagating wave will circulate in one film whereas a counterclockwise propagating wave will circulate in the other film. The counter-circulating waves can be picked up by two coaxial cables separated from each other and from the input, as shown in Figure 4.18b. When the cylinder is stationary, the circulating times are equal and the null condition can be observed on an oscilloscope. However, when the cylinder rotates, the circulating times are rendered unequal and readjustment by a calibrated phase shifter is required to renull the signal; the phase shift is a measure of the rotation rate. Further details about fabrication and experimental arrangements can be found in Sethares (1975) and Newburgh et al. (1974).

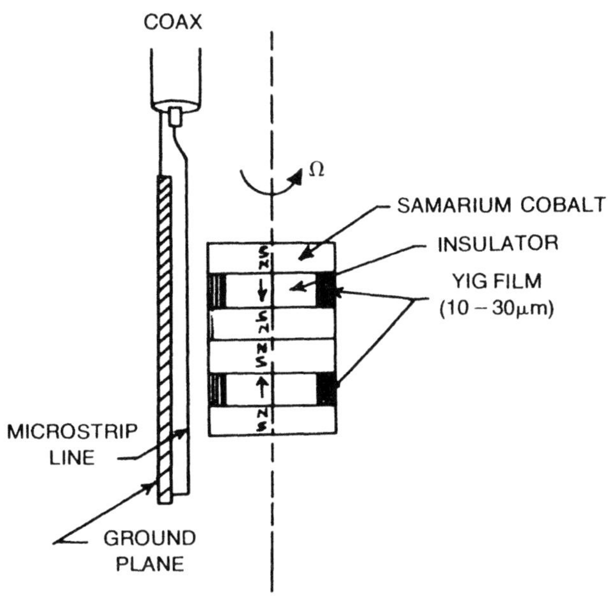

Figure 4.18a. Schematic diagram for sandwich of two YIG films, one carrying a clockwise propagating wave and the other carrying a counterclockwise propagating wave (after Newburgh et al., 1974).

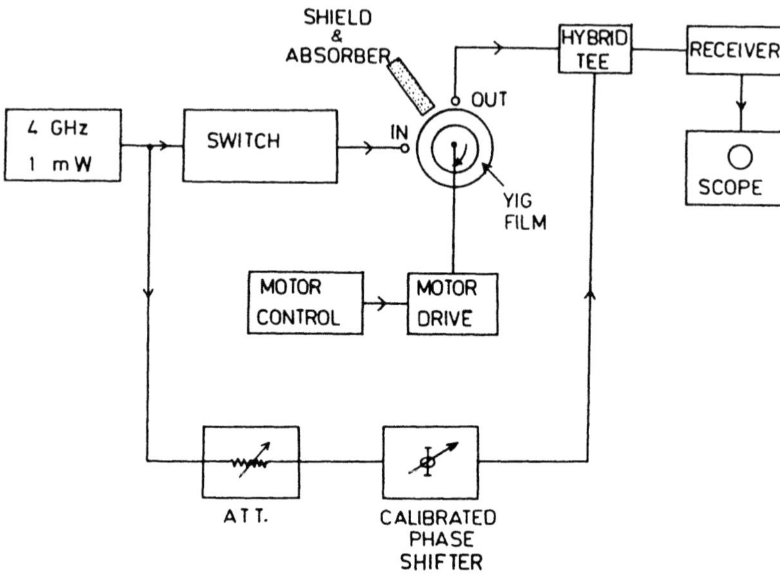

Figure 4.18b. Block diagram for magnetostatic wave ring interferometer for rotation rate sensing (after Newburgh et al., 1974).

An exact evaluation of the performance of a magnetostatic wave ring interferometer requires the analysis of propagation in curved moving films. It was stated earlier that the effect of curvature would not be significant in a practical situation since the film thickness would be much smaller than the radius of curvature of the cylinder. No theoretical analysis of magnetostatic propagation in moving films is presently available.

Some experimental work on magnetic steering of magnetostatic surface waves in YIG films grown by chemical vapor deposition (CVD) has been reported by Pizzarello et al. (1970). The electromagnetic energy was converted into magnetostatic surface wave energy through a periodic finger coupler (Figure 4.19). A magnetic probe scanned parallel to the direction of the fingers. The steering angle φ_0 corresponds to the state in which maximum output is obtained in the detector. Two separate experiments were carried out: (i) a probe scanning at $\theta = 5°$ and $10°$ and (ii) a probe using four parallel fingers, φ_0 is fixed at $32°$, whereas θ is experimentally determined to be $10°$ for $\beta = 350\,\text{cm}^{-1}$. The experimental results as well as the corresponding theoretical curves [obtained from equations (4.69) and (4.70)] are shown in Figure 4.20. Good agreement has been obtained between theory and experiment.

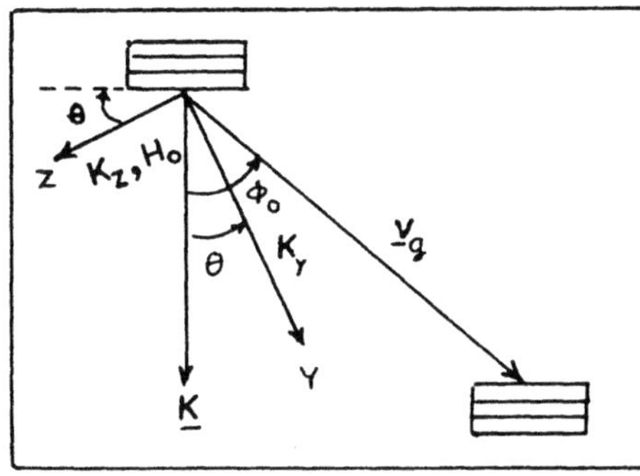

Figure 4.19. Relative orientations of the internal dc field, the direction of propagation, and the group velocity; the deflection angle φ_0 between the direction of propagation and group velocity depends on the magnitude of the biasing field as well as its orientation $\pi/2 - \theta$ with respect to the direction of propagation (after Pizzarello et al., 1970).

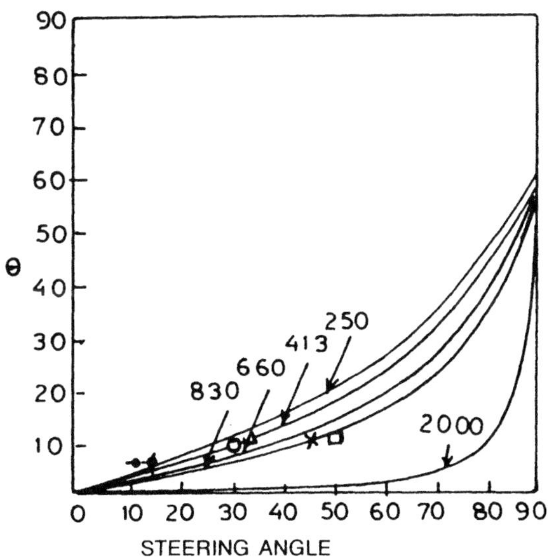

Figure 4.20. Comparison of experimental observations with theoretical curves of steering angle φ_0 as a function of θ at specific values of wavenumber β. The wave frequency is 3 GHz and the film thickness is 10 μm. The symbols △, ⊙, ×, □, –·–, and ⫶ represent experimental points corresponding to β = 250, 350, 660, 830, 413, and 660 cm^{-1} (after Pizzarello et al., 1970).

We now discuss some experimental results pertaining to magnetostatic wave resonances for finite ferrimagnetic slabs. The spectral resonances of a transversely magnetized slab were first observed by Brundle and Freedman (1968b) when they varied the biasing field at a fixed frequency of operation. These modes were appropriately identified by Sparks (1969). Since the extent of a real slab (Figure 4.21) is finite in the plane perpendicular to the thickness, we get a good approximation if we regard the magnetization at the edge surfaces as being pinned. Consequently, the variation of rf magnetization with y and z can be taken as $\sin k_y y \sin k_z z$. The pinning

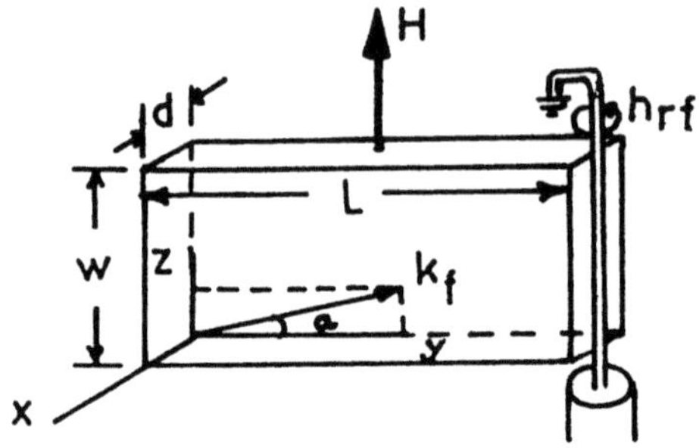

Figure 4.21. Sample geometry and coordinate system (after Sparks, 1969).

condition requires that

$$\left.\begin{aligned} k_y &= n_L \pi / L \\ k_z &= n_w \pi / w \end{aligned}\right\} \qquad (4.71)$$

where n_L and n_w are positive integers. The substitution of these values of k_y and k_z in equations (4.60) and (4.62) leads to the resonance frequencies for surface and bulk modes, respectively. Such modes, observed by Brundle and Freedman (1968b), Sparks (1969), and Adam et al. (1970), were labeled (n_L, n_w) modes by Sparks (1969). However, a comparison between theory and experiment was difficult owing to the nonuniformity of the internal dc magnetic field inherent in slab geometry (see Section 1.8). This difficulty was resolved by Adam et al. (1970). Their sample configuration is shown in Figure 4.22, in which the principal YIG sample is surrounded by metal-coated, dummy samples of a polycrystalline ferrite in

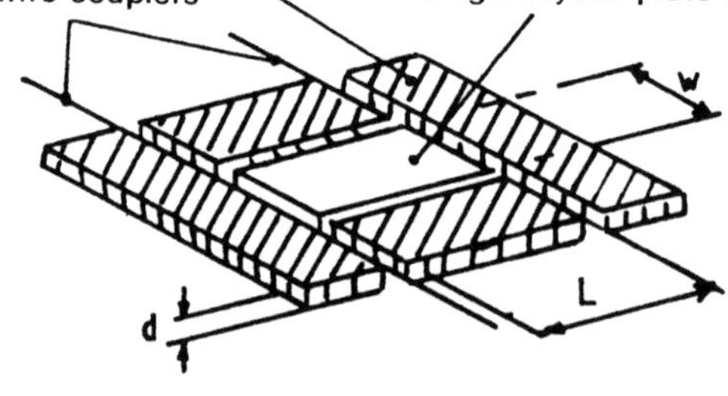

Figure 4.22. Composite YIG structure of Adam et al. (1970) used to study the magnetostatic wave resonances. The dimensions are $L = 1.0\,\text{cm}$, $w = 0.25\,\text{cm}$, and $d = 0.05\,\text{cm}$ (after Adam et al., 1970).

order to minimize the effect of nonuniform demagnetization. Figure 4.23 shows surface and volume mode resonances as functions of the biasing field, when the fine-wire couplers are in the plane of the slab. In this configuration, the driving field produced by the couplers is even about the central axis of the slab and, therefore, it couples only to the modes with odd n_w (i.e., $n_w = 1, 3, 5, \ldots$). The comparison of theoretical and experimental values of surface and bulk wave resonance frequencies is shown in Figure 4.24. The cut-off field (H_{bot}) between surface and bulk modes, which can be inferred from equation (4.61), is represented in the figure by a broken line. The $(n_L, 1)$ series of resonances ($H_0 < H_{bot}$) represents surface wave resonances. On the other hand, the series $(1, n_w)$ is obtained for $H_0 > H_{bot}$ and, therefore, represents volume wave resonances. In the dc field region between the mode (1, 1) and the field H_{bot}, there are a number of volume wave resonances in the $(n_L, 1)$ series, provided that k_x exceeds

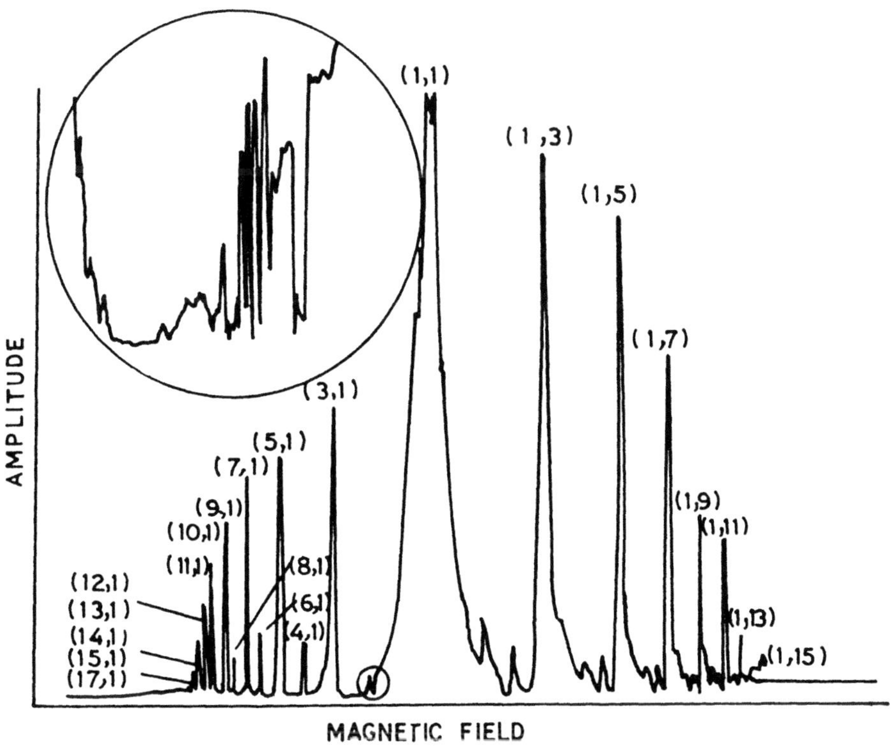

Figure 4.23. Experimentally observed surface and volume mode resonances corresponding to the configuration shown in Figure 4.22 (after Adam et al., 1970).

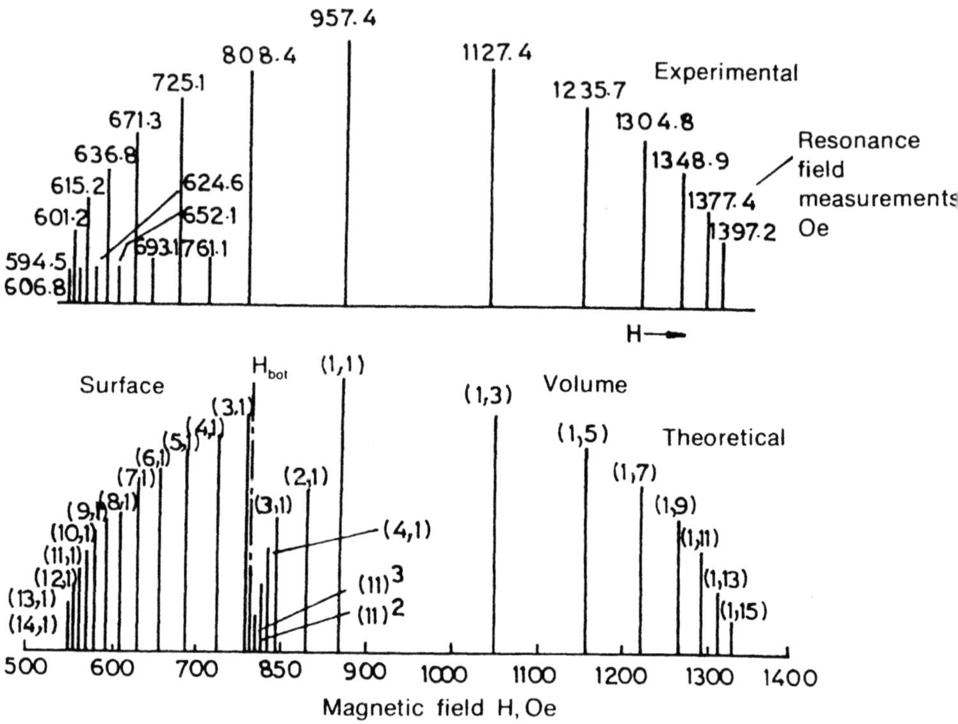

Figure 4.24. Comparison of experimental and theoretical field positions (after Adam et al., 1970).

π/d. When k_x lies between $(n-1)\pi/d$ and $n\pi/d$ such resonant modes are labeled $(n_L, n_w)^n$. It is seen from the figure that the general agreement between theory and experiment is good; the small difference is due to the combined effects of the anisotropy field and the residual demagnetizing field.

When the fine-wire couplers are normal to the plane of the slab (Sparks, 1969; Adam et al., 1970), the driving field is odd about the central axis of the slab. In this case, the modes with even n_w (i.e., $n_w = 2, 4, 6, \ldots$) do couple to the driving field. These modes are also volume modes.

4.2.2. *Normal Magnetization*

When a thin ferrimagnetic slab (or film) is subjected to a dc field directed perpendicular to its face, the internal field is smaller than the external field by an amount equal to the saturation magnetization (see Section 1.8). Thus a much larger external field may be required to obtain a given internal dc field. Nevertheless, the propagation of magnetostatic waves in such a configuration is of interest because it is possible to obtain isotropic propagation in the plane of the slab.

The results pertaining to the propagation of magnetostatic waves in a normally magnetized ferrimagnetic slab can be inferred from the works of Damon and van de Vaart (1965) and Gupta and Srivastava (1980b). The relatively recent theoretical and experimental investigations of guided wave propagation in normally magnetized films involve more complicated layered structures, which will be described later in this chapter. In what follows, we have carried out the magnetostatic field analysis (Section 4.2.2.1) to obtain the dispersion relation and group delay for magnetostatic waves in a normally magnetized slab. Energy distribution and power flow have been considered in Section 4.2.2.2, while Section 4.2.2.3 gives an approximate, zig-zag ray model that leads to increased physical insight regarding the path of rays in a bulk wave delay line.

4.2.2.1. *Magnetostatic Field Analysis*

The configuration to be analyzed is shown in Figure 4.25. The slab occupies the region $0 \leqslant z \leqslant d$ and is magnetized along the z-axis. The wave propagates along the y-axis and is assumed to have a uniform field distribution in the transverse direction (x-axis). The magnetostatic potential ψ and magnetic induction $\mathbf{b}$ in different regions can be obtained from equations (4.1) and (4.2) in the usual manner; thus

$$\left.\begin{aligned} \psi^{(1)} &= \exp(-\beta z)\exp j(\omega t - \beta y) \\ \psi^{f} &= [B\exp(j\alpha_v\beta z) + C\exp(-j\alpha_v\beta z)]\exp j(\omega t - \beta y) \\ \psi^{(2)} &= D\exp(\beta z)\exp j(\omega t - \beta y) \end{aligned}\right\} \tag{4.72}$$

$$\left.\begin{aligned} \mathbf{b}^{(1)} &= -\beta(j\hat{y} + \hat{z})\psi^{(1)} \\ \mathbf{b}^{f} &= \{\beta(\kappa\hat{x} - j\mu\hat{y}) + j\alpha_v\beta[B\exp(j\alpha_v\beta z) \\ &\quad + C\exp(-j\alpha_v\beta z)]\hat{z}\}\exp j(\omega t - \beta y) \\ \mathbf{b}^{(2)} &= -\beta(j\hat{y} - \hat{z})\psi^{(2)} \end{aligned}\right\} \tag{4.73}$$

where

$$\alpha_v^2 = -\mu \tag{4.74}$$

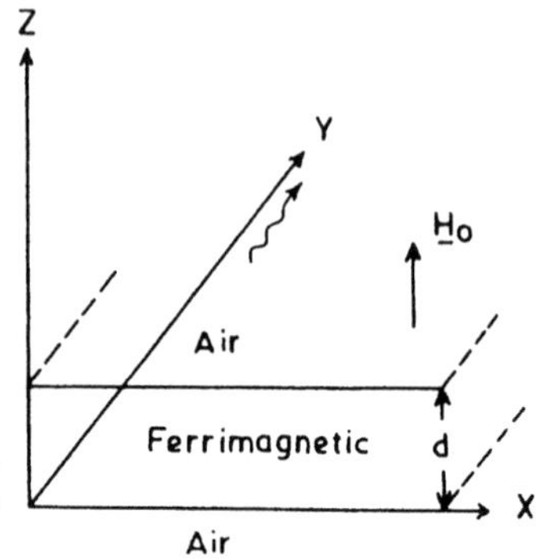

Figure 4.25. Normally magnetized ferrimagnetic slab. The wave propagates along the y-axis and has a uniform field distribution along the x-axis.

The application of the appropriate boundary conditions at $z = 0$ and $z = d$ leads to the following dispersion relation:

$$(\alpha_v^2 - 1)\tan \alpha_v \beta d = 2\alpha_v \tag{4.75a}$$

Equivalently,

$$\tan \alpha_v \beta d/2 = 1/\alpha_v \tag{4.75b}$$

from which the wavenumber is explicitly obtained as

$$\beta = (2/\alpha_v d)[n\pi + \arctan(1/\alpha_v)] \qquad n = 0, 1, 2, \ldots \tag{4.76}$$

The various unknown constants appearing in equation (4.72) are obtained as

$$\left.\begin{aligned} B &= \frac{j\alpha_v - 1}{2j\alpha_v}\exp[-(j\alpha_v + 1)\beta d] \\ C/B &= (j\alpha_v - 1)/(j\alpha_v + 1) \\ D/B &= 2j\alpha_v/(j\alpha_v + 1) \end{aligned}\right\} \tag{4.77}$$

When loss is neglected, μ is real, and it then follows from equation (4.74) that α_v is either purely real or purely imaginary. When μ is positive, α_v is imaginary and equation (4.75) does not allow any solution. As such, surface modes cannot be supported by a normally magnetized slab. However, when μ is negative, α_v is real and equation (4.75) necessarily yields a solution. Thus, the frequency range of allowed modes is given by $\mu < 0$, which occurs when $\omega_0 < \omega < [\omega_0(\omega_0 + \omega_m)]^{1/2}$. It can be inferred from the second of equations (4.72) that the guided waves are volume modes, which are reciprocal since the dispersion relation is independent of κ. Moreover, the rotational symmetry (about the z-axis) inherent in the configuration implies that the propagation characteristics are independent of the rotation of direction of propagation in the xy-plane; the propagation is isotropic in the plane of the slab. Equation (4.77) shows that $B, C,$ and D are all independent of κ, which implies that the field distribution is the same for $\pm y$-propagating waves. If we compare the operational bandwidth of bulk modes with surface modes discussed in Section 4.2.1, it is seen that if H_0 is taken as 3.25 kOe, the bulk modes are allowed in the range from 9.1 to 11.2 GHz. Comparison with the results presented in Section 4.2.1.1 shows that in the X-band the operational bandwidth of bulk modes is much larger than that of surface modes.

Figure 4.26 shows the dispersion curves as obtained from equation (4.75) for a typical set of data. It is seen that the wavenumber increases monotonically with frequency throughout the frequency range of allowed modes. Hence, the allowed modes are forward waves. As clearly shown in the figure, the successive higher-order modes which are characterized by $n = 1, 2, \ldots$ have larger wavenumbers at any frequency.

The group delay time is obtained, by differentiating equation (4.69), as

$$\tau_d = \frac{\beta\omega\omega_0\omega_m}{\alpha_v^2(\omega^2 - \omega_0^2)^2}\left(1 + \frac{2}{\alpha_v^2\beta d}\cos^2\frac{\alpha_v\beta d}{2}\right) \tag{4.78}$$

The frequency dependence of the group delay is shown in Figure 4.27 for a typical set of data. For the first-order mode, the time delay is finite at the lower cut-off frequency and it monotonically increases to infinity. However, for all higher-order modes, the delay time approaches infinity at both the frequency limits.

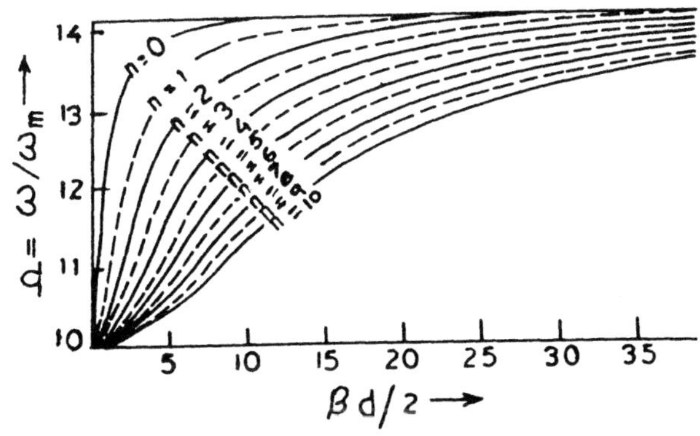

Figure 4.26. Magnetostatic mode spectrum for a normally magnetized slab. The phase velocity is smaller for the higher-order modes (after Damon and van de Vaart, 1965).

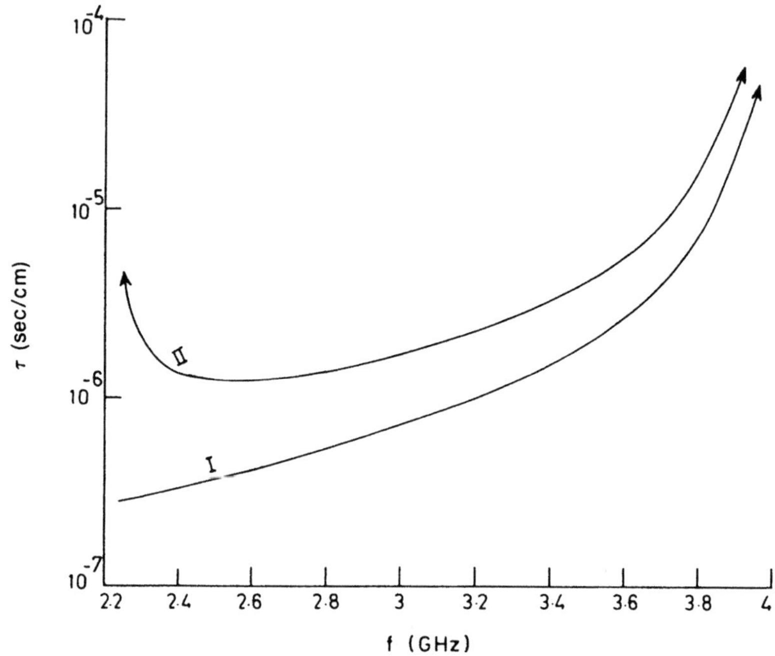

Figure 4.27. Variation of delay time with frequency for the first two modes in a normally magnetized film. The delay for the first-order mode is finite at the lower cut-off frequency and monotonically increases to infinity at the upper frequency limit. However, the delays for all higher-order modes approach infinity at both the frequency limits.

4.2.2.2. *Energy Distribution and Power Flow*

It is worth studying energy distribution and power flow in a delay line in order to form a better physical picture of guided wave phenomena. Since a biased ferrimagnetic medium is strongly dispersive, we must start from the generalized Poynting theorem for dispersive media. The theorem itself has been given in Appendix F, where it has also been shown that, under the magnetostatic approximation, the time-averaged power flow per unit area is given by

$$\mathbf{P} = \mathrm{Re}[-(j\omega/8\pi)\psi^*\mathbf{b}] \tag{4.79}$$

Using equations (4.72) and (4.73), we have, in dielectric regions,

$$\mathbf{P}_1 = \mathbf{P}_2 = -(\omega\beta/8\pi)\exp(-2\beta z)\hat{y} \tag{4.80}$$

Integrating this in the transverse direction, the unnormalized* power flow per unit width in dielectric regions is obtained as

$$\begin{aligned}\mathbf{P}_1^t = \mathbf{P}_2^t &= \int_d^{\infty} \mathbf{P}_1\,dz = \int_{-\infty}^{0} \mathbf{P}_2\,dz \\ &= -(\omega/16\pi)\exp(-2\beta d)\hat{y}\end{aligned} \tag{4.81}$$

On the other hand, in the ferrimagnetic region, we have (neglecting losses)

$$\begin{aligned}\mathbf{P}_f^t &= \int_0^d \mathbf{P}_f\,dz \\ &= (\omega/8\pi)\exp(-2\beta d)[1+((1+\alpha_v^2)/2)\beta d]\hat{y}\end{aligned} \tag{4.82}$$

It is not difficult to show from equation (4.82) that, in the frequency range of the allowed modes, the power flow in the ferrimagnetic region occurs in the $+y$-direction. However, it is evident from equation (4.81) that, in the dielectric region, the power flow occurs in the $-y$-direction, i.e., opposite to the direction of phase propagation. The situation is represented diagrammatically in Figure 4.28. The net power flow per unit width is

$$\begin{aligned}\mathbf{P}^t &= \mathbf{P}_1^t + \mathbf{P}_2^t + \mathbf{P}_f^T \\ &= (\omega/8\pi)\exp(-2\beta d)[(1+\alpha_v^2)/2]\beta d\hat{y}\end{aligned} \tag{4.83}$$

which is seen to be positive throughout the frequency range of guided modes. Thus, even though the net power flow occurs in the direction of phase propagation, the power flow in the dielectric regions occurs in the opposite direction; the importance of this result will be discussed later.

Now we consider the energy distribution in the transverse direction. As described in Appendix F, the energy density of the magnetostatic field in a lossless medium can be expressed as

$$U = \frac{1}{16\pi}\mathrm{Re}\left\{\mathbf{h}^*\left[\frac{\partial}{\partial\omega}(\omega\boldsymbol{\mu})\cdot\mathbf{h}\right]\right\} \tag{4.84}$$

With equations (4.72) and (4.84), the energy density in dielectric (U_1 and

* In equation (4.72), the amplitude of $\psi^{(1)}$ has been arbitrarily chosen as unity. In order to obtain a meaningful quantity, $\mathbf{P}_1^t$, $\mathbf{P}_2^t$, and $\mathbf{P}_f^t$ should be divided by the resultant power flowing down the slab.

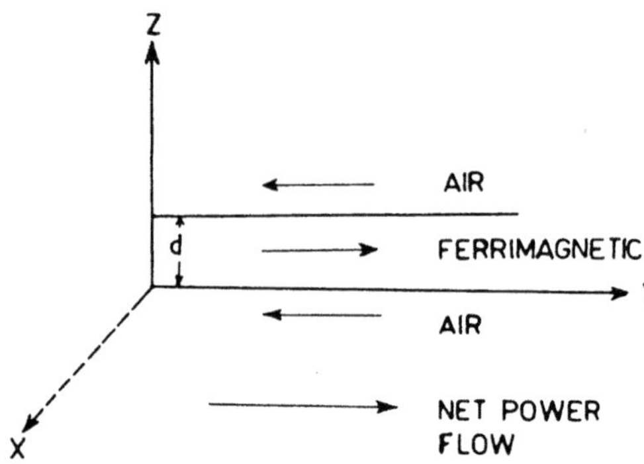

Figure 4.28. The direction of power flow in different regions, relative to the net power flow.

U_2) and ferrimagnetic (U_f) regions is obtained as

$$U_f = \frac{\beta^2}{16\pi}\left\{\frac{2\omega^2\omega_0\omega_m}{(\omega_0^2-\omega^2)^2}[|B|^2+|C|^2] + [BC^*\exp(2\alpha_v j\beta z) + B^*C\exp(-2j\alpha_v\beta z)]\left[2\mu + \frac{2\omega^2\omega_0\omega_m}{(\omega_0^2-\omega^2)^2}\right]\right\} \tag{4.85}$$

and

$$U_1 = U_2 = (\beta^2/8\pi)\exp(-2\beta d) \tag{4.86}$$

Integration over the transverse cross-section, followed by algebraic simplification, yields

$$U_f^t = \int_0^d U_f\,dz = \frac{\beta}{8\pi}\exp(-2\beta d)\frac{\omega^2\omega^0\omega_m}{\alpha_v^2(\omega_0^2-\omega^2)^2}\left[1 + \frac{\mu(\omega_0^2-\omega^2)^2}{\omega^2\omega_0\omega_m} + (1+\alpha_v^2)\beta d/2\right] \tag{4.87}$$

whereas, in the dielectric region, we obtain

$$U_1^t = U_2^t = \int_{-\infty}^0 U_2\,dz = \int_d^\infty U_1\,dz = (\beta/16\pi)\exp(-2\beta d) \tag{4.88}$$

Defining $U^t = U_1^t + U_2^t + U_f^t$ and using equations (4.87) and (4.88), we have

$$U^t = \frac{\beta}{8\pi}\exp(-2\beta d)\frac{\omega^2\omega_0\omega_m}{\alpha_v^2(\omega_0^2-\omega^2)^2}[1 + (1+\alpha_v^2)\beta d/2] \tag{4.89}$$

The average velocity of energy down the slab is given by

$$v_e = P^t/U^t \tag{4.90}$$

Using equations (4.83)–(4.85), we obtain the time duration for net energy flow through a unit distance as

$$\begin{aligned}\tau_e &= 1/v_e \\ &= \frac{\beta}{\omega[(1+\alpha_v^2)\beta d/2]} \\ &\quad \times \left\{1 + \frac{\omega^2\omega_0\omega_m}{\alpha_v^2(\omega_0^2-\omega^2)^2}\left[1 + (1+\alpha_v^2)\frac{\beta d}{2} + \frac{(\omega_0^2+\omega_0\omega_m-\omega^2)(\omega_0^2-\omega^2)}{\omega^2\omega_0\omega_m}\right]\right\}\end{aligned} \tag{4.91}$$

This expression appears to be quite complicated, but it can be simplified using the dispersion relation (Gupta and Srivastava, 1980b). It is seen that τ_e is equivalent to the group delay time as given by equation (4.78). Since the entire preceding analysis has been based on magnetostatic approximation, the equality of v_g and v_e in the present case is a manifestation of self-consistency and general validity of the magnetostatic approximation for guided waves. Moreover, this also suggests the possibility of analysis of magnetostatic bulk waves by ray optical techniques, as discussed below.

4.2.2.3. *Zig-Zag Ray Model*

In recent years, modes in dielectric optical waveguides have been analyzed by zig-zag ray models.* This description views the wave guidance as being a consequence of multiple total internal reflections of the "rays" by the two surfaces of the guiding structure. Such a model leads to increased physical insight into the nature of guided wave phenomena. Moreover, ray optic methods can easily be generalized to the cases of tapered sections and inhomogeneous and curved guides. An attempt to generalize the zig-zag ray model to guided magnetic waves presents several difficulties.

First, unlike the situation in dielectrics, the modes propagating in a ferrimagnetic medium in an arbitrary direction relative to the biasing field are characterized by complicated dispersion and polarization characteristics. Second, the strongly dispersive and dissipative nature of the magnetic media makes, in general, all the velocities (i.e., phase velocity, group velocity, energy velocity, and signal velocity) different from one another.

* See, for example, Kogelnik and Weber (1974).

Third, since the phase shift, on account of total reflection at a dielectric–ferrimagnetic interface, is a sensitive function of frequency and wavenumber (see Chapter 3), it is difficult to get reasonably correct expressions for lateral shift and the time delay associated with a single reflection,* because it is no longer possible to neglect (Brekhovskikh, 1960) the second- and higher-order derivatives of the phase shift.

In the present chapter, the first of these difficulties has been overcome by resorting to magnetostatic analysis. As to the second difficulty, it has already been shown in an earlier section that $v_e = v_g$ when losses are neglected. Therefore, knowledge of the path of the rays should lead to v_g and, hence, to the group delay. Unfortunately, no general solution to the last difficulty is presently available. We have obtained the lateral shift using an approximate method due to Renard (1964), which leads only to an approximate expression for group delay. The following analysis is based on the ad-hoc assumption that guided magnetostatic waves can be treated using a zig-zag ray model.

Consider the propagation of magnetostatic plane waves in the case when the field variation along one direction (say the x-axis) can be ignored. If $\mathbf{k}^f$ and $\mathbf{k}^d$ represent the wavevectors for the magnetostatic field in ferrimagnetic and dielectric regions, respectively, it can be shown from equation (4.1) that the components of $\mathbf{k}^f$ and $\mathbf{k}^d$ satisfy the following conditions:

$$\left.\begin{aligned} (k_y^d)^2 + (k_z^d)^2 &= 0 \\ \mu(k_y^f)^2 + (k_z^f)^2 &= 0 \end{aligned}\right\} \tag{4.92}$$

It follows from equation (4.92) that in the dielectric region the magnetostatic field can only be evanescent; writing $k_z^d = j|k_z^d|$, we have

$$|k_z^d| = k_y^d \tag{4.93}$$

When μ is positive, it follows from the second of equations (4.92) that the magnetostatic field in the ferrimagnetic region is also evanescent. However, when μ is negative, i.e., when $\omega_0 < \omega < [\omega_0(\omega_0 + \omega_m)]^{1/2}$, the field is a propagating plane wave for which

$$\left.\begin{aligned} k_z^f &= \alpha_v k_y^f \\ \alpha_v &= (-\mu)^{1/2} \end{aligned}\right\} \tag{4.94}$$

* The lateral shift and associated time delay are required to obtain the desired expression for the group velocity of magnetostatic waves propagating down the guide.

If we refer to Figure 4.29, in which a plane magnetostatic wave is incident at the ferrimagnetic–dielectric interface, the former being magnetized perpendicular to the interface, we see that the angle of incidence is given by

$$\tan\theta = k_y^f/k_z^f \tag{4.95}$$

The magnetostatic potential for incident, reflected, and transmitted waves is given by

$$\left.\begin{aligned} \psi^i &= \exp j(\omega t - k_y^f y - k_z^f z) \\ \psi^r &= R\exp j(\omega t - k_y^f y + k_z^f z) \\ \psi^t &= T\exp(-|k_z^d|z)\exp j(\omega t - k_y^d y) \end{aligned}\right\} \tag{4.96}$$

The magnetic induction is obtained as

$$\left.\begin{aligned} \mathbf{b}^i &= (\kappa k_y^f\hat{x} - j\mu k_y^f\hat{y} - jk_z^f\hat{z})\psi^i \\ \mathbf{b}^r &= (\kappa k_y^f\hat{x} - j\mu k_y^f\hat{y} + jk_z^f\hat{z})\psi^r \\ \mathbf{b}^t &= (-jk_y^d\hat{y} - |k_z^d|\hat{z})\psi^t \end{aligned}\right\} \tag{4.97}$$

Phase-matching at $z = 0$ yields

$$k_y^f = k_y^d (\equiv \beta) \tag{4.98}$$

The application of appropriate boundary conditions at $z = 0$ leads to

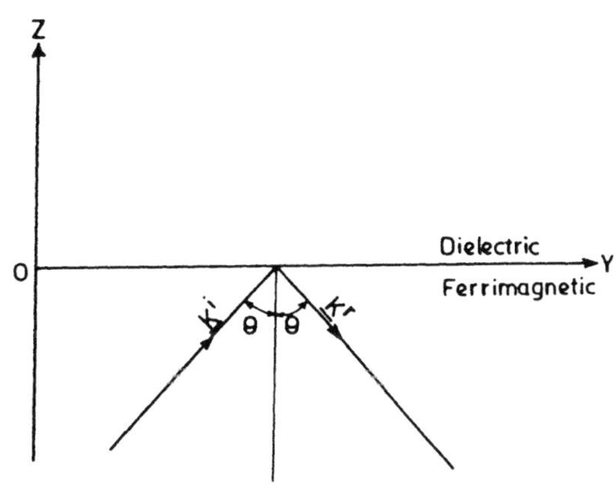

Figure 4.29. Reflection of a magnetostatic plane wave at a ferrimagnetic dielectric interface.

$$\left.\begin{aligned} 1 + R &= T \\ -jk_z^f + jk_z^f R &= -|k_z^d| T \end{aligned}\right\} \tag{4.99}$$

from which the transmission and reflection coefficients are obtained as

$$\left.\begin{aligned} |T| &= 2\alpha_v/(1+\alpha_v^2)^{1/2} \\ R &= (k_z^f + jk_y^f)/(k_z^f - jk_y^f) \equiv \exp(j\varphi_v) \end{aligned}\right\} \tag{4.100}$$

from which it follows that

$$\tan \varphi_v/2 = k_y^f/k_z^f \tag{4.101}$$

Thus, with equations (4.94), (4.95), and (4.101) the phase shift is obtained as

$$\left.\begin{aligned} \varphi_v &= 2 \arctan(k_y^f/k_z^f) \\ &= 2 \arctan(1/\alpha_v) \\ &= 2(m\pi + \theta) \qquad m = 0, 1, 2, \ldots \end{aligned}\right\} \tag{4.102}$$

The dispersion relation is now obtained by generalizing the well-known procedure (Tien, 1971, 1977). From Figure 4.30 it is required that the phases at points A and C should be the same [except for an integral multiple of 2π which has been accounted for in equation (4.102)]. Thus, we have

$$2k_z^f d = 2\varphi_v \tag{4.103}$$

With equations (4.94), (4.98), and (4.102), this leads to

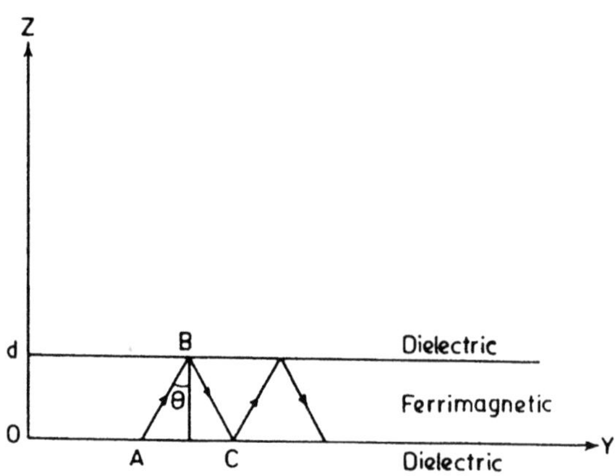

Figure 4.30. Phase propagation in a normally biased YIG film. The phase difference between points A and C should be an integral multiple of 2π (after Gupta and Srivastava, 1980b).

$$\alpha_v \beta d/2 = \arctan(1/\alpha_v) + m\pi$$

which transforms into

$$\tan(\alpha_v \beta d/2) = 1/\alpha_v \tag{4.104}$$

which is equivalent to equation (4.75).

In order to determine the path of the rays, refer again to Figure 4.30. Using equations (4.79), (4.84), (4.96), and (4.97), we obtain the time-averaged power and energy densities in the dielectric region as

$$\left.\begin{aligned} \mathbf{P}^d &= -(\omega/8\pi)k_y^d \exp(-2|k_z^d|z)|T|^2\hat{y} \\ U^d &= (1/8\pi)(k_y^d)^2|T|^2 \exp(-2|k_z^d|z) \end{aligned}\right\} \tag{4.105}$$

It is noted that in the dielectric region the power flow occurs in the negative y-direction, and this can be compared with the results of the last section (see Figure 4.29). The velocity of energy flow is obtained as

$$|v_e^d| = |\mathbf{P}^d|/U^d = \omega/\beta \tag{4.106}$$

which is, incidently, the same as the velocity of backward energy flow in the dielectric region of the normally biased delay line [which can be inferred from equations (4.81) and (4.88)].

In the ferrimagnetic region, the power and energy densities for the incident wave are obtained as

$$\left.\begin{aligned} \mathbf{P}^i &= (\omega\beta/8\pi)(-\mu\hat{y} - \alpha_v\hat{z}) \\ U^i &= (\beta^2/8\pi)\omega^2\omega_0\omega_m/(\omega_0^2 - \omega^2)^2 \end{aligned}\right\} \tag{4.107}$$

As for the reflected wave, it can be shown that the energy density is the same as that given by the second of equations (4.107), whereas the power flow per unit area is obtained as

$$\mathbf{P}^r = (\omega\beta/8\pi)(-\mu\hat{y} + \alpha_v\hat{z}) \tag{4.108}$$

It is easy to show that $\mathbf{k}^i \cdot \mathbf{P}^i = 0 = \mathbf{k}^r \cdot \mathbf{p}^r$, which means that the power flow for the incident and reflected plane waves occurs perpendicular to their respective wavevectors; the resulting situation is shown diagrammatically in Figure 4.31. In accordance with the discussion presented in Chapter 3, the present lateral shift on account of total reflection is clearly negative. The consequent path of rays* in total reflection is as shown in Figure 4.32,

* A ray is regarded as representing the direction of energy flow.

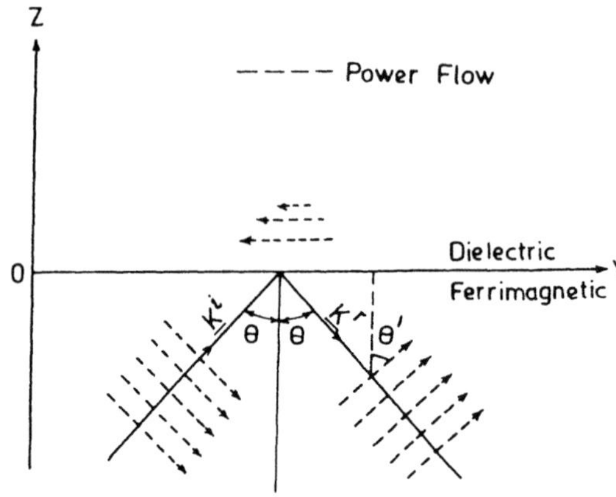

Figure 4.31. The direction of power flow in different regions in total reflection of a magnetostatic plane wave.

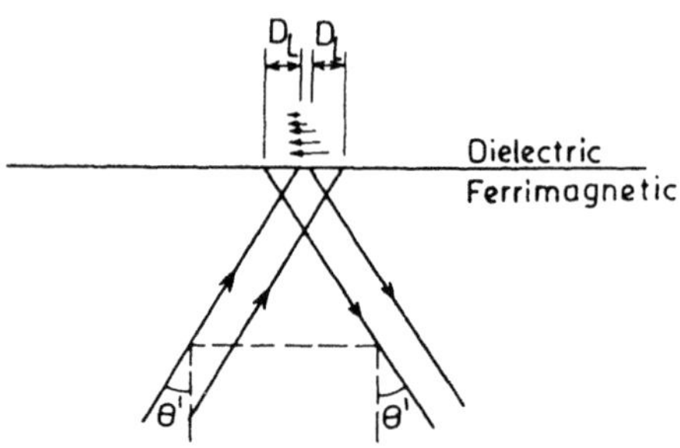

Figure 4.32. Lateral shift of a magnetostatic beam in total reflection at ferrimagnetic–dielectric interface (after Gupta and Srivastava, 1980b).

and the path of rays in a ferrimagnetic slab delay line is as shown in Figure 4.33.

In order to obtain an expression for group delay from the ray model, it is seen that the velocity of energy flow down the slab can be expressed as

$$v_e = (x \cos\theta - D_L)/(\tau_f + \tau_d)$$

where x and D_L are shown in Figure 4.33; τ_f and τ_d are time intervals associated with ray propagation from A to B and B to C, respectively. As noted earlier, it is difficult to derive an exact expression for D_L since φ_v is a sensitive function of ω and β. Therefore, in order to obtain an approximate result we apply Renard's procedure, which was discussed in Chapter 3. Thus

$$|\mathbf{P}^i| D_L \sin\theta = \int_0^\infty \mathrm{P}_y^r dz \tag{4.110}$$

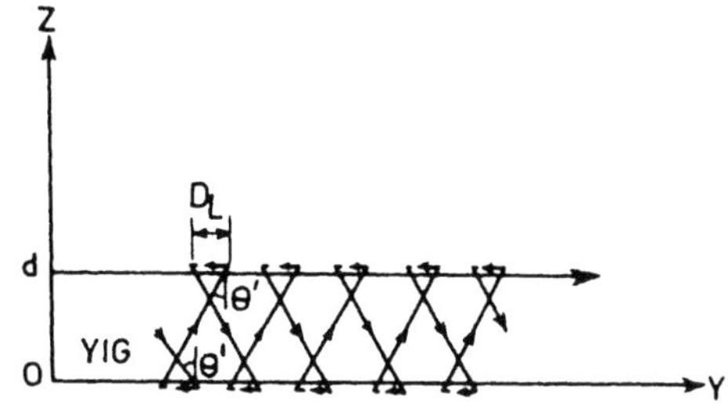

Figure 4.33. Path of rays in a YIG film delay line (after Gupta and Srivastava, 1980b).

Using equations (4.95), (4.105), and (4.107), we have

$$D_L = \frac{2}{\beta}\frac{(-\mu)^{1/2}}{(1-\mu)^{1/2}} \tag{4.111}$$

Since the speed of ray propagation in the dielectric region is ω/β, the time interval τ_d is obtained as

$$\tau_d = \frac{D_L}{\omega/\beta} = \frac{2}{\omega}\frac{(-\mu)^{1/2}}{(1-\mu)^{1/2}} \tag{4.112}$$

The speed of energy flow in the ferrimagnetic region (along the ray path) is obtained from equations (4.107) as

$$v_f = \frac{\sqrt{-\mu(1-\mu)}}{\beta\omega\omega_0\omega_m}(\omega_0^2-\omega^2)^2 \tag{4.113}$$

Hence τ_f is obtained as

$$\tau_f = \frac{x}{v_f} = \frac{d\beta}{\alpha_v}\frac{\omega\omega_0\omega_m}{(\omega_0^2-\omega^2)^2} \tag{4.114}$$

Substitution for x, D_L, τ_f, and τ_d in equation (1.109) leads to the group delay as

$$\tau_e = \frac{1}{v_e} = \frac{\omega\omega_0\omega_m}{\alpha_v^2(\omega_0^2-\omega^2)^2}\left\{\frac{\beta+[2\alpha_v^2(\omega_0^2-\omega^2)^2/\omega^2\omega_0\omega_m d][1/(1-\mu)]}{1-2/\beta d(1-\mu)}\right\} \tag{4.115}$$

In the limit when $\omega \to [\omega_0(\omega_0+\omega_m)]^{1/2}$, this τ_e approaches the exact expression for group delay, i.e., equation (4.78). The numerical results obtained from equation (4.115) have been presented by Gupta and Srivastava (1980b). There is good agreement between equations (4.115) and (4.78) throughout the frequency range of guided waves for the second- and higher-order modes. However, for $n = 1$, the two agree only near resonance. Further investigations are required to improve this model.

4.3. Metal-Backed Ferrimagnetic Slab

The study of magnetic wave propagation in a metal-backed ferrimagnetic slab is important because, under suitable biasing conditions, this configuration leads to nonreciprocal dispersion and delay characteristics. As regards the analysis, the presence of the metal can be accounted for by requiring the normal component of magnetic induction to vanish at the metal surface (assuming the conductivity to be infinite). In what follows, we have considered two specific configurations: (i) dc magnetization in the plane of the slab (Section 4.3.1), and (ii) dc magnetization perpendicular to the plane of the slab (Section 4.3.2).

4.3.1. Magnetization in Slab Plane

The configuration to be analyzed is shown in Figure 4.34. The slab occupies the region $0 \leq x \leq d$ and extends to infinity in the yz-plane. The region $x \leq 0$ is occupied by a metal. The biasing field is directed along the z-axis. First, in Section 4.3.1.1, we analyze the propagation of a magnetostatic surface wave transverse to the dc magnetization. In Section 4.3.1.2, we present a rigorous electromagnetic analysis of transverse propagation. The general magnetostatic analysis of propagation in an arbitrary direction relative to the biasing field is carried out in Section 4.3.1.3. The effect of the finite conductivity of metal on dispersion characteristics is considered in Section 4.3.1.4. Finally, some experimental results are discussed in Section 4.3.1.5.

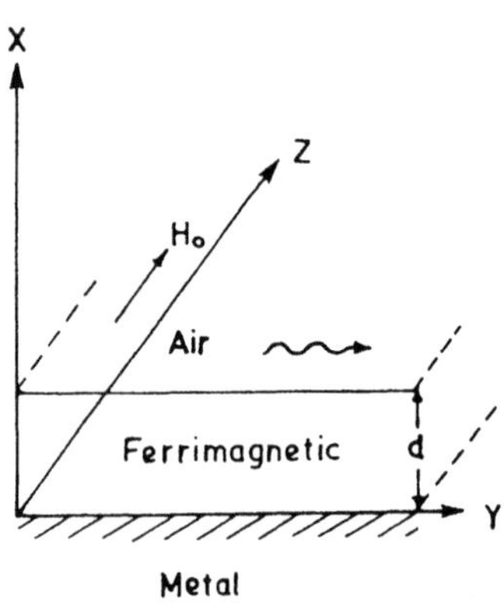

Figure 4.34. Transversely magnetized, metal-backed ferrimagnetic slab.

4.3.1.1. *Transverse Propagation: Magnetostatic Analysis*

The following analysis is based on a paper by Seshadri (1970). As shown in Figure 4.34, the direction of propagation is along the y-axis, whereas the wave field is uniform along the z-axis. The magnetostatic potential and normal component of the magnetic induction in ferrite and air regions can be expressed as

$$\left.\begin{aligned} \psi^{a} &= D\exp(-\beta x)\exp j(\omega t-\beta y) \\ \psi^{f} &= [B\exp(\beta x)+C\exp(-\beta x)]\exp j(\omega t-\beta y) \end{aligned}\right\} \tag{4.116}$$

$$\left.\begin{aligned} b_x^{a} &= -\beta\psi^{a} \\ b_x^{f} &= \beta[(\mu+\kappa)B\exp(\beta x) \\ &\qquad -(\mu-\kappa)C\exp(-\beta x)]\exp j(\omega t-\beta y) \end{aligned}\right\} \tag{4.117}$$

At $x=0$ (metal surface), it is required that $b_x^{f}=0$ and thus

$$(\mu+\kappa)B=(\mu-\kappa)C \tag{4.118}$$

Other equations are obtained by applying the usual boundary conditions at $x=d$. The elimination of constants leads to the following dispersion relation for propagation along the $+y$-direction:

$$\exp(2\beta d)=\frac{(\mu+\kappa)(\mu-\kappa-1)}{(\mu-\kappa)(\mu+\kappa+1)} \tag{4.119}$$

It is interesting that this dispersion relation is not invariant under reversal of the sign of κ, which implies nonreciprocal propagation characteristics. The wavenumber for magnetostatic surface wave propagation along the $\pm y$-direction is therefore

$$\beta_{\pm}=\frac{1}{2d}\ln\left[\frac{(\mu\pm\kappa)(\mu\mp\kappa-1)}{(\mu\mp\kappa)(\mu\pm\kappa+1)}\right] \tag{4.120}$$

As the medium is assumed to be lossless, substitution for μ and κ leads to

$$\beta_{\pm}=\frac{1}{2d}\ln\left[\frac{\frac{1}{2}\omega_m(\omega_0+\omega_m\mp\omega)}{(\omega_0+\omega_m\pm\omega)(\omega_0+\frac{1}{2}\omega_m\mp\omega)}\right] \tag{4.121}$$

It is easy to show from this that the range of allowed modes for propagation in the $+y$-direction is $[\omega_0(\omega_0+\omega_m)]^{1/2}\leqslant\omega\leqslant\omega_0+\frac{1}{2}\omega_m$, whereas that for propagation in the $-y$-direction is $[\omega_0(\omega_0+\omega_m)]^{1/2}\leqslant\omega\leqslant$

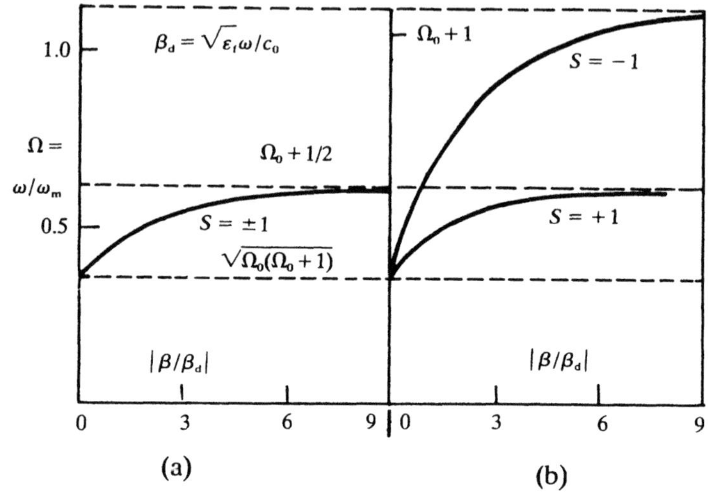

Figure 4.35. Magnetostatic dispersion curves for (a) free ferrimagnetic slab, and (b) metal-backed ferrimagnetic slab. In (b) $S = \pm 1$ corresponds to propagation along the $\pm y$-direction. The introduction of the metal plate thus removes the degeneracy and renders the propagation characteristics nonreciprocal (after Seshadri, 1970).

$(\omega_0 + \omega_m)$, with the respective upper limits corresponding to resonance. Figure 4.35 shows the dispersion curves for a typical set of data. For comparison, the dispersion curve corresponding to the unmetallized ferrimagnetic slab, as obtained from equation (4.48), has also been shown. It is seen that the introduction of a perfect conductor at $x = 0$ leaves the mode propagating along the $+y$-direction practically unaffected. However, the mode propagating along the $\mathbf{H}_0 \times \hat{n}$ direction, i.e., the $-y$-direction, is strongly influenced, which is understandable since the modification of boundary conditions at a particular ferrimagnetic surface (presently, $x = 0$) should influence only the surface wave associated with that surface. It is also seen that the phase velocity of the ferrite-metal (f.m.) mode is much larger than that of the ferrite-air (f.a.) mode.*

Direct differentiation of equation (4.121) yields the group delay for propagating a unit distance as

$$\tau_d^{\pm} = \frac{1}{d} \frac{\omega^2 \mp 2\omega(\omega_0 + \omega_m) + \omega_0(\omega_0 + \omega_m)}{[\omega^2 - (\omega_0 + \omega_m)^2](\pm 2\omega - 2\omega_0 - \omega_m)} \tag{4.122}$$

* The terms ferrite-air (f.a.) mode and ferrite-metal (f.m.) mode were introduced by Seidel and Fletcher (1959).

TABLE 4.1. Group Delay Time in Microseconds[a] (after Seshadri, 1970)

H_0	A	B	C
300	9.88×10^{-3}	9.40×10^{-3}	4.86×10^{-4}
290	1.24×10^{-2}	1.19×10^{-2}	4.92×10^{-4}
280	1.65×10^{-2}	1.60×10^{-2}	4.97×10^{-4}
270	2.46×10^{-2}	2.41×10^{-2}	5.03×10^{-4}
260	4.82×10^{-2}	4.76×10^{-2}	5.09×10^{-4}
250	8.95×10^{-1}	8.94×10^{-1}	5.15×10^{-4}

[a] H_0 is the applied field in oersteds, A is the ungrounded ferrite slab, B is the ferrite slab grounded at the bottom ($x = 0$), and C is the ferrite slab grounded at the top ($x = d$).

The delay computed from this relation for a typical set of data is shown in Table 4.1. For comparison, the delay obtainable from an unmetallized (but otherwise equivalent) slab has also been shown in Table 4.1. It is evident that the delay of the corresponding f.a. mode, associated with the surface $x = d$, is practically unaffected by the presence of the metal plate at $x = 0$. On the contrary, the f.m. mode is now characterized by a much smaller group delay. Thus, the delay characteristics of the metallized slab are nonreciprocal.

4.3.1.2. *Transverse Propagation: Electromagnetic Analysis*

The rigorous analysis of transverse propagation in a metal-backed ferrite slab has been carried out by Gerson and Nadan (1974) and Parekh (1975). Whereas Gerson and Nadan considered only surface modes, Parekh also investigated the dynamic bulk modes. The subsequent discussion follows these authors.

The analysis can be carried out on the same lines as in Sections 4.1.1 and 4.2.1.2. Referring to Figure 4.35, we see that the relevant components of electric and magnetic fields of the wave in ferrimagnetic and dielectric regions can be expressed as

$$\left.\begin{aligned} E_z^d &= D\exp(-k_d x)\exp j(\omega t - \beta y) \\ E_z^f &= [B\exp(k_f x) + C\exp(-k_f x)]\exp j(\omega t - \beta y) \end{aligned}\right\} \tag{4.123}$$

$$\left.\begin{aligned} h_y^d &= \frac{jc_0 k_d}{\omega} E_z^d \\ h_y^f &= -j\frac{c_0}{\omega\mu_{\text{eff}}}[(k_f - \beta\kappa/\mu)B\exp(k_f x) \\ &\qquad - (k_f + \beta\kappa/\mu)C\exp(-k_f x)]\exp j(\omega t - \beta y) \end{aligned}\right\} \tag{4.124}$$

where

$$\left.\begin{aligned} k_d^2 &= \beta^2 - \beta_d^2, \qquad & k_f^2 &= \beta^2 - \beta_f^2 \\ \beta_d^2 &= \varepsilon_d \omega^2 / c_0^2, \qquad & \beta_f^2 &= \varepsilon_f \mu_{eff} \omega^2 / c_0^2 \end{aligned}\right\} \tag{4.125}$$

It is required that E_z^f vanish at the metal surface ($x = 0$) whereas E_z^f and h_y^f should be continuous at $x = d$. The usual procedure yields the following dispersion relation for propagation in the $\pm y$-direction:

$$\exp(2k_f d) = \frac{\mu_{eff} k_d \mp \beta_\pm \kappa/\mu - k_f}{\mu_{eff} k_d \pm \beta_\pm \kappa/\mu + k_f} \tag{4.126}$$

which is essentially the same as that derived by Gerson and Nadan (1974) and Parekh (1975) in slightly different forms.

When μ_{eff} is negative, i.e., when $[\omega_0(\omega_0 + \omega_m)]^{1/2} < \omega < (\omega_0 + \omega_m)$, a solution to equation (4.126) leads to surface modes since k_f will be real. On the other hand, when μ_{eff} is positive, surface waves and volume waves may be obtained for $\beta > \beta_f$ and $\beta_d < \beta < \beta_f$, respectively. We shall first consider the surface modes.

When the slab is thick, the metal is far away from the ferrimagnetic dielectric interface. In the limit as $d \to \infty$, equation (4.126) reduces, as it should, to the surface wave dispersion relation for the ferrimagnetic half space. Also, when β is sufficiently large, we have $k_d \simeq \beta$ and $k_f \simeq \beta$, and equation (4.126) reduces to the magnetostatic dispersion relation, i.e., equation (4.120). Since the magnetostatic analysis predicts the upper limit of surface modes as $\omega = (\omega_0 + \omega_m)$, while the electromagnetic analysis (for a half space) predicts surface waves at frequencies above $\omega = (\omega_0 + \omega_m)$, it is worth solving equation (4.126) numerically in order to get a clear picture.

Figure 4.36 shows the resulting dispersion curves for different slab thicknesses. It is seen that the characteristics of surface waves for $\omega < (\omega_0 + \omega_m)$ are similar to those obtained from magnetostatic analysis. However, the cut-off wavenumber is finite rather than zero as predicted by magnetostatic analysis. For $\omega > (\omega_0 + \omega_m)$, surface modes do occur when the slab thickness exceeds a certain critical value. However, as shown in the inset, a small, thickness-dependent band gap exists at frequencies just above $\omega = (\omega_0 + \omega_m)$. This band gap decreases as the slab thickness increases. In the limit as $d \to \infty$, the dispersion curves reduce to the ones for a half space (see Figure 4.2). As pointed out by Parekh (1975), when the slab thickness is large (but finite), the surface wave branch above the band gap exhibits a backward wave characteristic (over a small frequency range), not shown explicitly in Figure 4.36.

Now consider bulk mode solutions to equation (4.126), which can be

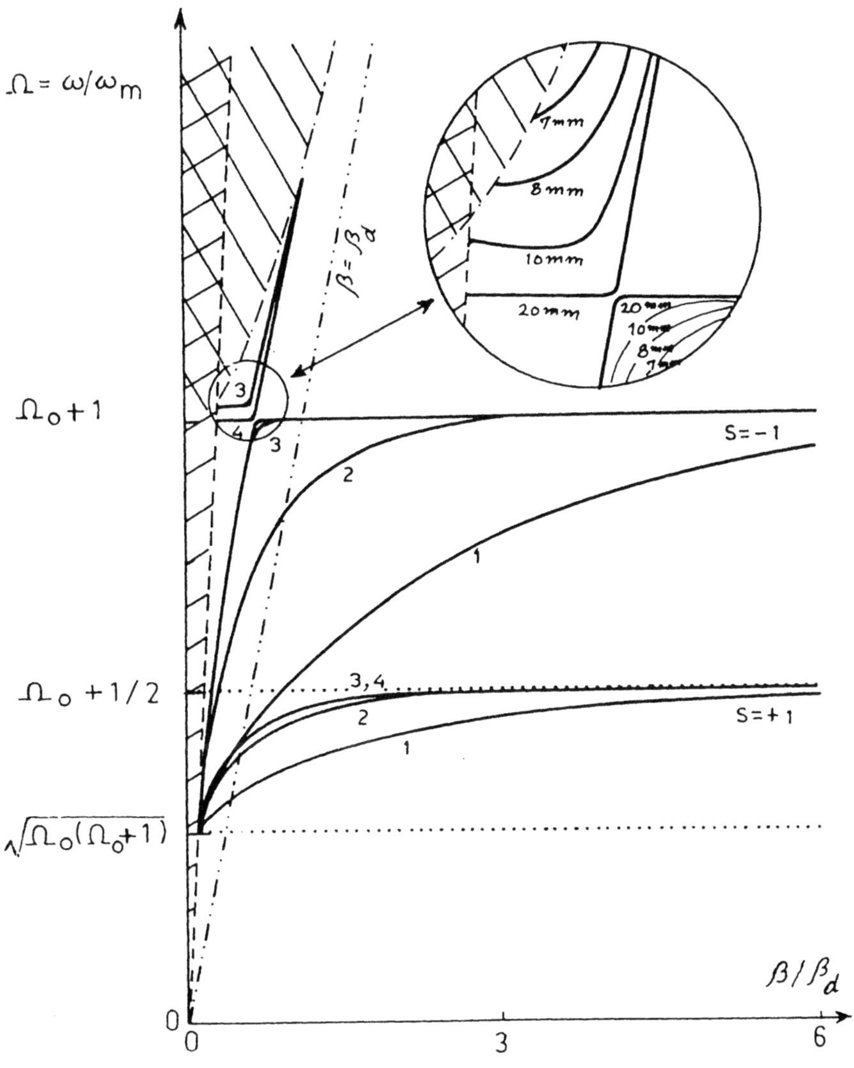

Figure 4.36. Surface electromagnetic mode dispersion curves for a transversely magnetized grounded ferrimagnetic slab. The parameters $\Omega_0 = 0.1124$ and $\varepsilon_d = 14$. The slab thickness d for the curves designated by numbers 1, 2, 3, and 4 is 0.5, 2.0, 10.0, and 20.0 mm, respectively (after Garson and Nadan, 1974).

expressed in a more convenient form as (Parekh, 1975)

$$k_f + (\mu_{eff} k_d \mp \beta_{\pm} \kappa/\mu) \tan k_f d = 0 \tag{4.127}$$

Figures 4.37 and 4.38 present the typical dispersion curves obtained from this relation. In comparison, the TE mode solutions for an equivalent

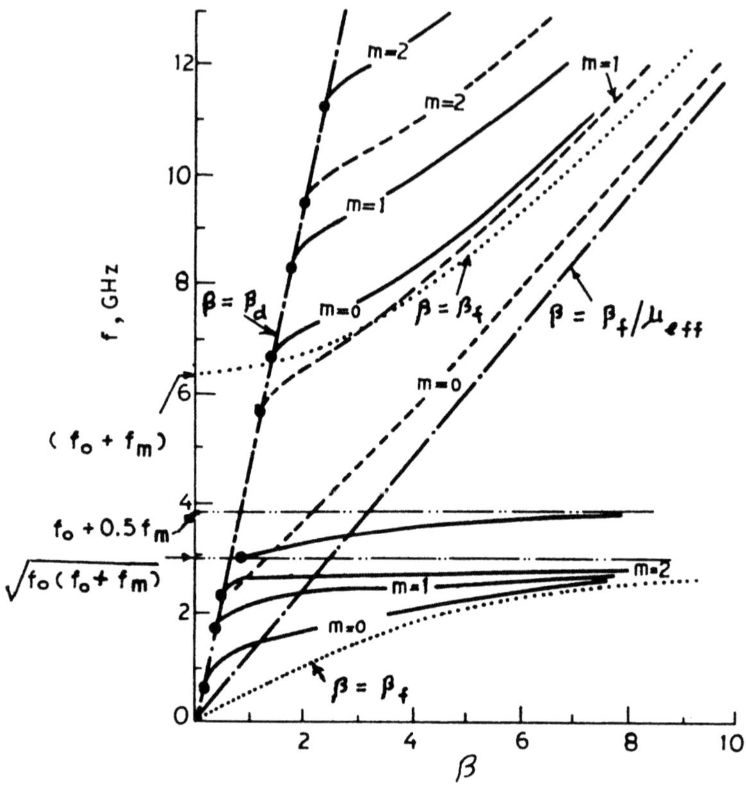

Figure 4.37. Dispersion curves for electromagnetic modes (of a metallized YIG plate) propagating in the $+y$-direction with a biasing field along the $+z$-direction. The dashed curves represent the TE mode solution, which would be obtained if all spins were clamped. The parameters are $4\pi M_0 = 1750$ G, $\varepsilon_f = 16$, $H_0 = 500$ Oe, and $d = 1$ cm (after Parekh, 1975).

dielectric plate [which can be obtained by substituting $\mu = 1$ and $\kappa = 0$ in equation (4.127)] have also been shown in Figure 4.37. The dynamic bulk modes, corresponding to each value of the mode number m, are characterized by two branches in the dispersion curves. The upper branches are the modified dielectric modes. The lower branches occur for $\omega < [\omega_0(\omega_0 + \omega_m)]^{1/2}$ and are induced purely by the magnetic nature of the slab. Surface wave branches have also been shown in Figures 4.37 and 4.38. The behavior of the upper surface wave branch in Figure 4.38 is quite interesting as it represents a continuum of the upper $m = 0$ dynamic mode and is characterized by backward wave characteristics over a small frequency range when the slab thickness is large but finite.

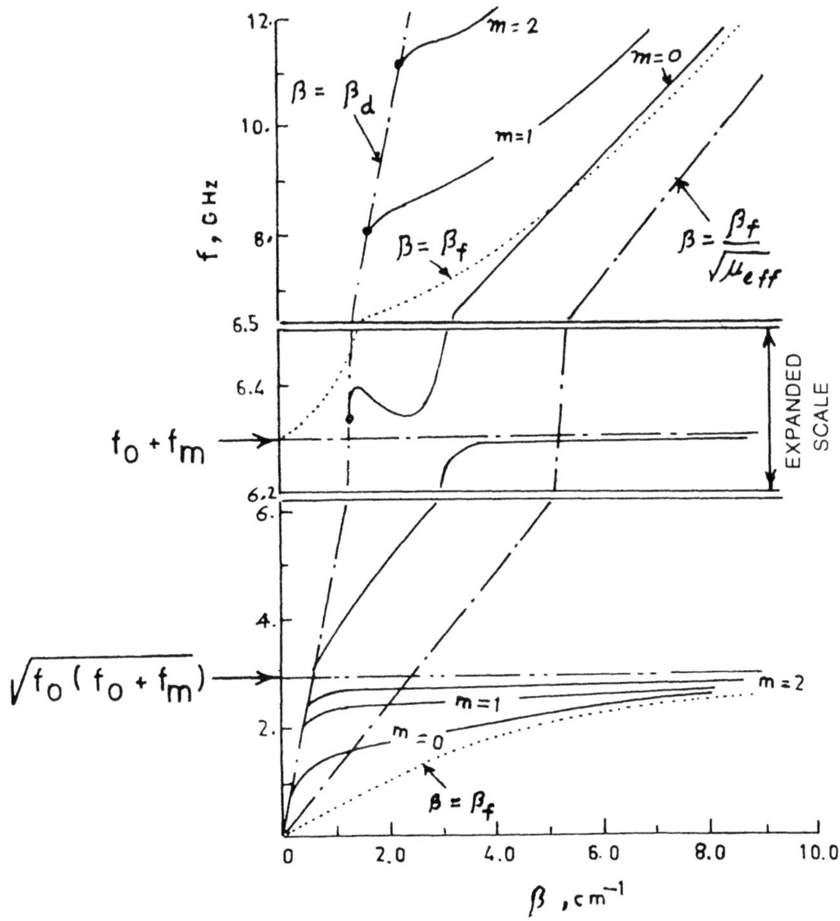

Figure 4.38. Dispersion curves for electromagnetic modes of a metallized YIG plate for propagation in the $-y$-direction. The parameters are the same as for Figure 4.37 (after Parekh, 1975).

4.3.1.3. *Arbitrary Propagation: Magnetostatic Analysis*

In the preceding sections, we investigated the dispersion characteristics for propagation perpendicular to the dc field. In what follows, we consider the general case of propagation at an oblique angle with respect to the biasing field. The results pertaining to this section have been obtained, in slightly different forms, by Young (1969), Bennett and Adam (1970), and Parekh (1973).

Referring again to Figure 4.34 and assuming a variation $\exp(\pm jk_z z)\exp j(\omega t - \beta y)$ for the magnetostatic potential, we have

$$\left.\begin{aligned} \psi^{d} &= D\exp(-px)\exp(\pm jk_z z)\exp j(\omega t - \beta y) \\ \psi^{i} &= [B\exp(qx) + C\exp(-qx)]\exp(\pm jk_z z)\exp j(\omega t - \beta y) \end{aligned}\right\} \quad (4.128)$$

where

$$\left.\begin{aligned} p^2 &= \beta^2 + k_z^2 \\ q^2 &= \beta^2 + k_z^2/\mu \end{aligned}\right\} \quad (4.129)$$

The usual procedure yields the following dispersion relation for propagation in the $\pm y$-directions:

$$\exp(2qd) = \frac{(\mu q \pm \kappa\beta)(\mu q \mp \beta\kappa - p)}{(\mu q \mp \beta\kappa)(\mu q \pm \beta\kappa + p)} \quad (4.130)$$

which is the same as the relations obtained by Bennett and Adam (1970) and Parekh (1973) in different forms. If ψ_{wb} represents the angle between the wavevector and biasing field, we have $\beta = p\sin\psi_{wb}$ and $k_z = p\cos\psi_{wb}$. The dispersion curves for $\psi_{wb} = 80°$ and $60°$ are shown in Figure 4.39. As

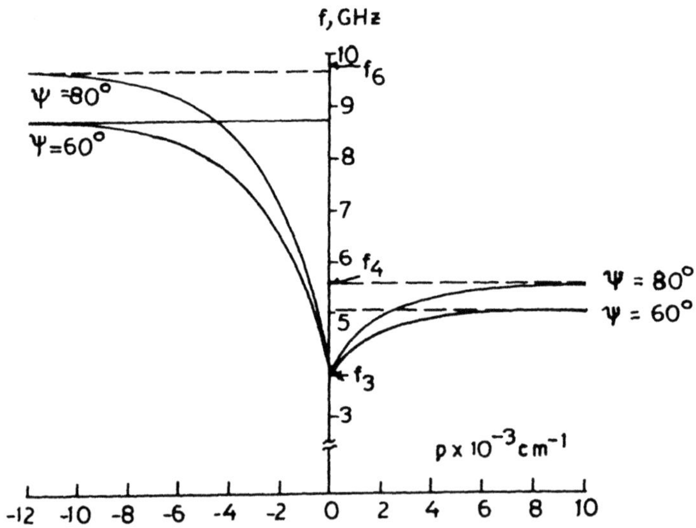

Figure 4.39. Computed dispersion curves for magnetostatic surface wave propagation in the configuration shown in Figure 4.34 with the difference that the wave propagates at an angle with respect to the biasing field. The curves for $\psi_{wb} = 60°$ and $80°$ have been explicitly shown in the figure. The parameters for a Ga–YIG film are $4\pi M_0 = 300\,\mathrm{G}$, $d = 2\,\mu\mathrm{m}$, and $H_0 = 50\,\mathrm{Oe}$ (after Parekh, 1973).

ψ_{wb} decreases, the frequency range of allowed modes decreases for f.a. ($+y$-propagating) as well as f.m. ($-y$-propagating) modes. The frequency range of f.m. modes just vanishes at $\psi_{wb}=0$, whereas that of f.a. modes vanishes when $\psi_{wb} < \psi_{crit} \equiv \tan^{-1}(\omega_0/\omega_m)^{1/2}$; the latter limit is the same as that for an unmetallized slab, as noted in Section 4.2.1. Furthermore, as noted by Parekh (1975), the wavevector and group velocity are collinear only when $\psi_{wb}=90°$ (transverse propagation). The angle between the wavevector and group velocity increases monotonically from 0° to 90° as ψ_{wb} decreases from 90° to its limiting value.

4.3.1.4. *Effect of Finite Conductivity of Metal*

In the preceding sections, the conductivity of the metal was regarded as infinite. As such, the normal component of **b** and tangential component of **E** were required to vanish at the metal surface. This has been termed an effective boundary condition model (EBCM) by DeWames and Wolfram (1970), who have also investigated the effect of finite conductivity of metal on the propagation characteristics of magnetostatic surface waves. We shall summarize below the method of analysis and important results.

Figure 4.40 shows the configuration: a transversely magnetized ferrimagnetic slab of thickness d is backed by a metallic layer of thickness t. The wave field in regions other than metal can still be obtained from magnetostatic analysis. In the metal region, however, the wave electric field cannot be neglected since the conductivity is finite. Neglecting the displacement current, we can write the relevant Maxwell equations as

$$\left.\begin{aligned} \nabla \times \mathbf{E} &= -j(\omega/c_0)\mathbf{h} \\ \nabla \times \mathbf{h} &= (4\pi\sigma_e/c_0)\mathbf{E} \end{aligned}\right\} \qquad (4.131)$$

where σ_e is the electrical conductivity of the metal. It is easy to show that

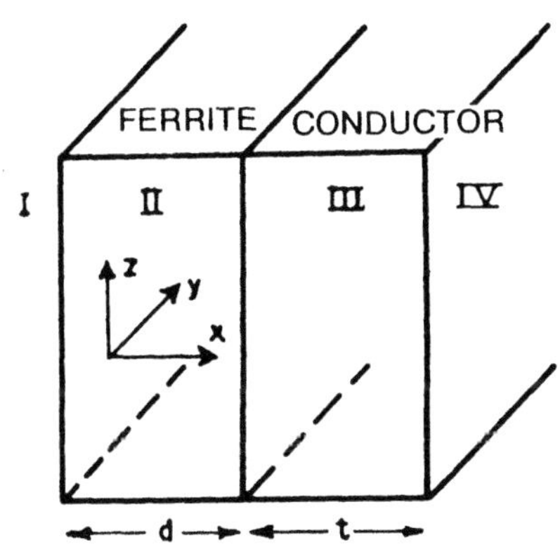

Figure 4.40. A ferrimagnetic layer of thickness d in contact with a metallic layer of thickness t. The conductivity of the metal is finite. The biasing field is along the z-axis, whereas the wave propagates along the y-axis (after DeWames and Wolfram, 1970).

the nonvanishing components of the field vectors, i.e., E_z, h_x, and h_y, satisfy the following equation [assuming a variation $\exp j(\omega t - \beta y)$]:

$$(\partial^2/\partial x^2 - \beta_m^2)U = 0 \tag{4.132}$$

where

$$\left.\begin{aligned} \beta_m^2 &= \beta^2 + 2j/\delta_p^2 \\ \delta_p &= (c_0^2/2\pi\sigma_e\omega)^{1/2} \end{aligned}\right\} \tag{4.133}$$

In equation (4.133), δ_p is the depth of penetration. The rest of the analysis is conventional, and the dispersion relation is obtained as

$$\exp(2\beta_\pm d) = \frac{\mu \mp \kappa - 1}{\mu \pm \kappa + 1} \times \frac{(1+\varepsilon_m)(\mu \pm \kappa - 1/\varepsilon_m)\exp(2\beta_m t) + (1-\varepsilon_m)(\mu \pm \kappa + 1/\varepsilon_m)}{(1+\varepsilon_m)(\mu \mp \kappa + 1/\varepsilon_m)\exp(2\beta_m t) + (1-\varepsilon_m)(\mu \mp \kappa - 1/\varepsilon_m)} \tag{4.134}$$

where $\varepsilon_m = \beta_m/\beta$. It is easy to show that equation (4.134) reduces to the Damon–Eshbach result, i.e., equation (4.47), for zero conductivity and to the Seshadri result, i.e., equation (4.120), for infinite conductivity. Since β_m is complex, a solution to equation (4.134) will also be complex. Equation (4.134) can be conveniently solved by regarding $\Omega_0 = \omega_0/\omega_m$ as a complex quantity; the imaginary part of Ω_0 represents the losses. Figures 4.41–4.43

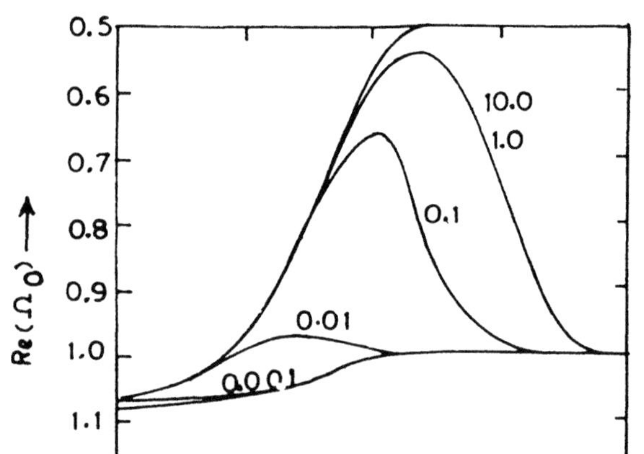

Figure 4.41. Variation of the real part of the applied resonant field Ω_0 with $\xi = \beta d$ for different values of t/δ_p, indicated by the side of each curve. The parameters are $\delta_p/d = 0.05$ and $\Omega = 1.5$. The dispersion curves approach the Damon–Eshbach curve when $t/\delta_p \to 0$ and the Seshadri curve when $t/\delta_p \to \infty$ (after DeWames and Wolfram, 1970).

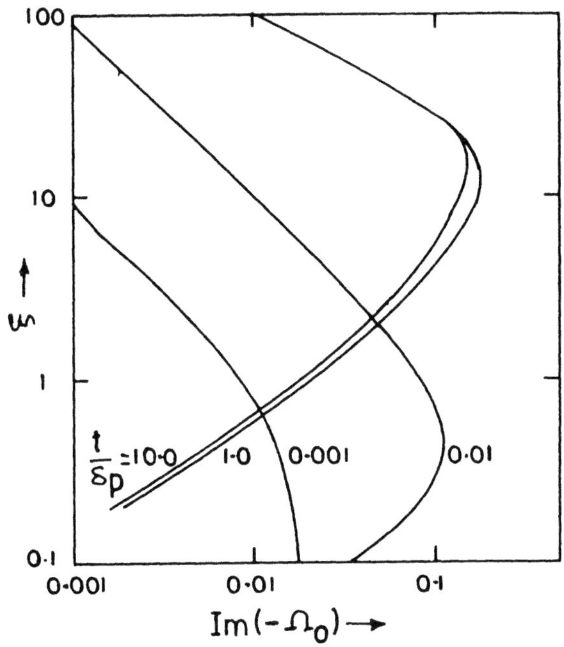

Figure 4.42. Variation of the imaginary part of Ω_0 with $\xi = \beta d$ for different values of t/δ_p. The parameters are the same as for Figure 4.41 (after DeWames and Wolfram, 1970).

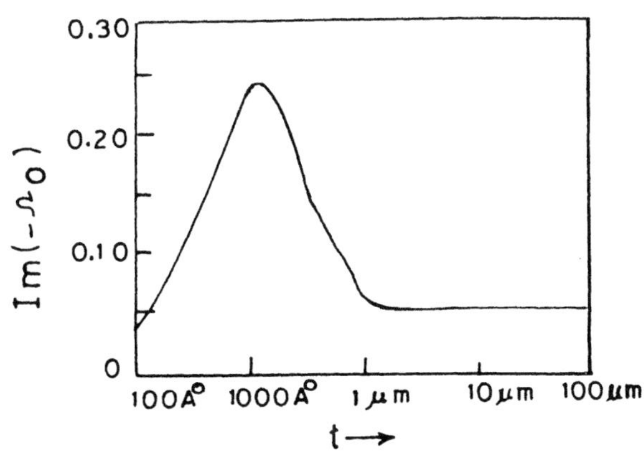

Figure 4.43. Variation of Im(Ω_0) with t for $\beta = 1.25 \times 10^3\ \mathrm{cm}^{-1}$, $\delta_p = 1$ mm, and $d = 20\ \mu$m (after DeWames and Wolfram, 1970).

show the variation of $\mathrm{Re}(\Omega_0)$ and $\mathrm{Im}(\Omega_0)$ with different parameters. The following general conclusions are drawn (DeWames and Wolfram, 1970):

1. The results of EBCM are valid for small $\beta\delta p$ provided that $t > \delta_p^2\beta$.
2. When the conductivity is finite (which is also true in practice), the asymptotic value of $\mathrm{Re}(\Omega_0)$ at high frequencies approaches the Damon–Eshbach limit rather than the limit deduced from the EBCM.
3. The modes are lossy, the degree of damping critically depending on βd, $\beta\delta_p$, and t/δ_p.
4. The group velocity can be negative for large β.

Since the conductivity can be controlled as desired by choosing an appropriate semiconductor, the importance of some of the aforementioned conclusions will be evident later in this chapter.

4.3.1.5. *Experimental Results*

Bennett and Adam (1970) have experimentally observed the magnetostatic surface wave resonances of a metal-backed ferrimagnetic slab and demonstrated the nonreciprocal propagation characteristics by an experiment at 4 GHz. The experimental configuration is essentially the same as in Figure 4.6a, except for the presence of a metal plate on one side of the YIG slab. Figures 4.44 and 4.45 display the variation of the transmitted amplitude as a function of the biasing field. As the biasing field is increased, a broad transmission region followed by a series of resonances is obtained (Figure 4.44). On the other hand, when the input and output terminals are interchanged (Figure 4.45), only the resonances are observed. This behavior can be understood in terms of the theory developed in Section 4.3.1.3. A resonance will occur provided that the total phase change for surface waves in one trip around the slab surface is an integral multiple of 2π; thus

$$(\beta_+ + \beta_-)L = 2n_L\pi \tag{4.135}$$

where L is the length of the slab and n_L is an integer. The wavenumber $\beta_\pm$ for the f.a. and f.m. modes can be obtained from equation (4.130) with $k_z = n_w\pi/w$ and $n_w = 1, 2, \ldots$. The resonances can be designated as (n_L, n_w). It is obvious that the occurrence of resonances requires that both the modes (f.a. and f.m.) should be of a propagating nature. Whereas the f.m. mode has no lower cut-off at 4 GHz, the f.a. mode is cut-off for $H_0 < H_{cr} \equiv (\omega/\gamma - 4\pi M_0)$. Thus, the resonances are observed for $H_0 > H_{cr}$ as in Figures 4.44 and 4.45. When $H_0 < H_{cr}$ only the f.m. mode can propagate, which leads to a broad transmission region as obtained in Figure 4.44 but not in Figure 4.45.

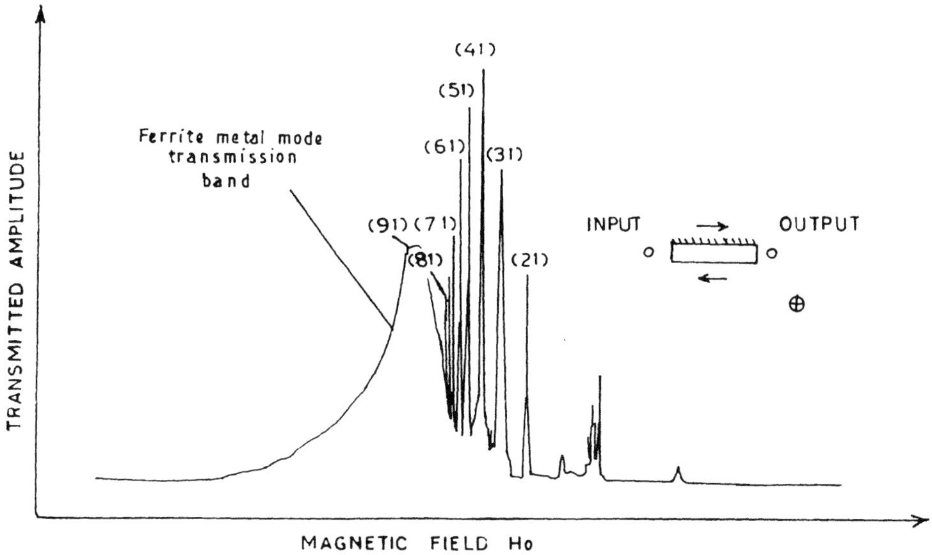

Figure 4.44. Measured transmitted power vs. biasing field for magnetostatic surface waves in a YIG plate with top surface metallized (after Bennett and Adam, 1970).

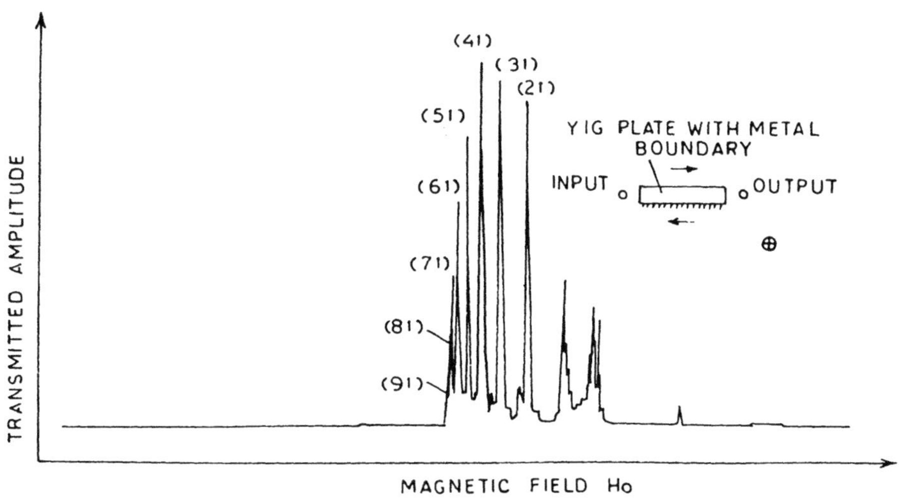

Figure 4.45. Measured transmitted power vs. biasing field for magnetostatic surface waves in a YIG plate with bottom surface metallized (after Bennett and Adam, 1970).

4.3.2. Normal Magnetization

Figure 4.46 shows the configuration; the ferrimagnetic slab occupies the region $0 \leqslant z \leqslant d$ and is backed by a perfect conductor at $z = 0$. If the variation of the wave field along the x-axis is ignored, the magnetostatic potential in air and ferrimagnetic regions can be expressed as

$$\left.\begin{aligned} \psi^{a} &= D \exp(-\beta z) \exp j(\omega t - \beta y) \\ \psi^{f} &= [B \exp(j\alpha_{v}\beta z) + C \exp(-j\alpha_{v}\beta z)] \exp j(\omega t - \beta y) \end{aligned}\right\} \quad (4.136)$$

where

$$\alpha_{v} = (-\mu)^{1/2} \quad (4.137)$$

The following dispersion relation is obtained by the usual procedure:

$$\beta = (1/\alpha_{v} d)[n\pi + \arctan(1/\alpha_{v})] \quad (4.138)$$

where n is a nonnegative integer. A comparison of this relation with equation (4.70) shows that, for each n, the wavenumber for a metallized slab is half of that for an unmetallized slab with the same parameters. This can be readily understood on the basis of the zig-zag ray model presented in Section 4.2.2.3. It is easy to show that the phase shift of a plane magnetostatic wave due to total reflection from a metal is zero. Thus, in equation (4.100), $2\varphi_{v}$ should be replaced by φ_{v} in order to obtain equation

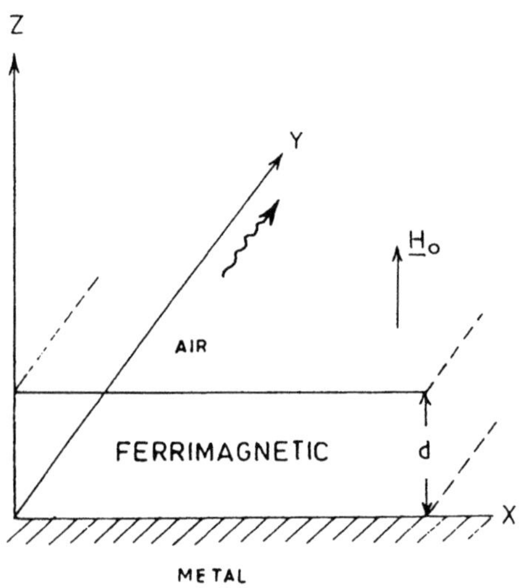

Figure 4.46. Normally biased, metallized ferrimagnetic slab, supporting a magnetostatic wave propagation along the y-axis.

(4.138). Since the phase shift in one complete "cycle" is reduced to half of its value, the overall reduction of the wavenumber by a factor of two is understandable.

Further discussion of magnetostatic waves in normally magnetized films will be taken up in the next section.

4.4. *Ferrimagnetic–Dielectric–Metal Structure*

In the preceding sections, the utility of magnetostatic waves in microwave delay devices was discussed. Apart from lower loss above 3 GHz (see Figure 4.8), an important advantage of a magnetostatic wave delay device over other [e.g., surface acoustic wave (SAW)] devices is its magnetic tunability; the frequency of operation can be varied by changing the strength of the biasing field. The price paid for magnetic tunability is the greatly reduced effective bandwidth of operation on account of the strong dispersion of the group velocity which, in turn, leads to distortion of a wave packet as it propagates down the line. It would be ideal to have a configuration wherein the, apparently, conflicting requirements of tunability and nondispersive delay over a reasonable bandwidth are simultaneously met. The desired characteristics can be obtained in the configuration involving a ferrimagnetic film separated from a perfect conductor by a dielectric layer, as demonstrated by Bongianni (1972). The occurrence of nondispersive delay over a small frequency range is explained as follows.

It is clear from the discussion of Section 4.3 that, at a given frequency, the wavenumber and group delay of magnetostatic waves are smaller for the f.m. mode than for the f.a. mode. The curves A and D in Figure 4.47 and 4.48 show the typical dispersion and delay characteristics for surface f.m. and f.a. modes, respectively. As the metal plate is separated from the ferrimagnetic slab and moved to infinity, curve A is expected to change gradually and approach curve D (in each figure). Now consider the dispersion curves in the case when the distance from the slab to the metal plate is finite. If the wavelength is large (small β), the extent of the field distribution away from the slab is relatively large, the wave interacts strongly with the metal plate, and, therefore, the dispersion curve approaches that of the f.m. mode, i.e., curve A. On the other hand, when the wavelength is small (large β), the field distribution is confined to the surface of the slab and the wave hardly "feels" the presence of the metal plate. Therefore, in the large-wavenumber limit, the dispersion curve approaches curve D. Curves B and C in Figure 4.47 display this behavior. Curve C corresponds to a dielectric thickness larger than that corresponding to curve B. Consequently, curve C approaches curve D at a frequency much higher than that at which curve B approaches curve D.

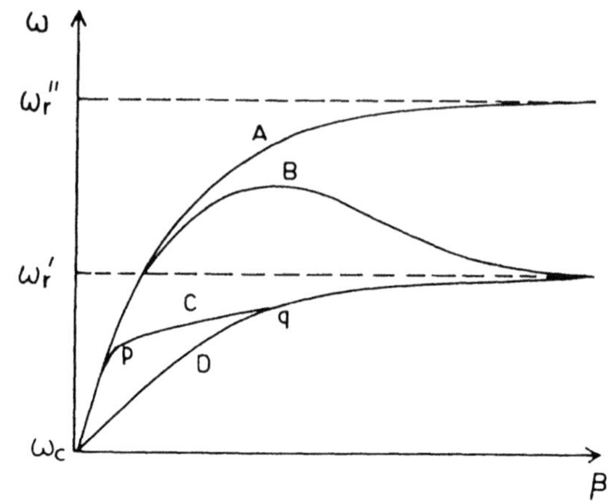

Figure 4.47. Qualitative dispersion curves for magnetostatic surface waves in a transversely magnetized ferrimagnetic slab. The curves designated by D and A correspond to free and metallized slabs, respectively. The curves B and C correspond to finite separation of the metal plate from the ferrite slab.

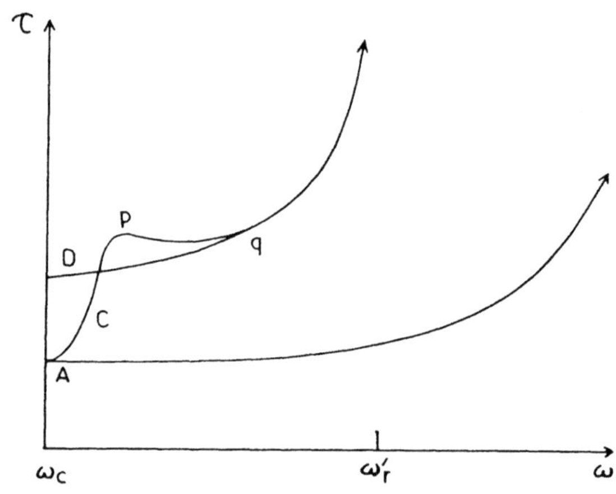

Figure 4.48. The time delay τ vs. frequency corresponding to the dispersion curves in Figure 4.47.

Now consider the delay curves in Figure 4.48. It has already been noted that the f.a. and f.m. modes exhibit monotonic increases of delay with frequency, as shown clearly in Figure 4.48. As noted above, curve C in Figure 4.47 approaches curve A near the lower cut-off frequency. Consequently, the corresponding delay curve in Figure 4.48 also approaches curve A at low frequencies. However, at slightly higher frequencies, it is seen in Figure 4.47 that there is a steep rise in β with ω (because of the tendency to move away from curve A and approach curve D) for curve C. As such, in this frequency range, the slope of β with ω is larger for curve C than for curve D. Hence, the corresponding group delay is larger for curve C than for curve D. The point p in Figure 4.48 represents this behavior. Beyond this point, the tendency of curve C in Figure 4.47 is to approach curve D which finally occurs, say, at point q. The positions of points p and q (on the frequency scale) in Figure 4.48 can be varied slightly by changing the thickness of the dielectric layer so as to get more or less constant delay over a small bandwidth. Clearly, the region of nondispersive delay occurs in a frequency range slightly above the lower cut-off frequency. Since the cut-off frequency limit can be changed by varying the biasing field, it is possible to shift the region of nondispersive delay around a desired frequency. Similar arguments hold good for forward bulk modes, and it is not very difficult to show, on the same lines, that there is no such region of nondispersive delay for backward waves.

In what follows, we present an analysis of magnetostatic wave propagation in a ferrimagnetic–dielectric–metal structure* under three biasing conditions: (i) transverse magnetization in the plane of the film, (ii) normal magnetization, and (iii) arbitrary magnetization.

4.4.1. *Transverse Magnetization in the Film Plane*

Figure 4.49 shows the configuration in which a ferrimagnetic film of thickness d occupies the region $0 \leq x \leq d$ and is separated from a perfect conductor at $x = d + t$ by a dielectric layer of thickness t. The slab is

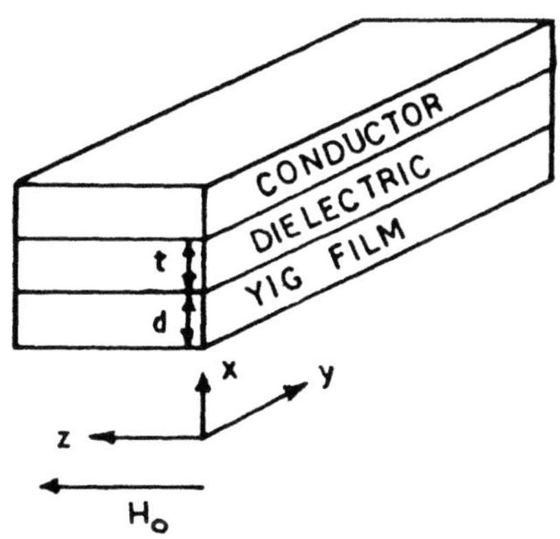

Figure 4.49. The dielectric layered structure: a YIG film of thickness d is separated from a perfect conductor by a dielectric layer of thickness t. The biasing field is along the z-axis while the magnetostatic surface wave propagates along the y-axis (after Bongianni, 1972).

* This structure was introduced by Bongianni (1972) who called it a "dielectric layered structure".

magnetized along the z-axis and supports a magnetostatic surface wave propagating in the y-direction. Following the magnetostatic field analysis described in preceding sections, the dispersion relation for propagation in the $\pm y$-direction is obtained as

$$\exp(2\beta_{\pm}d)=\frac{(\mu\mp\kappa-1)[(\mu\pm\kappa+1)\exp(-2\beta_{\pm}t)+(\mu\pm\kappa-1)]}{(\mu\pm\kappa+1)[(\mu\mp\kappa-1)\exp(-2\beta_{\pm}t)+(\mu\mp\kappa+1)]} \tag{4.139}$$

which is essentially the same as that derived by Bongianni (1972). In the limits when $t=0$ or $t\to\infty$, this equation reduces to the correct dispersion relation for a metal-backed or an unbacked ferrimagnetic slab, as discussed in Sections 4.2 and 4.3. If losses are neglected, the substitution for μ and κ in equation (4.139) leads to

$$\exp(2\beta_{\pm}d)=\frac{(\frac{1}{2}\omega_m)[\frac{1}{2}\omega_m+(\omega_0+\frac{1}{2}\omega_m\mp\omega)\exp(-2\beta_{\pm}t)]}{(\omega_0+\frac{1}{2}\omega_m\mp\omega)[\omega_0+\frac{1}{2}\omega_m\pm\omega+\frac{1}{2}\omega_m\exp(-2\beta_{\pm}t)]} \tag{4.140}$$

It follows that, in general, the propagation characteristics are nonreciprocal. It is easy to show from this dispersion relation that the resonance and cut-off limits for propagation in each direction are the same as those for the unbacked ferrimagnetic slab. This is physically understandable since, when β is large, the decay of the wave field away from the ferrimagnetic surface is very fast and, therefore, the presence of a metal plate (which is some distance away from the ferrimagnetic surface) does not affect the propagation characteristics.

Figure 4.50 displays the dispersion curves for propagation in the $\pm y$-direction with the thickness of the dielectric as the parameter. The nature of the curves is the same as was discussed in connection with Figure 4.47. Figure 4.51 shows the dispersion curves with the biasing field as the parameter, and the tunability of the frequency range of interest is evident.

The group delay time is obtained from equation (4.140) as

$$\left.\begin{aligned}
\tau_d^{\pm}&=N_\tau/D_\tau\\
N_\tau&=\exp[-2\beta_{\pm}(t+d)]\mp\exp(-2\beta_{\pm}t)-4\omega/\omega_m\\
D_\tau&=2t(\omega_0+\tfrac{1}{2}\omega_m\mp\omega)\exp(-2\beta_{\pm}t)-d\omega_m\exp(-2\beta_{\pm}d)\\
&\quad-2(d+t)(\omega_0+\tfrac{1}{2}\omega_m-\omega)\exp[-2\beta_{\pm}(t+d)]
\end{aligned}\right\} \tag{4.141}$$

This expression is based on magnetostatic analysis and is, therefore, expected to lead to approximate results. Vaslow (1973) has carried out rigorous electromagnetic analysis of waves in the dielectric layered structure. His calculations indicate that the departure of rigorous group delay from the corresponding magnetostatic result, i.e., equation (4.141), is

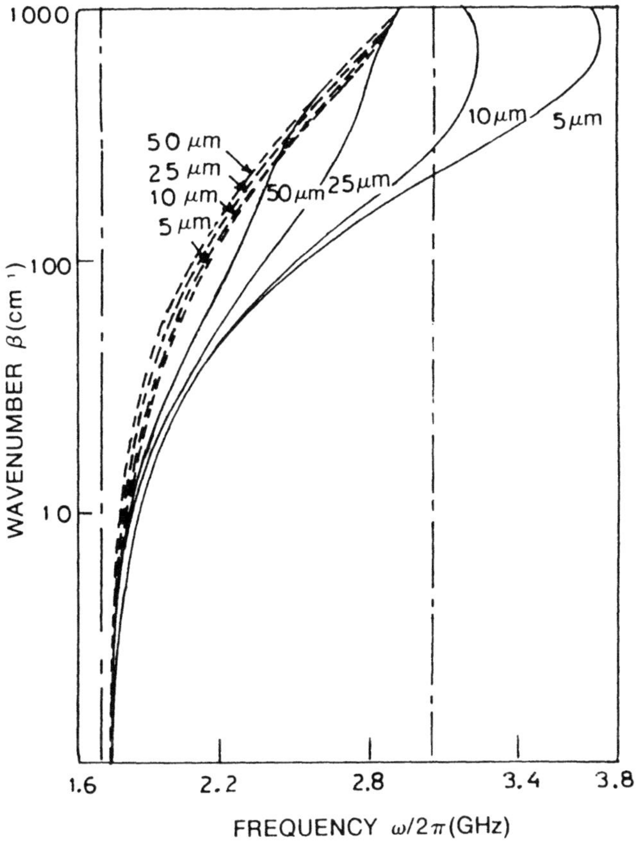

Figure 4.50. Dispersion curves for magnetostatic surface waves in a dielectric layered structure for different values of the dielectric layer thickness (shown alongside each curve). The solid and dashed curves correspond to propagation along the $+y$- and $-y$-directions, respectively. The parameters are $H_0 = 200\,\text{Oe}$, $d = 10\,\mu\text{m}$, and $4\pi M_0 = 1750\,\text{G}$ (after Bongianni, 1972).

negligible except in the region very close to the cut-off. Bongianni (1972) introduced the effective electrical thickness of the dielectric layer as

$$t_{\text{eff}} = (\varepsilon_f/\varepsilon_d)^{1/2}\,t \tag{4.142}$$

where t is the actual thickness of the dielectric layer. It has been suggested that t appearing in equation (1.141) should be replaced by t_{eff} for the purpose of calculations. Figure 4.52a compares a group delay measured by the standard pulse technique with that computed from equation (4.141)

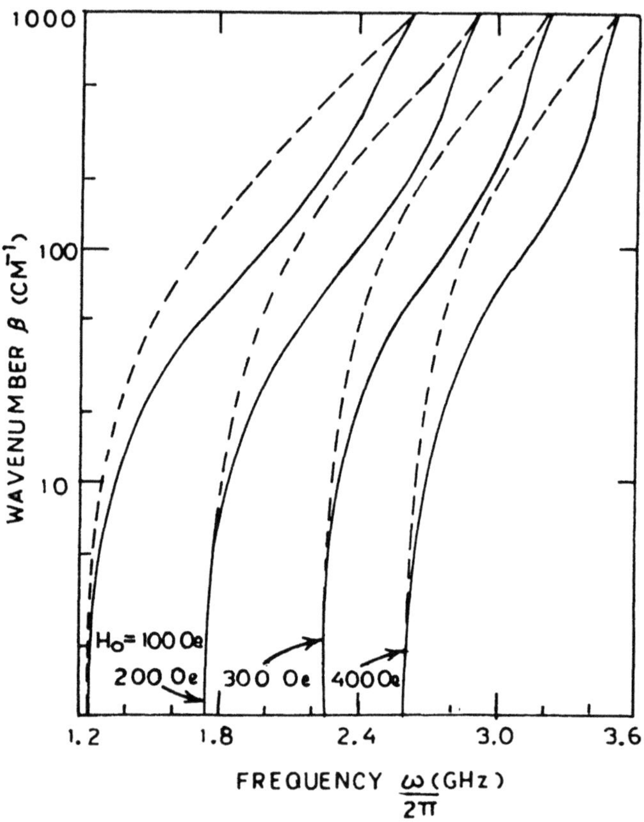

Figure 4.51. Dispersion curves for magnetostatic surface waves in a dielectric layered structure as a function of the biasing field H_0. The solid and dashed curves correspond to propagation in the $+y$- and $-y$- directions, respectively. In the calculations, the thickness of the dielectric layer has been taken as 25 μm while other parameters are the same as for Figure 4.50 (after Bongianni, 1972).

with modified t. The oscilloscope traces of pulse response are shown in Figure 4.52b. It is seen from Figure 4.52a that the agreement between calculated and measured values is excellent since the nonuniform demagnetization, which makes it difficult to compare theory and experiment in the case of slabs (see Section 4.1), is automatically eliminated in the case of films. The effect of dielectric thickness on delay characteristics is shown in Figure 4.53. As the dielectric thickness decreases, the time delay of nondispersive regions increases. The delay characteristics can be controlled by adjusting the dielectric and ferrimagnetic film thicknesses and the biasing field.

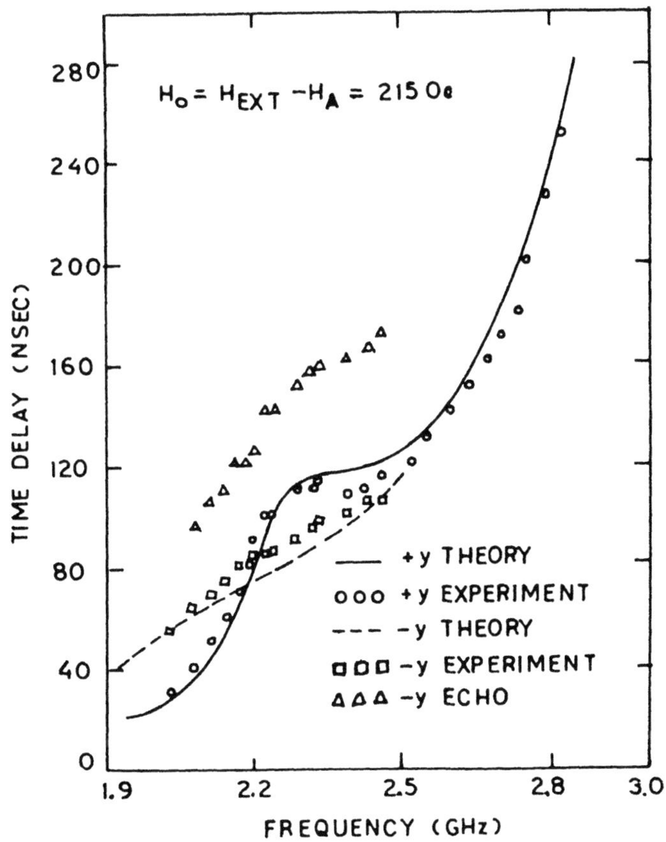

Figure 4.52a. Comparison of calculated and measured variation of delay time with frequency for 10-μm-thick YIG film separated from the grounded plane by a 125-μm-thick glass dielectric (after Bongianni, 1972).

The explicit effect of the finite linewidth on the attenuation of magnetostatic surface waves in the dielectric layered structure has been theoretically investigated by Vittoria and Wilsey (1974). When μ and κ in equation (4.139) are replaced by their appropriate expressions from equations (1.131) and (1.32), β turns out to be complex. The group velocity may then be defined as

$$v_g = \mathrm{Re}(\partial\omega/\partial\beta) \tag{4.143}$$

and the attenuation factor is given by

$$\alpha_p = 20\log_{10} e \times \mathrm{Im}(\beta) \tag{4.144}$$

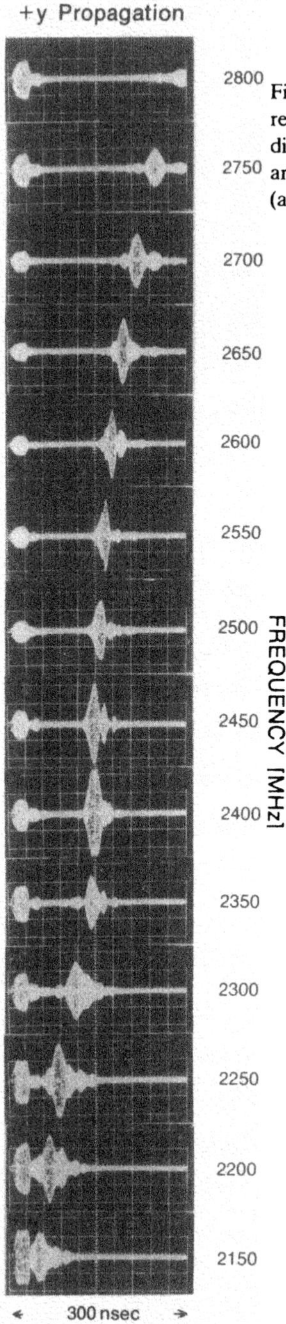

Figure 4.52b. Oscilloscope traces of the pulse response of magnetostatic surface waves in a dielectric layered structure. The parameters are $d = 10\,\mu\text{m}$, $t = 75\,\mu\text{m}$, and $H_0 = 215\,\text{Oe}$ (after Bongianni, 1972).

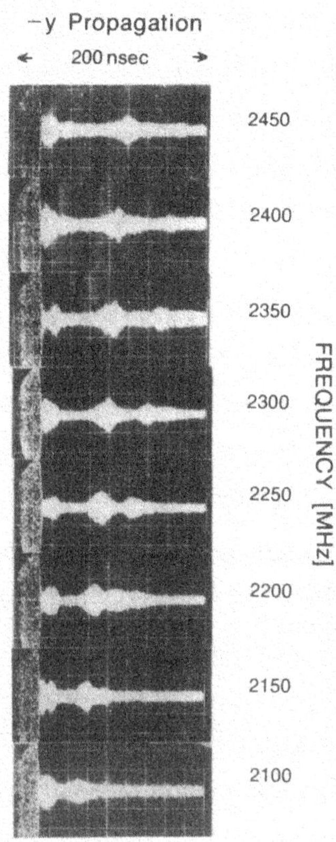

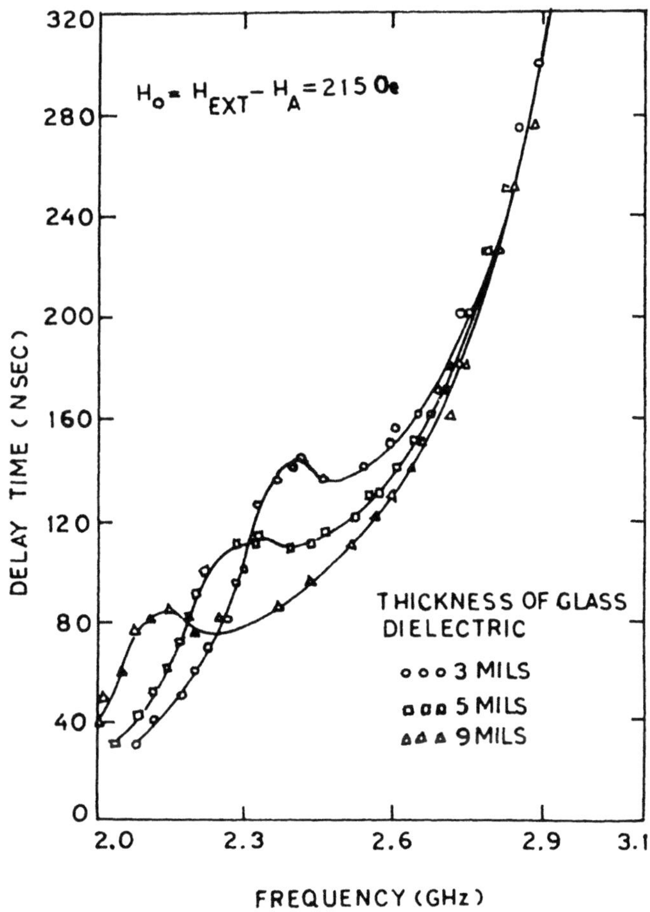

Figure 4.53. Frequency dependence of the delay time for different thicknesses of the glass dielectric layer (after Bongianni, 1972).

Vittoria and Wilsey (1974) have defined the loss factor as

$$L_p = v_g \alpha_p \times 10^{-6} \qquad [\mathrm{dB}/\mu\mathrm{sec}] \tag{4.145}$$

This factor is regarded as the inverse figure of merit of a magnetostatic wave device. Figure 4.54 shows the variation of L_p with frequency for a typical set of parameters. At low frequencies, the loss factor is more or less a constant, whereas, at relatively high frequencies, it is proportional to the frequency and is given by (Adam, 1970; Vittoria and Wilsey, 1974):

$$L_p = 4\pi \times 76.4\alpha f/\gamma \tag{4.146}$$

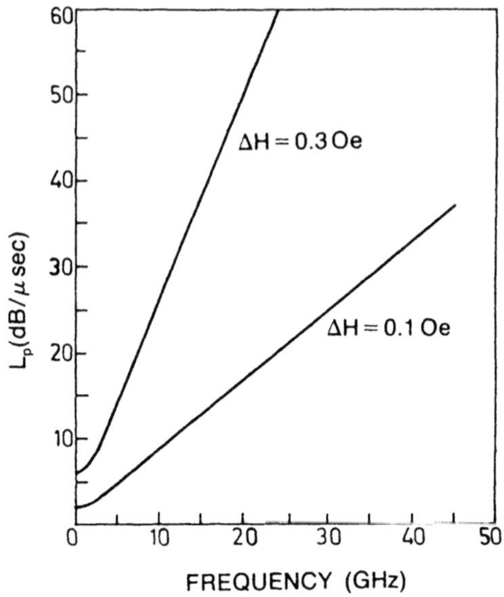

Figure 4.54. Variation of the magnetostatic wave loss factor L_p with frequency for $\Delta H = 0.1$ and 0.3 Oe. At high frequencies the loss factor increases linearly with the wave frequency. At low frequencies, however, the loss factor attains a constant value. The damping parameter is assumed to be frequency-independent (after Vittoria and Wilsey, 1974).

where α is given by equation (1.36), with the aid of which this reduces to equation (4.39), i.e.,

$$L_p = 76.4\,\Delta H \qquad [\text{dB}/\mu\text{sec}] \qquad (4.147)$$

Figure 4.55 shows the frequency dependence of experimentally measured attenuation of magnetostatic surface waves. Whereas the high-frequency behavior seems to be consistent with equation (4.147), it is seen that below about 3 GHz, the attenuation increases sharply with decreasing frequency. Webb et al. (1975) have shown that this is due to the increase of the damping parameter λ below 3 GHz which, in turn, leads to increased linewidth and hence to increased loss. Figure 4.56 shows the comparison of measured attenuation with that calculated from equations (4.139), (4.144), and (4.145), after accounting for the increase in λ. Clearly, at least above 3 GHz, the magnetostatic surface waves have a tolerably low attenuation rate.

It is clear from this discussion that the consideration of losses and bandwidth of the region of nondispersive delay leads to the conclusion that magnetostatic surface wave devices are particularly useful in the 3–6 GHz

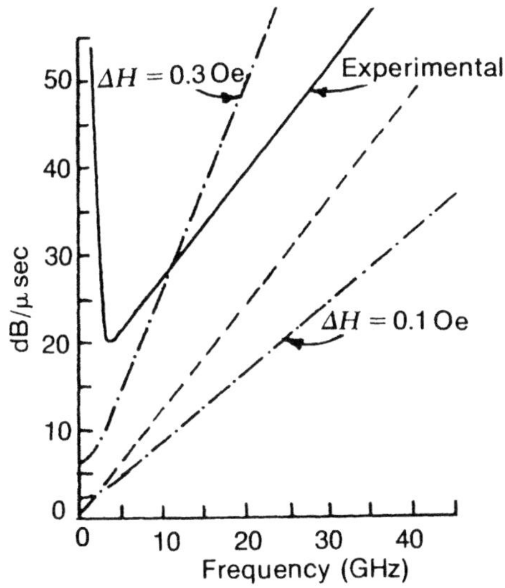

Figure 4.55. An experimentally measured loss factor compared with the calculated one. The general nature in the high-frequency region seems to be explained by the theory of Vittoria and Wilsey (1974). However, below about 3 GHz, there is a drastic increase in the loss factor, which is not accounted for by the theory based on the assumption of frequency-independent damping parameter (after Sethares and Stiglitz, 1974).

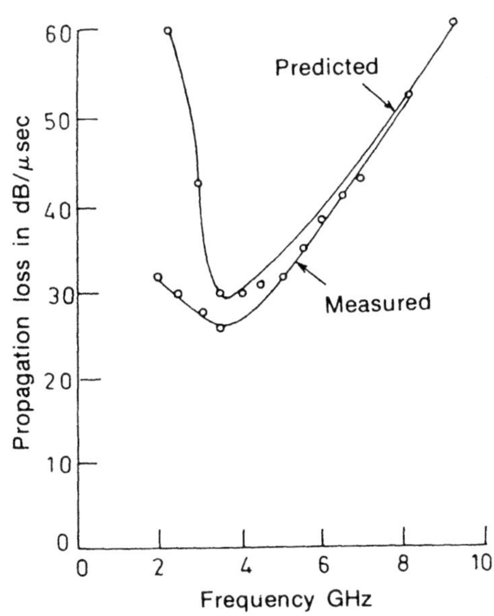

Figure 4.56. Comparison of measured magnetostatic wave propagation loss with that calculated from the theory of Vittoria and Wilsey (1974) after accounting for the increase in the damping parameter at low frequencies (after Webb et al., 1975).

range. Adam et al. (1975) have experimentally investigated spatially tapped magnetostatic surface wave delay lines, which form the basis for real-time microwave signal processing. Figure 4.57 shows such a delay line. The S-band experimental plot of delay vs. frequency for the tapped delay line is shown in Figure 4.58. Both tap and full delay are seen to be nondispersive over a bandwidth of about 225 MHz.

The analysis leading to equation (4.139) was carried out under the assumption that the wave field is uniform in the direction of the biasing field. Kawasaki et al. (1974a) have theoretically and experimentally investigated the magnetostatic surface wave propagation in the configuration shown in Figure 4.59. The variation of the wave field in the z-direction, i.e., the direction of dc magnetization, has also been considered. It has been shown that continuous pass-band control is obtained by changing the spacing from zero to infinity; the upper limit of the pass-band varies

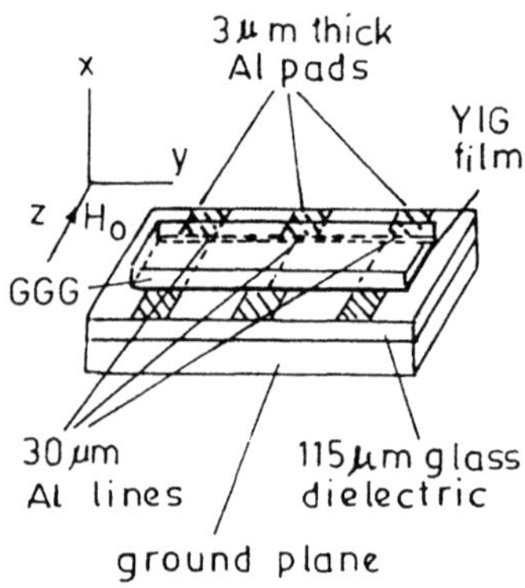

Figure 4.57. Construction of a trapped, nondispersive YIG film delay line (after Adam et al., 1975).

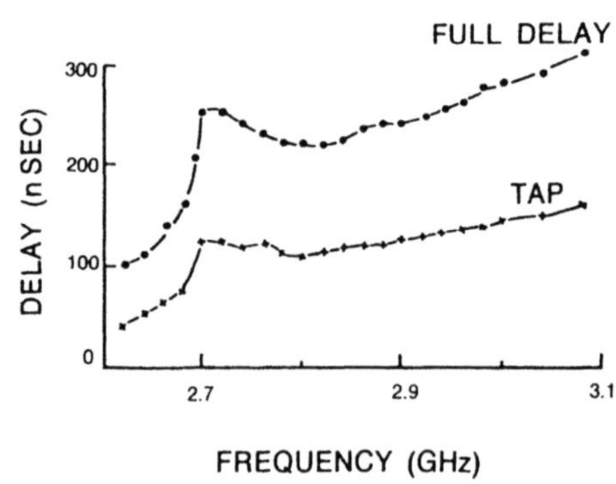

Figure 4.58. Experimental plot of delay vs. frequency for the tapped delay line shown in Figure 4.57. The parameters are YIG film thickness $d = 4.5\,\mu$m, glass dielectric thickness $t = 115\,\mu$m, and biasing field $H_0 = 500$ Oe (after Adam et al., 1975).

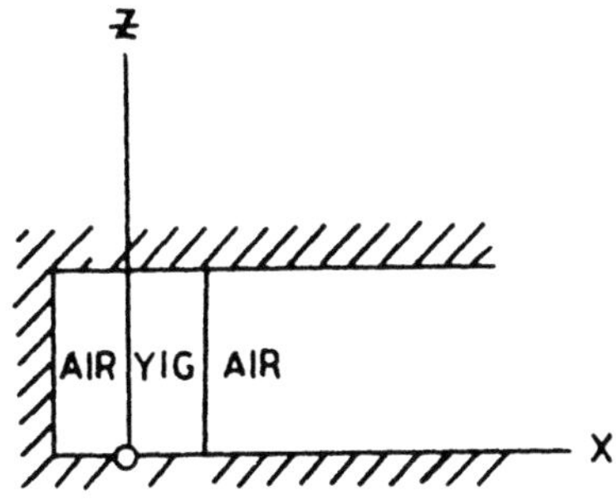

Figure 4.59. Air–YIG–air system shielded on three sides. The biasing field is along the z-axis, whereas the magnetostatic wave propagates in the y-direction (after Kawasaki et al., 1974a).

continuously from $\omega = \omega_0 + \omega_m$ to $\omega = \omega_0 + \frac{1}{2}\omega_m$ as the spacing varies from zero to infinity. The nonreciprocity disappears from a finite spacing and a partial stop-band also exists from the interchange of input and output ports. Broad transmission can be obtained by exciting multiple modes. The detailed results can be found in Kawasaki et al. (1974a). Recently, O'Keeffe and Patterson (1978) presented a detailed theoretical analysis accounting for the finite sample width. The numerical results compare extremely well with experimental results, thus removing the discrepancy between theory and experiment.

4.4.2. *Normal Magnetization*

This configuration has been theoretically and experimentally investigated, from the viewpoint of magnetostatic bulk wave propagation, by Bardai et al. (1976), Tsai et al. (1976), Miller (1976a, b), and Gupta and Srivastava (1980b). This is, perhaps, the most promising configuration for X-band delay devices since, as noted earlier, only the forward bulk waves have a significant bandwidth at the X-band and lead to a region of nondispersive delay over a reasonable frequency range. In what follows, we have summarized theoretical analyses and experimental results regarding dispersion characteristics (Section 4.4.2.1). The energy distribution and power flow are discussed briefly in Section 4.4.2.2.

4.4.2.1. *Dispersion Characteristics*

The configuration is shown in Figure 4.60. The magnetization is in the z-direction, whereas the wave propagates in the y-direction. The magnetostatic potential and induction in air, ferrimagnetic, and dielectric regions can be expressed as:

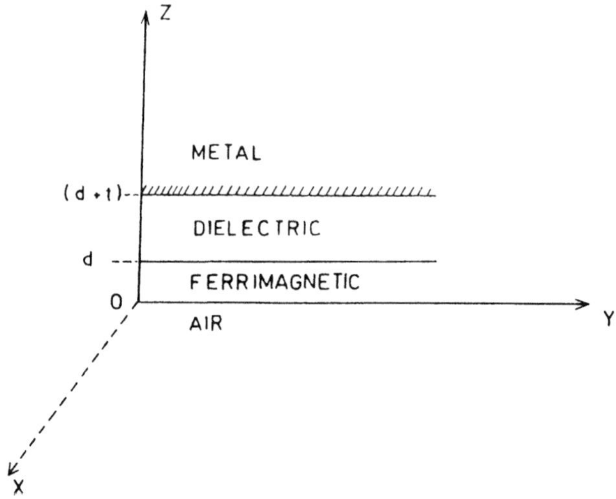

Figure 4.60. A normally biased ferrimagnetic film in a dielectric layered structure.

$$\left.\begin{aligned}
\psi^{a} &= \exp(\beta z)\exp(-j\beta y) \\
\psi^{f} &= [A\exp(j\alpha_v\beta z) + B\exp(-j\alpha_v\beta z)]\exp(-j\beta y) \\
\psi^{d} &= [C\exp(\beta z) + D\exp(-\beta z)]\exp(-j\beta y) \\
\mathbf{b}^{a} &= -\beta(j\hat{y} - \hat{z})\psi^{a} \\
\mathbf{b}^{f} &= \beta\{\kappa\psi^{f}\hat{x} - j\mu\psi^{f}\hat{y} + j\alpha_v\hat{z} \\
&\qquad \times [A\exp(j\alpha_v\beta z) - B\exp(-j\alpha_v\beta z)]\exp(-j\beta y)\} \\
\mathbf{b}^{d} &= -\beta\{j\psi^{d}\hat{y} - [C\exp(\beta z) - D\exp(-\beta z)]\exp(-j\beta y)\hat{z}\}
\end{aligned}\right\} \qquad (4.148)$$

where $\alpha_v = (-\mu)^{1/2}$. The standard procedure leads to the following dispersion relation (Bardai et al., 1976; Miller, 1976a):

$$2\alpha_v \cot \alpha_v\beta d = (\alpha_v^2 + 1)\exp(-2\beta t) + (\alpha_v^2 - 1) \qquad (4.149)$$

The group velocity is obtained as (Miller, 1976a):

$$\left.\begin{aligned}
v_g &= N_v/D_v \\
N_v &= (\omega_0^2 - \omega^2)^2\alpha_v^2\left[d - \left(\frac{\alpha_v^2 + 1}{\alpha_v^2}\right)t\sin^2\alpha_v\beta d\exp(-2\beta t)\right] \\
D_v &= \omega\omega_0\omega_m\left\{\beta d + \frac{\sin^2\alpha_v\beta d}{2\alpha_v^2}[(\alpha_v^2 - 1)\exp(-2\beta t) + (\alpha_v^2 + 1)]\right\}
\end{aligned}\right\} \qquad (4.150)$$

The delay characteristics computed for a typical set of data from equation (4.150) are shown in Figure 4.61. The delay curves have been shown for the first two modes. Only the first mode exhibits a region of nondispersive delay for a finite thickness of the dielectric layer. Figures 4.62 and 4.63 compare measured and calculated delay times as obtained by Bardai et al. (1976) and Miller (1976a), respectively. In Figure 4.62, there is a constant delay of about 350 nsec ±5% over a 200 MHz bandwidth (in the S-band). Figure 4.63 shows the results for the X-band, in which almost constant delay is obtained over a bandwidth of about 150 MHz. More experimental results have been presented by Tsai et al. (1976).

4.4.2.2. *Energy and Power*

The energy distribution and power flow associated with magnetostatic volume wave propagation in a free, normally magnetized ferrimagnetic

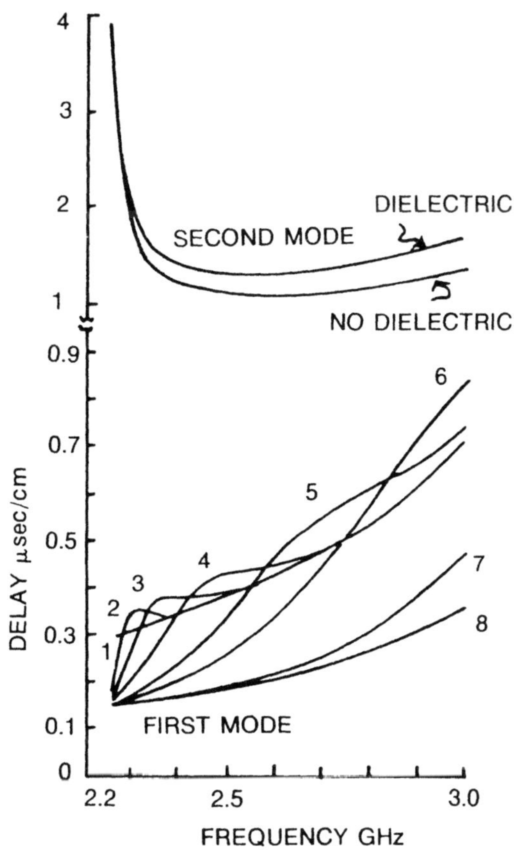

Figure 4.61. Magnetostatic forward bulk wave delay vs. frequency for the first two modes for different thicknesses of the dielectric layer. The curves for the first mode, designated by numbers from 1 to 8 in increasing order, correspond to dielectric thickness (in micrometers) = ∞, 100, 50, 25, 10, 5, 2, and 0, respectively. Other parameters are $H_0 = 800$ Oe, $4\pi M_0 = 1750$ G, and $d = 4.5\,\mu$m (after Bardai et al., 1976).

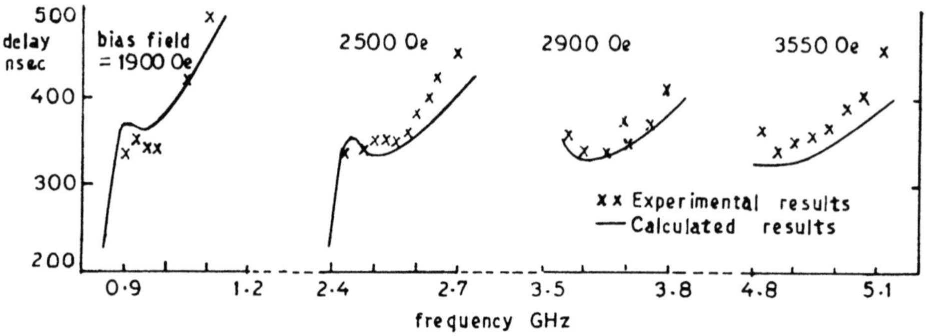

Figure 4.62. Comparison of measured and calculated magnetostatic forward bulk wave delay for four different values of the biasing field. The parameters are $d = 4.5\,\mu\text{m}$ and $4\pi M_0 = 1750\,\text{Oe}$ (after Bardai et al., 1976).

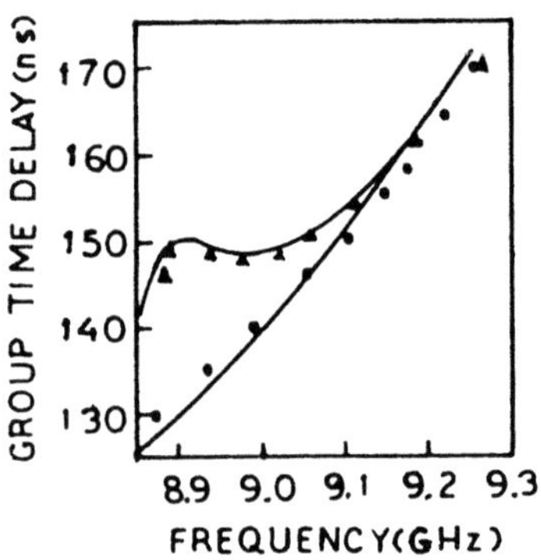

Figure 4.63. Comparison of experimental and theoretical group delay for magnetostatic forward bulk waves in the X-band. The parameters are: propagation path length = 8 mm, internal biasing field = 4850 Oe; $\varepsilon_d = 2.35$, $t_{eff} = 178\,\mu\text{m}$ for triangles, whereas $\varepsilon_d = 9.6$, $t_{eff} = 635\,\mu\text{m}$ for dots (after Miller, 1976a).

slab was discussed in Section 4.2.2.2. The analysis presented in that section has been generalized to the dielectric layered structure by Gupta and Srivastava (1980b). The group velocity and the velocity of energy flow are once again found to be equal. Using equations (4.79) and (4.148), we can express the power flowing through the different regions as

$$\left.\begin{aligned}
\mathbf{P}_a^t &= -\frac{\omega}{16\pi}\,\hat{y} \\
\mathbf{P}_f^t &= \frac{\omega\beta}{8\pi}\left(\frac{\alpha_v^2+1}{2}\right) \\
&\quad \times \left\{ d + \left(\frac{\alpha_v^2+1}{2\beta\alpha_v^2}\right)\sin^2\alpha_v\beta d\left[1+\frac{\alpha_v^2-1}{\alpha_v^2+1}\exp(-2\beta t)\right]\right\}\hat{y} \\
\mathbf{P}_d^t &= -\frac{\omega}{16\pi}\,\hat{y}\left\{\left[1+\left(\frac{\alpha_v^2-1}{\alpha_v}\right)\cot\alpha_v\beta d\right.\right. \\
&\quad \left.\left. + \left(\frac{\alpha_v^2+1}{\alpha_v}\right)^2\beta t\exp(-2\beta t)\right]\sin^2\alpha_v\beta d - \cos^2\alpha_v\beta d\right\}
\end{aligned}\right\} \tag{4.151}$$

The net power flowing (per unit width in the x-direction) in the direction of phase propagation is obtained as

$$\left.\begin{aligned}\mathbf{P}^{T} &= \mathbf{P}_{a}^{t} + \mathbf{P}_{f}^{t} + \mathbf{P}_{d}^{t} \\ &= \frac{\omega\beta(\alpha_{v}^{2}+1)}{16\pi}\left[d - \left(\frac{\alpha_{v}^{2}+1}{\alpha_{v}}\right) t \sin^{2}\alpha_{v}\beta d \exp(-2\beta t)\right]\hat{y}\end{aligned}\right\} \quad (4.152)$$

It is easy to show from the last of equations (4.151) that the power flow in dielectrics is also opposite to the direction of phase propagation. Figures 4.64a and b show the dependence on the wave frequency of relative power

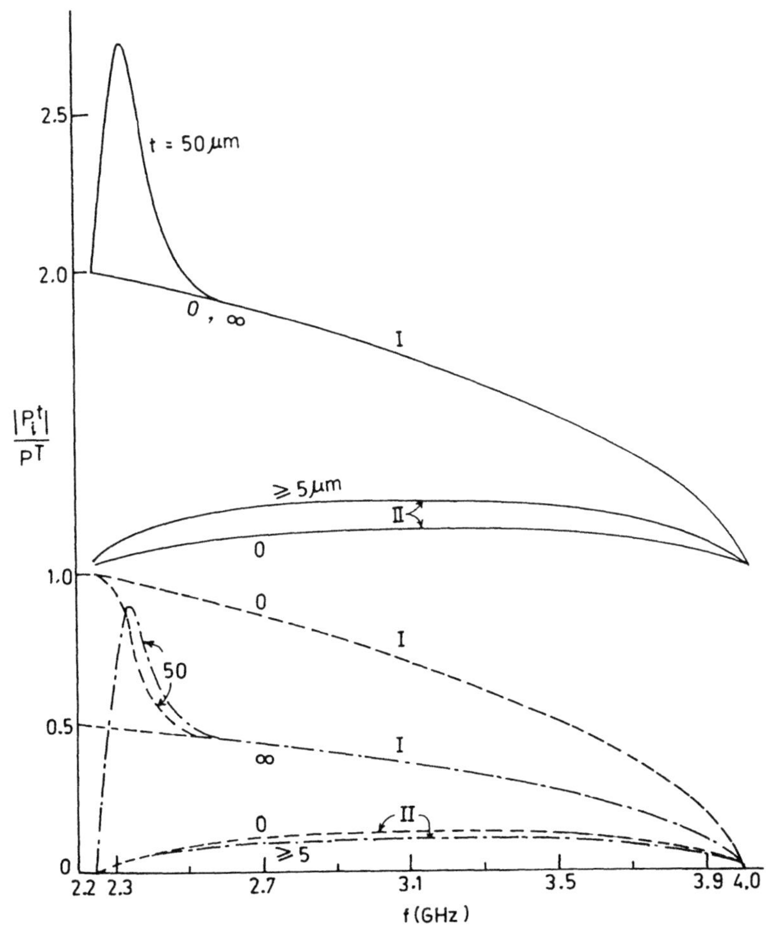

Figure 4.64a. Frequency dependence of the magnitudes of fractional powers flowing in ferrimagnetic (———), dielectric (—·—·—) and air (- - - -), for the first two modes for $t = \infty$, 50 μm, and 0. Other parameters are $H_0 = 800$ Oe, $4\pi M_0 = 1750$ G, and $d = 4.50$ μm.

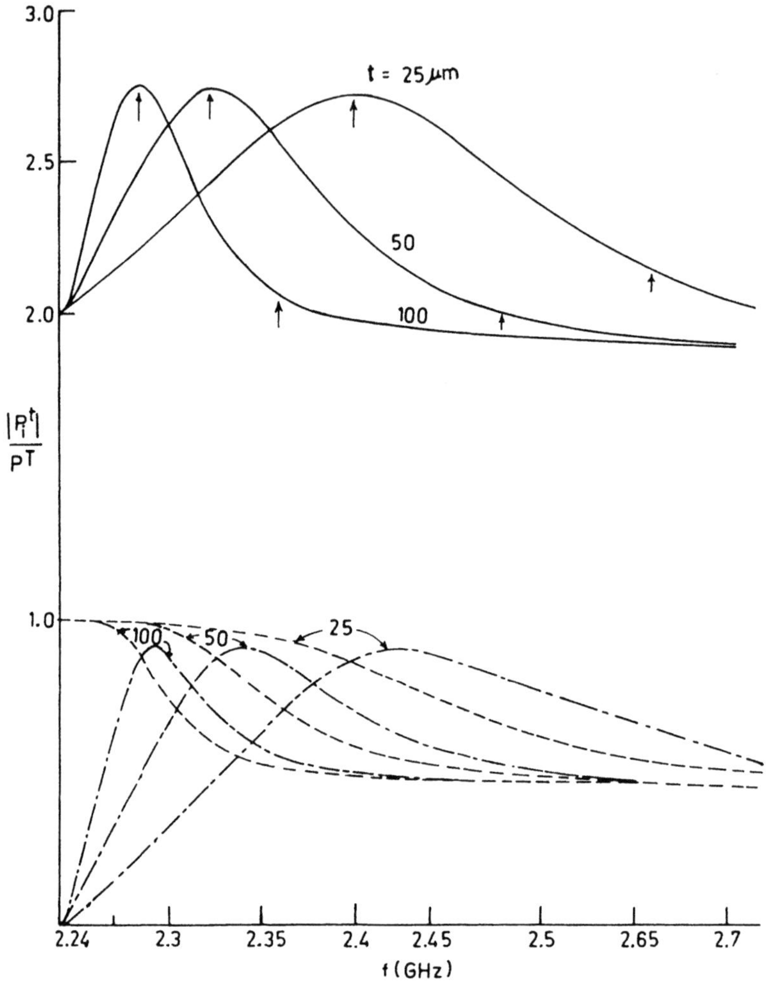

Figure 4.64b. Frequency dependence of the magnitudes of fractional power flow in different regions, on an expanded scale, for different dielectric thicknesses shown alongside each curve. The regions of nondispersive delay are marked by arrows under each curve. Other parameters are the same as for Figure 4.64a.

flow through ferrite, air, and dielectric regions. Figure 4.64a shows the fractional powers through the three regions for the first two modes. It is seen that the fractional power through the ferrimagnetic region exceeds unity, which is understandable since, as noted above, the power flow in a ferrimagnetic region exceeds the net power flow (due to power flow in the opposite direction in dielectrics). The frequency limits corresponding to the region of nondispersive delay have also been shown in Figure 4.64a. Clearly, the region of nondispersive delay is characterized by the sharp decrease of fractional power through the ferrimagnetic region. The threshold frequency of the region of nondispersive delay practically coincides with the frequency at which the fractional power through the ferrimagnetic region is maximum. For the second mode, the effect of the presence of the metal plate is rather insignificant. In contrast to the first mode, all the power for the second mode flows through the ferrimagnetic region at the cut-off limit as well as at resonance, and this seems to be related to the fact that the group delay for the second mode is infinite at both the frequency limits. The qualitative power flow behavior for still higher modes is the same as that for the second-order mode (results not shown). In Figure 4.64b, the critical regions of Figure 4.64a have been presented on an expanded frequency scale for different thicknesses of the dielectric layer. As the dielectric thickness increases, there is a slight decrease in the bandwidth of the region of nondispersive delay. Also, this region shifts to lower frequencies. Further details regarding power flow and energy distribution can be found in Gupta and Srivastava (1980b).

It is worth mentioning that the zig-zag ray model presented in Section 4.2.2.3 can also be generalized to the dielectric layered structure.

4.4.3. *Arbitrary Magnetization*

This section follows the papers by Miller (1977, 1978) and Bajpai and Srivastava (1980a). The configuration is shown in Figure 4.65. The dc magnetization is assumed to be directed along the z'-axis of the primed coordinate system, which is oriented with respect to the unprimed one through the first two Eulerian angles, φ and θ, as in Figure C.1 of Appendix C. In the present case, since the dc field is along the z'-axis, it is

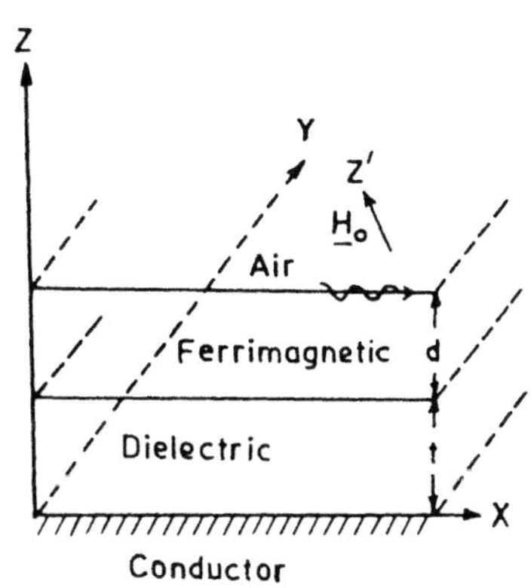

Figure 4.65. A ferrimagnetic film magnetized along the z' (arbitrary) direction in a dielectric layered structure.

$\boldsymbol{\mu}'$ which has the form given by equation (1.25). The permeability tensor $\boldsymbol{\mu}$ (in an unprimed coordinate system) can be related to $\boldsymbol{\mu}'$ (see Appendix C) by

$$\boldsymbol{\mu}' = R\boldsymbol{\mu}\tilde{R} \tag{4.153}$$

where R is the same as that given by equation (C.16). Equivalently,

$$\begin{aligned}\boldsymbol{\mu} &= \tilde{R}\boldsymbol{\mu}'R \\ &= \begin{pmatrix} \mu^{a}_{11} & \mu^{a}_{12} & \mu^{a}_{13} \\ \mu^{a}_{21} & \mu^{a}_{22} & \mu^{a}_{23} \\ \mu^{a}_{31} & \mu^{a}_{32} & \mu^{a}_{33} \end{pmatrix}\end{aligned} \tag{4.154}$$

where

$$\left.\begin{aligned}
\mu^{a}_{11} &= \mu(\cos^2\varphi + \cos^2\theta\sin^2\varphi) + \sin^2\theta\sin^2\varphi \\
\mu^{a}_{12} &= (\mu - 1)\sin^2\theta\sin\varphi\cos\varphi + j\kappa\cos\theta \\
\mu^{a}_{13} &= [(1-\mu)\sin\varphi\cos\theta + j\kappa\cos\varphi]\sin\theta \\
\mu^{a}_{21} &= (\mu - 1)\sin^2\theta\sin\varphi\cos\varphi - j\kappa\cos\theta \\
\mu^{a}_{22} &= \mu(\sin^2\varphi + \cos^2\theta\cos^2\varphi) + \sin^2\theta\cos^2\varphi \\
\mu^{a}_{23} &= [(\mu - 1)\cos\theta\cos\varphi + j\kappa\sin\varphi]\sin\theta \\
\mu^{a}_{31} &= [(1-\mu)\sin\varphi\cos\theta - j\kappa\cos\varphi]\sin\theta \\
\mu^{a}_{32} &= [(\mu - 1)\cos\theta\cos\varphi - j\kappa\sin\varphi]\sin\theta \\
\mu^{a}_{33} &= \mu\sin^2\theta + \cos^2\theta
\end{aligned}\right\} \tag{4.155}$$

It is of interest that the dc field appearing in the expressions for μ and κ is the internal field $\mathbf{H}_0$ which, in turn, is related to the external field $\mathbf{H}_e$ by

$$\left.\begin{aligned}
H_0\sin\theta &= H_e\sin\psi \\
H_0\cos\theta &= H_e\cos\psi - 4\pi M_0\cos\theta
\end{aligned}\right\} \tag{4.156}$$

where ψ is the angle between $\mathbf{H}_e$ and the z-axis.

Consider the propagation of a magnetostatic wave in the x-direction, assuming a uniform field distribution in the transverse y-direction. The standard magnetostatic analysis (Bajpai and Srivastava, 1980a) leads to the following dispersion relation:

$$\begin{aligned}\mu^{a}_{33}\alpha_a\cot\alpha_a\beta d(1+\coth\beta t) = \{&\kappa\sin\theta\cos\varphi(1-\coth\beta t) \\ &+ \coth\beta t[(\mu^{a}_{33}\alpha_a)^2 + (\kappa\sin\theta\cos\varphi)^2] - 1\}\end{aligned} \tag{4.157}$$

where

$$\left.\begin{aligned} \alpha_a &= (p_a^2 - n_a^2)^{1/2}/\beta \\ p_a &= \beta(1-\mu)\sin\theta\sin\varphi\cos\theta/\mu_{33}^a \\ n_a &= \beta(\mu_{11}^a/\mu_{33}^a)^{1/2} \end{aligned}\right\} \tag{4.158}$$

In the case of guided bulk waves, α_a is required to be real. This leads to the frequency range of allowed modes as $\omega_1 < \omega < \omega_3'$, where

$$\left.\begin{aligned} \omega_1' &= [\omega_0(\omega_0 + \omega_m \sin^2\theta\cos^2\varphi)]^{1/2} \\ \omega_3' &= [\omega_0(\omega_0 + \omega_m)]^{1/2} \end{aligned}\right\} \tag{4.159}$$

The dispersion relation for guided surface modes can be obtained from equation (4.157) by replacing α_a by $j\alpha_a$. The expression for group velocity* is obtained from equation (4.157) as

$$v_g = \partial\omega/\partial\beta = (A_a + B_a)/(P_a + Q_a + R_a + S_a) \tag{4.160}$$

where

$$\left.\begin{aligned} A_a &= \mu_{33}^a d\alpha_a^2(1+\coth\beta t)\operatorname{cosech}^2\beta\alpha_a d \\ B_a &= t\operatorname{cosech}^2\beta t(\mu_{33}^a\alpha_a\cot\beta\alpha_a d + \kappa\sin\theta\cos\varphi \\ &\qquad - \mu_{33}^a\alpha_a^2 - \kappa^2\sin^2\theta\cos^2\varphi) \\ P_a &= -\beta\alpha_a' d\mu_{33}^a\alpha_a(1+\coth\beta t)\operatorname{cosec}^2\beta\alpha_a d \\ Q_a &= (1+\coth\beta t)(\mu_{33}^a\alpha_a' + \mu_{33}^{a\prime}\alpha_a)\cot\beta\alpha_a d \\ R_a &= -\kappa'\sin\theta\cos\varphi(1-\coth\beta t) \\ S_a &= -2[\mu_{33}^a\alpha_a(\mu_{33}^a\alpha_a' + \mu_{33}^{a\prime}\alpha_a) + \kappa\kappa'\cos^2\varphi\sin^2\theta]\coth\beta t \end{aligned}\right\} \tag{4.161}$$

In the expressions given above, the primes denote derivatives with respect to ω. Thus

$$\left.\begin{aligned} \mu' &= 2\omega_0\omega_m\omega/(\omega_0^2-\omega^2)^2 \\ \kappa' &= \omega_m(\omega^2+\omega_0^2)/(\omega_0^2-\omega^2)^2 \\ (\mu_{11}^a)' &= \mu'(\cos^2\varphi + \cos^2\theta\sin^2\varphi) \\ (\mu_{33}^a)' &= \mu'\sin^2\theta \\ \alpha_a' &= \frac{1}{2\alpha_a}\left\{\frac{(\mu-1)\sin^2\theta\cos^2\theta\sin^2\varphi}{\mu_{33}^a}\left[\frac{(1-\mu)\mu_{33}^{a\prime}}{(\mu_{33}^a)^2} - \frac{\mu'}{\mu_{33}^a}\right]\right. \\ &\qquad \left. + \frac{\mu_{11}^a\mu_{33}^{a\prime}}{(\mu_{33}^a)^2} - \frac{\mu_{11}^{a\prime}}{\mu_{33}^a}\right\} \end{aligned}\right\} \tag{4.162}$$

* This velocity is in fact the inverse of the time required for a wave group to propagate between a set of parallel microstrips (separated by unit distance), each of which is perpendicular to the x-axis. It may be noted, however, that the actual direction of propagation may deviate from the x-axis owing to steering.

Figure 4.66 shows typical backward wave dispersion curves for $\theta = 45°$ and $\varphi = 60°$. It is seen that at low (high) frequencies, the guided bulk waves are backward (forward) waves. For $\omega > \omega_3'$, surface modes are also allowed provided that θ exceeds $\theta_c = \cos^{-1}[(\omega_m/\omega_0)^{1/2}]$, as discussed in Section 4.2. Surface and bulk modes for the special case of magnetization in the transverse plane ($\varphi = 0$, variable θ) have been investigated by Miller (1978); Figure 4.67 shows such dispersion curves for some selected values of θ. When θ is small, only bulk modes are allowed. When θ exceeds θ_c, surface as well as bulk modes propagate. Finally, when $\theta = \pi/2$, only the surface modes are allowed.

The delay characteristics for different combinations of θ and φ have been shown in Figures 4.68–4.70. Figure 4.68 displays the frequency dependence of delay time for $\varphi = \pi/2$ and different values of θ; this corresponds to rotation of the direction of dc magnetization in the

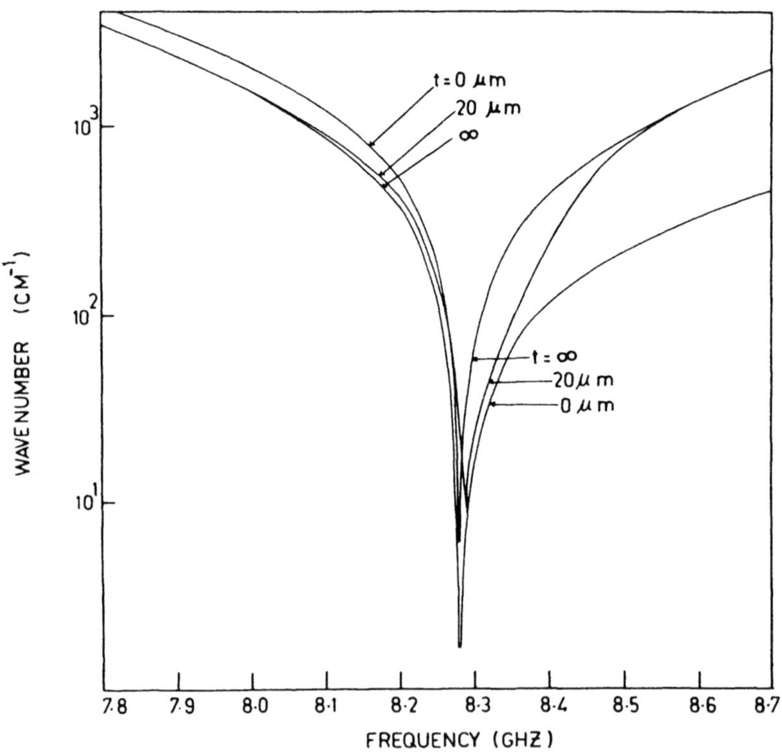

Figure 4.66. Dispersion curves for the configuration shown in Figure 4.65 with the orientation of the z'-axis characterized by the Eulerian angles $\varphi = 60°$ and $\theta = 45°$. The dielectric thickness is shown alongside each curve. Other parameters are $4\pi M_0 = 1750$ G, $H_0 = 2550$ Oe, and $d = 4.5\ \mu$m (S.N. Bajpai, private communication).

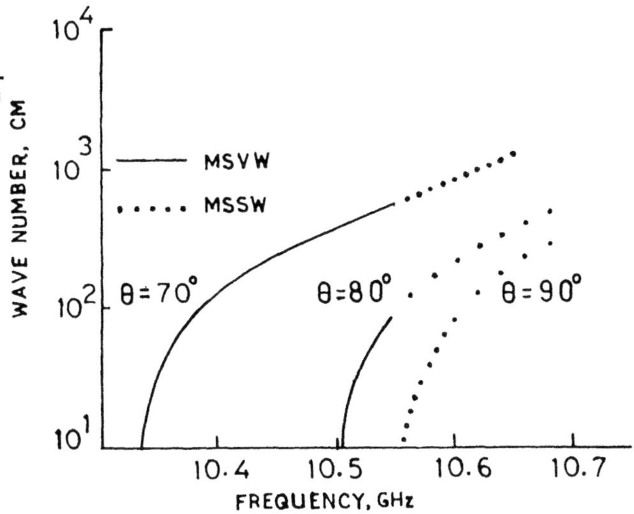

Figure 4.67. Computed dispersion curves for $\varphi = 0$ and variable θ. The parameters are $H_0 = 3000$ Oe, $t = 1000\,\mu$m, $4\pi M_0 = 1750$ G, and $d = 10\,\mu$m (after Miller, 1978).

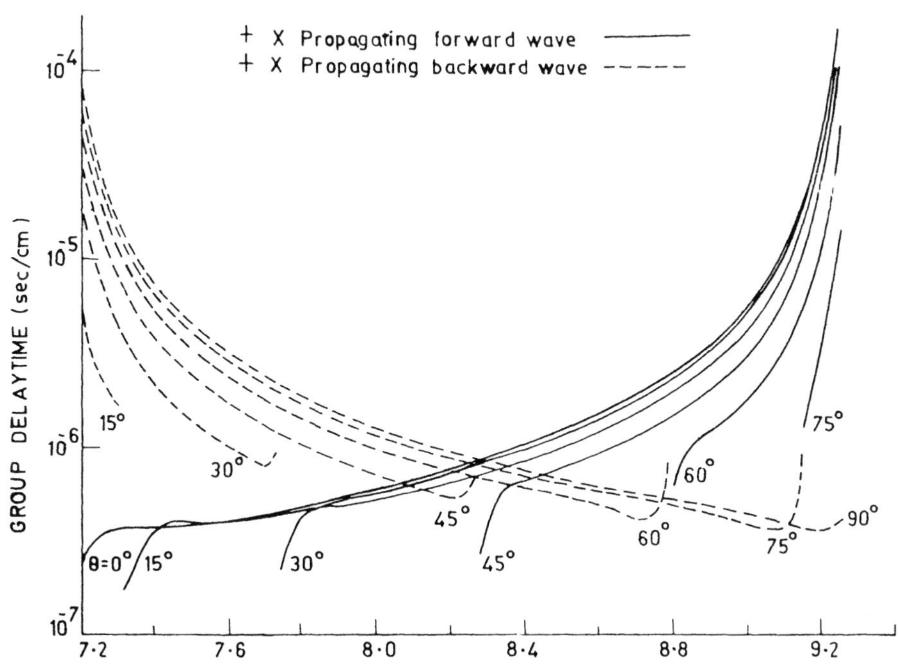

Figure 4.68. Variation of group delay time with frequency for $\varphi = 90°$ and different values of θ. Other parameters are $t = 40\,\mu$m, $d = 4.5\,\mu$m, $H_0 = 2550$ Oe, and $4\pi M_0 = 1750$ G (after Bajpai and Srivastava, 1980a).

xz-plane, i.e., the plane containing the direction of propagation and the normal (to the film). In this case, the delay characteristics are reciprocal since equation (4.160) is independent of κ when $\varphi = \pi/2$. The general nature of the curves can be understood in terms of the specific cases investigated earlier. In the case of normal magnetization ($\theta = 0$), the guided waves are forward waves with resonance frequency $\omega = \omega_3'$, whereas, in the case of longitudinal magnetization ($\varphi = \pi/2$, $\theta = \pi/2$), the guided waves are backward waves with resonance frequency $\omega = \omega_0$. It is seen from Figure 4.68 that when θ differs from 0 or $\pi/2$, the delay curves contain forward (solid lines) as well as backward (broken lines) wave branches. The backward (forward) waves propagate in the low (high) frequency region. The transition from backward to forward waves is abrupt (since losses have been neglected) and occurs at a frequency $\omega = \omega_2'$, where

$$\omega_2' = \sqrt{\omega_0(\omega_0 + \omega_m \sin^2\theta)} \tag{4.163}$$

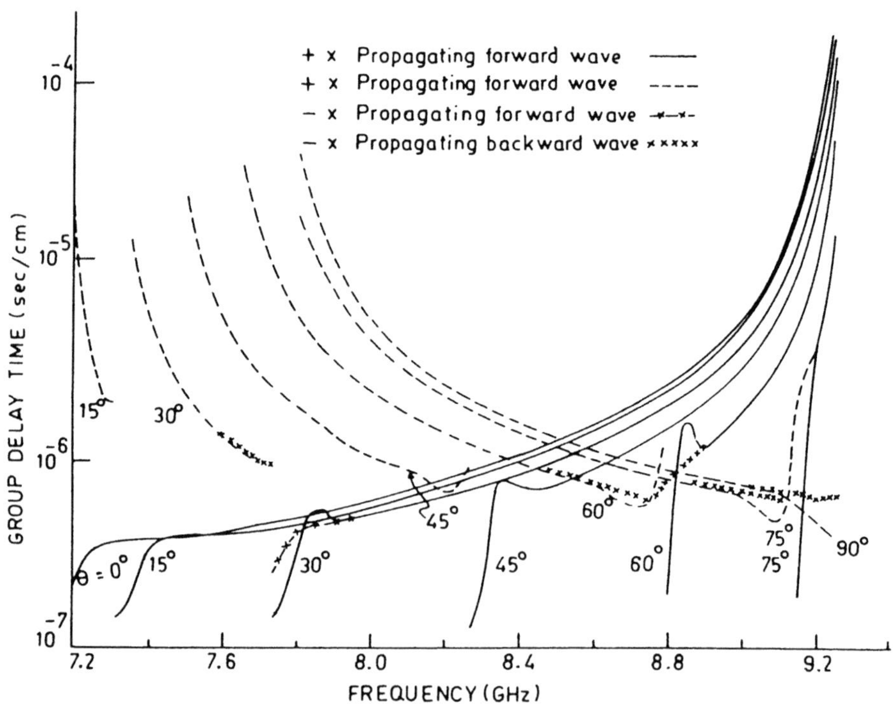

Figure 4.69. Variation of group delay time with frequency for $\varphi = 60°$ and different values of θ. Other parameters are the same as for Figure 4.68. To avoid confusion, the delay curves for $-x$-propagating waves have not been shown for all values of θ (after Bajpai and Srivastava, 1980a).

The backward- and forward-wave branches are obtained in the ranges $\omega_0 < \omega < \omega_2'$ and $\omega_2' < \omega < \omega_3'$, respectively. It follows from equation (4.163) that as θ increases from 0 to $\pi/2$, ω_2' varies from ω_0 to ω_3'. Consequently, the backward-wave region grows at the expense of the forward-wave region. It is also seen from the figure that the forward-wave branches contain a region of nondispersive delay. No such region is obtained for backward waves. Figures 4.69 and 4.70 are plotted for $\varphi = 60°$ and 30°, respectively. It is seen that, for $\theta \neq 0$, the lower cut-off frequency limit ω_1' depends on θ as well as φ. Thus, as θ increases, the dc magnetization deviates from the normal and the overall range of the allowed modes decreases. Consequently, there is a decrease in the frequency range of nondispersive delay for forward waves. It is interesting to note that when $\varphi \neq \pi/2$, the propagation characteristics are nonreciprocal. The delay curves for reverse propagation are also shown in Figures 4.69 and 4.70. The nonreciprocity is small except when ω is close to ω_2'.

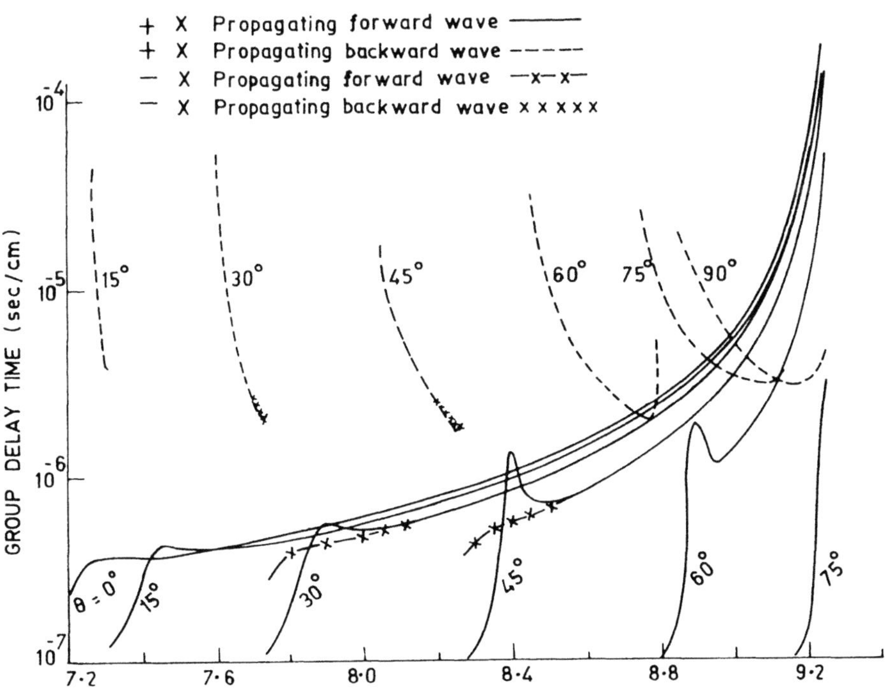

Figure 4.70. Variation of group delay time with frequency for $\varphi = 30°$ and different values of θ. Other parameters are the same as for Figure 4.68. The delay curves for $-x$-propagating waves have not been shown for all values of θ (after Bajpai and Srivastava, 1980a).

Experimental results pertaining to the case of arbitrary magnetization have been reported by Schilz (1973) and Miller (1977). Figures 4.71 and 4.72 compare the theoretical and experimental group delay for the case of magnetization in the plane of the film ($\theta = \pi/2$, different $\varphi' = \pi/2 - \varphi$). Good agreement between theory and experiment is obtained when the physical thickness of the dielectric layer, as appearing in equations (4.157) and (4.161), is replaced by the effective thickness, as given by equation (4.142). It is seen from Figure 4.72 that the observed delay characteristics are nonreciprocal and a small region of nondispersive delay also exists.

Miller (1978) has explained the experimental results obtained by Schilz (1973) in the case of magnetization in the transverse plane [$\varphi = 0$ and different θ in equations (4.157)–(4.162)]. Figure 4.73 shows the dependence of the frequency of a magnetostatic wave on the applied magnetic field for a constant wave number. In the calculations, the internal field H_0 and its orientation θ have been obtained from equation (4.156). The agreement between measured and calculated delay is generally good. It is also evident from the figure that there are regions in which the wave frequency is independent of the biasing field and in which it increases on decreasing the biasing field. These features may prove to be important in temperature-stable magnetostatic wave oscillators and narrow-bandwidth filters.

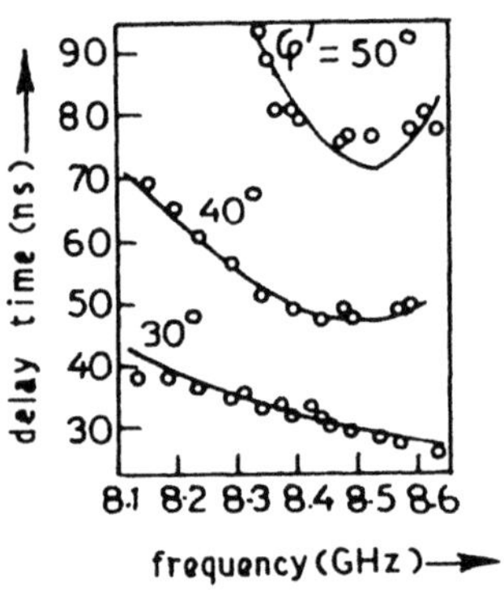

Figure 4.71. Comparison of experimentally measured and theoretically calculated group delay time as a function of the propagation angle $\varphi' = \pi/2 - \theta$ with $\theta = \pi/2$. The parameters are $H_0 = 2350\,\text{Oe}$, $\varepsilon_d = 9.9$, $t_{eff} = 635\,\mu\text{m}$, $d = 38\,\mu\text{m}$, and propagation path length $= 3\,\text{mm}$ (after Miller, 1977).

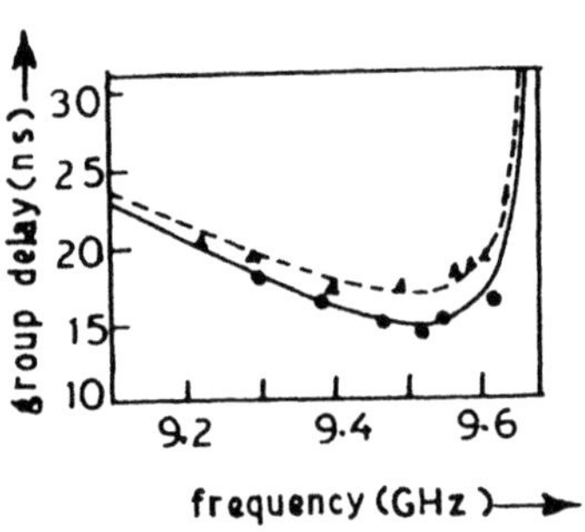

Figure 4.72. Comparison of experimentally measured and calculated group delay time for propagation along $+x$ (triangles) and $-x$ (dots) directions. The parameters are $H_0 = 2690\,\text{Oe}$, $\varepsilon_d = 2.35$, $t = 110\,\mu\text{m}$, $d = 38\,\mu\text{m}$, $\varphi = 80°$, $\theta = \pi/2$, and propagation path length $= 3\,\text{mm}$ (after Miller, 1977).

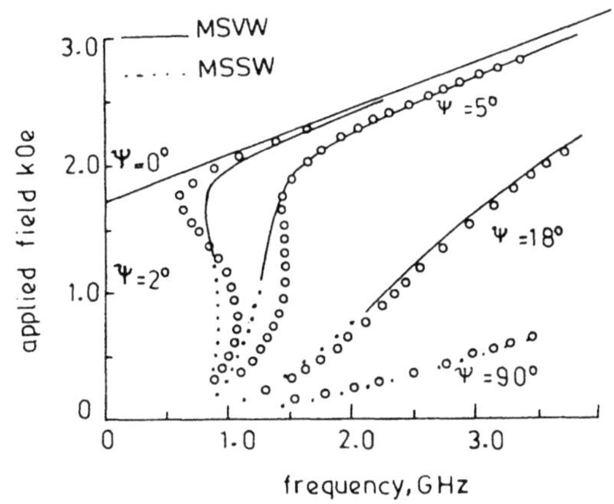

Figure 4.73. Comparison of theoretical and experimental applied field vs. frequency curves for a constant β. The experimental points represented by circles are due to Schilz (1973). The parameters are $d = 3\,\mu$m, $t = 254\,\mu$m, and $\beta = 157\,\text{cm}^{-1}$. The values of ψ are explicitly shown in the figure (after Miller, 1978).

4.5. *Metal–Ferrimagnetic–Metal Structure*

In this section, we shall discuss magnetostatic wave propagation in a structure consisting of a ferrimagnetic slab (or film) supported on both sides by metal-backed dielectrics. The resulting configuration has been extensively studied in the past in connection with TE_{n0} mode propagation in a ferrite-slab loaded rectangular waveguide. Earlier studies were concerned mainly with phase-shift characteristics, but recently, the mode spectrum of magnetostatic waves in this structure was reported by Tsai et al. (1976) and Yukawa et al. (1977, 1978a). In what follows, we shall consider seaparately the cases of transverse and normal magnetizations.

4.5.1. *Transverse Magnetization*

This section follows Yukawa et al. (1977). Figure 4.74 shows the configuration: a ferrimagnetic slab of thickness d occupies the region $0 \leqslant x \leqslant d$ and is separated from perfect conductors placed at $x = (A_d + 1)d$ and $x = -B_d d$ by means of dielectrics. The dc magnetization is along the z-axis. The magnetostatic wave propagates in the yz-plane. In the magnetostatic approximation, the magnetostatic potential can be expressed as

$$\left.\begin{aligned} \psi^{(1)} &= [A\exp(k_x^e x) + B\exp(-k_x^e x)]\exp(-j\beta y)\exp(-jk_z z) \\ \psi^{f} &= [C\exp(jk_x^i x) + D\exp(-jk_x^i x)]\exp(-j\beta y)\exp(-jk_z z) \\ \psi^{(2)} &= [E\exp(k_x^e x) + F\exp(-k_x^e x)]\exp(-j\beta y)\exp(-jk_z z) \end{aligned}\right\} \qquad (4.164)$$

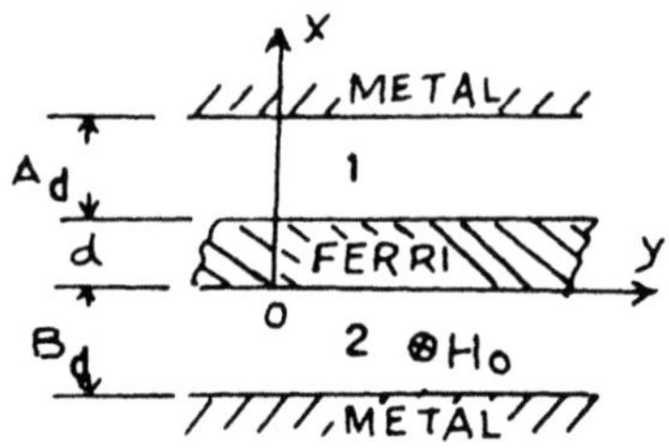

Figure 4.74. Metal–air–ferrimagnetic–air–metal structure and the coordinate system. The biasing field is along the z-axis, whereas the wave propagates along the y-axis (after Yukawa et al., 1977).

Substitution of these ψ's in the magnetostatic equations (4.1) and (4.2) yields

$$\left.\begin{aligned} \mu(k_x^{i2}+\beta^2)+k_z^2=0 \\ k_x^{e2}-\beta^2-k_z^2=0 \end{aligned}\right\} \qquad (4.165)$$

When $k_z=0$, we have $k_x^e=\beta$ and $k_x^i=j\beta$. In this case, the magnetic induction can be easily obtained from equations (4.164). The application of appropriate boundary conditions yields the following dispersion relation:

$$\mu^2-[\kappa+\tanh(A_d\beta d)][\kappa-\tanh(B_d\beta d)] + \mu[\tanh(A_d\beta d)+\tanh(B_d\beta d)]\coth\beta d=0 \qquad (4.166)$$

In the limit when A_d, $B_d\rightarrow\infty$, equation (4.166) reduces to equation (4.44). When either $A_d=0$ and $B_d\rightarrow\infty$ or $A_d\rightarrow\infty$ and $B_d\rightarrow 0$, the dispersion relations (4.120) (for a metallized ferrimagnetic slab) are obtained. Finally, when $B_d\rightarrow\infty$ and A_d is finite, equation (4.166) reduces, as expected, to equation (4.139), which applies to the dielectric layered structure.

The cut-off limit for magnetostatic surface waves can be obtained from equation (4.166) on requiring that $\beta\rightarrow 0$, which leads to

$$\mu^2-\kappa^2+\mu(A_d+B_d)=0 \qquad (4.167)$$

which finally yields

$$\omega_{cs}=\left[(\omega_0+\omega_m)\left(\omega_0+\frac{\omega_m}{A_d+B_d+1}\right)\right]^{1/2} \qquad (4.168)$$

The variation of ω_{cs} with A_d+B_d is shown in Figure 4.75 in which Ω_1, Ω_2, and Ω_3 are, respectively, the normalized resonance frequencies for volume

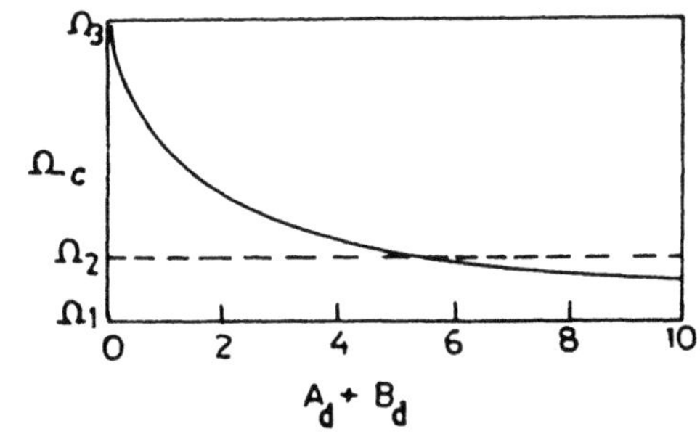

Figure 4.75. Variation of $\Omega_c \equiv \omega_c/\omega_m$ with $(A_d + B_d)$ for a magnetostatic surface wave propagation in the configuration shown in Figure 4.74; $\Omega_0 \equiv \omega_0/\omega_m = 0.5$ (after Yukawa et al., 1977).

mode, f.a. mode, and f.m. mode, and are given by

$$\left.\begin{aligned} \Omega_1\omega_m &= \sqrt{\omega_0(\omega_0+\omega_m)} \\ \Omega_2\omega_m &= \omega_0 + \tfrac{1}{2}\omega_m \\ \Omega_3\omega_m &= \omega_0 + \omega_m \end{aligned}\right\} \qquad (4.169)$$

The resonance frequency can be obtained by requiring that $\beta \to \infty$, which yields

$$\Omega \to \Omega_2 \qquad \text{for } A_d, B_d \neq 0 \qquad (4.170)$$

and

$$\Omega \to \Omega_3 \qquad \text{for } A_d = B_d = 0 \qquad (4.171)$$

Clearly, the range of magnetostatic modes vanishes as $A_d = B_d = 0$, i.e., when the metal slabs are in contact on both sides of the slab.

Figure 4.76 shows the dispersion curves for different combinations of A_d and B_d, and the nonreciprocity of magnetostatic waves is evident. When $A_d = B_d$, however, the nonreciprocity vanishes and reciprocal propagation

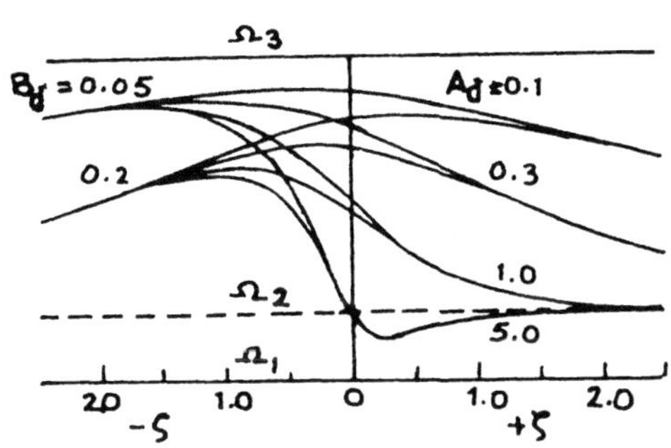

Figure 4.76. Dispersion curves: $\Omega \equiv \omega/\omega_m$ vs. $\zeta = \beta d$ for the modes with $\Omega_0 \equiv \omega_0/\omega_m = 0.5$ (after Yukawa et al., 1977).

is obtained; the resulting dispersion curves are shown in Figures 4.77a and b. The dispersion curves for the case when one of the metal plates touches the ferrite ($B_d = 0$) and the other is at a finite distance away ($A_d \neq 0$) are shown in Figure 4.78a and b. Interestingly, the guided wave becomes a backward wave over a wide range of wavenumbers.

Yukawa et al. (1977) have also experimentally observed the mode spectrum for the present configuration. The absorption spectrum at a fixed frequency $f = 4.8054$ GHz (C-band) and varying dc field is shown in Figure 4.79. In this figure, the dc fields H_0, H_1, and H_2 correspond to $\Omega = \Omega_0, \Omega_1$, and Ω_2, respectively. Table 4.2 shows the comparison of the experimental results with the ones calculated without accounting for anisotropy and demagnetization effects. The effects of anisotropy and demagnetization can be almost eliminated by comparing the quantities $(H_0 - H_1)$ and $(H_1 - H_2)$. In this case, good agreement between theoretical and

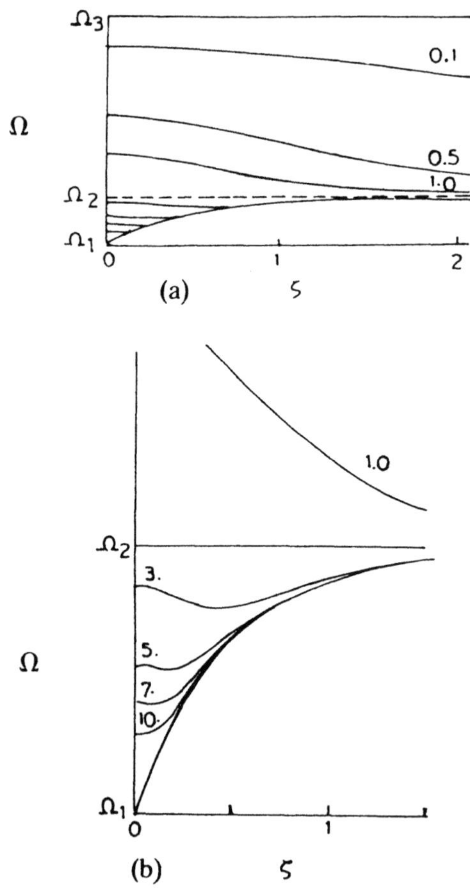

Figure 4.77. Dispersion curves for the (A_d, A_d) mode for $\Omega_0 = 0.5$. The value of the parameter A_d has been shown alongside each curve. The figure (a) shows general dispersion curves, whereas the region $\Omega_1 < \Omega < \Omega_2$ has been shown in (b) on an expanded scale (after Yukawa et al., 1977).

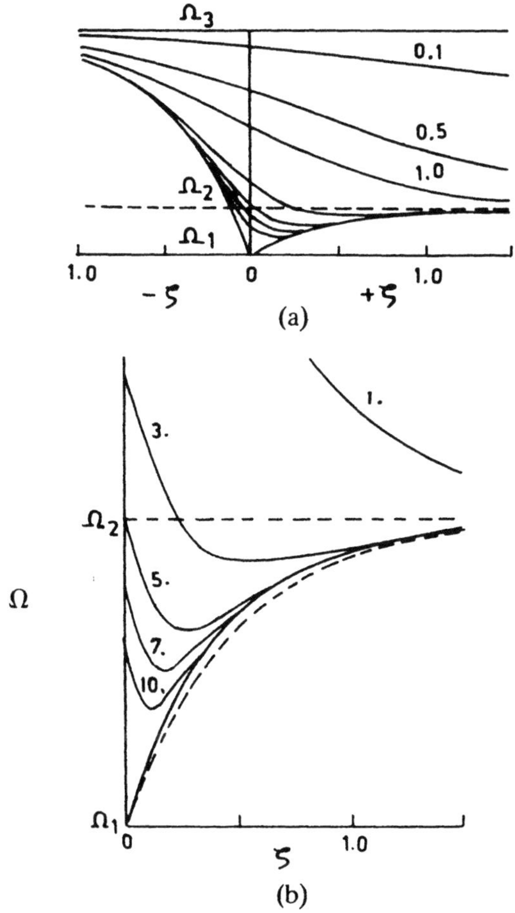

Figure 4.78. Dispersion curves for the $(A_d, 0)$ mode. The value of A_d is shown alongside each curve. The figure (a) shows the general dispersion curves whereas the region $\Omega_1 < \Omega < \Omega_2$ has been shown in (b) on an expanded scale (after Yukawa et al., 1977).

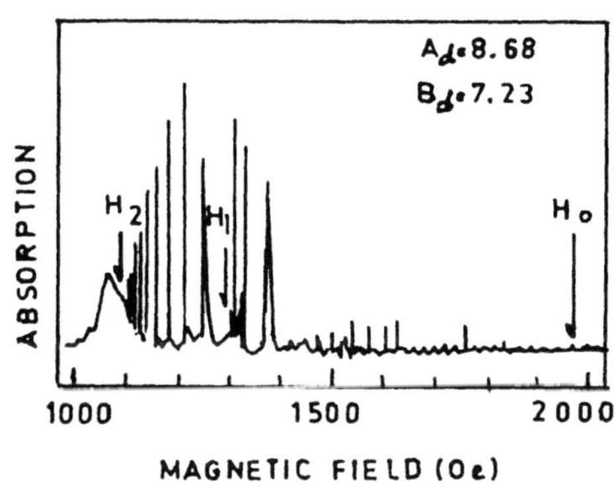

Figure 4.79. Absorption spectrum of the (8.68, 7.23) mode. The excitation frequency is 4.8054 GHz while the measured values of the applied fields H_0, H_1, and H_2 are 1976 Oe, 1295.7 Oe, and 1083.6 Oe, respectively (after Yukawa et al., 1977).

TABLE 4.2. Comparison of the Experimental Results with Calculated Ones Neglecting the Anisotropy and Demagnetizing Fields for H_0, H_1, and H_2 (after Yukawa et al., 1977)

	Theoretical value (Oe)	Experimental value (Oe)
H_0	1716.2	1976.0
H_1	1043.3	1295.7
H_2	826.2	1083.6
$H_0 - H_1$	672.9	680.3
$H_1 - H_2$	217.1	212.1

experimental results is obtained (see Table 4.2). Further discussion of the absorption spectrum and standing wave resonances can be found in Yukawa et al. (1977).

4.5.2. *Normal Magnetization*

The relevant analyses have been carried out by Yukawa et al. (1978a) and Tsai et al. (1976). The configuration is shown in Figure 4.80 and can be analyzed in, essentially, the same way as the dielectric layered structure in Section 4.4.2. The dispersion relation, in the magnetostatic approximation, is obtained as

$$\tanh(A_d \beta d)\tanh(B_d \beta d) - \alpha_v^2 + \alpha_v[\tanh(A_d \beta d) + \tanh(B_d \beta d)]\cot(\alpha_v \beta d) = 0 \qquad (4.172)$$

where $\alpha_v = (-\mu)^{1/2}$. This dispersion relation implies reciprocal propagation. It reduces to the appropriate forms in the various limits. It is easy to show that surface modes are not allowed. However, bulk modes may possibly be allowed when $\mu < 0$, i.e., when $\omega_0 < \omega < [\omega_0(\omega_0 + \omega_m)]^{1/2}$. The limiting procedure shows that $\beta = 0$ when $\omega = \omega_{cv} = \{\omega_0[\omega_0 + \omega_m/(A_d + B_d + 1)]\}^{1/2}$, whereas $\beta \to \infty$ when $\omega = \omega_3'$. Thus, the range of allowed (bulk) modes is $\omega_{cv} < \omega < \omega_3'$.

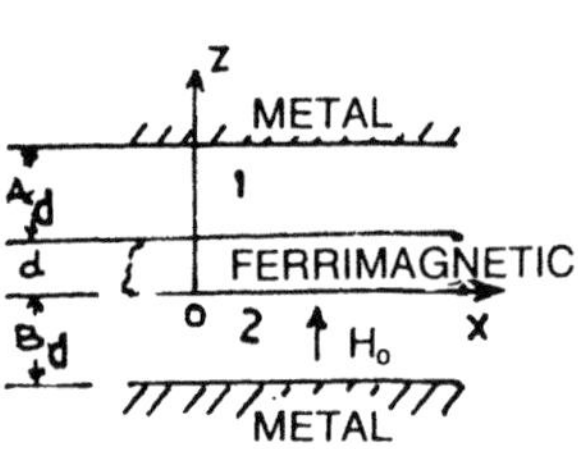

Figure 4.80. Metal–air–ferrimagnetic–air–metal structure and the coordinate system. The magnetostatic wave propagates along the y-axis whereas the biasing field is along the z-axis, i.e., normal to the ferrimagnetic (after Yukawa et al., 1978a).

Figure 4.81 shows the variation of ω_{cv} with $(A_d + B_d)$ for $\omega_0 = \omega_m$, and the propagating region for magnetostatic volume waves is explicitly shown in the figure. The dispersion curves for $A_d = B_d$ and for $B_d = 0$, $A_d \neq 0$ are shown in Figures 4.82 and 4.83, respectively. The relevant experimental results have been obtained and discussed by Yukawa et al. (1978a).

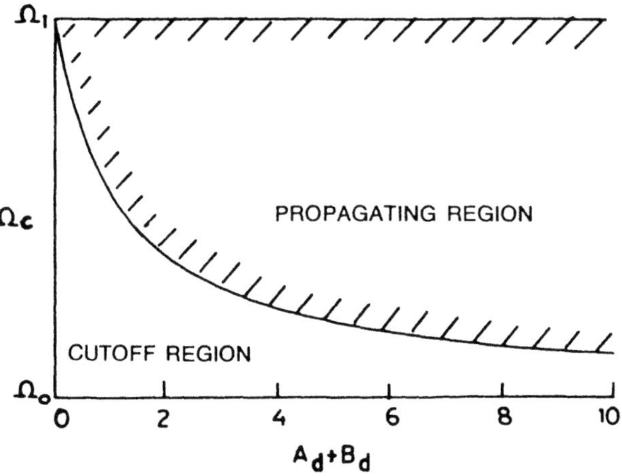

Figure 4.81. Variation of $\Omega_c \equiv \omega_{cv}/\omega_m$ with $(A_d + B_d)$ for $\Omega_0 = \omega_0/\omega_m = 1.0$ in the case of magnetostatic bulk waves propagating in the structure shown in Figure 4.80. The first-order (dominant) mode exists in the shaded region while higher-order modes exist in the entire region (after Yukawa et al., 1978a).

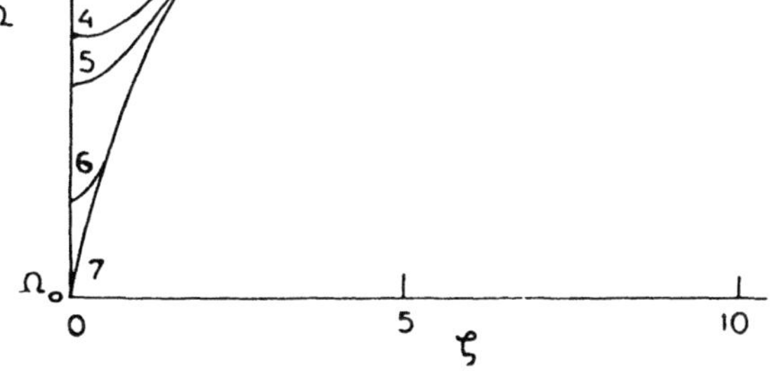

Figure 4.82. Dispersion curves for the (A_d, A_d) mode. The value of the parameter A_d is shown alongside each curve (after Yukawa et al., 1978a).

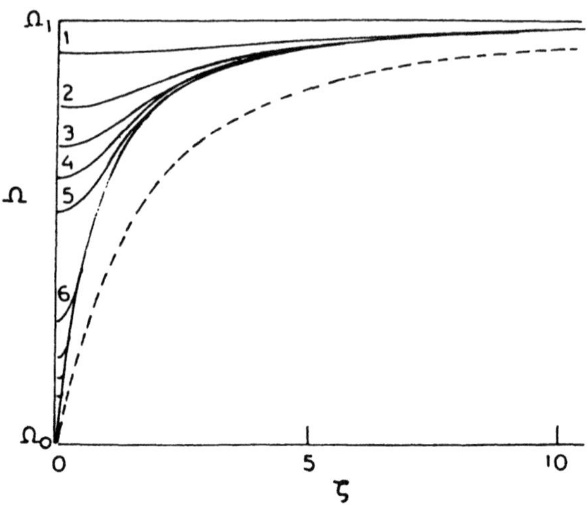

Figure 4.83. Dispersion curves for the $(A_d, 0)$ mode. The value of the parameter A_d is shown alongside each curve. The broken line indicates the dispersion curve for the (∞, ∞) mode (after Yukawa et al., 1978a).

4.6. *Double Magnetic Layered Structures*

Magnetostatic surface wave propagation in more than one adjacent layer of transversely magnetized ferrimagnetics has been investigated by Wolfram (1970), Hurd (1970), Tsutsumi (1974), Ganguly et al. (1974). Tsutsumi et al. (1976), and Srivastava (1978c). We shall discuss magnetostatic propagation in the following configurations: (i) the gap between two adjacent ferrimagnetic substrates (Section 4.6.1), and (ii) a ferrimagnetic slab in contact with another ferrimagnetic substrate (Section 4.6.2).

4.6.1. *Air Gap between Two Ferrimagnetic Substrates*

This section follows papers by Tsutsumi (1974) and Srivastava (1978c). Figure 4.84 shows the configuration: the ferrites 1 and 2 occupy the regions $x > d/2$ and $x < -d/2$, respectively, and are separated by an air gap $-d/2 \leq x \leq d/2$. The saturation magnetizations of the substrates are $4\pi M_1$ and $4\pi M_2$, respectively, and are assumed to be directed along the z-axis. The wave propagates in the y-direction. The magnetostatic potential and

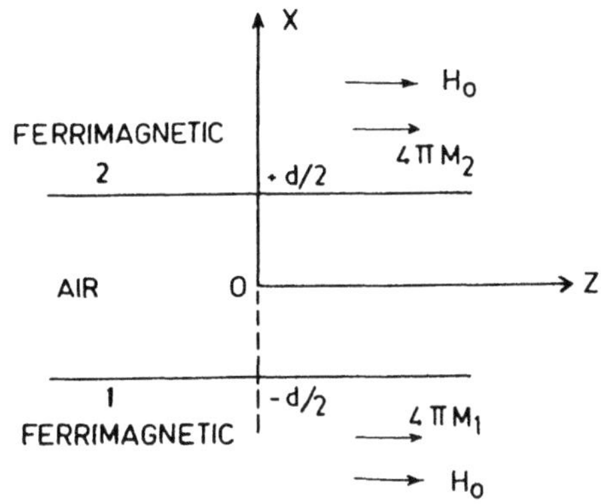

Figure 4.84. Air gap of width d between two similarly magnetized ferrimagnetics. The magnetizations of the ferrimagnetics 1 and 2 (along the z-axis) are $4\pi M_1$ and $4\pi M_2$, respectively. The waves propagate in the $\pm y$-directions.

induction can be expressed as

$$\left.\begin{aligned} \psi^{(1)} &= C\exp(\beta x)\exp j(\omega t-\beta y) \\ \psi^{a} &= [A\exp(\beta x)+B\exp(-\beta x)]\exp j(\omega t-\beta y) \\ \psi^{(2)} &= \exp(-\beta x)\exp j(\omega t-\beta y) \\ b_x^{(1)} &= \beta(\mu_1-\kappa_1)\psi^{(1)} \\ b_x^{a} &= \beta[A\exp(\beta x)-B\exp(-\beta x)]\exp j(\omega t-\beta y) \\ b_x^{(2)} &= -\beta(\mu_2-\kappa_2)\psi^{(2)} \end{aligned}\right\} \tag{4.172}$$

where μ_i and κ_i refer to the ith ferrimagnetic region. The application of boundary conditions at the interfaces, followed by elimination of constants, yields the following dispersion relation for surface wave propagation in the $\pm y$-direction:

$$\exp(2\beta_\pm d)=\frac{(\mu_1\pm\kappa_1-1)(\mu_2\mp\kappa_2-1)}{(\mu_1\pm\kappa_1+1)(\mu_2\mp\kappa_2+1)} \tag{4.173}$$

If losses are neglected, equation (4.173) leads to

$$\exp(2\beta_\pm d)=\frac{(\frac{1}{2}\omega_{m1})(\frac{1}{2}\omega_{m2})}{(\omega_0+\frac{1}{2}\omega_{m1}\mp\omega)(\omega_0+\frac{1}{2}\omega_{m2}\pm\omega)} \tag{4.174}$$

where the expressions for μ_i and κ_i have been used. It follows that, in general, the propagation characteristics are nonreciprocal. However, when $M_1=M_2$, the dispersion relation is rendered reciprocal and is, in fact, equivalent to the dispersion relation for a free ferrimagnetic slab of thickness equal to d [cf. equation (4.45)]. This is quite interesting since the thickness of a given ferrimagnetic slab cannot be changed easily, whereas the thickness of the air gap can be controlled as desired. The dispersion curves for different magnetization ratios are shown in Figure 4.85 for a typical set of data. The magnetostatic propagation in the $+y$- and $-y$-directions occurs in the ranges $\omega_c^+ < \omega < \omega_r^+$ and $\omega_c^- < \omega < \omega_r^-$, respectively, where

$$\left.\begin{aligned} \omega_r^+ &= \omega_0 + \tfrac{1}{2}\omega_{m1} \\ \omega_r^- &= \omega_0 + \tfrac{1}{2}\omega_{m2} \\ \omega_c^\pm &= \mp(\omega_{m2} - \omega_{m1})/4 + \tfrac{1}{2}\sqrt{[2\omega_0 + (\omega_{m1} + \omega_{m2})/2]^2 - \omega_{m1}\omega_{m2}} \end{aligned}\right\} \tag{4.175}$$

It is seen that the waves are forward waves since ω increases monotonically with β throughout the frequency range of allowed modes.

It would be interesting to consider the case when the magnetization of one of the substrates is reversed (Srivastava, 1978c): the resulting configuration is shown in Figure 4.86. As noted in Chapter 1, the desired dispersion relation can be obtained from equation (4.173) by changing the sign of κ_2, which yields

$$\exp(2\beta_\pm d) = \frac{(\mu_1 \pm \kappa_1 - 1)(\mu_2 \pm \kappa_2 - 1)}{(\mu_1 \pm \kappa_1 + 1)(\mu_2 \pm \kappa_2 + 1)} \tag{4.176}$$

Assuming $M_1 = M_2$ and substituting for μ_i and κ_i, we obtain

$$\exp(2\beta_\pm d) = \frac{(\tfrac{1}{2}\omega_m)^2}{(\omega_0 + \tfrac{1}{2}\omega_m \mp \omega)^2} \tag{4.177}$$

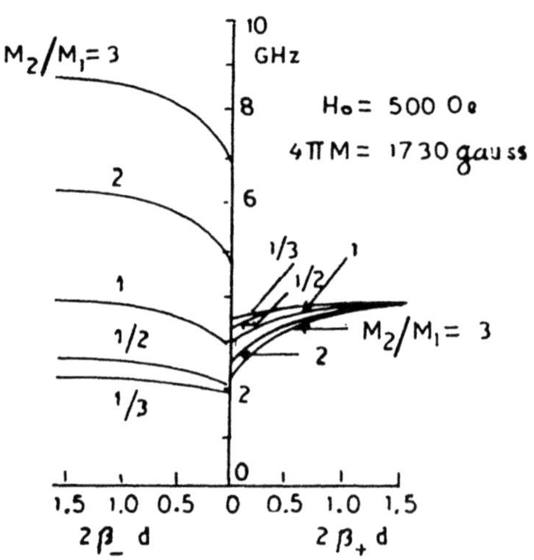

Figure 4.85. Dispersion curves for magnetostatic surface waves propagating in the configuration shown in Figure 4.84. The values of the parameters are shown in the curves (after Tsutsumi, 1974).

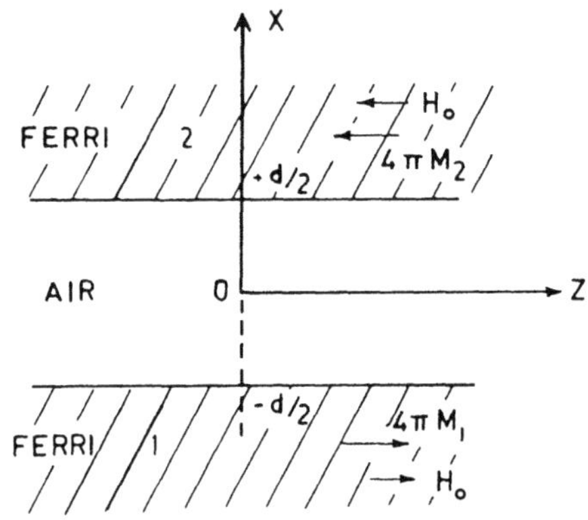

Figure 4.86. Air gap between two oppositely magnetized ferrimagnetic substrates. The direction of magnetization in ferrimagnetic 1(2) is the $+z(-z)$-axis. The wave propagates in the y-direction (after Srivastava 1978c).

It is easy to show that surface wave propagation occurs only in the $+y$-direction. This is understandable since the application of $\mathbf{H}_0 \times \hat{n}$ criteria* to the configuration of Figure 4.86 shows that the surface wave associated with each interface, i.e., $x = -d/2$ and $x = d/2$, propagates only in the $+y$-direction. The resonance frequency for this mode is given by $\omega = (\omega_0 + \frac{1}{2}\omega_m)$ whereas there are two cut-off limits, i.e., $\omega = \omega_0$ and $\omega = \omega_0 + \omega_m$. It is easy to show that the guided waves in the ranges $\omega_0 < \omega < (\omega_0 + \frac{1}{2}\omega_m)$ and $(\omega_0 + \frac{1}{2}\omega_m) < \omega < (\omega_0 + \omega_m)$ are forward and backward waves, respectively. The group velocity is obtained as

$$v_g = (\omega_0 + \tfrac{1}{2}\omega_m - \omega)d \qquad (4.178)$$

i.e., the group velocity varies linearly with the wave frequency as well as the biasing field.

Srivastava (1978c) carried out a rigorous electromagnetic analysis of surface wave propagation in the configuration of Figure 4.86. Although the general results are in consonance with those of the magnetostatic analysis discussed above, there are some notable differences. When ω is slightly smaller than $[\omega_0(\omega_0 + \omega_m)]^{1/2}$, there exists a small "forbidden gap" (not brought out by the magnetostatic analysis) in the dispersion curves, irrespective of whether the wavenumber is small or large. This is a unique situation where the results of magnetostatic analysis prove to be qualitatively incorrect even in the large-wavenumber domain. Furthermore, rigorous analysis shows that the cut-off wavenumber is finite for forward as well as backward waves.

It is worth mentioning that the unidirectionality of wave propagation [which can be shown to be inconsistent with thermodynamics, e.g., Lax and

* The outward normal to the medium 2(1) is in the $-x$ $(+x)$-direction.

Button (1956)] is due to the neglect of losses. A proper consideration of losses in a realistic structure would definitely lead to bidirectional propagation (Gardiol, 1967).

4.6.2. *Ferrimagnetic Layer Over Another Ferrimagnetic Substrate*

The configuration is shown in Figure 4.87 and was first analyzed by Wolfram (1970). A ferrimagnetic layer ($0 \leqslant x \leqslant d$) of magnetization M_1 is in contact with another ferrimagnetic substrate $x \geqslant d$ of magnetization M_2. The dc magnetic field is along the z-axis. Following the conventional magnetostatic analysis, the dispersion relation for $\pm y$-propagating magnetostatic surface waves is obtained as

$$\exp(2\beta_{\pm} d) = \frac{(\mu_1 \mp \kappa_1 - \mu_2 \pm \kappa_2)(\mu_1 \pm \kappa_1 - 1)}{(\mu_1 \pm \kappa_1 + \mu_2 \mp \kappa_2)(\mu_1 \mp \kappa_1 + 1)} \qquad (4.179)$$

where subscripts 1 and 2 refer to the first and second ferrimagnetic substrates, respectively. In the case when $M_2 = 0$, equation (4.179) reduces, as it should, to equation (4.44), i.e., the dispersion relation for a single ferrimagnetic slab. Figures 4.88 and 4.89 show the dispersion curves as computed from equation (4.179). The various parameters shown in the

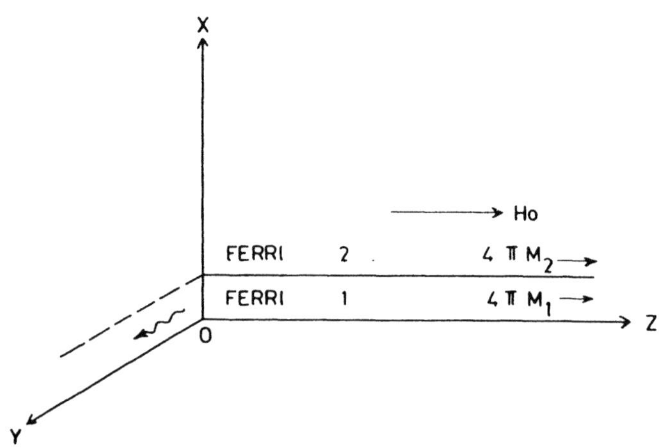

Figure 4.87. A ferrimagnetic film of thickness d is in contact with another ferrimagnetic substrate. The biasing field is along the z-axis, whereas the magnetostatic surface wave propagates along the y-axis.

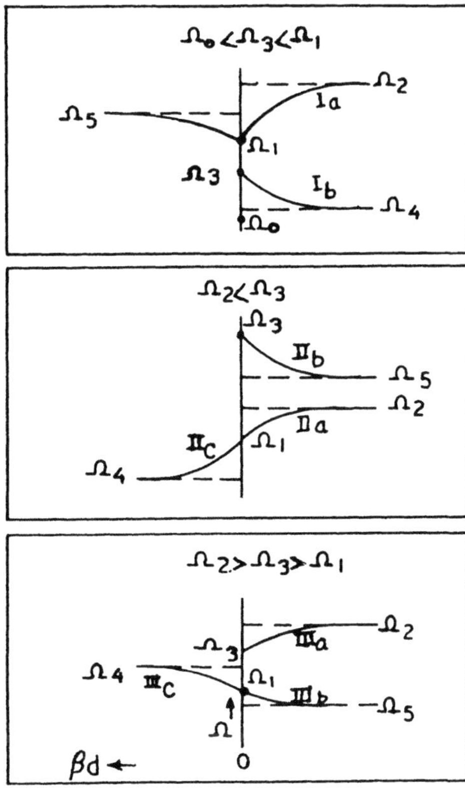

Figure 4.88. Dispersion relation for three different ranges of $r = M_2/M_1$ for a magnetostatic surface wave propagation in the structure shown in Figure 4.87 (after Wolfram, 1970).

curves are defined as follows:

$$\left.\begin{aligned}
\Omega_0 &= H_0/4\pi M_1 \\
\Omega &= \omega/4\pi M_1 \gamma \\
r &= M_2/M_1 \\
R &= (r-1)/4 \\
\Omega_1 &= [\Omega_0(\Omega_0+1)]^{1/2} \\
\Omega_2 &= \Omega_0 + 0.5 \\
\Omega_3 &= \Omega_0 + 0.5r \\
\Omega_4 &= R - (R^2 + 2\Omega_0 R + \Omega_1^2)^{1/2} \\
\Omega_5 &= R + (R^2 + 2\Omega_0 R + \Omega_1^2)^{1/2}
\end{aligned}\right\} \qquad (4.180)$$

Figure 4.88 distinguishes three specific cases corresponding to different ranges of magnetization ratio. In each case, branches a and c represent $\pm y$

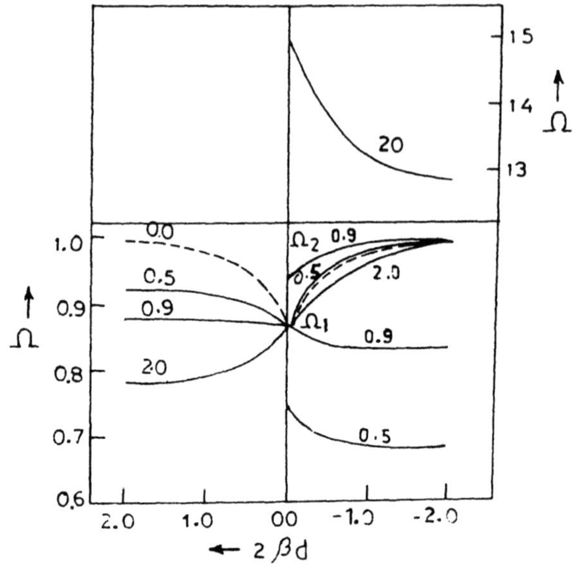

Figure 4.89. Dispersion curves for different values of $r = M_2/M_1$ shown alongside each curve. The broken lines represent a Damon–Eshbach curve for a single-film case (after Wolfram, 1970).

propagating magnetostatic waves, associated with medium 1 and modified by the presence of medium 2. Branch b essentially corresponds to a $-y$ propagating wave, associated with medium 2 and modified by the presence of medium 1. In case I, branches a and c correspond to forward waves, whereas branch b corresponds to a backward wave. In case II, only branch a corresponds to a forward wave; the remaining two correspond to backward waves. Lastly, in case III, only branch b represents a backward wave, while the remaining two represent forward cases.

Figure 4.89 illustrates each of the cases for $\Omega_0 = 0.5$, and cases I, II, and III are represented by $r = 0.5$, 2.0, and 0.9, respectively. The dotted curves correspond to $r = 0.0$, i.e., to the case of a single ferrimagnetic slab as discussed in Section 4.2. It is evident that the dispersion, and hence the delay characteristics, can be controlled by suitable choice of magnetic film and substrate.

4.7. *Periodic Structures*

The study of magnetostatic wave propagation in periodic structures is important because of its potential applications to magnetostatic wave filters, distributed feedback (DFB) magnetic wave amplifiers and oscillators, etc. The theoretical investigations of the propagation of magnetic waves in periodic structures were carried out by Elachi (1975), Tsutsumi and Yuki (1975), Tsutsumi et al. (1977a, b), and Seshadri (1978). The first experimental evidence of magnetostatic wave propagation in a periodic structure was reported by Sykes et al. (1976), and in what follows, we shall first discuss some of their experimental results and then summarize the analysis and results of magnetostatic wave propagation in a corrugated YIG slab (Tsutsumi et al., 1977b).

The experimental work of Sykes et al. (1976) involves a periodic shallow grooved structure as shown in Figure 4.90. Specifically, grooves 30 μm wide by 4 mm long and separated by 120 μm were etched in a 9-μm thick YIG film, grown by liquid phase epitaxy on a ⟨111⟩-oriented GGG substrate. The periodic structure consisted of 20 grooves which were defined photolithographically; the sputtered SiO_2 masked the YIG during etching in hot orthophosphoric acid at 270 °C. A groove depth of 1 μm was obtained in 1-sec etching. Figure 4.91 shows the insertion loss through the structure as a function of magnetostatic wave frequency. Three "rejection notches" are clearly visible in the figure, which are a consequence of periodicity. It is interesting to note that the frequencywise position of the rejection notches remains unchanged even when the biasing field is reversed in order to cause field displacement of magnetostatic surface waves from the top to the bottom surface. It is interesting that the experimental results given in Figure 4.91 are in reasonable agreement (Sykes et al., 1976) with the theory of a repetitively mismatched transmission line (Sittig and Coquin, 1968).

As regards the theoretical analysis of magnetostatic waves in periodic structures, we shall specifically consider the case of a transversely magne-

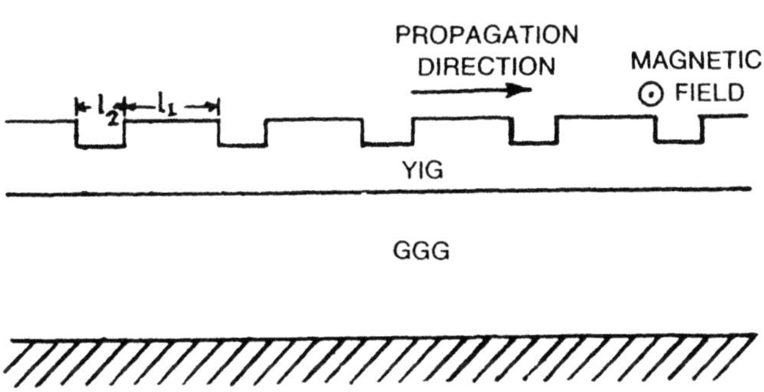

Figure 4.90. Schematic cross-section for the periodic shallow-grooved structure prepared by selective etching in an epitaxial YIG film (after Sykes et al., 1976).

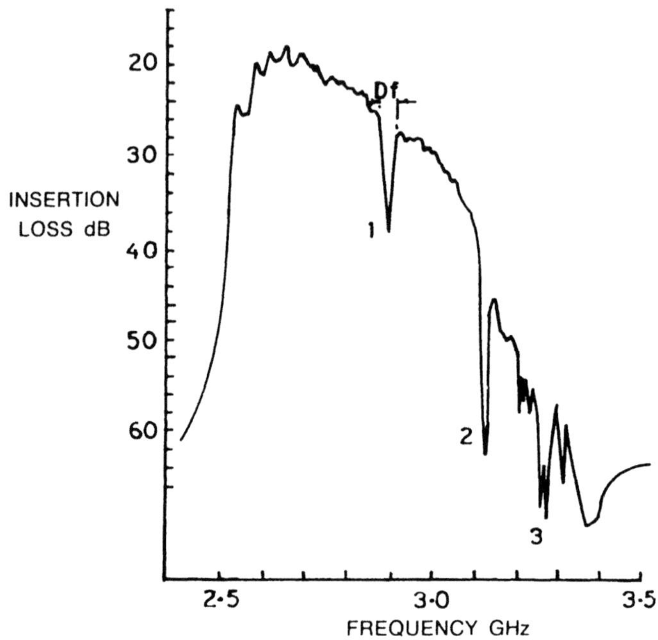

Figure 4.91. Variation of measured insertion loss with frequency for magnetostatic surface wave propagation through the periodic structure shown in Figure 4.90. The biasing field is 380 Oe (after Sykes et al., 1976).

tized, sinusoidally corrugated ferrimagnetic slab. The analysis follows Tsutsumi et al. (1977b), and the configuration is shown in Figure 4.92. There is a periodically (specifically, sinusoidally) corrugated ferrimagnetic slab with an average thickness $2d$ and periodicity h. The surfaces 1 and 2 are given, respectively, by

$$x_1 = d - \Delta \cos(2\pi/h)y \tag{4.181}$$

and

$$x_2 = -d + \Delta \cos(2\pi/h)y \tag{4.182}$$

where Δ is a measure of the strength of modulation in the x-direction. Neglecting the presence of the incident wave in the corrugated region,* we

* Neglect of the incident wave field in the corrugated region is justified only for small modulations, as discussed by Peng et al. (1975), Neviere et al. (1973), Tsutsumi et al. (1977a, b), etc. Thus, the present formulation would be applicable only to the case of small modulations.

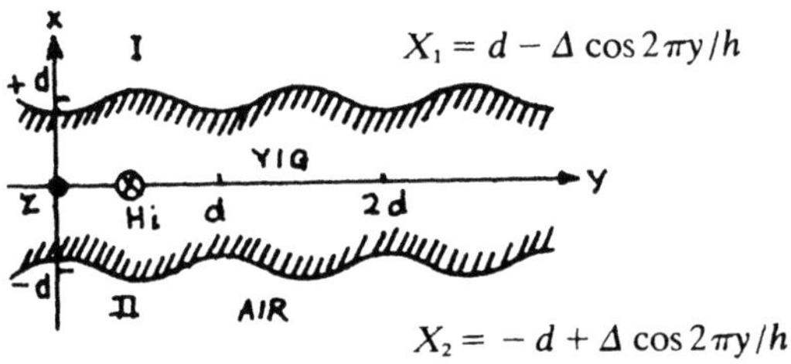

Figure 4.92. The corrugated YIG film structure. The average thickness of the film is $2d$, the strength of modulation and periodicity are characterized by Δ and h, respectively. The biasing field is along the z-axis while the wave propagates along the y-axis (after Tsutsumi et al., 1977b).

can express the magnetostatic potential in the three regions in Floquet form as follows:

$$\left.\begin{aligned} \psi^{(1)} &= \sum_{m=-\infty}^{+\infty} C_m \exp(-\gamma_m x)\exp(j\alpha_m y) \\ \psi^{f} &= \sum_{m=-\infty}^{+\infty} [A_m \exp(\gamma_m x) + B_m \exp(-\gamma_m x)]\exp(j\alpha_m y) \\ \psi^{(2)} &= \sum_{m=-\infty}^{+\infty} D_m \exp(\gamma_m x)\exp(j\alpha_m y) \end{aligned}\right\} \tag{4.183}$$

where superscripts 1 and 2 refer to the air regions in the positive and negative x-directions, respectively. The quantities A_m, B_m, C_m, and D_m are the coefficients of space harmonics of nth order, whereas α_m and γ_m are given by

$$\left.\begin{aligned} \mathrm{Re}(\gamma_m) &= |\mathrm{Re}(\alpha_m)| \\ \alpha_m &= \beta + (2\pi/h)m \qquad m = 0, \pm 1, \pm 2, \ldots \end{aligned}\right\} \tag{4.184}$$

The magnetic field and magnetic induction can be obtained from equations (4.183) in the usual manner. It is necessary to match the tangential component of the magnetic field and the normal component of induction at the interfaces. The resulting conditions, for the first interface, can be expressed as

$$h_{1y} \cos\theta + h_{1x} \sin\theta = h_{fy} \cos\theta + h_{fx} \sin\theta \tag{4.185}$$

and

$$b_{1y} \sin\theta + b_{1x} \cos\theta = b_{ty} \sin\theta + b_{tx} \cos\theta \tag{4.186}$$

respectively. The boundary conditions for the lower surface can be obtained from equations (4.185) and (4.186) on replacing θ by $-\theta$. These conditions can be simplified by noting that the slope of the surface is given by

$$\tan\theta = (2\pi/h)\Delta \sin(2\pi/h)y \tag{4.187}$$

Substitution of the field quantities in the boundary conditions leads to an infinite set of homogeneous algebraic equations involving the coefficients A_m to D_m. The condition for the existence of a nontrivial solution yields the desired dispersion relation which has a rather complicated form. It is seen that the mth space harmonic gets coupled to the $(m-1)$th space harmonic directly, and hence to other space harmonics indirectly. If the strength of modulation is assumed to be very small, it is possible to neglect second- and higher-order harmonics. As such, the dispersion relation can be approximated to (Tsutsumi et al., 1977b):

$$\cosh 4(\beta - \pi/h)d = \frac{1}{2}\left\{\frac{\mu_p\bar{\mu}_p}{\mu_q\bar{\mu}_q}\exp\left(\frac{4\pi d}{h}\right) + \frac{\mu_q\bar{\mu}_q}{\mu_p\bar{\mu}_p}\exp\left(-\frac{4\pi d}{h}\right) - 4\left(\frac{\pi\Delta}{h}\right)^2\left[2\exp\left(\frac{4\pi d}{h}\right) + \frac{\mu_q\bar{\mu}_q}{\mu_p\bar{\mu}_p}\right]\right\} \tag{4.188}$$

where

$$\left.\begin{aligned} \mu_p &= 1 + \mu - \kappa, & \bar{\mu}_p &= 1 + \mu + \kappa \\ \mu_q &= 1 - \mu - \kappa, & \bar{\mu}_q &= 1 - \mu + \kappa \end{aligned}\right\} \tag{4.189}$$

Figures 4.93a and b display typical dispersion diagrams. The dispersion curve in Figure 4.93a clearly exhibits a coupling between the $m = 0$ forward wave and the $m = 1$ backward wave. The region close to the coupling range is expanded in Figure 4.93b for different strengths of modulation. The broken lines represent the imaginary value of the propagation constant within the stop-band. The simultaneous presence of a pass-band and a stop-band, which is evident in Figure 4.93a, is also seen in Figure 4.91.

4.8. *Effect of Magnetocrystalline Anisotropy*

It was seen from equation (4.147) and subsequent discussions that the propagation loss for magnetostatic waves is prohibitively high unless ΔH is

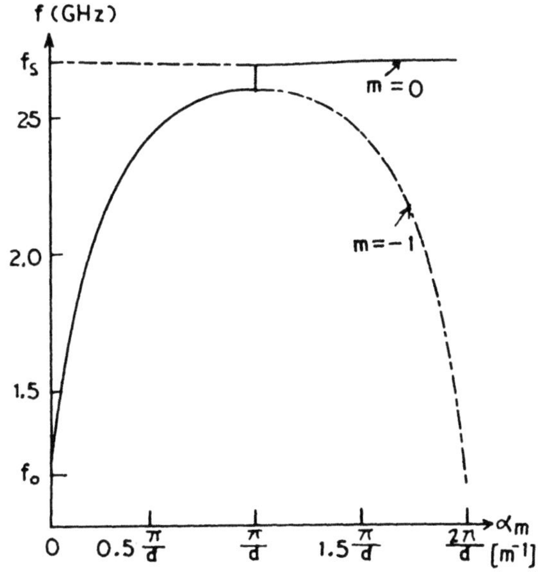

Figure 4.93a. Brillouin diagram for magnetostatic surface wave propagation through the corrugated film shown in Figure 4.92 (after Tsutsumi et al., 1977b).

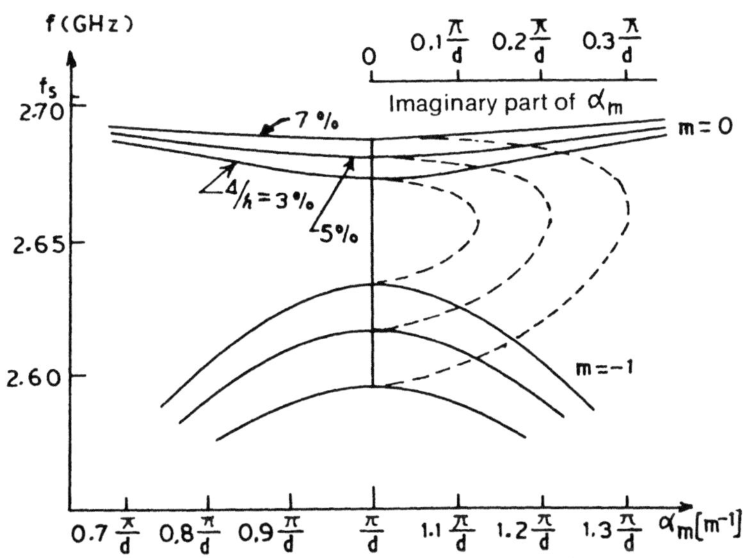

Figure 4.93b. The details of the Brillouin diagram shown in Figure 4.93a near the coupling region (after Tsutsumi et al., 1977b).

sufficiently small. Since ΔH is small only for certain single crystals, the studies of magnetostatic propagation are usually confined only to these crystals. The crystals in which magnetostatic wave propagation has been theoretically and experimentally investigated include pure YIG, Ge–YIG (Tsutsumi et al., 1976), lithium ferrite (Vittoria and Wilsey, 1974; Ganguly et al, 1975), Zn_2Y magnetoplumbite (Srivastava, 1976b; Bajpai et al., 1979). All the single crystals are characterized by magnetocrystalline anisotropy: the first three are cubic and the last is planar hexagonal. Therefore, it is worth investigating the effect of magnetocrystalline anisotropy on the propagation characteristics of magnetostatic waves.

In what follows, we discuss* the effect of cubic as well as hexagonal planar anisotropy on propagation in a normally magnetized dielectric layered structure. The configuration and the appropriate coordinate system are shown in Figures 4.94 and 4.95, respectively. In the case of the cubic material, the primed coordinate axes are assumed to be parallel to the {100}-axes, whereas, in the case of the hexagonal material, the y-axis of the unprimed coordinate system is parallel to the hard axis. The sample is cut in such a way that the z-axis, which is also the direction of magnetization, is perpendicular to the face of the slab. As such, the permeability tensors (in the unprimed coordinate system) are given by equations (1.75) and (1.76) for the cubic case and equations (1.87) and (1.88) for the hexagonal planar case. The wave propagates in the xy-plane in a direction making an angle δ with the y-axis. Following the conventional magnetostatic analysis (Bajpai et al., 1979), we obtain the dispersion relation as

$$\cot \alpha_n \beta d = \frac{\alpha_n^2 - \tanh \beta t}{\alpha_n(1 + \tanh \beta t)} \tag{4.190}$$

where

$$\left.\begin{aligned} \alpha_n &= (-\mu'_{22})^{1/2} \\ \mu'_{22} &= \mu_{xx} \sin^2 \delta + \mu_{yy} \cos^2 \delta \end{aligned}\right\} \tag{4.191}$$

It is seen that this dispersion relation is essentially the same as that for the case of an isotropic film [cf. equation (4.149)], except for the modification in α_n, resulting from magnetocrystalline anisotropy. The group velocity is obtained from equation (4.190) as

$$v_g = P_n / Q_n \tag{4.192}$$

* The present discussion is based on a paper by Bajpai et al. (1979). Other investigations of the effect of magnetocrystalline anisotropy on the dispersion characteristics of magnetostatic waves have been carried out by Schneider (1972, 1974), Vittoria and Wilsey (1974), Ganguly and Vittoria (1974), Ganguly et al. (1974), and Srivastava (1976b).

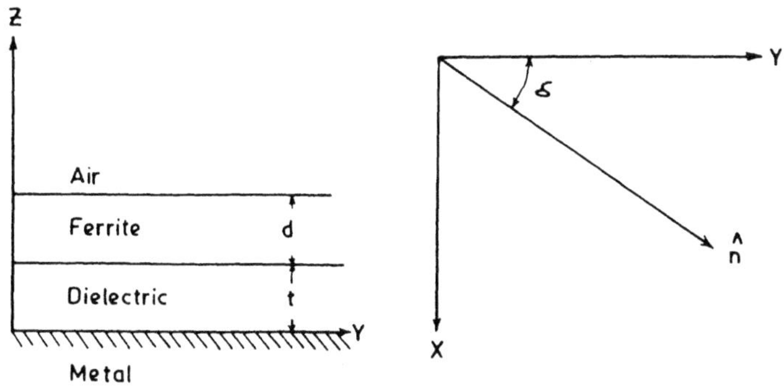

Figure 4.94. The dielectric layered structure analyzed in the present section. The magnetostatic wave propagates in a direction making an angle δ with the y-axis (after Bajpai et al., 1979).

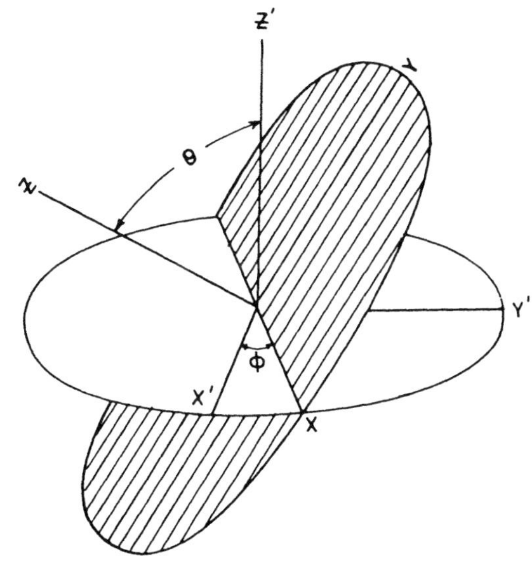

Figure 4.95. The unprimed and primed coordinate systems (after Bajpai et al., 1979).

where

$$\left.\begin{aligned} P_n &= \alpha_n d(\tanh\beta t - \alpha_n^2)\sec^2\alpha_n\beta d + t(\alpha_n + \tanh\alpha_n\beta d)\,\mathrm{sech}^2\,\beta t \\ Q_n &= \alpha_n'(2\alpha_n\tan\alpha_n\beta d - 1 - \tanh\beta t) \\ &\qquad + \alpha_n'\beta d(\alpha_n^2 - \tanh\beta t)\sec^2\alpha_n\beta d \\ \alpha_n' &= (\partial\alpha_n/\partial\omega) = -\omega(\mu_{22}' - 1)/\alpha_n D \end{aligned}\right\} \quad (4.193)$$

The group delay time per unit distance of propagation is, of course, the inverse of the group velocity. The specific cases of YIG (cubic) and Zn_2Y* (planar) materials are discussed below.

4.8.1. *YIG*

We first consider the case in which the [001]-axis (or one of its crystallographic equivalents) is oriented perpendicular to the plane of the YIG slab (or film), which evidently corresponds to $\theta = 0$. It follows from equations (1.76), (1.74), and (1.66) that, in this case, the permeability tensor is independent of δ, which implies that the magnetostatic wave propagation characteristics will be unaffected by magnetocrystalline anisotropy.

The case when the [111]-axis (or one of its crystallographic equivalents) is oriented perpendicular to the plane of the slab (or film) corresponds to $\varphi = 3\pi/4$ and $\theta = \cos^{-1}(1/\sqrt{3})$. It follows from equations (1.74) and (1.76) that the elements of the permeability tensor do not depend on k_1 or δ explicitly. However, it follows from equation (1.66) that the effective dc field is modified by the anisotropy:

$$H_0 = H - 4k_1/3M_0 \tag{4.194}$$

Thus, the characteristics of magnetostatic wave propagation would be the same as in the case of an isotropic ferrimagnetic film subjected to a field given by equation (4.194), other parameters in the configuration remaining unchanged. The propagation characteristics would therefore be unaffected by a rotation of the direction of propagation in the plane of the slab.

When the [101]-axis (or one of its crystallographic equivalents) is oriented perpendicular to the plane of the slab, we have $\varphi = \pi/2$ and $\theta = \pi/4$. It follows from equations (1.76) and (1.74) that, in this case, α_n depends on k_1 as well as δ, which implies that the delay characteristics are influenced by the rotation of the direction of propagation in the plane of the slab. Figure 4.96 shows the variation of the group delay with frequency for a typical set of parameters. When $\delta = 0$, i.e., when the wave propagates along the [$\bar{1}$01]-axis, the group delay time is comparatively large, although the band of nondispersive delay is small. As δ increases, the direction of propagation rotates in the plane containing the [$\bar{1}$01]- and [010]-axes and the group delay time continuously decreases, while the frequency range of nondispersive delay increases. When $\delta = \pi/2$, i.e., when the wave propagates along the [010]-axis, the delay time is least, whereas the frequency range of the region of nondispersive delay is maximum. It follows that the group delay time is large in the case of propagation along the easy axis and

* Zn_2Y magnetoplumbite is the only planar ferrimagnetic material, the single crystals of which have a sufficiently narrow linewidth for low-loss magnetostatic wave propagation.

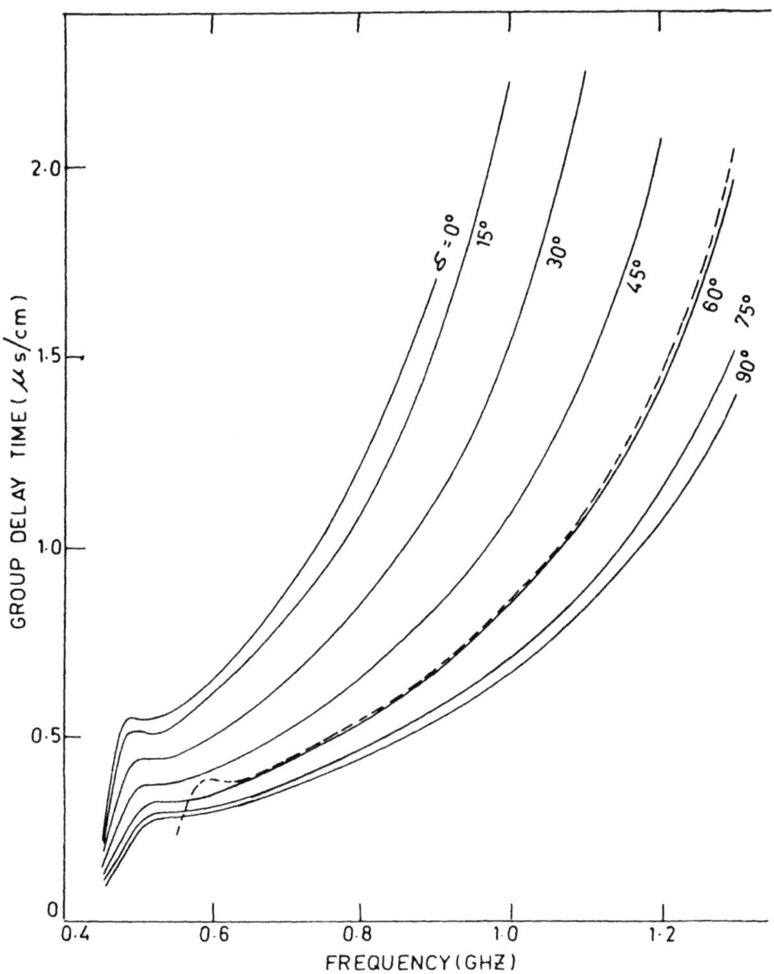

Figure 4.96. Frequency dependence of the group delay time for a ⟨101⟩ oriented YIG film with δ as the parameter. Other parameters are $\varphi = 90°$, $\theta = 45°$, $t = 115\,\mu m$, $d = 4.5\,\mu m$, $4\pi M_0 = 1750\,G$, $H_0 = 150\,Oe$, and $2k_1/M_0 = -80\,Oe$. The broken line corresponds to isotropic ferrite with the same parameters (Bajpai et al., 1979).

smaller for propagation along the hard axis. It is seen from the figure that the direction sensitivity of delay characteristics is smaller when the direction of propagation is closer to either the $[\bar{1}01]$- or the [010]-axis. It is also evident that for a given thickness of the dielectric the frequency range of nondispersive delay and the magnitude of delay time are inverse functions of each other, which is an important conclusion.

When θ and φ are arbitrary, $\mathbf{H}_0$ and $\mathbf{H}$ are only approximately parallel, as can be concluded from equation (1.66). Figures 4.97–4.99 show the frequency dependence of magnetostatic wave group delay time for some selected combinations of θ and φ: (i) $\theta = 30°$, $\varphi = 30°$ (Figure 4.97); (ii) $\theta = 60°$, $\varphi = 30°$ (Figure 4.98); and (iii) $\theta = 90°$, $\varphi = 30°$ (Figure 4.99). Other combinations of θ and φ which lead to the same physical situations

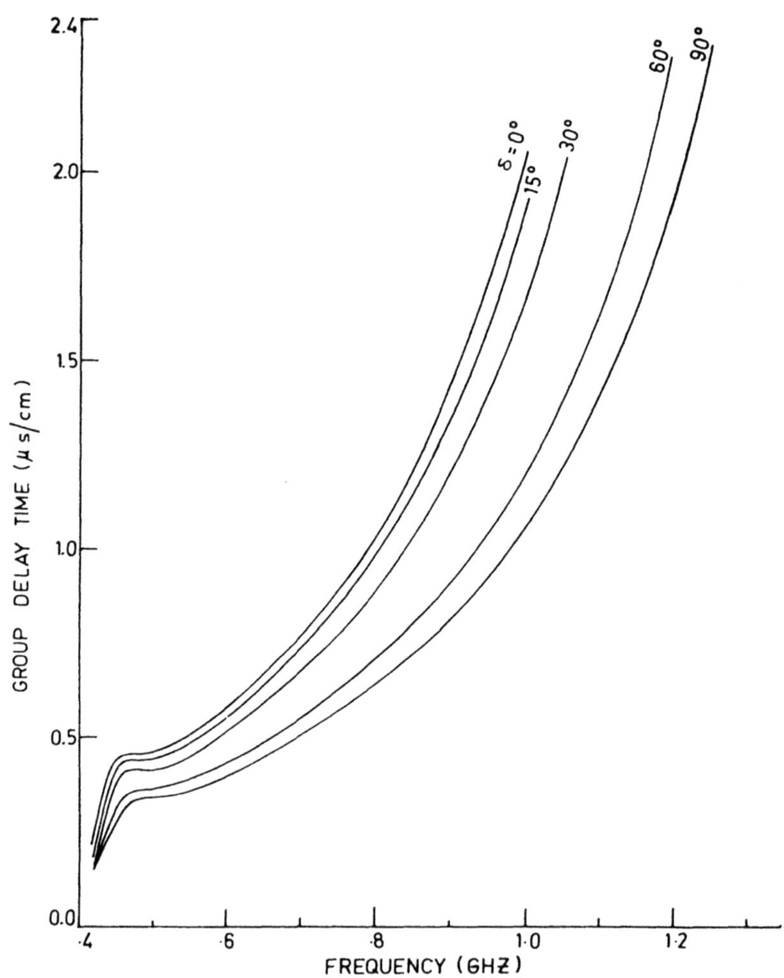

Figure 4.97. Variation of group delay time with frequency for different values of δ when $\varphi = 30°$ and $\theta = 30°$. Other parameters are the same as for Figure 4.96 (S. N. Bajpai, private communication).

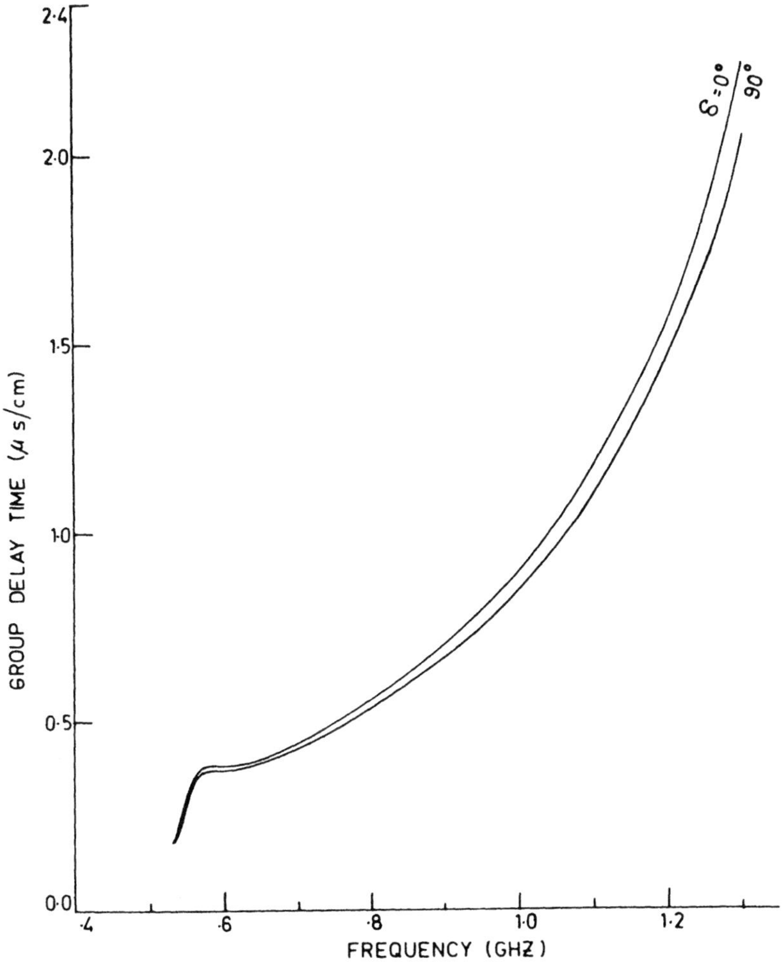

Figure 4.98. Variation of group delay time with frequency for different values of δ when $\varphi = 30°$ and $\theta = 60°$. Other parameters are the same as for Figure 4.96 (S. N. Bajpai, private communication).

as these three cases are:

(i) (a) $\theta = 30°, \varphi = 30°, \delta = \psi$; (b) $\theta = 30°, \varphi = 60°, \delta = \psi$
(ii) (a) $\theta = 60°, \varphi = 30°, \delta = \psi$; (b) $\theta = 60°, \varphi = 60°, \delta = \psi$
(iii) (a) $\theta = 90°, \varphi = 30°, \delta = \psi$; (b) $\theta = 30°, \varphi = 0, \quad \delta = 90° - \psi$;
(c) $\theta = 30°, \varphi = 90°, \delta = 90° - \psi$; (d) $\theta = 60°, \varphi = 0, \quad \delta = 90° - \psi$;
(e) $\theta = 60°, \varphi = 90°, \delta = 90° - \psi$; (f) $\theta = 90°, \varphi = 60°, \delta = \psi$

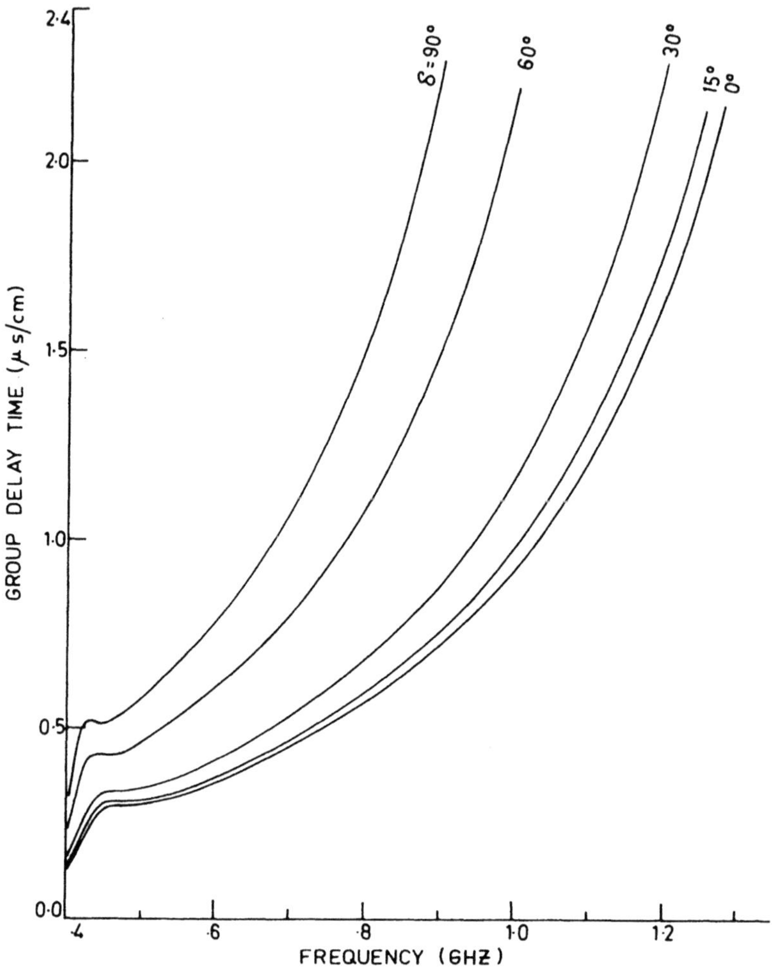

Figure 4.99. Variation of group delay time with frequency for different δ when $\varphi = 30^\circ$ and $\theta = 90^\circ$. Other parameters are the same as for Figure 4.96 (after S. N. Bajpai, private communication).

The general nature of the curves in Figures 4.97–4.99 can be understood on the basis of the fact that it is the dependence of α_n and δ which is responsible for the modification of delay characteristics when the direction of propagation is rotated in the plane of the slab. It follows from equations (1.78), (1.79) and (4.191) that

$$-\alpha_n^2 = \mu'_{22} = 1 + \omega_m\omega_0/D - \omega_m\omega_a/D$$

$$\times\{3\sin^2\delta\sin^2\theta[(1+\sin^4\varphi+\cos^4\varphi)\cos^2\theta - \tfrac{1}{2}\sin^2 2\varphi] + \tfrac{3}{2}\sin^2\theta\sin^2 2\varphi - 1\}$$

It is evident from this relation that α_n is independent of δ when either $\theta = 0$ or $\theta = \theta_c$, where

$$\cos^2\theta_c = 2\sin^2\varphi\cos^2\varphi/(1+\sin^4\varphi+\cos^4\varphi) \tag{4.195}$$

Figure 4.100 shows the variation of θ_c with φ. When $\varphi = 0$ or $\pi/2$, we have $\theta_c = \pi/2$, which corresponds to magnetization along a cube edge. When $\varphi = \pi/4$, θ_c is a minimum and is equal to $\cos^{-1}(1/\sqrt{3})$; this combination corresponds to magnetization along the body diagonal. In general, for a given φ, it is possible to find a $\theta = \theta_c$ from Figure 4.100 such that if the normal to the plane of the slab is characterized by Eulerian angles φ, θ_c (see Figure 4.95), the the wave delay characteristics will be unaffected by rotation of the direction of propagation in the plane of the slab. Figures 4.97–4.99 are plotted for different values of θ but a fixed $\varphi = 30°$ for which $\theta_c \simeq 61°$, which explains why the variation of the delay time with δ is least significant for Figure 4.98, which corresponds to $\theta = 60°$. It is also seen from Figures 4.97–4.99 that when $\theta < \theta_c$, the overall delay time is larger and the frequency range of nondispersive delay is smaller for $\delta = 0$ than for $\delta = \pi/2$. However, the converse is true for $\theta > \theta_c$.

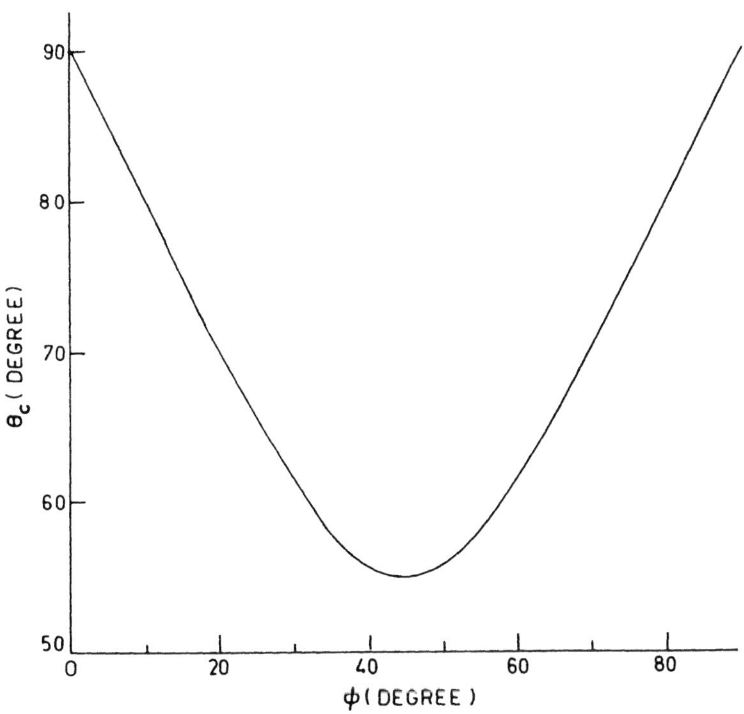

Figure 4.100. Variation of θ_c with φ (after Bajpai et al., 1979).

4.8.2. Zn_2Y Planar Magnetoplumbite

Figures 4.101–4.104 illustrate the frequency dependence of the group delay time for various thicknesses of the dielectric layer. The different values of δ correspond to different orientations of the direction of propagation with respect to the hard axis. When δ changes from 0 to $\pi/2$, the direction of propagation rotates from the y-axis (hard axis) to the x-axis. Figure 4.101 shows the variation of group delay with wave

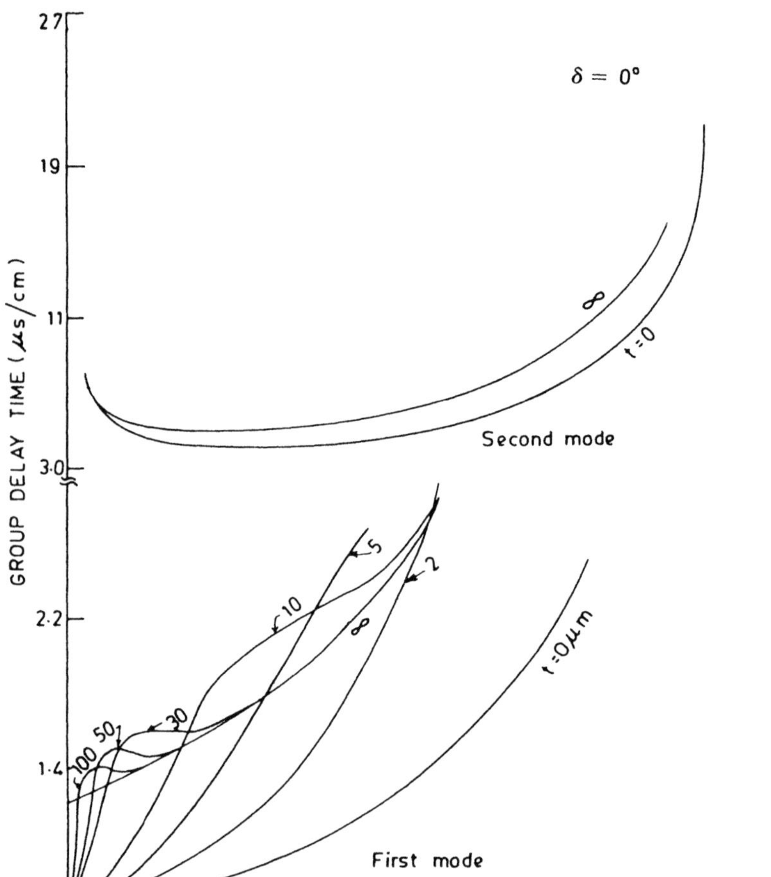

Figure 4.101. Group delay time vs. frequency with dielectric thickness as the parameter for the first- and second-order modes of magnetostatic bulk waves in a Zn_2Y planar magnetoplumbite–dielectric–metal structure. The parameters are $\delta = 0$, $H_a = 9000$ Oe, $H_0 = 200$ Oe, $4\pi M_0 = 2850$ G, and $d = 4.5\,\mu$m (after Bajpai et al., 1979).

frequency for the first two modes, in the case of propagation along the hard axis ($\delta = 0$). The general delay characteristics are similar to those for YIG. However, there are important quantitative differences. The magnitude of the delay time is quite large (more than 1 μsec/cm) in the frequency region of nondispersive delay, while the maximum frequency range in which nondispersive delay is obtained is very small (only about 50 MHz). When δ increases to 15° (Figure 4.102), there is a sharp fall in the delay time, but the frequency range of nondispersive delay increases significantly. In particular, a nondispersive group delay of about 0.4 μsec/cm is obtainable over a

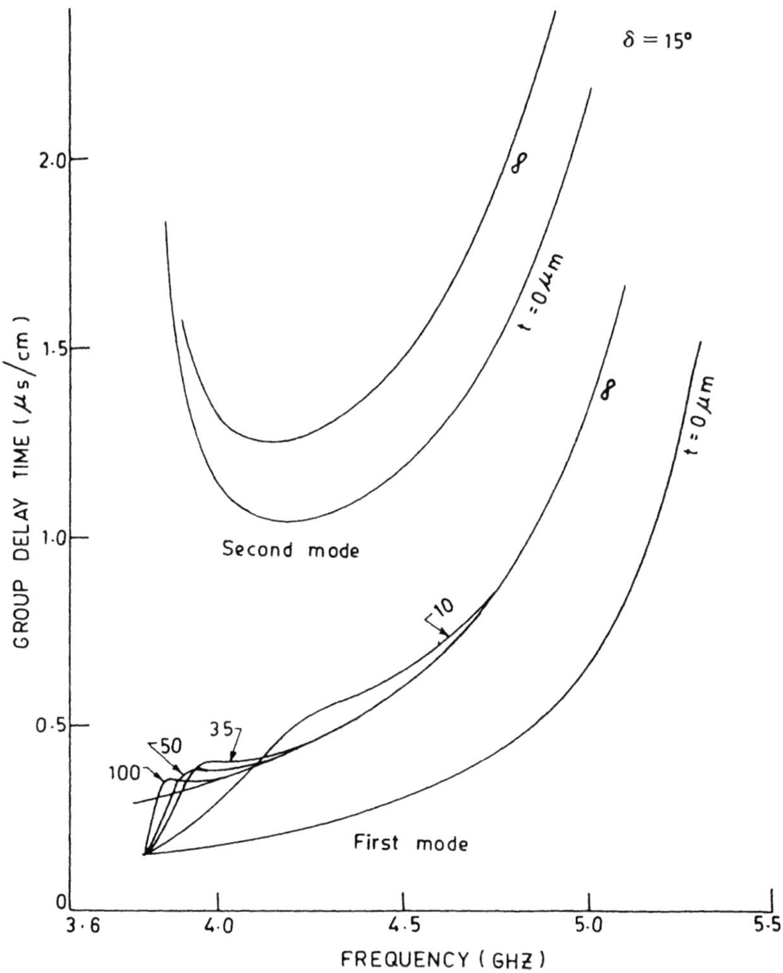

Figure 4.102. Group delay time vs. frequency for $\delta = 15°$. Other parameters are the same as for Figure 4.101 (after Bajpai et al., 1979).

frequency range of about 200 MHz around the central frequency of 4 GHz. It is also seen that the optimum thickness of the dielectric layer, corresponding to maximum nondispersive delay, increases with δ. The central frequency around which nondispersive delay is obtained also increases slightly with δ. When δ increases further (Figures 4.103 and 4.104), the overall delay time continues to decrease, while the frequency range of nondispersive delay increases. It follows from Figure 4.104 that, for $\delta = \pi/2$, the frequency range of nondispersive delay increases to about 100 MHz around 4.5 GHz, which is quite significant. However, in this

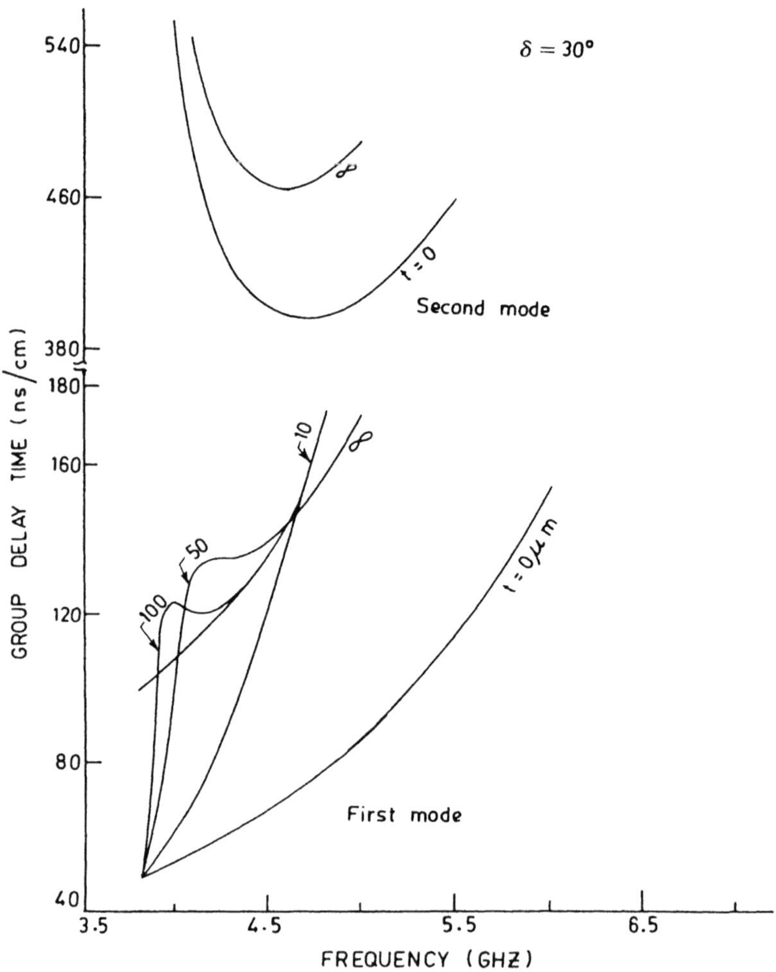

Figure 4.103. Group delay time vs. frequency for $\delta = 30°$. Other parameters are the same as for Figure 4.101 (after Bajpai et al., 1979).

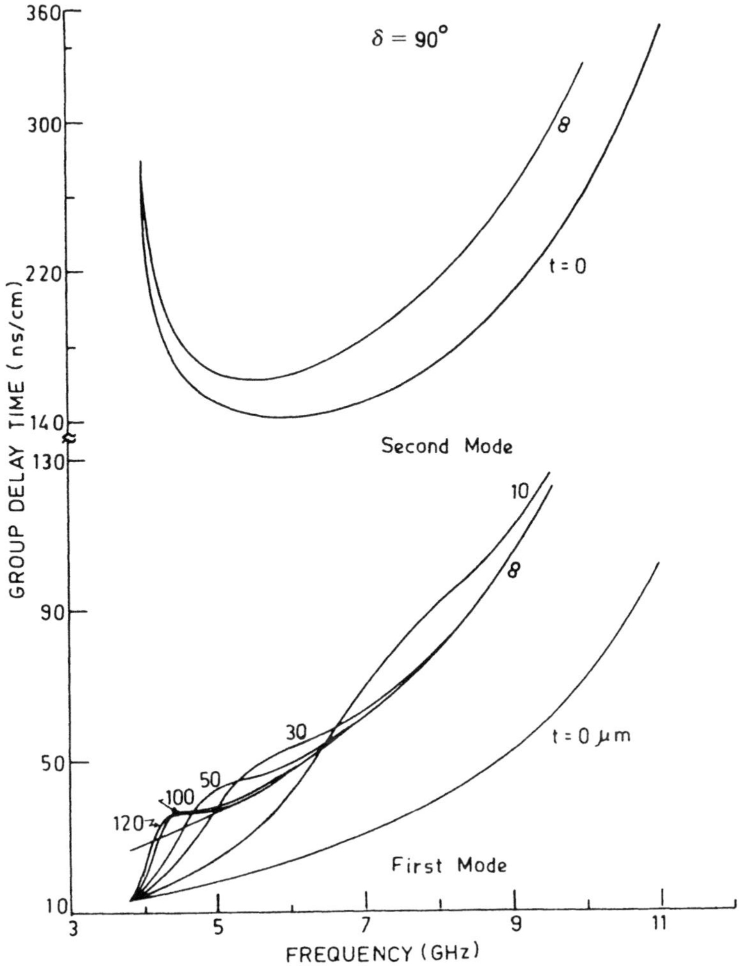

Figure 4.104. Group delay time vs. frequency for $\delta = 90°$. Other parameters are the same as for Figure 4.101 (after Bajpai et al., 1979).

range, the delay time is rather small (about 35 nsec/cm). Thus the frequency range of nondispersive delay and the magnitude of the group delay time can be controlled by changing the orientation of the direction of propagation with respect to the hard axis. It is also seen from Figures 4.101–4.104 that the delay time is more sensitive to changes in δ when it is closer to 0, i.e., when the direction of propagation is closer to the hard axis. The frequency range of allowed modes is obtained from equations (1.93), (1.94), and (4.191) as

$$[\omega_0(\omega_0 + \omega_a)]^{1/2} < \omega < [\omega_0(\omega_0 + \omega_a + \omega_m) + \omega_a\omega_m \sin^2 \delta]^{1/2}$$

When ω_0 is small and ω_a is large (which is so for Zn_2Y magnetoplumbite), this frequency range increases significantly with δ. For a numerical appreciation, it is seen that when $\delta = \pi/2$, magnetostatic waves can propagate in a frequency range from about 3 to 12 GHz with a biasing field as low as 200 Oe, which is important from the standpoint of practical application.

4.9. *Interaction of Magnetic Waves with Drifting Carriers*

In Sections 4.1 and 4.2, we saw that the propagation loss for magnetostatic waves is significantly high even for the materials with low linewidths. Moreover, coupling losses are also present in all practical situations. Consequently, even in carefully performed experiments, with the best available materials (e.g., YIG), magnetic wave signals get heavily damped before reaching the output terminals. Thus in order to compensate for the propagation and coupling losses the magnetic waves must be amplified. This can be achieved, under favorable conditions, by interaction with drifting charge carriers.

The literature on the interaction of magnetic waves with drifting carriers is sizable. It includes investigations by Trivelpiece et al. (1961), Akhiezer et al. (1963), Vural (1966), Vural and Thomas (1968), Chang and Matsuo (1968, 1977), Steele and Vural (1969), Schlömann (1969), Robinson et al. (1970), Vashkovskii et al. (1972), Szustakowski et al. (1973), Masuda et al. (1974), Kawasaki et al. (1974a, b), Chang, Yamada and Matsuo (1975, 1976a, b, 1977), Awai et al. (1976), Yamada et al. (1977), Bini et al. (1976, 1977, 1978a, b), Yukawa et al. (1978b), etc.

In spite of all the theoretical and experimental investigations already carried out, there has been some confusion in the literature (see Bini et al., 1976) regarding the interpretation of results, which is partly responsible for the difficulties in realizing the desired net amplification. However, some recent investigations (Bini et al., 1976) have clarified most of the aspects of the interaction of magnetic waves with drifting carriers.

In what follows, we explain the process of amplification physically in Section 4.9.1. The air–ferrimagnetic semiconductor configuration is considered in Section 4.9.2. Multilayered structures comprising metals, dielectrics, semiconductors and YIG are discussed in Section 4.9.3.

4.9.1. *Physical Explanation of Amplification*

Schlömann (1969) has explained the amplification of magnetic waves by interaction with drifting carriers physically. Consider the configuration shown in Figure 4.105. The magnetic field is applied along the z-axis, whereas the carrier drift current flows along the y-axis, which is also the direction of propagation. It follows from the $\mathbf{H}_0 \times \hat{n}$ criterion that a magnetic surface wave will propagate along the $+y$-direction.

The rf magnetization has been shown in the figure at three points along the y-direction half a wavelength apart. Now consider the rf magnetic field inside the semiconductor. Its direction is obviously the same as the direction of magnetization in the ferrimagnetic region. Since there is a dc electric field along the y-direction (which causes carrier drift), the rf magnetic field gives rise to rf currents due to the Hall effect. In the figure, the directions of these currents are shown by vertical arrows. The rf currents produce an rf magnetic field $\mathbf{h}_S$, which is represented by broken lines. This additional magnetic field is perpendicular to the direction of rf magnetization at each point shown in the figure. The field vector precedes the magnetization vector and has a tendency to enhance it. Thus, if $\mathbf{h}_S$ is sufficiently strong, it can compensate for propagation losses, in which case the wave can be amplified. If the direction of the dc current is reversed, the wave will be damped.

Although this physical explanation refers to the specific configuration shown in Figure 4.105, it can be easily extended to other configurations involving ferrimagnetics and semiconductors.

4.9.2. *Air–Ferrimagnetic Semiconductor Composite*

The configuration shown in Figure 4.105 can be realized in practice by placing thick ferrimagnetic and semiconductor substrates in contact. However, in the case of a ferrimagnetic half space, the magnetic surface wave is

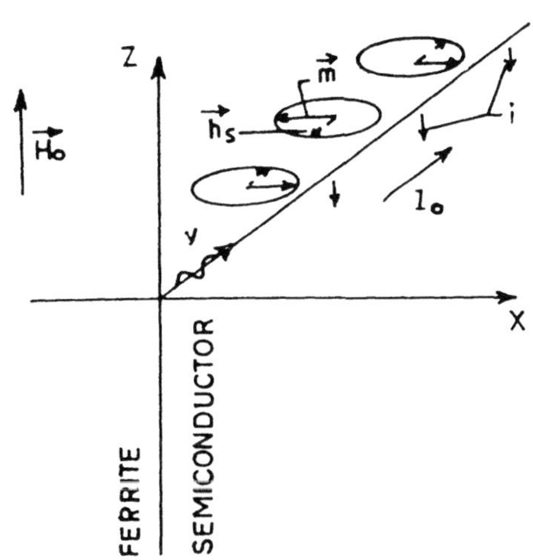

Figure 4.105. Ferrimagnetic-semiconductor composite; the biasing field is along the z-axis whereas the carriers drift along the y-axis which is also the direction of wave propagation (after Schlömann, 1969).

characterized by a single frequency (see Section 4.1). As such, the active bandwidth for amplification would be negligible for the configuration in question.

It is well known that the mode spectrum in the case of a ferrimagnetic film is strongly thickness-dependent. Specifically, the active bandwidth is significantly large when the film is thin. Therefore, it is worth considering the interaction of magnetic waves with drifting charge carriers in a ferrimagnetic film (rather than a half space) placed in contact with a semiconductor substrate.

Figure 4.106 shows the configuration; a ferrimagnetic slab ($0 \leqslant x \leqslant d$) is placed in contact with a semi-infinite semiconductor ($x \geqslant d$). The magnetization is along the z-axis, whereas the wave propagates along the y-axis. Magnetic loss is neglected, and exchange effects are ignored. The carrier mobility is assumed to be a scalar and independent of the applied dc electric field, and the carriers are of a single kind. The analysis of the interaction between the magnetic wave and the drifting charges under these assumptions has been carried out by Masuda et al. (1974) and Awai et al. (1976). We shall closely follow the latter.

(a) *Air Region* ($x \leqslant 0$). The field equations for the air region, under the magnetostatic approximation, are expressed as

$$\left.\begin{aligned} \nabla \times \mathbf{h}^a &= 0 \\ \nabla \cdot \mathbf{h}^a &= 0 \\ \nabla \times \mathbf{E}^a &= -j(\omega/c_0)\mathbf{h}^a \end{aligned}\right\} \tag{4.196}$$

The magnetic potential and the magnetic and electric fields are obtained as

$$\left.\begin{aligned} &\psi^a = A \exp(\beta x) \exp j(\omega t - \beta y) \\ &h_x^a = \beta\psi^a, \qquad h_y^a = -j\beta\psi^a \\ &E_z^a = (\omega/c_0)\psi^a \end{aligned}\right\} \tag{4.197}$$

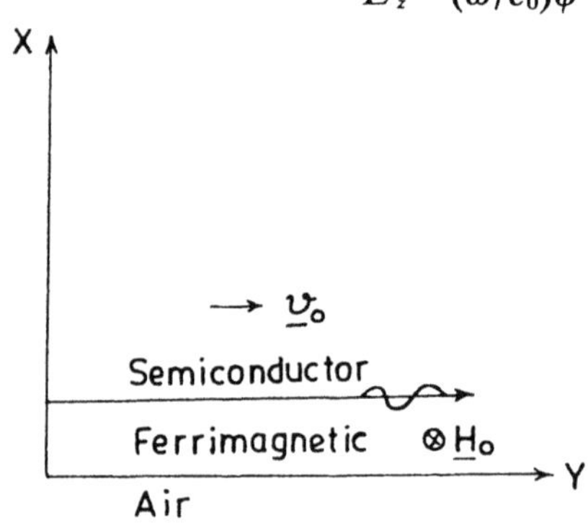

Figure 4.106. Air–ferrite–semiconductor structure.

(b) *Ferrimagnetic Region* $(0 \leqslant x \leqslant d)$. The field equations are expressed as

$$\left.\begin{aligned} \nabla \times \mathbf{h}^{\mathrm{f}} &= 0 \\ \nabla \cdot \mathbf{b}^{\mathrm{f}} &= 0 \\ \mathbf{b}^{\mathrm{f}} &= \boldsymbol{\mu} \cdot \mathbf{h}^{\mathrm{f}} \\ \nabla \times \mathbf{E}^{\mathrm{f}} &= -j(\omega/c_0)\mathbf{b}^{\mathrm{f}} \end{aligned}\right\} \tag{4.198}$$

The field quantities are obtained as

$$\left.\begin{aligned} &\psi^{\mathrm{f}} = [B\exp(\beta x) + C\exp(-\beta x)]\exp j(\omega t - \beta y) \\ &h_x^{\mathrm{f}} = \beta[B\exp(\beta x) - C\exp(-\beta x)]\exp j(\omega t - \beta y) \\ &h_y^{\mathrm{f}} = -j\beta\psi^{\mathrm{f}} \\ &b_x^{\mathrm{f}} = \beta[(\mu + \kappa)B\exp(\beta x) - (\mu - \kappa)C\exp(-\beta x)]\exp j(\omega t - \beta y) \\ &E_z^{\mathrm{f}} = (\omega/\beta c_0)b_x^{\mathrm{f}} \end{aligned}\right\} \tag{4.199}$$

(c) *Semiconductor Region* $(x \geqslant d)$. In this region, the electric field cannot be ignored (since mobile charges are present) and, therefore, the magnetostatic approximation is not valid (see Section 4.3.1.4). As such, the field equations are expressed as

$$\left.\begin{aligned} \nabla \times \mathbf{h}^{\mathrm{S}} &= j(\omega/c_0)\varepsilon_{\mathrm{d}}\boldsymbol{\varepsilon} \cdot \mathbf{E}^{\mathrm{S}} \\ \nabla \times \mathbf{E}^{\mathrm{S}} &= -j(\omega/c_0)\mathbf{b}^{\mathrm{S}} \end{aligned}\right\} \tag{4.200}$$

where ε_{d} is the background (lattice) dielectric constant and the relative permittivity $\boldsymbol{\varepsilon}$ for the case of drifting carriers can be expressed as (Steele and Vural, 1969):

$$\boldsymbol{\varepsilon} = \begin{pmatrix} \varepsilon_x & j\varepsilon_y & 0 \\ -j\varepsilon_y & \varepsilon_x & 0 \\ 0 & 0 & \varepsilon_z \end{pmatrix} \tag{4.201}$$

where

$$\varepsilon_z = 1 - \frac{\omega_{\mathrm{p}}^2(\omega - \beta v_0)}{\omega^2(\omega - \beta v_0 - j\nu_{\mathrm{c}})} \tag{4.202}$$

In this expression ω_{p}, v_0, and ν_{c} are the plasma frequency, drift velocity, and carrier collision frequency, respectively.

Since the magnetic waves under study are TE waves (see Section 4.1), the relevant component of the electric field E_z can be shown, from equations (4.200), to satisfy the following equation:

$$(\partial^2/\partial x^2 - k_S^2)E_z^S = 0 \tag{4.203}$$

where

$$k_S^2 = \beta^2 - \varepsilon_d(\omega^2/c_0^2)\varepsilon_z \tag{4.204}$$

The electric and magnetic field vectors are thus obtained as

$$\begin{aligned} E_z^S &= G \exp(-k_S x) \exp j(\omega t - \beta y) \\ h_x^S &= c_0\beta/\omega E_z^S \\ h_y^S &= j(\omega/c_0)k_S E_z^S \end{aligned} \tag{4.205}$$

Matching of E_z^S and h_y^S at $x = 0, d$ and elimination of constants yields the following dispersion relation (Awai et al., 1976):

$$\exp(-2\beta d) = \frac{(1+\mu \mp \kappa)[1+\gamma_3(\mu \pm \kappa)]}{(1-\mu \mp \kappa)[1-\gamma_3(\mu \mp \kappa)]} \tag{4.206}$$

where

$$\gamma_3^2 = 1 - \frac{\varepsilon_d \omega^2}{\beta^2 c_0^2} + \frac{\omega_p^2 \varepsilon_d(\omega - \beta v_0)}{c_0^2 \beta^2(\omega - \beta v_0 - j\nu_c)} \tag{4.207}$$

If it is assumed that $\beta \gg (\varepsilon_d)^{1/2}\omega/c$ and $\nu_c \gg \omega_1, \beta\nu_0$ (which hold for magnetic waves and for ordinary semiconductors) γ_3 reduces to

$$\gamma_3 \simeq \sqrt{1 + j(4\pi\sigma_e/c_0^2\beta^2)(\omega - \beta v_0)} \tag{4.208}$$

It is easy to show that when the carrier density approaches zero, γ_3 approaches unity and equation (4.206) assumes the Damon–Eshbach form, i.e., equation (4.44). On the other hand, when the carrier density approaches infinity (i.e., when the semiconductor is replaced by a perfect conductor), the dispersion relation (4.206) assumes, as expected, the Seshadri form, i.e., equation (4.119).

For arbitrary carrier concentration, equation (4.206) is a transcendental equation involving complex quantities and requires numerical solution for the study of the interaction. We shall first consider the case when there is no carrier drift ($v_0 = 0$) and then take up the case of drifting carriers.

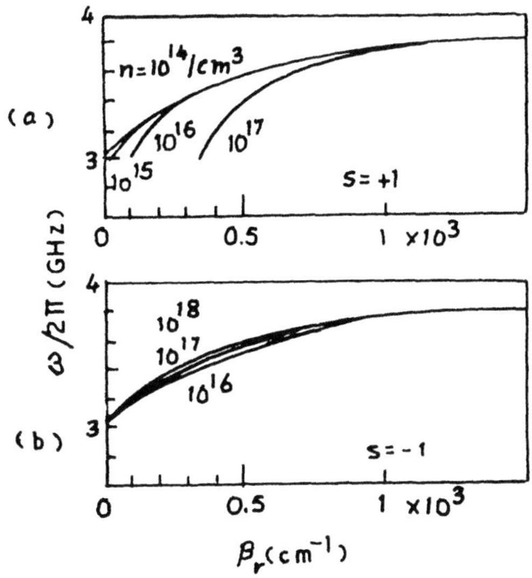

Figure 4.107. Dispersion curves in the case when there is no carrier drift ($v_0 = 0$). The carrier density n is the variable parameter and is shown alongside each curve. Other parameters used in the calculations are $4\pi M_0 = 1750\,G$, $H_0 = 500\,Oe$, $d = 10\,\mu m$, $\varepsilon_d = 17.9$, and electron mobility $\mu_e = ne/m\nu_c = 5 \times 10^4\,cm^2/V$ sec. The cases (a) and (b) correspond to magnetization along the $+z$- and $-z$-directions, respectively (after Awai et al., 1976).

Case I: *No Drift* ($v_0 = 0$). In this case we have $v_0 = 0$ in equation (4.208). Out of the infinite solutions to the complex equation (4.206), it is required to choose only those which: (i) reduce to Damon–Eshbach and Seshadri forms as the carrier density approaches zero and infinity, respectively, and (ii) have negative imaginary parts since, in the case when there is no carrier drift, the wave always decays as it propagates. The proper solutions obtained with the help of these criteria are shown in Figures 4.107–4.109. Figure 4.107 shows the dispersion curves for a typical set of data with carrier density as the parameter. It is seen that only the $S = +1$ ($+z$ magnetization) mode wavenumbers depend significantly on the carrier density, which is understandable, since the wave energy for this mode is concentrated at the ferrite–semiconductor interface.* The other ($S = -1$) mode is characterized by field concentration at the ferrite–air interface and hence it interacts rather weakly with the drifting carriers. Figure 4.108 shows the dispersion curves for the $S = +1$ mode in the case when the carrier density is very large. It is seen that a new branch appears in the low-wavenumber region, which has been designated as branch 2 in

* This can be inferred from the $\mathbf{H}_0 \times \hat{n}$ criterion.

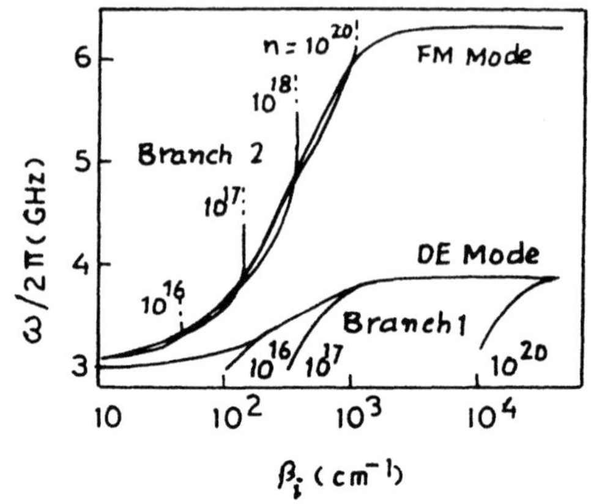

Figure 4.108. Dispersion curves for the $S = +1$ mode without the carrier drift in the case when the carrier concentration n is large. The parameters are the same as for Figure 4.107. It is clear that, as the carrier concentration increases, branch 2 dominates over branch 1, thereby explaining the transition from the Damon–Eshbach mode to the Seshadri or the ferrite–metal mode when a metal plate is brought into contact with the ferrite slab (after Awai et al., 1976).

the figure. The dispersion curves in Figure 4.107a correspond to branch 1. As the carrier density increases, branch 2 grows in wavenumber as well as frequency domains at the expense of branch 1. In the limit when the carrier density approaches infinity, only branch 2 remains, which is then, in fact, the Seshadri curve (ferrimagnetic–metal mode). Figure 4.109 shows the frequency dependence of the imaginary part of the propagation constant for $S = \pm 1$ modes. As expected, the loss factor is, in general, smaller for the $S = -1$ mode than for the $S = +1$ mode.

Case II: *Drifting Carriers.* In this case, β_i can be either positive or negative. The positive values of β_i could imply either amplifying or even evanescent modes; the distinction can be made using Briggs' criteria (Briggs, 1964). The procedure for the choice of proper solutions is then as follows.

1. Consider several solutions which are comparable to the Damon–Eshbach mode when the carrier concentration n is small (less than about 10^{17}/cc; see Figures 4.108 and 4.109) and to the Seshadri mode when n is large (above 10^{17}/cc).

2. Apart from the solutions with positive β_i, retain also one solution for which β_i is negative and smallest in magnitude (which will obviously dominate over the solutions with larger negative absolute β_i because of least attenuation).

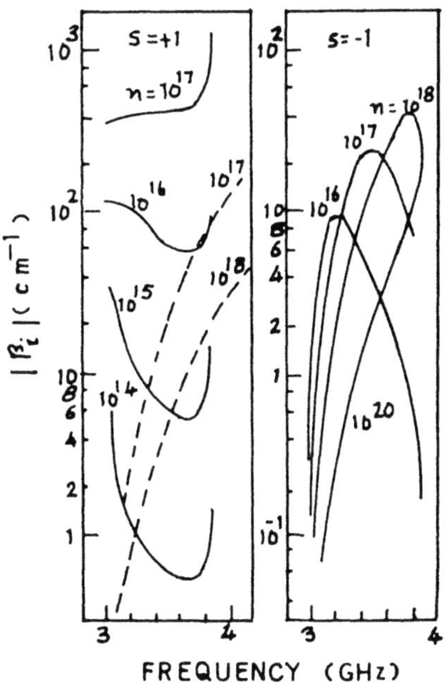

Figure 4.109. Variation of the attenuation factor with frequency for different values of the carrier concentration. The dotted lines in (a) are for branch 2 in Figure 4.108. The parameters are the same as for Figure 4.107 (after Awai et al., 1976).

3. Subject the solutions with positive β_i to Briggs' criteria, and determine whether β_i has a different sign when the frequency is assumed to have a large negative imaginary part. If it does then the wave is amplifying, otherwise it is an evanescent wave.

The dispersion curves obtained from the above criteria have been shown in Figures 4.110 and 4.111 for a typical set of data. A comparison of Figure 4.110 with Figure 4.107 shows that the curves for $S = +1$ $(S = -1)$ are significantly (insignificantly) altered when the drift field is applied. It is seen from Figure 4.111 and $\beta_i \gtrless 0$ when $v_0 \gtrless \omega/\beta_r$, which implies that the wave is amplified when the drift velocity of the carrier exceeds the phase velocity of the waves. Further analysis and calculations based on equations (4.206) lead (Awai et al., 1976) to the following conclusions: (i) The amplification factor β_i is a maximum at a certain optimum value of the carrier concentration. (ii) When the carrier concentration is low, β_i is a linearly increasing function of the electric field. (iii) The amplification factor β_i strongly depends on the wave frequency.

4.9.3. *Metal–Dielectric–Semiconductor–YIG Structure*

Consider the layered structure shown in Figure 4.112, which is the most general composite that can be formed with metals, dielectrics, semiconductor, and YIG for the purpose of guiding (specifically, amplifying) the magnetic waves. The analysis of the amplification of magnetic surface waves due to the interaction with drifting carriers in the present configuration has been carried out by Chang and Matsuo (1975) and Chang

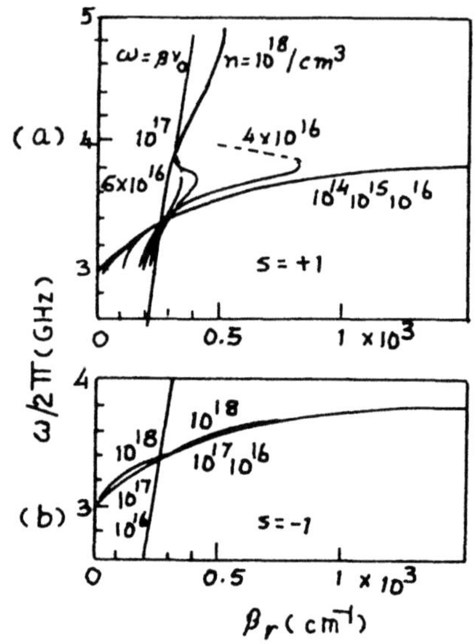

Figure 4.110. Dispersion curves for the case of drifting carriers. The drift velocity $v_0 = 8 \times 10^7$ cm/sec while other parameters are the same as for Figure 4.107 (after Awai et al., 1976).

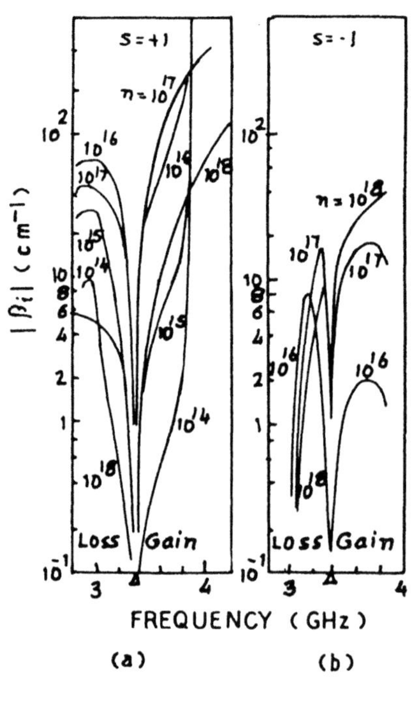

Figure 4.111. Frequency dependence of the amplification factor for different carrier concentration for the $S = \pm 1$ modes. The crossover between attenuation and amplification occurs at $\omega = \beta_r/v_0$. The parameters are the same as for Figure 4.110 (after Awai et al., 1976).

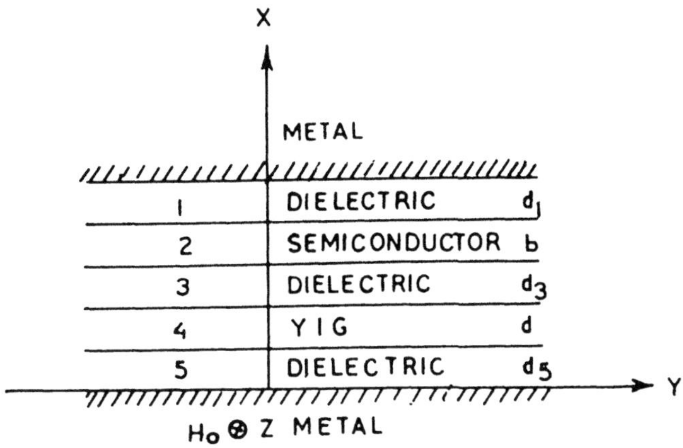

Figure 4.112. The layered structure consisting of metals, dielectrics, semiconductor, and YIG (after Chang et al., 1976a).

et al. (1976a). The dispersion relation can be obtained as

$$\begin{aligned}\exp(2fd) = \{&\mu_+[(\mu_-\xi - 1)\tanh fd_3 + (\mu_- - \xi)] \\ &+ \tanh fd_5[(\xi - \mu_-) + (1 - \mu_-\xi)\tanh fd_3]\} \\ &\times \{\mu_-[(\mu_+\xi + 1)\tanh fd_3 + (\mu_+ + \xi)] \\ &\quad + \tanh fd_5[(\xi + \mu_+) + (1 + \mu_+\xi)\tanh fd_3]^{-1}\}\end{aligned} \tag{4.209}$$

where

$$\left.\begin{aligned}\xi &= \frac{1}{\gamma_S'}\frac{\gamma_S'\tanh fd_1 + \tanh\gamma_S b}{\gamma_S'\tanh fd_1\tanh\gamma_S b + 1} \\ \mu_\pm &= \mu \pm s\kappa \\ \gamma_S^2 &= \beta^2 - \beta_d^2[1 - j(\omega_p^2/\omega^2)\zeta(\omega - \beta v_0)] \\ \gamma_S' &= \gamma_S/f \\ \zeta &= \omega - \beta v_0 - j\nu_c \\ f &= s\beta \\ \beta &= \beta_r + j\beta_i \\ \beta_d^2 &= \varepsilon_d\omega^2/c_0^2 \\ \beta_0 &= \omega/v_0\end{aligned}\right\} \tag{4.210}$$

The results of numerical calculations for a typical set of data have been presented in Figures 4.113–4.115. Figure 4.113 shows the dispersion curves for $S = \pm 1$ modes. The gain parameter has been plotted in Figure 4.114. It

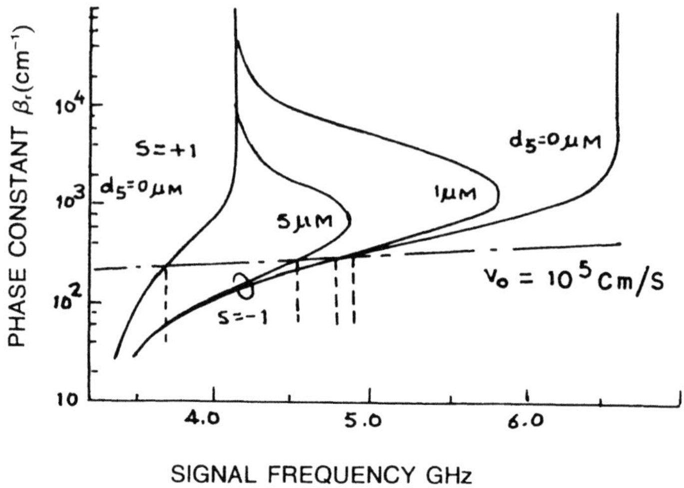

Figure 4.113. The dispersion curves for waves propagating in the layered structure shown in Figure 4.112. The solid lines are for the magnetic surface waves whereas the chain dotted line is for the carrier wave in a semiconductor. The parameters used in calculations are: $4\pi M_0 = 1750$ G, $H_0 = 600$ Oe, $d_1 = \infty$, $d_3 = 0$, $d = 10\,\mu$m, and $b = 1\,\mu$m. The values of d_5 are shown in the figure (after Chang et al., 1976a).

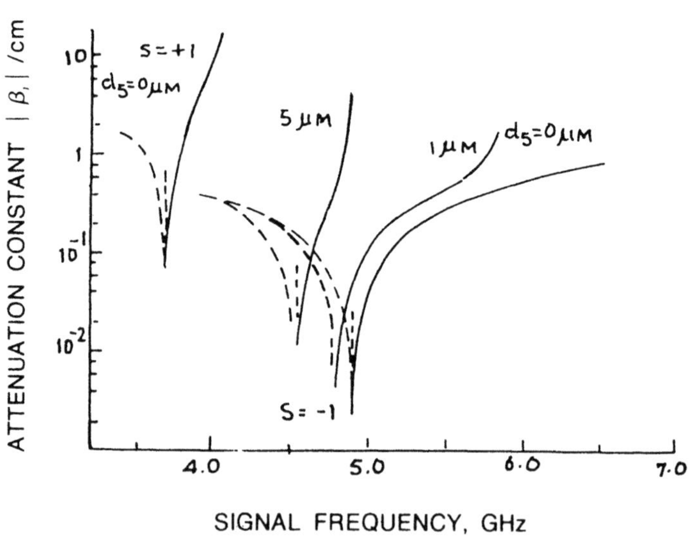

Figure 4.114. Frequency dependence of the gain parameter for different values of d_5 shown alongside each curve. The solid and broken lines correspond to gain and loss, respectively. The parameters are the same as for Figure 4.113 (after Chang et al., 1976a).

is seen that, in the active region, the gain is larger for the $S = +1$ mode than for the $S = -1$ mode. However, the bandwidth over which net gain occurs is larger for the $S = -1$ mode. Figure 4.115 shows the frequency dependence of the gain parameter on the magnetic linewidth of YIG. It is seen from the figure that the bandwidth over which net gain occurs decreases with increasing linewidth. Thus, in order to achieve significant gain over a reasonable bandwidth, it is necessary to have the smallest

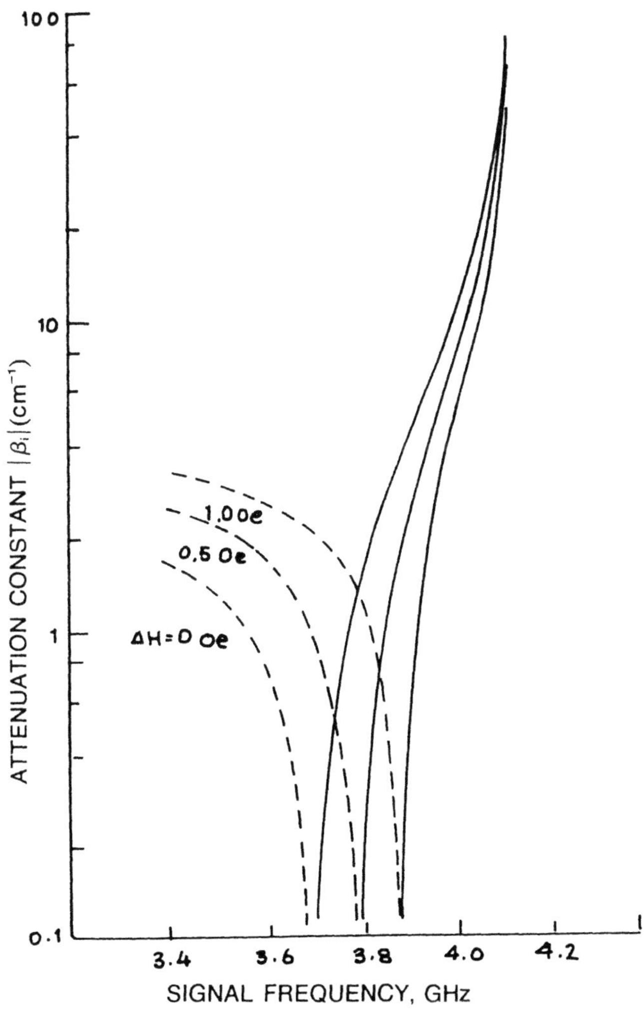

Figure 4.115. Frequency dependence of the gain parameter for different values of the linewidth ΔH. Other parameters are the same as for Figure 4.113 (after Chang et al., 1976a).

possible value of ΔH for the ferrimagnetic material and the largest feasible value for the drift velocity in the semiconductor.

It is interesting to note that several investigations (e.g., Chang et al., 1976b; Bini et al., 1976) have been carried out to study the interaction of forward, bulk magnetic waves with drifting carriers. However, in general, it seems to be easier to achieve amplification in the case of surface waves than in bulk waves (Bini et al., 1978).

Several experimental investigations, including those by Vural and Thomas (1968), Vaskovskii et al. (1972), and Kawasaki et al. (1974a, b), have exhibited reduced loss or even net gain as a result of the convective instability. However, a working amplifier with the desired characteristics has yet to be fabricated.

4.10. *Magnetostatic Wave Oscillators*

Howarth (1975), Miller and Brown (1976), Castéra (1978), and Castéra and Hartemann (1978) have described tunable, magnetostatic surface wave oscillators. Such an oscillator uses a magnetostatic wave delay line in the feedback loop of an amplifier. The block diagram is shown in Figure 4.116. It follows from the general principles that oscillations occur at frequency f, provided that the loop gain exceeds unity and the phases satisfy the condition (Howarth, 1975; Miller and Brown, 1976)

$$2\pi\tau_f f + \varphi_e = 2n\pi \tag{4.211}$$

where φ_e is the phase shift associated with the amplifier and transmission lines, τ_f is the delay time at frequency f, and n is an integer. Consequently, the frequency at which oscillations set in is obtained as

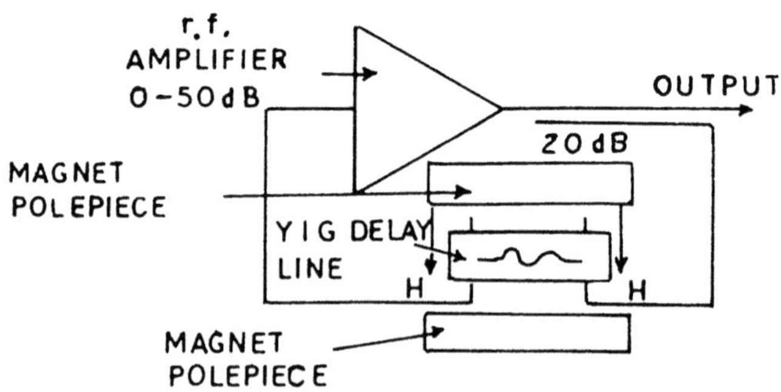

Figure 4.116. Block diagram of the magnetostatic wave oscillator (after Miller and Brown, 1976).

$$f = n/(\tau_f + \varphi_e/2\pi) \tag{4.212}$$

The system shown in Figure 4.116 is thus able to support a comb of frequencies given by equation (4.212).

We now describe some experimental results obtained by Castéra (1978). The experiments involve a magnetostatic surface wave delay line consisting of a film of YIG ($\sim 10\,\mu$m thick) grown over a GGG substrate by LPE. The waves are excited or detected by means of microstrip couplers about $50\,\mu$m wide and spaced over 1 cm. A typical mode spectrum of such an oscillator is shown in Figure 4.117.

It is possible to obtain a single-mode oscillator by narrow-banding the delay line, which is, in turn, achieved by using multibar microstrip transducers (Wu et al., 1977). The wavevector of the magnetostatic surface wave thus excited depends on the mutual spacing of the successive microstrips. The resulting insertion loss has been shown in Figure 4.118 as a function of frequency. A 3 dB bandwidth of 35 MHz is obtained at a frequency of 3.2 GHz, which corresponds to a wavelength of $419\,\mu$m (which is precisely equal to the interstrip spacing). The insertion loss is about 12 dB. This response is tunable over more than 1 GHz with no appreciable

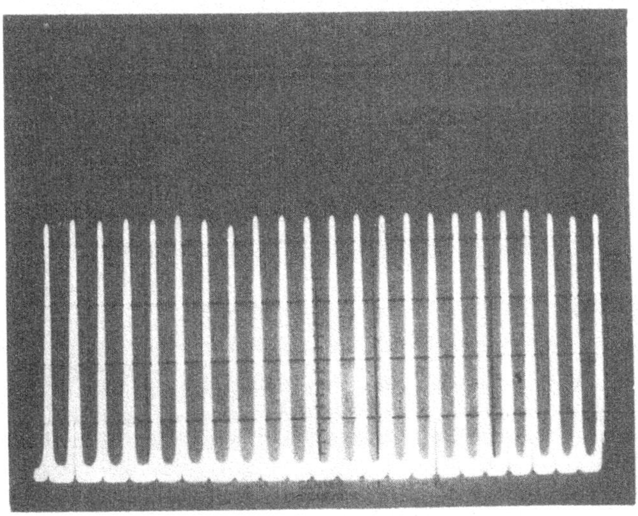

Figure 4.117. Mode spectrum of an oscillator using the circuit shown in Figure 4.116. The YIG film delay line has a single-strip transducer. The parameters are $H_0 = 200$ Oe, central frequency $f = 2.1$ GHz, and mode spacing $\Delta f = 8.55$ MHz (after Castéra, 1978).

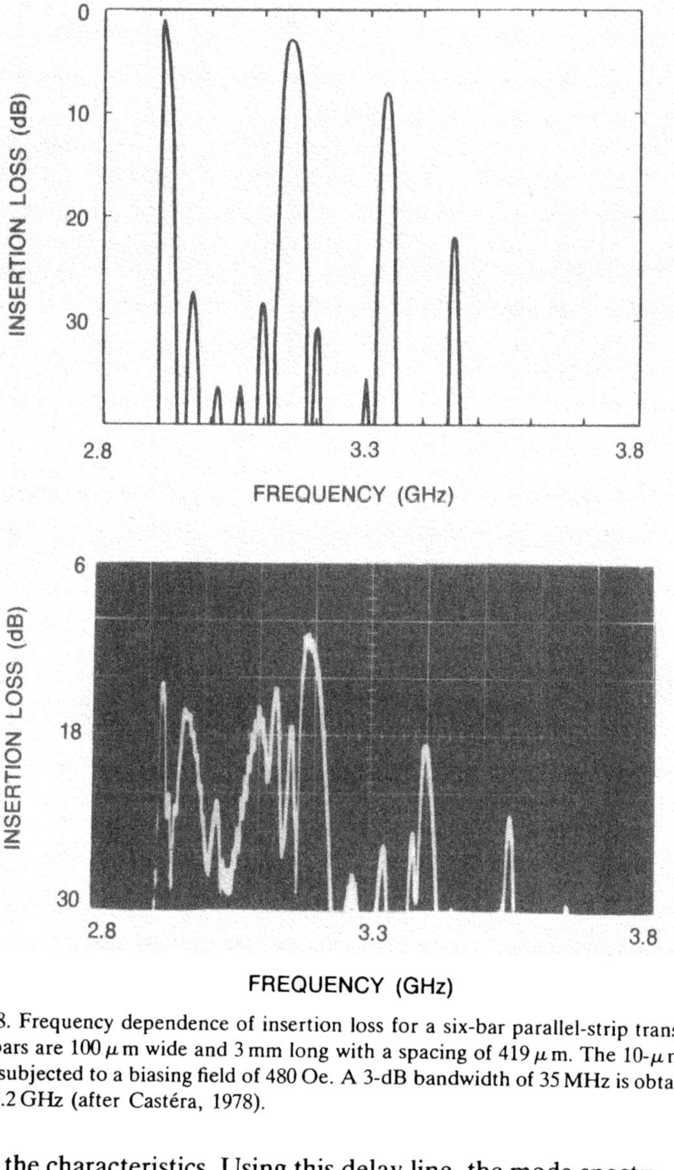

Figure 4.118. Frequency dependence of insertion loss for a six-bar parallel-strip transducer. Individual bars are 100 μm wide and 3 mm long with a spacing of 419 μm. The 10-μm-thick YIG film is subjected to a biasing field of 480 Oe. A 3-dB bandwidth of 35 MHz is obtained at frequency 3.2 GHz (after Castéra, 1978).

change in the characteristics. Using this delay line, the mode spectrum of a single-mode oscillator has been shown in Figure 4.119. The oscillator is tunable over a wide range (from 2 to 3.8 GHz) with restrictions only due to limited amplifier bandwidth. Castéra (1978) and Castéra and Hartemann (1978) have improved the performance of such oscillators by using two delay paths in the feedback loop. One- and two-part resonators (Collins et al., 1977) using reflective arrays have also been studied in recent years.

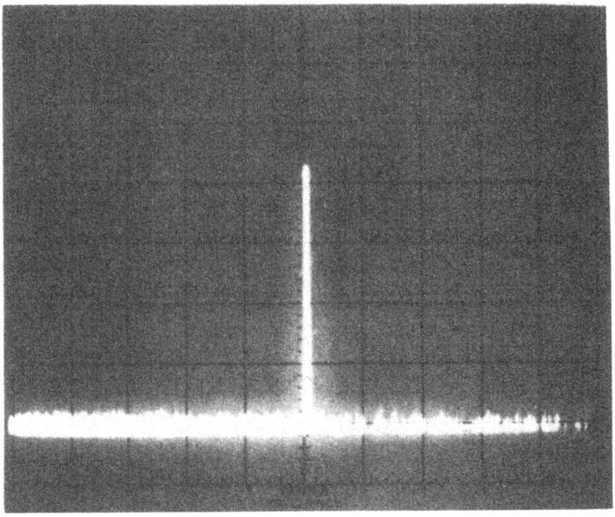

CENTER FREQUENCY: 2.5 GHz
20 MHz/div

Figure 4.119. Mode spectrum of an oscillator using a delay line with a six-bar parallel-strip transducer, the insertion loss of which has been shown in Figure 4.118 (after Castéra, 1978).

4.11. *Transduction of Magnetostatic Waves*

In a guided magnetic wave device, higher figures of merit can be achieved by minimizing the insertion loss which, in turn, is a sum of the propagation loss and the coupling loss. While the propagation loss (measured in decibels) is approximately proportional to the magnetic linewidth, and can be minimized by choosing a well-prepared sample of a narrow linewidth material like YIG, the coupling loss depends on the details of the transducer geometry and configuration.

In earlier investigations on magnetostatic waves in planar structures, the transduction was achieved using cables with (grounded) extended central conductors. It was seen later that efficient excitation of magnetostatic waves in slabs and films could be achieved through the use of microstrip couplers. In recent years, more sophisticated couplers, specifically, meander lines (Schilz, 1973), multibar parallel strips and π-transducers (Wu et al., 1977), and interdigital transducers (Adam et al., 1978) consisting of open-circuited microstrip fingers have also been fabricated.

The first exhaustive analysis of magnetostatic wave excitation was reported by Ganguly and Webb (1975), and further analyses dealing with the various couplers in different configurations were published by Sethares et al. (1978), Parekh and Tuan (1979b).

In what follows, we present a simplified, quasi-static analysis of microstrip excitation of magnetostatic bulk waves in a normally magnetized YIG film, and then take up the case of surface wave excitation.

Figure 4.120a shows a ferrimagnetic film ($-d \leqslant z \leqslant 0$). A thin microstrip is deposited on the top surface $z = 0$ so as to occupy the region $-b/2 \leqslant x \leqslant b/2$. Magnetostatic bulk waves propagating along the x-axis are excited by means of a surface current distribution $J_y(x)\exp(j\omega t)$ in the microstrip. The surface current, and hence the excited wave field, is uniform along the y-axis. Magnetocrystalline anisotropy and magnetic loss are neglected.

The magnetostatic wave field is required to satisfy equations (4.1) and (4.2), with $\partial/\partial y = 0$, in ferrimagnetic and free-space regions, respectively. The general solution is then expressed as the sum (in fact, integral) of particular solutions over the entire wavenumber spectrum. Thus, ignoring the time dependence, the magnetostatic potentials in different regions are expressed as follows*:

$$\left.\begin{aligned} \psi^{\mathrm{I}} &= \int_{-\infty}^{+\infty} A \exp(\beta z)\exp(-j\beta x)\,d\beta \\ \psi^{\mathrm{f}} &= \int_{-\infty}^{+\infty} [C\exp(j\alpha_v\beta z) + D\exp(-j\alpha_v\beta z)]\exp(-j\beta x)\,d\beta \\ \psi^{\mathrm{II}} &= \int_{-\infty}^{+\infty} E\exp(-\beta z)\exp(-j\beta x)\,d\beta \end{aligned}\right\} \tag{4.213}$$

where $\alpha_v = \sqrt{-\mu}$. The relevant components of the wave magnetic field and induction are obtained as

$$\left.\begin{aligned} h_x^{\mathrm{I}} &= -j\int_{-\infty}^{+\infty} \beta A \exp(\beta z)\exp(-j\beta x)\,d\beta \\ h_x^{\mathrm{f}} &= -j\int_{-\infty}^{+\infty} \beta[C\exp(j\alpha_v\beta z) + D\exp(-j\alpha_v\beta z)]\exp(-j\beta x)\,d\beta \\ h_x^{\mathrm{II}} &= -j\int_{-\infty}^{+\infty} \beta E\exp(-\beta z)\exp(-j\beta x)\,d\beta \end{aligned}\right\} \tag{4.214}$$

* The "constants" A, C, D, and E may depend on β.

$$\left.\begin{aligned} b_z^{\mathrm{I}} &= \int_{-\infty}^{+\infty} \beta A \exp(\beta z)\exp(-j\beta x)d\beta \\ b_z^{\mathrm{f}} &= \int_{-\infty}^{+\infty} j\alpha_v\beta[C\exp(j\alpha_v\beta z) - D\exp(-j\alpha_v\beta z)]\exp(-j\beta x)d\beta \\ b_z^{\mathrm{II}} &= -\int_{-\infty}^{+\infty} \beta E \exp(-\beta z)\exp(-j\beta x)d\beta \end{aligned}\right\} \tag{4.215}$$

The boundary conditions are as follows:

At $z = -d$:

$$b_z^{\mathrm{f}} = b_z^{\mathrm{I}} \tag{4.216}$$

$$h_x^{\mathrm{f}} = h_x^{\mathrm{I}} \tag{4.217}$$

At $z = 0$:

$$b_z^{\mathrm{f}} = b_z^{\mathrm{II}} \tag{4.218}$$

$$h_x^{\mathrm{II}} - h_x^{\mathrm{f}} = J_y(x) \tag{4.219}$$

Using equations (4.214)–(4.218), we can relate the unknown quantities appearing in equations (4.213):

$$\left.\begin{aligned} C &= T\exp(2j\alpha_v\beta d)D \\ A &= (T+1)\exp[(j\alpha_v+1)\beta d]D \\ E &= j\alpha[1 - T\exp(2j\alpha_v\beta d)]D \end{aligned}\right\} \tag{4.220}$$

where

$$T = (\alpha_v - j)/(\alpha_v + j) \tag{4.221}$$

Substitution from equations (4.214) and (4.220) into equation (4.219) leads to

$$-j\int_{-\infty}^{+\infty} \beta F_T(\omega, \beta)\exp(2j\alpha_v\beta d)\exp(-j\beta x)D\,d\beta = J_y(x) \tag{4.222}$$

where

$$F_T(\omega, \beta) - F(\omega, \beta)\exp(-2j\alpha_v\beta d) \tag{4.223}$$

$$F(\omega, \beta) = j\alpha_v(1 - T\exp(2j\alpha_v\beta d)] - [1 + T\exp(2j\alpha_v\beta d)] \tag{4.224}$$

Multiplying both sides of equation (4.222) by $\exp(j\beta' x)$, integrating over x from $-\infty$ to $+\infty$, and using the condition

$$\int_{-\infty}^{+\infty} \exp j(\beta' - \beta)x dx = 2\pi\delta(\beta' - \beta)$$

we obtain

$$D = \frac{j\tilde{J}(\beta)\exp(-2j\alpha_v\beta d)}{2\pi\beta F_T(\omega, \beta)} \tag{4.225}$$

where

$$\tilde{J}(\beta) = \int_{-\infty}^{+\infty} J_y(x)\exp(j\beta x)dx \tag{4.226}$$

The mode fields can exist even in the absence of the exciting current, but this is possible only when the integrands in the right-hand sides of equations (4.213) are singular. It is seen from equations (4.220) and (4.225) that this condition leads to

$$F_T(\omega, \beta) = 0 \tag{4.227}$$

Thus, the dispersion relation is obtained from equations (4.223), (4.224), and (4.227) as

$$\cot \beta\alpha_v d = (\alpha_v^2 - 1)/2\alpha_v \tag{4.228}$$

Equivalently,

$$\cot \alpha_v\beta d/2 = \alpha_v \tag{4.229}$$

which, as expected, is the same as equation (4.75). Substitution from equations (4.220) and (4.225) into equations (4.213) followed by integration by the method of residues leads to (Ganguly and Webb, 1975; Sethares et al., 1978):

$$\left.\begin{aligned} \psi^{\mathrm{I}} &= -[(T+1)G/\beta]\exp[\beta(d+z)]\exp(-j\beta x) \\ \psi' &= -(G/\beta)\{T\exp[j\beta\alpha_v(z+d)] + \exp[-j\alpha_v\beta(z+d)]\}\exp(-j\beta x) \\ \psi^{\mathrm{II}} &= (j\alpha_v G/\beta)[T\exp(j\alpha_v\beta d) - \exp(-j\alpha_v\beta d)]\exp(-\beta z)\exp(-j\beta x) \end{aligned}\right\} \tag{4.230}$$

where

$$\left.\begin{aligned} G &= \tilde{J}(\beta)\exp(-j\alpha_v\beta d)/(\partial F_T/\partial\beta) \\ \partial F_T/\partial\beta &= F'_{T\mathrm{i}} + jF'_{T\mathrm{i}} \\ F'_{T\mathrm{r}} &= 2\alpha_v d\sin 2\alpha_v\beta d + 2\alpha_v^2 d\cos 2\alpha_v\beta d \\ F'_{T\mathrm{i}} &= 2\alpha_v d\cos 2\alpha_v\beta d - 2\alpha_v^2 d\sin 2\alpha_v\beta d \end{aligned}\right\} \tag{4.231}$$

With equation (4.79), the x-component of the Poynting vector in the dielectric or free-space region can be written as

$$P_x = -\omega\beta/8\pi\,|\psi|^2$$

Substituting from equation (4.230), we obtain the total power flowing (in the x-direction) in region I per unit width of the guiding structure as

$$\begin{aligned} P^{\mathrm{I}} &= \int_{-\infty}^{-d} (-\omega\beta/8\pi)|\psi|^2\,dz \\ &= -(\omega/8\pi)|G|^2 M_1 \end{aligned} \tag{4.232}$$

$$M_1 = 2\alpha_v^2/\beta^2(\alpha_v^2+1) \tag{4.233}$$

Similarly, power flow per unit width in the remaining two regions is expressed as

$$\left.\begin{aligned} P^{\mathrm{II}} &= (-\omega/8\pi)|G|^2 M_2 \\ P^{\mathrm{f}} &= (-\omega/8\pi)|G|^2 M_{\mathrm{f}} \end{aligned}\right\} \tag{4.234}$$

where

$$\left.\begin{aligned} M_2 &= 2\alpha_v^2/\beta(\alpha_v^2+1) \\ M_{\mathrm{f}} &= -\frac{2\alpha_v^2[\beta d(\alpha_v^2+1)+2]}{\beta^2(\alpha_v^2+1)} \end{aligned}\right\} \tag{4.235}$$

Thus, total power flowing per unit width is obtained as

$$\begin{aligned} P &= P^{\mathrm{I}} + P^{\mathrm{f}} + P^{\mathrm{II}} \\ &= (-\omega/8\pi)|G|^2(M_1 + M_{\mathrm{f}} + M_2) \end{aligned}$$

Substitution from equations (4.231), (4.233), and (4.235) yields

$$P = \frac{\omega}{16\pi}\frac{|\tilde{J}(\beta)|^2}{\beta d(\alpha_v^2+1)} \tag{4.236}$$

Though it is not obvious, this is essentially the same as the result derived by Parekh (1979) and Bajpai and Srivastava (1980b).

In the specific case of uniform current I_0 in a microstrip of width b, we have

$$\tilde{J}(\beta) = I_0 \frac{\sin \beta b/2}{\beta b/2} \tag{4.237}$$

The excitation efficiency of a microstrip can be inferred from equations (4.236) and (4.237). Figure 4.120b shows the magnetostatic forward

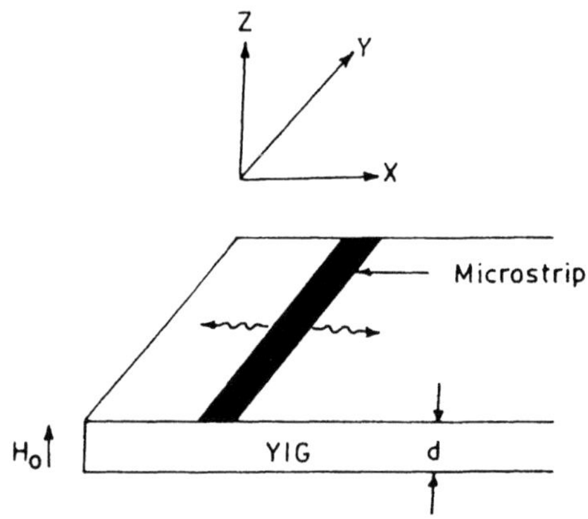

Figure 4.120a. The geometry of the configuration for excitation of magnetostatic forward volume waves in a normally magnetized ferrimagnetic film $-d \leq z \leq 0$ by means of a microstrip of width b.

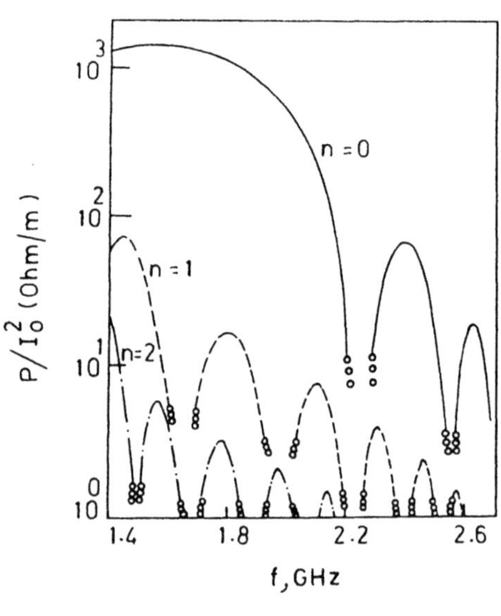

Figure 4.120b. Frequency variation of power transduced into magnetostatic forward volume waves by a single microstrip coupler. The open circles denote a zero crossing of the function (after Parekh, 1979).

wave excitation characteristics of a single-microstrip transducer for the $n = 0, 1,$ and 2 modes for a typical set of data. The modes above $n = 2$ are too weakly excited to be shown on the ordinate scale of the figure. It is seen that over the significant frequency range, the excitation efficiency for the $n = 1$ mode is about two orders of magnitude smaller than that for the $n = 0$ mode. Evidently, the excitation of $n = 1$ and higher-order modes can generally be ignored.

The preceding analysis has been generalized to the case of multilayered structures by Bajpai and Srivastava (1980b), who also presented results for meander line and grating couplers, as well as a general analysis of magnetostatic surface wave excitation. In the specific case of the dielectric layered structure (Figure 4.121), the power flow per unit width can be expressed as*:

$$\frac{P_\pm}{|I_0|^2} = -\frac{s\omega}{8\pi\beta^2}\frac{\exp(-2\beta d)}{|F'_T(\omega,\beta)|^2}(M_1^S + M_f^S + M_2^S)\left(\frac{\sin\beta b/2}{\beta b/2}\right)^2 \tag{4.238}$$

where

$$\left.\begin{aligned}
M_1^S &= (T+1)^2/2 \\
M_2^S &= \frac{(\mu + s\kappa)T\exp(\beta d) - (\mu - s\kappa)\exp(-\beta d)}{\sinh\beta t}\left(\tfrac{1}{4}\sinh 2\beta t + \tfrac{1}{2}\beta t\right) \\
M_f^S &= \{(T^2/2)(\mu + s\kappa)[\exp(2\beta d) - 1] \\
&\qquad - \tfrac{1}{2}(\mu - s\kappa)[\exp(-2\beta d) - 1]2\beta\mu d T\} \\
T &= (\mu - s\kappa)/(\mu + s\kappa) \\
F'_T(\omega,\beta) &= Q_1 + Q_2 \\
Q_1 &= [(\mu + s\kappa)T - (\mu - s\kappa)\exp(-2\beta d)]t\,\mathrm{cosech}\,\beta t \\
Q_2 &= 2d\exp(-2\beta d)[1 - (\mu - s\kappa)\coth\beta t]
\end{aligned}\right\} \tag{4.239}$$

In these expressions, $S = \pm 1$ for propagation along the $\pm y$-axis.

The total radiated power is expressed as

$$P_T = P_+ + P_- = \tfrac{1}{2}R_m|I_0|^2$$

The input resistance is given by (Ganguly and Webb, 1975)

$$R_i = (l/2)R_m$$

where l is the length of the microstrip line.

* Though it is not obvious, this result is essentially the same as that of Ganguly and Webb (1975), and also the same as that of Sethares et al. (1978) with $l \to \infty$.

Figures 4.122 and 4.123 compare the calculated and measured values of the input resistance for typical sets of data, and agreement between the two is satisfactory. The results suggest that: (i) At a fixed biasing field strength and given coupler geometry, the increase in excitation bandwidth is approximately proportional to the thickness of the YIG film. The

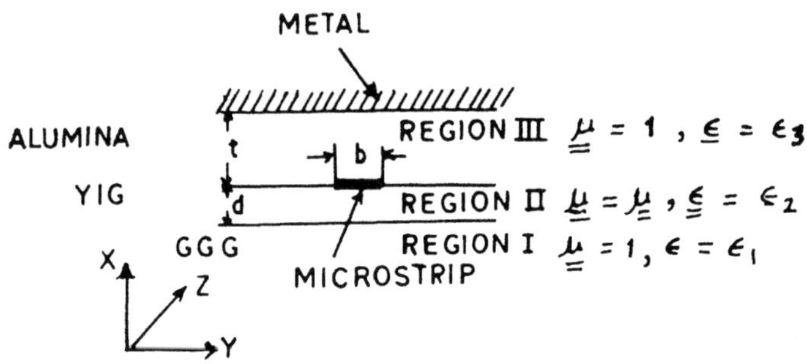

Figure 4.121. The configuration for excitation of magnetostatic surface waves in a dielectric layered structure. A microstrip of width b extends from $-b/2 \leq y \leq b/2$ at the interface between the dielectric and YIG ($x = 0$). The biasing field is along the z-axis while the wave propagates along the y-axis (after Ganguly and Webb, 1975).

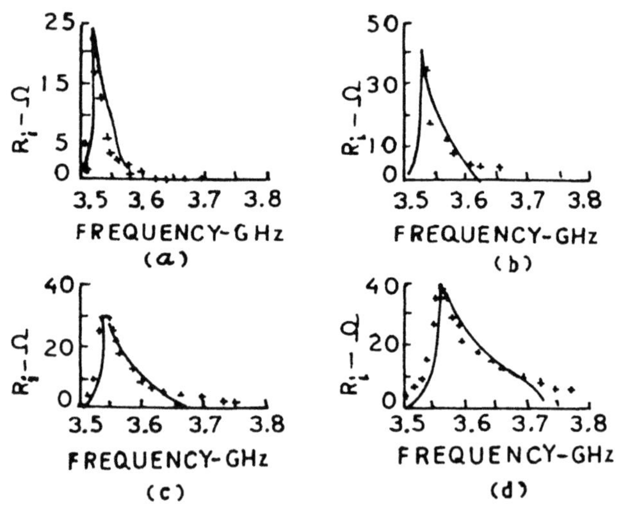

Figure 4.122. Comparison of measured and computed values of input resistance. The YIG film thickness d and the propagation path length l for different cases are as follows: (a) $d = 1.7\,\mu$m, $l = 1.53$ mm, (b) $d = 2.6\,\mu$m, $l = 2.54$ mm, (c) $d = 4.0\,\mu$m, $l = 1.85$ mm, and (d) $d = 6.25\,\mu$m, $l = 2.28$ mm. The common parameters are $H_0 = 650$ Oe, $b = 177.8\,\mu$m, and $t = 254\,\mu$m (after Ganguly and Webb, 1975).

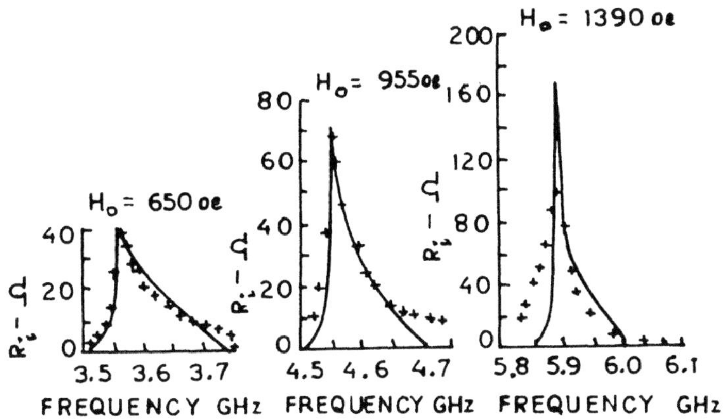

Figure 4.123. Comparison of measured and computed values of input resistance with the biasing field as the parameter. Other parameters are also shown in the figure (after Ganguly and Webb, 1975).

maximum value of R_m, however, increases only slightly with the film thickness. (ii) For a given geometry, the excitation bandwidth narrows and maximum R_m increases when the biasing field is increased (Figure 4.123).

The reader interested in the various aspects of the excitation efficiency of different configurations and coupler geometries should refer to Schilz (1973), Ganguly and Webb (1975), Wu et al. (1977), Ganuly et al. (1978), Sethares et al. (1978), Sethares (1978), Adam et al. (1978), Weinberg and Sethares (1978), Parekh (1979), Parekh and Tuan (1979a, b), Sethares et al. (1979a, b), Sethares (1979), Volluet (1980), and Bajpai and Srivastava (1980a, b).

5

Magnetoelastic Waves in Layered Planar Structures

The first observations of guided magnetoelastic bulk waves were carried out in ferrimagnetic rods, ellipsoids, etc. (LeCraw and Comstock, 1965; Strauss, 1966; Auld, 1971). However, in later investigations, magnetoelastic surface waves were excited on planar structures such as flat ferrimagnetic substrate and multilayers consisting of ferrimagnetic films and other media.

Voltmer et al. (1969) were the first to observe magnetoelastic surface waves on a YIG substrate. Subsequently, several experimental and theoretical investigations of magnetoelastic surface waves were published. Most of these studied delay characteristics and fabricated the associated devices in the UHF and microwave region. Incidently, long time delays are obtainable even with surface elastic waves on piezoelectric substrates (White, 1970). However, in piezoelectrics, the delay characteristics are essentially reciprocal and cannot be varied electronically. On the other hand, magnetoelastic surface waves can be nonreciprocal and are electronically tunable. Thus, where electronic tunability and/or nonreciprocity are required, magnetoelastic surface waves in planar ferrimagnetics provide the only viable alternative to conventional elastic surface waves in piezoelectric and other media.

In spite of extensive theoretical investigations, many of the experimental observations in this area are still unexplained. Except for the excellent works of Ganguly et al. (1976a, b), there are few publications which report satisfactory agreement between experimental and theoretical results. Thus, in this chapter, we first discuss, in Sections 5.1–5.3, theoretical analyses pertaining to magnetoelastic waves in different biasing configurations, and then in Section 5.4 summarize the experimental results and describe some of the more useful devices.

5.1. *Transverse Magnetization*

In this section, we analyze guided magnetoelastic wave propagation in planar structures when the dc magnetic field is parallel to the direction of propagation. Specifically, the cases of a ferrimagnetic half space, a nonmagnetic layered ferrimagnetic substrate, a free ferrimagnetic slab, and a ferrimagnetic layered nonmagnetic substrate are discussed.

5.1.1. *Ferrimagnetic Half Space*

The following analysis of magnetoelastic surface wave propagation on free and metallized ferrimagnetic substrates is based on papers by Parekh (1969a, b, 1970). Figure 5.1a shows the configuration in which the region $x \leqslant 0$ is occupied by a magnetic medium while the region $x > 0$ is free space. The dc magnetic field is directed along the z-axis. It is required to

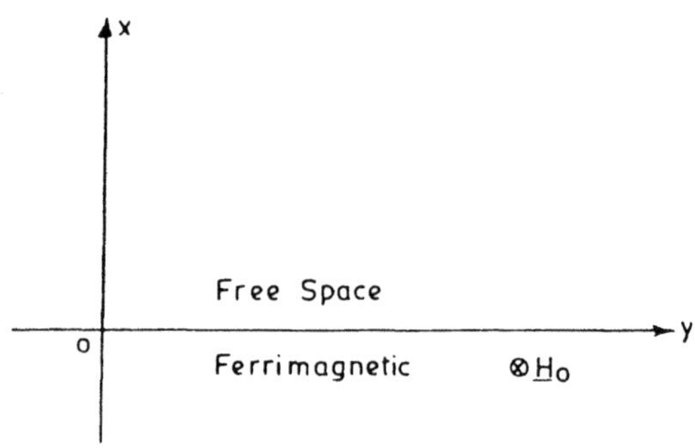

Figure 5.1a. Ferrimagnetic half space: dc magnetic field is along the z-axis and the wave propagates along the $\pm y$-axis.

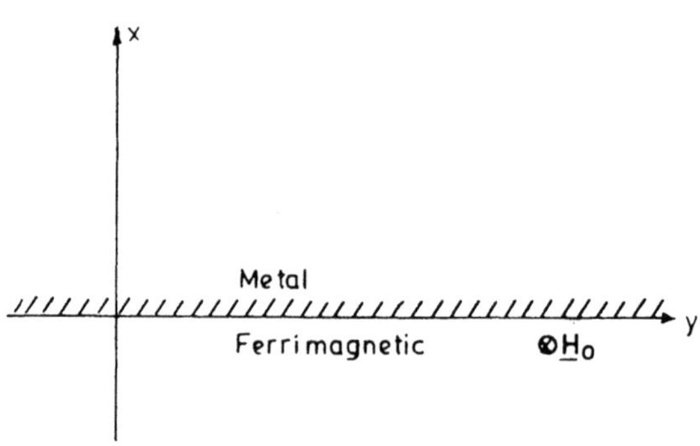

Figure 5.1b. Metallized ferrimagnetic half space.

obtain the dispersion characteristics of a magnetoelastic surface wave (with elastic polarization along the z-axis) propagating along the $\pm y$-axis. The elastic and magnetic fields of the wave are assumed to be uniform along the z-axis.

Since there is no variation of the wave field along the direction of the dc magnetic field ($\partial/\partial z = 0$), it can be shown (following the results of Chapter 2, Part B, Chapter 3, Part B) that the elastic and magnetic field quantities will be governed by equations (3.69) and (3.71) of Chapter 3. The wave field varies as $\exp j(\omega t - s\beta y)$, where $s = \pm 1$ for wave propagation along the $\pm y$-axis. Hence using equation (3.69), one obtains

$$(\partial^2/\partial x^2 - \xi_m^2)u_z^m = 0 \tag{5.1}$$

where

$$\left.\begin{aligned} \xi_m^2 &= \beta^2 - \omega^2/c_t^2\delta_\perp \\ c_t^2 &= c_{44}^m/\rho_m \\ \delta_\perp &= (\omega_2'^2 - \omega^2)/\Delta_1 \\ \Delta_1 &= \omega_3'^2 - \omega^2 \\ \omega_3'^2 &= \omega_0(\omega_0 + \omega_m) \\ \omega_2'^2 &= \omega_3'^2 - \omega_0\sigma \end{aligned}\right\} \tag{5.2}$$

The solution of equation (5.1), which decays away from the interface, is expressed as

$$u_z^m = u_0^m \exp(\xi_m x)\exp j(\omega t - s\beta y) \tag{5.3}$$

where u_0^m is an arbitrary constant. Substitution for u_z^m into equation (3.71) yields

$$\left(\frac{\partial^2}{\partial x^2} - \beta^2\right)\psi^m = \frac{4\pi\gamma b_2\omega_0(\xi_m^2 - \beta^2)}{\Delta_1}\, u_0^m \exp(\xi_m x)\exp j(\omega t - s\beta y) \tag{5.4}$$

The general solution to this equation, which decays away from the interface, is expressed as

$$\psi^m = \left[\frac{4\pi\gamma b_2\omega_0}{\Delta_1}\, u_0^m \exp(\xi_m x) + D\exp(\beta x)\right]\exp j(\omega t - s\beta y) \tag{5.5}$$

where D is an arbitrary constant. We are also interested in the expressions for rf magnetization m_x and the normal component of magnetic induction

b_x^{m}. Using the first of equations (3.66) of Chapter 3 and equations (5.3) and (5.55), we obtain

$$m_x = \left[-\frac{\gamma b_2(\omega_0 \xi_{\mathrm{m}} + \omega\beta)}{\Delta_1} u_0^{\mathrm{m}} \exp(\xi_{\mathrm{m}} x) + \frac{\omega_{\mathrm{m}}\beta}{4\pi(\omega_0 - s\omega)} D \exp(\beta x) \right] \exp j(\omega t - s\beta y) \tag{5.6}$$

and using $b_x^{\mathrm{m}} = \partial\psi^{\mathrm{m}}/\partial x + 4\pi m_x$, we have

$$b_x^{\mathrm{m}} = \left[-\frac{4\pi\gamma b_2 \omega\beta s}{\Delta_1} u_0^{\mathrm{m}} \exp(\xi_{\mathrm{m}} x) + \frac{\beta(\omega_0 + \omega_{\mathrm{m}} - \omega s)}{(\omega_0 - \omega s)} D \exp(\beta x) \right] \exp j(\omega t - s\beta y) \tag{5.7}$$

In a free space region ($x > 0$), the elastic displacement is zero whereas the magnetostatic potential is expressed as

$$\psi^{\mathrm{S}} = C \exp(-\beta x) \exp j(\omega t - s\beta y) \tag{5.8}$$

where C is an arbitrary constant. The normal component of magnetic induction is obtained as

$$b_x^{\mathrm{S}} = -\beta C \exp(-\beta x) \exp j(\omega t - s\beta y) \tag{5.9}$$

The boundary conditions to be applied at the interface $x = 0$ can be inferred from equations (3.74)–(3.77) of Chapter 3 and are as follows:

$$\psi^{\mathrm{m}} = \psi^{\mathrm{S}} \tag{5.10}$$

$$b_x^{\mathrm{m}} = b_x^{\mathrm{S}} \tag{5.11}$$

$$c_{44}^{\mathrm{m}} \left(\frac{\partial u_z^{\mathrm{m}}}{\partial x} \right) + \frac{b_2 m_x}{M_0} = 0 \tag{5.12}$$

The application of these boundary conditions at $x = 0$ leads to the following:

$$\frac{4\pi\gamma b_2 \omega_0}{\Delta_1} u_0^{\mathrm{m}} + D = C \tag{5.13}$$

$$-\frac{4\pi\gamma b_2 \omega s}{\Delta_1} u_0^{\mathrm{m}} + \frac{\omega_0 + \omega_{\mathrm{m}} - \omega s}{\omega_0 - \omega s} D = -C \tag{5.14}$$

$$c_{44}^{m}\left(\delta_{\perp}\xi_{m}-\frac{\sigma\omega\beta s}{\Delta_{1}}\right)u_{0}^{m}+\frac{b_{2}\omega_{m}\beta}{4\pi M_{0}(\omega_{0}-\omega s)}D=0 \tag{5.15}$$

Elimination of C from equations (5.13) and (5.14) yields

$$\frac{2(\omega_{0}+\frac{1}{2}\omega_{m}-\omega s)}{\omega_{0}-\omega s}D=\frac{4\pi\gamma b_{2}}{\Delta_{1}}(\omega s-\omega_{0})u_{0}^{m} \tag{5.16}$$

Equations (5.15) and (5.16) are solved to obtain ξ_m as

$$\xi_{m}=\frac{\beta\sigma[\omega^{2}-s\omega\omega_{0}-\frac{1}{2}\omega_{0}\omega_{m}]}{(\omega_{0}+\frac{1}{2}\omega_{m}-s\omega)(\omega^{2}-\omega_{2}^{\prime 2})} \tag{5.17}$$

The dispersion relation can be obtained by eliminating ξ_m from equations (5.2) and (5.17).

At this stage, it is easy to obtain the dispersion relation for guided magnetostatic surface wave propagation in the configuration shown in Figure 5.1b, which is obtained from that shown in Figure 5.1a by overlaying a thin, perfectly conducting metallic film; this metallic layer provides an electrical short but leaves the surface practically stress-free. In this case the field quantities in the ferrimagnetic region are still described by equations (5.2), (5.3), and (5.5)–(5.7). The boundary condition (5.12) continues to be applicable while (5.11) is modified to $b_x^m = 0$ at $x = 0$ and (5.10) is no longer required.

Equating b_x^m given by equation (5.7) to zero (at $x = 0$), we have

$$\frac{\omega_{0}+\omega_{m}-s\omega}{\omega_{0}-s\omega}D=\frac{4\pi\gamma b_{2}s\omega}{\Delta_{1}}u_{0}^{m} \tag{5.18}$$

Eliminating D and u_0^m from equations (5.15) and (5.18) we obtain the expression for ξ_m as

$$\xi_{m}=\beta\frac{\sigma s\omega}{\Delta_{1}-\sigma\omega_{0}}\,\frac{\omega_{0}-s\omega}{\omega_{0}+\omega_{m}-s\omega} \tag{5.19}$$

The explicit dispersion relation for a guided magnetoelastic surface wave on a metallized substrate is then obtained by eliminating ξ_m from equations (5.2) and (5.19).

The computed dispersion diagrams for a typical set of data are displayed in Figure 5.2a for $s = +1$ and in Figure 5.2b for $s = -1$. The dispersion curves for both metallized and unmetallized configurations have been shown. The strong influence of metallization on dispersion characteristics is evident from the figures.

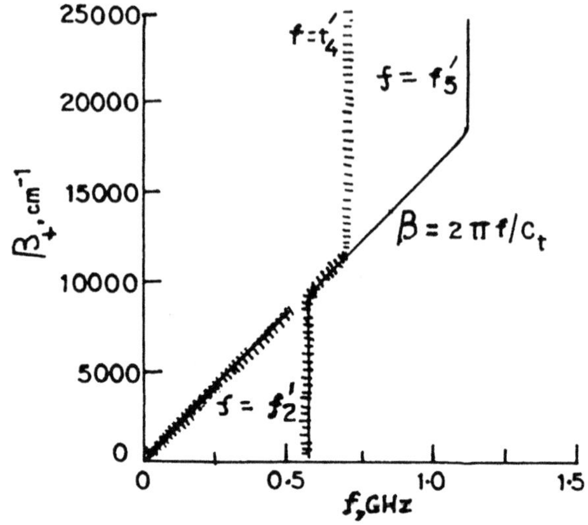

Figure 5.2a. Dispersion curves for propagation along the $+y$-axis. Solid and dashed curves are for metallized and unmetallized configurations. The parameters for Ga-doped YIG are $4\pi M_0 = 300\,\mathrm{G}$, $b_2 = 2.9 \times 10^6$ ergs/cc, $c_{44} = 7.74 \times 10^{11}$ ergs/cc, and $H_0 = 100\,\mathrm{Oe}$ (after Parekh, 1970).

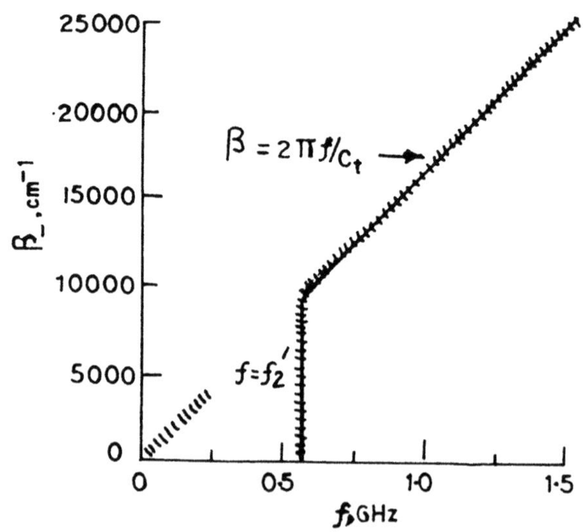

Figure 5.2b. Dispersion curves for propagation along the $-y$-axis. The parameters are the same as for Figure 5.2a (after Parekh, 1970).

The dispersion behavior can be understood by considering the fact that the resonance frequency of magnetostatic bulk waves in a transversely magnetized ferrimagnetic medium is $\omega = \omega_3' \equiv [\omega_0(\omega_0 + \omega_m)]^{1/2}$ (see Chapter 2, Part A). Also, as discussed in Section 4.1 of Chapter 4, a transversely magnetized ferrimagnetic half space supports only a unidirectional (slow) surface magnetostatic wave which propagates along the $\mathbf{H}_0 \times \hat{n}$ direction (the $+y$-axis in the present configuration); the resonance frequency for this wave is $\omega = \omega_4' \equiv \omega_0 + \frac{1}{2}\omega_m$ for an unmetallized half space and $\omega = \omega_5' \equiv \omega_0 + \omega_m$ for a metallized half space.

Now consider propagation along the $+y$-axis (Figure 5.2a). The dispersion curves result from the interaction of the transverse elastic wave (dispersion relation: $\beta = \omega/c_t$) with the bulk and surface magnetostatic waves. The consequence of the magnetoelastic coupling is that the bulk wave frequency is shifted from ω_3' to ω_2'. Another consequence of this coupling is the existence of a small band gap just below $\omega = \omega_2'$. The upper resonance frequencies are, as expected, ω_4' for the unmetallized and ω_5' for the metallized configuration. It should be noted that the band gap below $\omega = \omega_2'$ is larger for the metallized geometry than for the unmetallized one.

Next, consider the case of propagation along the $-y$-axis. Since there is no magnetostatic surface wave along the $-y$-axis in a half space configuration, the form of the dispersion curves shown in the figure results from the coupling of the transverse elastic wave to the magnetic bulk wave of frequency $\omega = \omega_3'$. Again, the consequences of the magnetoelastic coupling are the modification of the wave resonance frequency to ω_2' and the existence of a small band gap below this frequency.

The dispersion curves, shown in Figure 5.2, hold for a fixed biasing field. Since the frequencies ω_2', ω_4', and ω_5' depend on the biasing field, their positions in the diagram can be horizontally displaced by making appropriate changes in H_0. This implies that, for a fixed operating frequency, an appropriate change in H_0 can shift the operating point from one branch of the dispersion curves to another, which leads to magnetic tunability of the wave velocity.

While the foregoing discussion is restricted to a surface wave on a ferrimagnetic substrate ($x \leq 0$), with free space above it, Tsutsumi et al. (1975) and Bhattacharyya et al. (1976) have investigated the case when the region $x > 0$ in Figures 5.1a and 5.1b is occupied by a piezoelectric medium.

5.1.2. *Nonmagnetic Layer Over Ferrimagnetic Substrate*

A nonmagnetic slab or film of an isotropic elastic medium supports a variety of modes. When the elastic displacement is parallel to the interfaces but transverse to the direction of propagation, the resulting guided waves are called "Love waves" (Auld, 1973). The Love waves are modified when one side of the slab/film is backed by a ferrimagnetic substrate. The coupling of magnetostatic waves to Love waves has been investigated by Matthews and van de Vaart (1969) and van de Vaart (1971).

Figure 5.3a shows the configuration in which the region $0 \leqslant x \leqslant d$ is occupied by a nonmagnetic cubic medium whereas the region $x<0$ is a cubic ferrimagnetic medium. The dc magnetic field is parallel to the z-axis along which there is no variation of the wave field. It is required to obtain the dispersion relation for guided bulk waves propagating along the $\pm y$-axis.

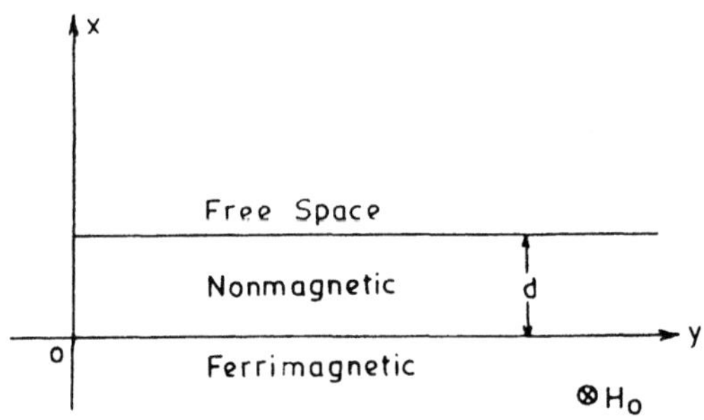

Figure 5.3a. Nonmagnetic layered ferrimagnetic substrate: the layer thickness is d and the dc field H_0 is along the z-axis.

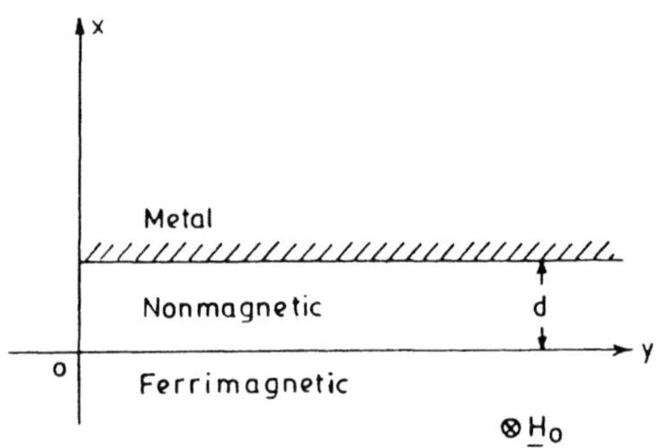

Figure 5.3b. Nonmagnetic layered ferrimagnetic substrate with a thin metal film at $x = d$.

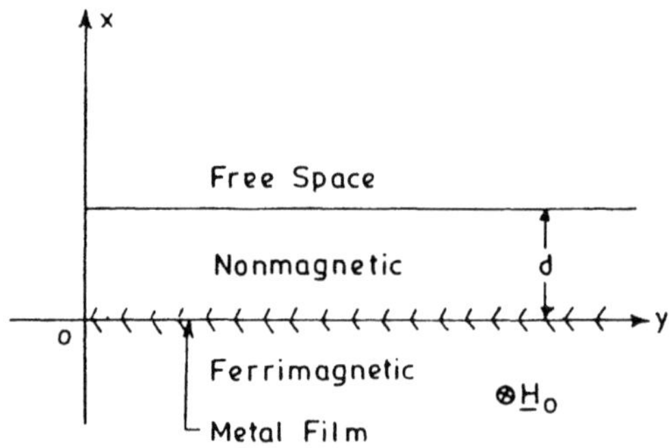

Figure 5.3c. Nonmagnetic layered ferrimagnetic substrate with a thin metal film between the layer and the substrate.

The elastic displacement u_z^e in the nonmagnetic medium satisfies equation (5.1) with ξ_m^2 replaced by $-\xi_e^2$ such that

$$\left.\begin{aligned} \xi_e^2 &= \beta^2 - \omega^2/c_t^{e2} \\ c_t^e &= \sqrt{c_{44}^e/\rho_e} \end{aligned}\right\} \tag{5.20}$$

Thus the elastic displacement and magnetostatic potential in the nonmagnetic region $(0 \leqslant x \leqslant d)$ are expressed as

$$\left.\begin{aligned} u_z^e &= [A \exp(j\xi_e x) + B \exp(-j\xi_e x)] \exp j(\omega t - s\beta y) \\ \psi^e &= [F \exp(\beta x) + G \exp(-\beta x)] \exp j(\omega t - s\beta y) \end{aligned}\right\} \tag{5.21}$$

where $A, B, F,$ and G are arbitrary constants.

The elastic displacement and magnetostatic potential in the ferrimagnetic region $(x < 0)$ are the same as those given by equations (5.3) and (5.5), respectively, whereas the magnetostatic potential in the free space region $(x > d)$ is given by equation (5.8). The boundary conditions are as follows:

At $x = 0$:

$$\left.\begin{aligned} \psi^m &= \psi^e \\ b_x^m &= \partial\psi^e/\partial x \\ u_z^m &= u_z^e \\ c_{44}^m \frac{\partial u_z^m}{\partial x} + \frac{b_2 m_x}{M_0} &= c_{44}^e \frac{\partial u_z^e}{\partial x} \end{aligned}\right\} \tag{5.22}$$

At $x = d$:

$$\left.\begin{aligned} \psi^e &= \psi^S \\ \partial\psi^e/\partial x &= \partial\psi^S/\partial x \\ c_{44}^e \frac{\partial u_z^e}{\partial x} &= 0 \end{aligned}\right\} \tag{5.23}$$

The application of the aforementioned boundary conditions followed by the standard procedure of eliminating the arbitrary constants leads to the following dispersion relation:

$$c_{44}^m \delta_\perp \xi_m - c_{44}^e \xi_e \tan \xi_e d + \frac{s c_{44}^m \sigma\beta}{-\Delta_1} \frac{(\omega - s\omega_+)(\omega + s\omega_-)}{\omega - s\omega_4'} = 0 \tag{5.24}$$

where

$$\left.\begin{aligned} \omega_4' &= \omega_0 + \tfrac{1}{2}\omega_m \\ 2\omega_\pm &= \pm\,\omega_0 + \omega_0(1 + 2\omega_m/\omega_0)^{1/2} \end{aligned}\right\} \tag{5.25}$$

When $d = 0$, equation (5.24) reduces, as expected, to equation (5.17). As discussed earlier, it would be interesting to obtain the dispersion characteristics in the case when a thin, perfectly conducting metal film is deposited at $x = d$ (Figure 5.3b). In this case the boundary conditions at $x = 0$ remain unchanged, i.e., the same as in equations (5.22). However, the boundary conditions at $x = d$ are modified to the following:

$$\partial\psi^e/\partial x = 0$$

$$c_{44}^e\,\partial u_z^e/\partial x = 0$$

Using these boundary conditions, we obtain the dispersion relation for the configuration shown in Figure 5.3b as

$$c_{44}^m\delta_\perp\xi_m - c_{44}^e\xi_e\tan\xi_e d$$

$$+\frac{c_{44}^m\sigma\omega\beta s}{-\Delta_1}\,\frac{\omega(\omega - s\omega_0) + [\omega^2 - s\omega_0(\omega + s\omega_m)]\tanh\beta d}{\omega - s(\omega_0 + \omega_m) + (\omega - s\omega_0)\tanh\beta d} = 0 \tag{5.26}$$

When $d = 0$, this dispersion relation reduces, as expected, to equation (5.19).

Another related configuration is shown in Figure 5.3c in which the conducting film is deposited at the interface between the nonmagnetic layer and the magnetic substrate. In this case the magnetostatic potential in the nonmagnetic layer is zero because of the presence of the conducting film at $x = 0$, which isolates the two regions. Thus, the boundary conditions are modified to the following:

At $x = 0$:

$$b_x^m = 0$$

$$u_z^m = u_z^e$$

$$c_{44}^m\frac{\partial u_z^m}{\partial x} + \frac{b_2}{M_0}m_x = c_{44}^e\frac{\partial u_z^e}{\partial x}$$

At $x = d$:

$$c_{44}^e\frac{\partial u_z^e}{\partial x} = 0$$

Using these boundary conditions, we obtain the dispersion relation as

$$c_{44}^{m}\delta_{\perp}\xi_{m} - c_{44}^{e}\xi_{e}\tan\xi_{e}d + \frac{s\sigma\beta c_{44}^{m}}{-\Delta_{1}}\frac{\omega(\omega - s\omega_{0})}{\omega - s(\omega_{0} + \omega_{m})} = 0 \tag{5.27}$$

The computed dispersion diagrams based on equations (5.25)–(5.27) are shown in Figure 5.4 for the configurations shown in Figure 5.3.

First consider the unmetallized geometry (Figure 5.3a). It is better to start with the constituent uncoupled modes. Substituting $\sigma = 0$ in equation (5.25), we have

$$c_{44}^{m}\xi_{m} - c_{44}^{e}\xi_{e}\tan\xi_{e}d = 0 \tag{5.28}$$

$$-\Delta_{1} \equiv \omega^{2} - \omega_{3}^{\prime 2} = 0 \tag{5.29}$$

$$\omega - s\omega_{4}^{\prime} = 0 \tag{5.30}$$

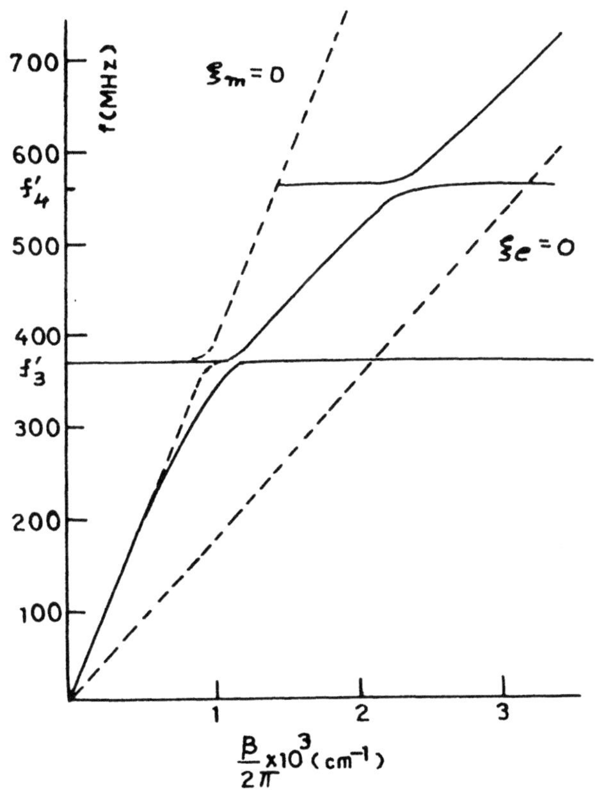

Figure 5.4a. Dispersion curves corresponding to the configuration shown in Figure 5.3a for + y-propagation. The parameters are: $c_t^e = 1.87 \times 10^5$ cm/sec and $\rho_e = 4.83$ g/cc for CdS, $d = 1.0\,\mu$m, and $H_0 = 50$ Oe. Other parameters are the same as for Figure 5.2a (after Matthews and van de Vaart, 1969).

While the first equation is the dispersion relation for Love waves (Auld, 1973) in the nonmagnetic layer, the second and third represent, respectively, the resonance frequencies for a magnetostatic plane (bulk) wave and a surface wave on a half space (for $s = +1$ only).

Figure 5.4a shows the dispersion curves computed from equation (5.25) for the $s = +1$ mode. As a consequence of magnetoelastic coupling, small band gaps exist at the cross-over points of Love wave branch(es) and

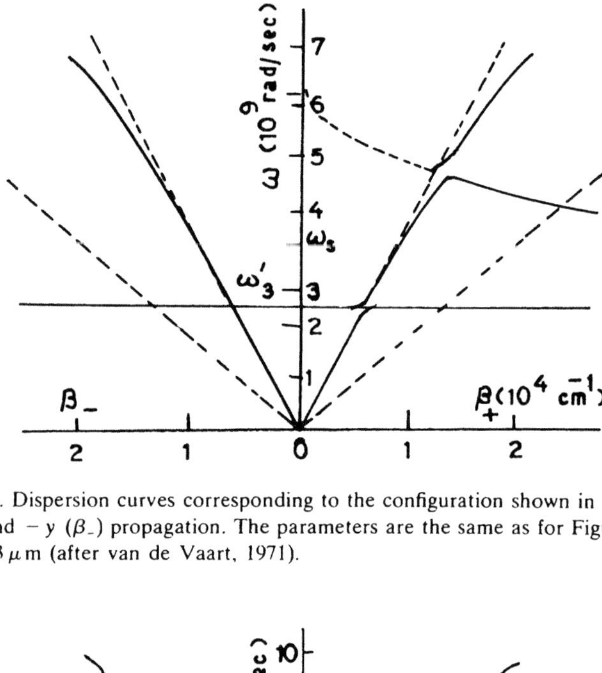

Figure 5.4b. Dispersion curves corresponding to the configuration shown in Figure 5.3b for $+y$ (β_+) and $-y$ (β_-) propagation. The parameters are the same as for Figure 5.4a except that $d = 0.3\,\mu$m (after van de Vaart, 1971).

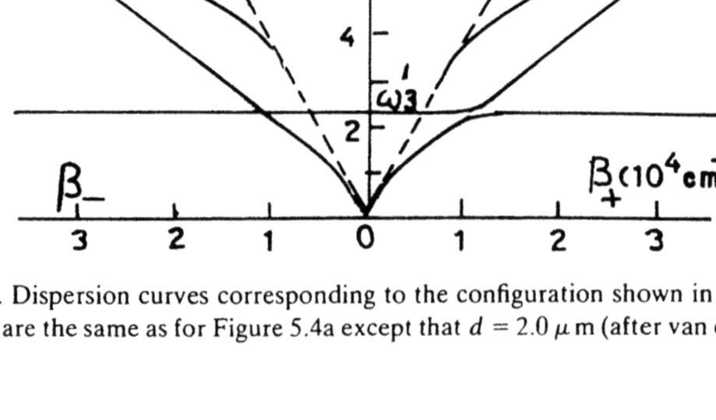

Figure 5.4c. Dispersion curves corresponding to the configuration shown in Figure 5.3c. The parameters are the same as for Figure 5.4a except that $d = 2.0\,\mu$m (after van de Vaart, 1971).

the magnetostatic bulk and surface waves. The curves for the $s = -1$ mode (not shown) would be similar to the ones for $s = +1$ except for the absence of the $f = f_4'$ branch and the consequent splitting of the elastic wave branch in that region.

Next we discuss the configuration shown in Figure 5.3b, the dispersion relation for which is given by equation (5.26). We first examine the uncoupled modes. Substituting $\sigma = 0$ in equation (5.26), we obtain equations (5.28) and (5.29) as before; however, instead of equation (5.30) we have

$$\omega = s\,[\omega_0 + \omega_m/(1 + \tanh \beta d)] \tag{5.31}$$

As can be inferred from equation (4.40) of Chapter 4, this is the magnetostatic surface wave resonance frequency of the $\mathbf{H}_0 \times \hat{n}$ surface wave on a ferrimagnetic substrate when there is a perfect conductor at a distance d from the substrate. Figure 5.4b displays the dispersion curves for a typical set of data. The curves for $s = +1$ and $s = -1$ are more or less alike except that, for $s = +1$, there is additional splitting at the frequency given by equation (5.31).

It should be noted that the thickness d of the nonmagnetic layer corresponding to the curves shown in Figure 5.4b is too small to allow coupling of more than one Love wave mode to the magnetostatic mode. If the thickness of the nonmagnetic layer were larger, higher-order Love wave modes would also be guided which would then couple to the magnetostatic modes (van de Vaart, 1971).

When the metal film exists in between the nonmagnetic layer and magnetic substrate (Figure 5.3c), the dispersion curves, computed using equation (5.27), are shown in Figure 5.4c. Since the thickness of the nonmagnetic layer is larger than that corresponding to Figure 5.4b, the magnetostatic surface mode characterized by the frequency $\omega = \omega_5' \equiv \omega_0 + \omega_m$ (for $s = +1$) couples to two Love wave modes, causing splitting at two distinct points at this frequency.

5.1.3. *Ferrimagnetic Slab*

The analysis of propagation of magnetoelastic waves in a transversely magnetized plate has been carried out by van de Vaart and Matthews (1970a) and van de Vaart (1971). When the slab is very thick, the analysis carried out in Section 5.1.1 is generally applicable. However, when the slab is thin, it is necessary to solve the magnetoelastic equations and apply boundary conditions on both the surfaces in order to obtain the dispersion relation. Instead of carrying out the complicated analysis, we shall only explain the results qualitatively.

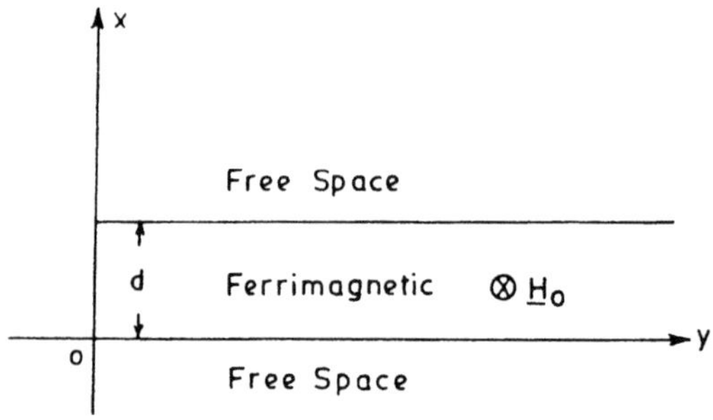

Figure 5.5. Ferrimagnetic slab of thickness d.

Figure 5.5 shows the configuration. The dispersion curves* for a typical set of data are shown in Figure 5.6a. As expected, there are three constituent uncoupled modes: the bulk magnetostatic mode at $\omega = \omega_3'$, the magnetostatic surface mode of a transversely magnetized slab† for $\omega_3' < \omega < \omega_4'$, and bulk elastic wave modes of the slab, characterized by $m = 0, 1, 2, \ldots$. As shown in Figure 5.6a, the magnetoelastic coupling leads to splitting in the dispersion curves wherever the uncoupled magnetostatic and elastic wave branches cross over. For clarity, the details of the dispersion curves near $\omega = \omega_3'$ are shown in Figure 5.6b.

The dispersion curves corresponding to a ferrimagnetic slab coated with a metallic film at $x = d$ (Figure 5.5) are shown in Figure 5.7 for a typical set of data. The nonreciprocity in dispersion characteristics is due to the nonreciprocity of the constituent surface magnetostatic mode. The dispersion characteristics can be understood on the basis of the fact that the resonance frequencies of the magnetostatic surface modes associated with unmetallized and metallized surfaces are ω_4' and ω_5', respectively (see Section 4.3). The reader interested in further details is referred to van de Vaart (1971).

5.1.4. *Ferrimagnetic Layered Nonmagnetic Substrate*

The guided propagation of magnetoelastic waves in an unmetallized and metallized ferrimagnetic layered nonmagnetic substrate has been investigated by van de Vaart and Matthews (1970b) and van de Vaart (1971). As in the preceding section, we shall only summarize the results instead of carrying out the analysis.

* See van de Vaart and Matthews (1970a) for details.

† Magnetostatic surface modes of a transversely magnetized slab have been discussed in Section 4.2 of Chapter 4.

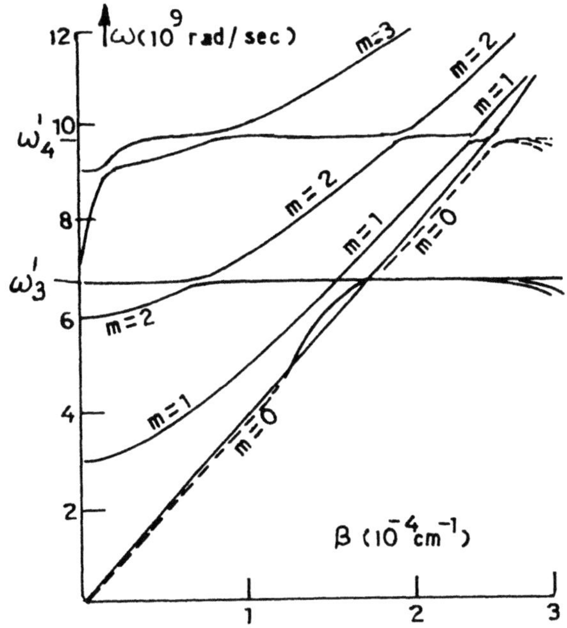

Figure 5.6a. Dispersion curves corresponding to the configuration shown in Figure 5.5. The parameters are: $d = 2.0\,\mu$m, $4\pi M_0 = 800$ G, and $H_0 = 150$ Oe (after van de Vaart and Matthews, 1970a).

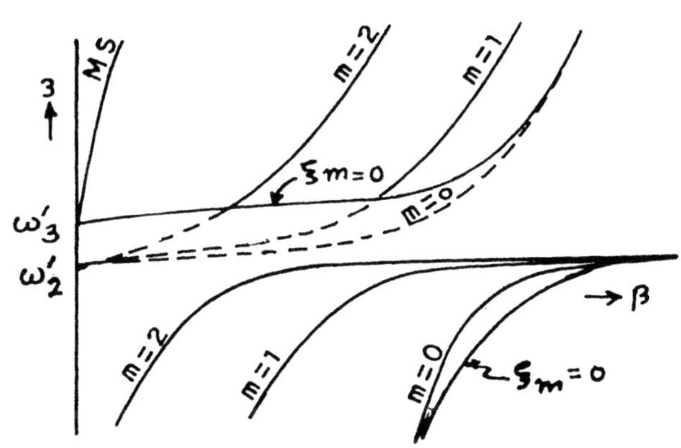

Figure 5.6b. Details of the dispersion curves shown in Figure 5.6a near $\omega = \omega_3'$ (after van de Vaart and Matthews, 1970b).

The configuration is shown in Figure 5.8. The dispersion curves for propagation along the $\pm y$-axis are displayed in Figure 5.9 for a typical set of data. The various branches of the dispersion curves are bounded by the curves $\xi_m = 0$ and $\xi_e = 0$. It is seen that the magnitude of splitting at $\omega = \omega_4'$ is different for propagation in opposite directions.

Figure 5.10 shows the dispersion curves for propagation along the $\pm$-direction when the surface $x = d$ is coated with a thin metallic film. As in earlier cases, the inequality of the resonance frequencies for $+y$ and $-y$ propagating magnetostatic surface waves is responsible for the nonreciprocity.

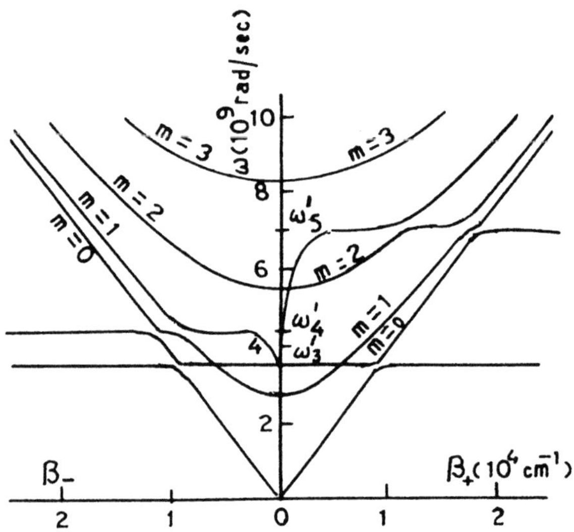

Figure 5.7. Dispersion curves corresponding to the configuration of Figure 5.5 with a thin metal film at $x = d$. The parameters are $4\pi M_0 = 300\,\mathrm{G}$, $H_0 = 100\,\mathrm{Oe}$, and $d = 2.0\,\mu\mathrm{m}$ (after van de Vaart, 1971).

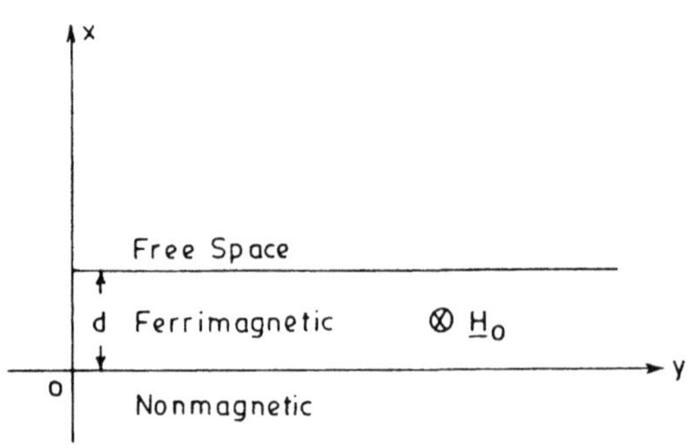

Figure 5.8. Ferrimagnetic layered nonmagnetic substrate.

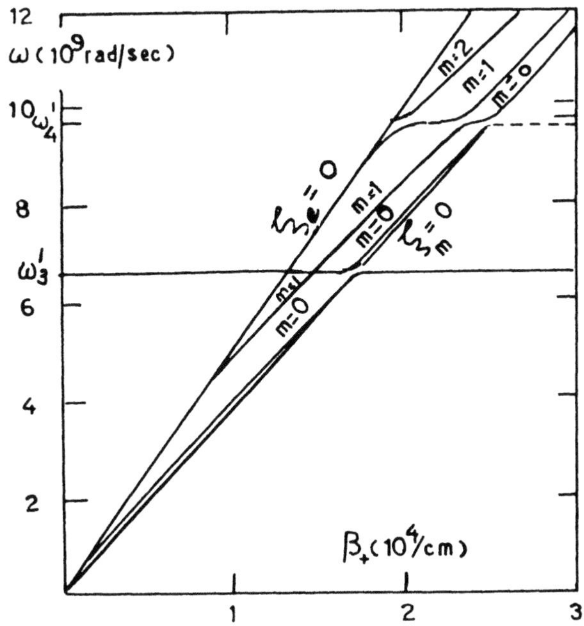

Figure 5.9. Dispersion curves corresponding to the $+y$-propagating wave in the configuration shown in Figure 5.8. The parameters are $d = 2.1\ \mu m$, $4\pi M_0 = 800\,G$, and $H_0 = 150\,Oe$ (after van de Vaart and Matthews, 1970b).

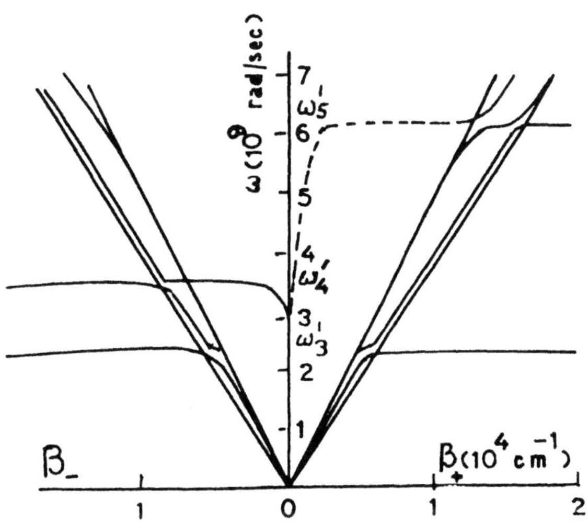

Figure 5.10. Dispersion curves corresponding to the configuration shown in Figure 5.8 with a thin metal film at $x = d$. The parameters are: $d = 4.0\ \mu m$, $4\pi M_0 = 300\,G$, and $H_0 = 50\,Oe$ (after van de Vaart, 1971).

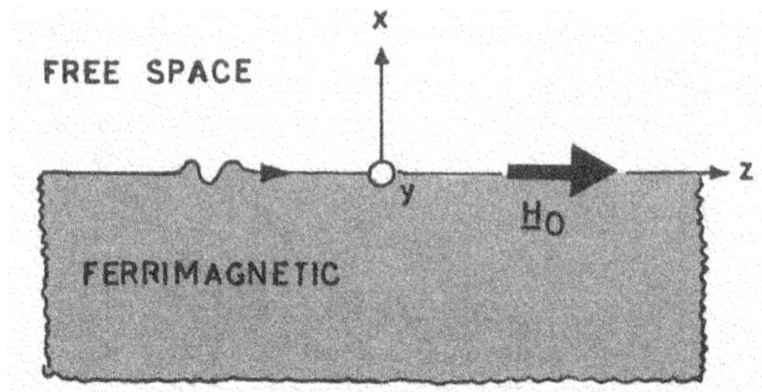

Figure 5.11. Ferrimagnetic half space: the dc magnetic field is along the z-axis which is also the direction of surface wave propagation (after Parekh and Bertoni, 1974a).

5.2. *Longitudinal Magnetization*

In this section we discuss the guided propagation of magnetoelastic Rayleigh waves* on a ferrimagnetic substrate (half space) which is magnetized parallel to the direction of wave propagation. This configuration has been analyzed by Parekh and Bertoni (1972, 1974a) and Emtage (1976). The following discussion is based on the work of Parekh and Bertoni (1974a).

5.2.1. *The Dispersion Relation*

Figure 5.11 shows the configuration wherein the region $x \leq 0$ is occupied by an isotropic ferrimagnetic half space while the region $x > 0$ is free space. The dc magnetic field is along the z-axis, which is parallel to the direction of wave propagation. There is no variation of the wave field along the y-axis; thus, the wavevector is restricted to the xz-plane and wave amplitude is independent of y. Consequently, the elastic displacement and magnetostatic field and potential will be characterized by space–time variation $\exp j(\omega t - \beta\hat{n} \cdot \mathbf{r})$, where $\beta\hat{n} = k_x\hat{x} + k_z\hat{z}$. Following the development of Section 2.6.1 we can easily show that k_x and k_z will be governed by equation (2.95) with $\beta = (k_x^2 + k_z^2)^{1/2}$ and $\theta = \arctan(k_x/k_z)$. If a value of k_z is assumed, equation (2.95) will yield (for a given set of values of H_0, $4\pi M_0$,

* On a nonmagnetic substrate, the Rayleigh wave is a guided surface wave with particle displacement confined to the plane containing the direction of propagation and normal to the substrate. In the presence of magnetoelastic coupling the Rayleigh wave is modified and the particle displacement has all the three components.

γ, σ, and ω) four values* of k_x^2, designated by $[k_x^{(q)}]^2$, where $q = 1, 2, 3, 4$. For each $[k_x^{(q)}]^2$, two values of the transverse wavenumber, $k_x^{(q)}$ and $-k_x^{(q)}$, are obtained which could be real (positive or negative) or complex. When $[k_x^{(q)}]^2$ is real and positive, let the root $k_x^{(q)}(-k_x^{(q)})$ correspond to the wavenumber of a propagating wave which carries energy in the positive (negative) x-direction; the direction of energy flow is normal to the refractive index diagrams shown in Section 2.6.3.2. If $[k_x^{(q)}]^2$ is negative (real) or complex, the wavenumber $k_x^{(q)}$ is chosen to be only that root of $[k_x^{(q)}]^2$ whose imaginary part is negative. Thus, $k_x^q(-k_x^{(q)})$ is the root which corresponds to the decay of a wave field in the positive (negative) x-direction. In view of this, the elastic displacement, rf magnetization, and magnetostatic potential can be expressed as

$$\begin{aligned}
\mathbf{u} &= \left[\sum_{q=1}^{4} Q_q \mathbf{u}_n^{(q)} \exp(jk_x^{(q)}x)\right] \exp j(\omega t - k_z z) \\
\mathbf{m} &= \left[\sum_{q=1}^{4} Q_q \mathbf{m}_n^{(q)} \exp(jk_x^{(q)}x)\right] \exp j(\omega t - k_z z) \\
\psi_{\mathrm{m}} &= \left[\sum_{q=1}^{4} Q_q \psi_{\mathrm{m}}^{(q)} \exp(jk_x^{(q)}x)\right] \exp j(\omega t - k_z z)
\end{aligned} \tag{5.32}$$

where

$$\begin{aligned}
\mathbf{u}_n^{(q)} &= u_{nx}^{(q)} \hat{x} + u_{ny}^{(q)} \hat{y} + u_{nz}^{(q)} \hat{z} \\
\mathbf{m}_n^{(q)} &= m_{nx}^{(q)} \hat{x} + m_{ny}^{(q)} \hat{y} + m_{nz}^{(q)} \hat{z}
\end{aligned} \tag{5.33}$$

In the foregoing expressions, $\mathbf{u}_n^{(q)}$ and $\mathbf{m}_n^{(q)}$ are the normalized eigenvectors of equation (2.94) of Chapter 2, while the Q_q represent the weight factors.

Using the condition $\nabla^2 \psi_{\mathrm{m}} = -4\pi \nabla \cdot \mathbf{m}$, we obtain

$$\psi_{\mathrm{m}}^{(q)} = 4\pi j k_x^{(q)} m_{nx}^{(q)} / [(k_x^{(q)})^2 + k_z^2] \tag{5.34}$$

The magnetostatic potential in the region $x > 0$ can be expressed as†

$$\psi_{\mathrm{S}} = -Q_5 \exp(-k_z x) \exp j(\omega t - k_z z) \tag{5.35}$$

where Q_5 is an arbitrary constant.

* In fact, these four solutions give rise to the four branches of the dispersion curves as discussed in Chapter 2, Part B.

† The introduction of a negative sign before Q_5 does not affect the final result.

The boundary conditions at the stress-free surface $x = 0$ requires that

$$\mathbf{T}_x \equiv T_1\hat{x} + T_6\hat{y} + T_5\hat{z} = 0 \tag{5.36}$$

Considering elastic as well as magnetoelastic contributions, we see that the condition $T_1 = 0$ is equivalent to

$$c_{11}\partial u_z/\partial x + c_{12}\partial u_x/\partial z = 0 \tag{5.37}$$

If the ferrimagnetic medium is regarded as elastically isotropic, we have $c_{12} = c_{11} - c_{44}$ (see Appendix D). If the ratio of bulk longitudinal and bulk shear wave velocities is taken as 2 (which is a good approximation for YIG), we have $c_{11} = 2c_{44}$. Thus, $c_{12} = \frac{1}{2}c_{11}$ and equation (5.37) can be rewritten as

$$\partial u_x/\partial x + \tfrac{1}{2}\partial u_z/\partial z = 0 \tag{5.38}$$

Equating T_6 and T_5 to zero, we have

$$\partial u_y/\partial x = 0 \tag{5.39}$$

and

$$\partial u_x/\partial z + \partial u_z/\partial x + (\sigma/\gamma b_2)m_x = 0 \tag{5.40}$$

Besides equations (5.38)–(5.40), the remaining boundary conditions involve magnetostatic potential and a normal component of magnetic induction at $x = 0$:

$$\psi_m = \psi_S \tag{5.41}$$

$$\partial\psi_m/\partial x + 4\pi m_x = \partial\psi_S/\partial x \tag{5.42}$$

Substitution from equations (5.32)–(5.35) into (5.38)–(5.42) leads to a set of five inhomogeneous, linear algebraic equations involving the five unknowns Q_β ($\beta = 1, 2, 3, 4$, and 5). The dispersion relation is obtained by equating the determinant of the coefficients to zero. Thus

$$\mathcal{D} \equiv [a_{\alpha\beta}]_{5\times5} = 0 \tag{5.43}$$

where each of the α and β vary from 1 to 5. The coefficients $a_{\alpha\beta}$ can be expressed as

$$\left.\begin{aligned}
a_{1\beta} &= -2k_x^{(\beta)}u_{nx}^{(\beta)} + k_z u_{nz}^{(\beta)}, && \beta \neq 5 \\
a_{15} &= 0 \\
a_{2\beta} &= k_x^{(\beta)}u_{ny}^{(\beta)}, && \beta \neq 5 \\
a_{25} &= 0 \\
a_{3\beta} &= jk_x^{(\beta)}u_{nz}^{(\beta)} - jk_z u_{nx}^{(\beta)} + (\sigma/\gamma b_2)m_{nx}^{(\beta)}, && \beta \neq 5 \\
a_{35} &= 0 \\
a_{4\beta} &= 4\pi m_{nx}^{(\beta)} k_z^2/[(k_x^{(\beta)})^2 + k_z^2], && \beta \neq 5 \\
a_{45} &= 0 \\
a_{5\beta} &= 4\pi j m_{nx}^{(\beta)} k_x^{(\beta)}/[(k_x^{(\beta)})^2 + k_z^2], && \beta \neq 5 \\
a_{55} &= 1
\end{aligned}\right\} \tag{5.44}$$

It is worth obtaining the dispersion relation for the case when a thin metallic film is deposited at $x = 0$ to provide electrical short, retaining the stress-free character of the surface at the same time. In this case, the foregoing procedure can be modified as follows. Equation (5.35) is no longer required. The boundary condition (5.41) is redundant while the right-hand side in the boundary condition (5.42) should be replaced by zero. The resulting set of equations leads to the following dispersion relation:

$$\mathscr{D} \equiv [a_{\alpha\beta}]_{4\times 4} = 0 \tag{5.45}$$

where the coefficients $a_{\alpha\beta}$ $(\alpha, \beta = 1, 2, 3, 4)$ are the same as those defined by equations (5.44).

Before discussing the formal solution to the dispersion relations (5.43) and (5.45), we consider some general features of the characteristic waves. The characteristic wave has four components corresponding to the four values of the transverse wavenumber—$k_x^{(q)}$ $(q = 1, 2, 3, 4)$—which, in turn, correspond to the four branches shown in refractive index diagrams discussed in Section 2.6.3.2. It was also pointed out there that the quasi-magnetic branch exists in the bulk wave spectrum in the range $(\omega_0 - \sigma) < \omega < \omega_2$; this essentially means that for any feasible, real k_z, one obtains real k_x (or $-k_x$) implying that at least the quasi-magnetostatic wave component is not guided even if the remaining component waves decay into the substrate. Thus, within the range $\omega_0 - \sigma < \omega < \omega_2'$, one can expect only a quasi-guided wave (Leaky wave). When $\omega < \omega_0 - \sigma$ or $\omega > \omega_2'$, the quasi-magnetic component wave is always inhomogeneous. If the other three quasi-elastic component waves are also inhomogeneous in some range, a bound surface wave results.

The existence or otherwise of a bound surface wave is ascertained only by numerical calculations and is discussed below.

5.2.2. *Numerical Solution of the Dispersion Relation*

For a given set of parameters such as $4\pi M_0$, γ, σ, and H_0, we must obtain the curves for k_z vs. ω. For a given ω, one can assume a value of k_z and solve equation (2.95) of Chapter 2 for k_x (with this set of ω and k_z), in order to obtain the transverse wavenumbers $-k_x^{(q)}$ ($q = 1, 2, 3, 4$). The polarization vectors $\mathbf{m}_n$ and $\mathbf{u}_n$ are then obtained from the equations in Section 2.6.3.3. These values are then substituted in the dispersion relation (5.43) or (5.45). If the resulting $\mathcal{D}$ is equal to zero, the assumed value of k_z represents the solution.

In general $\mathcal{D}$ will be different from zero. In this case the assumed value of k_z is suitably increased or decreased and the foregoing procedure is repeated to compute the corresponding $\mathcal{D}$. The procedure is repeated till $\mathcal{D}$ changes its sign. If $\mathcal{D}$ changes sign between $k_z = k_z^{(1)}$ and $k_z = k_z^{(2)}$, it follows that the solution lies in between these values of k_z. Again, by an iterative procedure (e.g., the bisection method) one obtains a value of k_z at which $\mathcal{D}$ is close to zero within specified accuracy. The entire procedure is then repeated for different values of ω to construct the dispersion diagram for a bound surface wave.

It is possible that in this search procedure no change in the sign of $\mathcal{D}$ will be obtained, which would imply the absence of a bound surface wave and the possibility of the existence of a leaky wave. Exact calculation of the complex wavenumber of a leaky wave being difficult, Parekh and Bertoni (1974a) have adopted the following approximate procedure.

Consider a series expansion of the function $\mathcal{D}(\omega, k_z)$ for the values of k_z close to $k_R = \text{Re}(k_R) + j\,\text{Im}(k_R)$. Assuming that $|\text{Im}(k_R)| \ll |\text{Re}(k_R)|$ (to ensure that energy leakage is small), and k_z is close to k_R, we can write

$$\mathcal{D}(\omega, k_z) = A(k_z - k_R) \tag{5.46}$$

where A depends on ω, k_z, etc. For real k_z, we have

$$|\mathcal{D}|^2 = |A|^2\{[k_z - \text{Re}(k_R)]^2 + [\text{Im}(k_R)]^2\} \tag{5.47}$$

and

$$\text{Phase}(\mathcal{D}) = \text{Phase}(A) - \arctan\left[\frac{\text{Im}(k_R)}{k_z - \text{Re}(k_R)}\right] \tag{5.48}$$

It is evident from equation (5.47) that, in the vicinity of $k_z = \text{Re}(k_R)$, a plot of $|\mathscr{D}|^2$ with real k_z is a parabola, the minimum of which is located at $k_z = \text{Re}(k_R)$; the value of $|\mathscr{D}|^2$ at this minimum is $|A|^2[\text{Im}(k_R)]^2$. It is also evident from equation (5.48) that the quantity $[\text{Phase}(\mathscr{D}) - \text{Phase}(A)]$ changes its sign at $k_z = \text{Re}(k_R)$. Figure 5.12 shows typical computations which confirm these features.

The foregoing discussion indicates the procedure used to obtain the wavenumber, $k_R = \text{Re}(k_R) + j\,\text{Im}(k_R)$, of the leaky waves which decay slowly. The function $|\mathscr{D}(f, k_z)|^2$ is computed as a function of real k_z to obtain the parabolic curve as shown in Figure 5.12; this directly yields $\text{Re}(k_R)$ and $|A|\,\text{Im}(k_R)$. Since $|A|$ and $\text{Im}(k_R)$ are both unknown, one more numerical relation between the two is needed to obtain $\text{Im}(k_R)$. This is achieved by using equation (5.47) for a point other than, but in the vicinity of, the minimum in Figure 5.12. A simultaneous solution of the two equations involving $|A|$ and $\text{Im}(k_R)$ yields $\text{Im}(k_R)$; the sign of $\text{Im}(k_R)$ is negative for $+z$-propagation, since the wave cannot grow as it propagates.

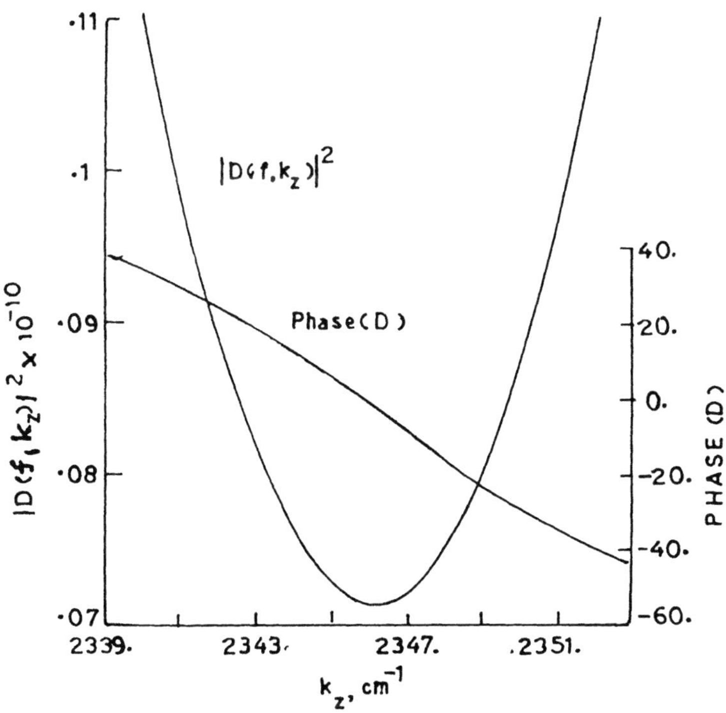

Figure 5.12. Variation of $|\mathscr{D}|^2$ and phase $(\mathscr{D})$ for the metallized configuration as a function of real k_z. The parameters are $f = 134$ MHz, $f_0 = 140$ MHz, $f_m = \gamma' 4\pi M_0 = 840$ MHz, and $f_\eta = 1.272$ MHz (after Parekh and Bertoni, 1974a).

The numerical calculations for a set of typical parameters for Ga–YIG have been carried out by Parekh and Bertoni (1974a). The computed dispersion diagrams for magnetoelastic Rayleigh waves in unmetallized and metallized configurations are presented in Figures 5.13a and 51.3b for $H_0 = 50$ and 100 Oe, respectively. The quantity $\Delta k = k_R - k_{R0}$ rather than k_R is plotted, where k_{R0} is the wavenumber of the corresponding Rayleigh wave when the spins are all clamped. Both the real and the imaginary parts of $\Delta k / k_{R0}$ have been shown. It is seen that, at high frequencies, the dispersion characteristics of magnetoelastic surface waves are asymptotic to those of the ordinary Rayleigh wave. As $f \to 0$, $\mathrm{Re}(\Delta k / k_{R0})$ tends to a small constant value.

The diagrams clearly indicate that the surface wave is bound [$\mathrm{Im}(k_R) = 0$] outside the frequency range $f_L < f < f_2'$, whereas $\mathrm{Im}(k_R) \neq 0$ within this range, thereby rendering the surface wave leaky. The frequency f_L is

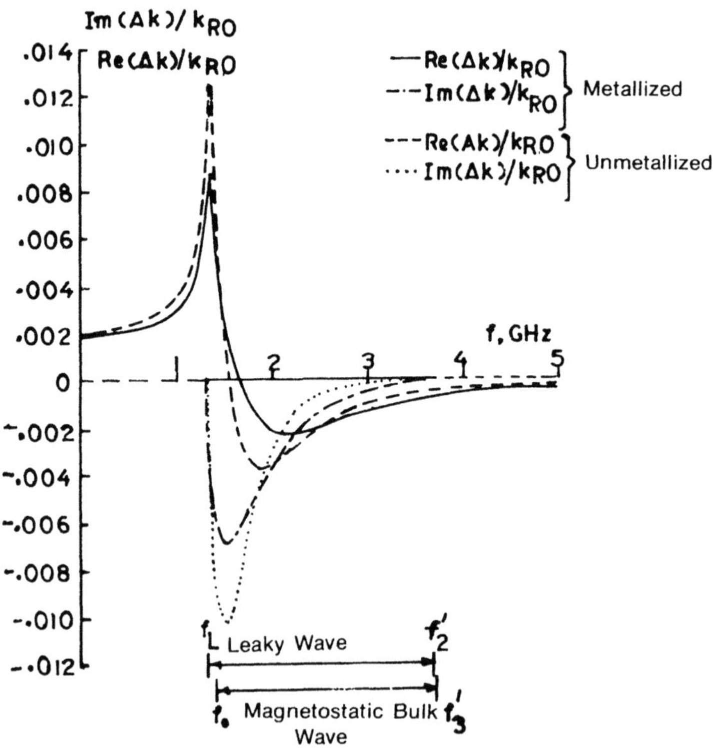

Figure 5.13a. Dispersion curves for magnetoelastic Rayleigh waves in metallized and unmetallized parallel bias configuration. The parameters are $f_m = 840$ MHz, $f_n = 1.272$ MHz, and $H_0 = 50$ Oe (after Parekh and Bertoni, 1974a).

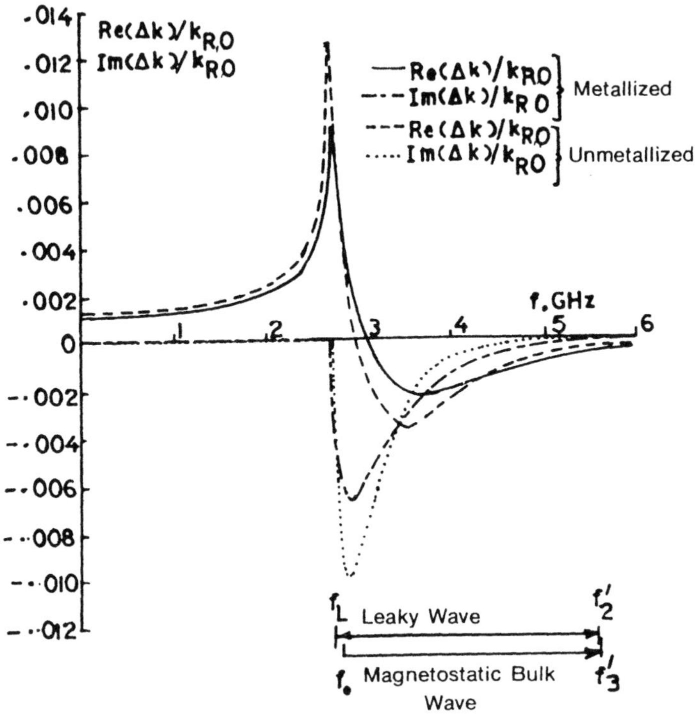

Figure 5.13b. Dispersion curves similar to those in Figure 5.13a except that $H_0 = 100$ Oe (after Parekh and Bertoni, 1974a).

slightly smaller than $f_0 - f_\sigma$, where $f_\sigma = \sigma/2\pi$; f_L is slightly larger for the unmetallized configuration than for the metallized one. The extent of the leaky wave spectrum, i.e., $f_L < f < f'_2$, is slightly greater than that of quasi-magnetic plane waves along the biasing field, which is $f_0 < f < f'_3$.

Whereas the nature of the dispersion curves for the metallized and unmetallized configuration is similar, the peak values of $\mathrm{Im}(k_R)$ and $[\mathrm{Re}(k_R) - k_{R0}]$ are significantly larger for the unmetallized configuration than for the metallized one. On the other hand, the form of the curves is similar for Figures 5.13a and 5.13b, implying that a dc magnetic field mainly influences the locations (values) of the characteristic frequencies f_L, $f'_2, f'_3, \ldots$, and not the strength of magnetoelastic coupling. Thus, the frequency of operation corresponding to specific characteristics is magnetically tunable.

It is further seen in Figures 5.13 that as the wave frequency increases above f_L, the leak rate suddenly rises to attain a maximum and then falls off rather slowly. This sharp rise in the leak rate just above f_L could be useful in designing sharp cut-off devices.

Figure 5.14 displays the explicit magnetic tuning characteristics; the variation of k_R with $f_0 = \gamma' H_0$ has been plotted at a fixed frequency $f = 500$ MHz. A comparison with Figure 5.13 shows that the dispersion behavior at higher (lower) frequencies is similar to that at lower (higher) dc bias.

5.2.3. Energy Loss for Leaky Waves

As discussed above, only leaky waves exist in the frequency range $f_L < f < f_2'$. It was also noted that at least one of the component waves is of a propagating rather than an inhomogeneous nature. The existence of leaky waves and the direction of energy leakage can be understood from the frequency dependence of the outermost branch (K_4 branch) of the refractive index diagrams discussed in Section 2.6.3.2. The relevant portions of this branch are shown in Figure 5.15 for different magnitudes of the wave frequency designated as $f_I < f_L < f_{II} < f_{III}$ such that $f_L < f_{II} < (f_0 - f_\sigma)$

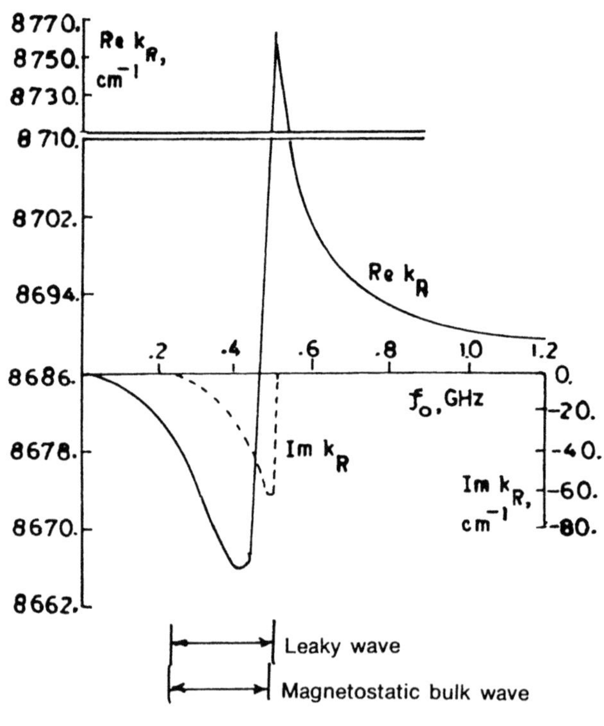

Figure 5.14. Biasing field dependence of the complex wavenumber k_R in the metallized configuration. The parameters are the same as in Figure 5.13 except that f is fixed at 500 MHz (after Parekh and Bertoni, 1974a).

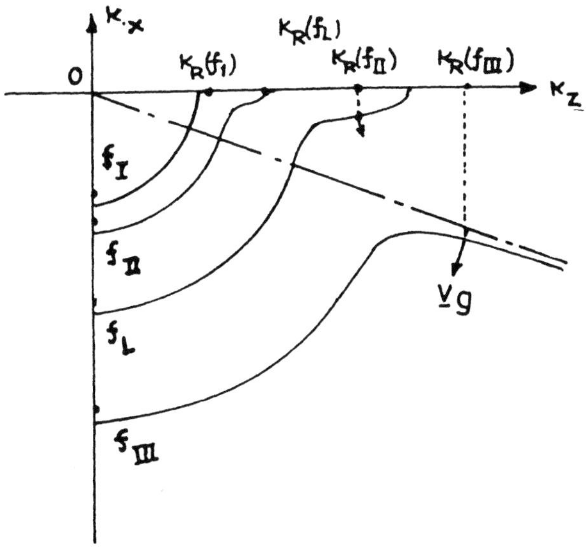

Figure 5.15. Qualitative frequency dependence of the outermost branch of the refractive index diagram (after Parkeh and Bertoni, 1974a).

and $(f_0 - f_\sigma) < f_{III} < f_2'$. It is seen that when $f = f_I$, the K_4 branch crosses the k_z-axis at a point at which $k_z < k_R$. Consequently, the component wave corresponding to the K_4 branch decays in the ferrimagnetic substrate. Since the K_4 branch is the outermost branch, the remaining three component waves (which correspond to the inner branches) also decay along the negative x-axis. Thus, one obtains a bound surface wave at $f = f_I$.

As $f \rightarrow f_L$, it is seen from Figure 5.15 that a bulge along the k_z-axis appears in the K_4 branch which crosses the k_z-axis exactly at the point $k_z = k_R(f_L)$. Thus, the K_4 component wave (which is in fact the quasi-shear component wave; see Section 2.6.3.2) propagates along the z-axis without decaying along the x-axis. This is the threshold of the leaky wave region.

At $f = f_{II}$, the K_4 branch cuts the k_z-axis at a point $k_z > \text{Re}[k_R(f_{II})]$, thereby leading to a component wave which propagates (instead of decaying) along the negative x-axis. Evidently, the surface wave at $f = f_{II}$ is of a leaky nature. As shown in Figure 5.15, the direction of the group velocity vector $\mathbf{v}_g$ is, in fact, the direction of energy leakage in the ferrimagnetic substrate. Since the z-component of $\mathbf{v}_g$ is positive, the surface wave is called a forward-shedding-type leaky wave. Figure 5.16a shows the equi-phase and equi-amplitude lines; the latter also represents the direction of energy flow. The increasing spacing of the lines of constant amplitude corresponds to the decreasing wave intensity with z due to leakage.

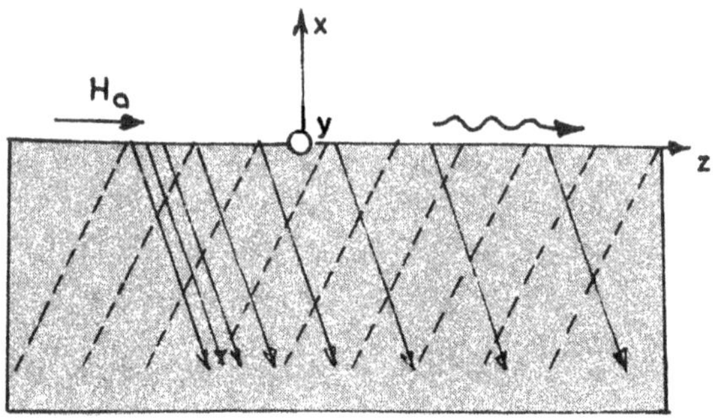

Figure 5.16a. Qualitative diagram of equi-phase and equi-amplitude planes for a forward-shedding leaky wave. The direction of the arrow represents the direction of energy flow (after Parekh and Bertoni, 1974a).

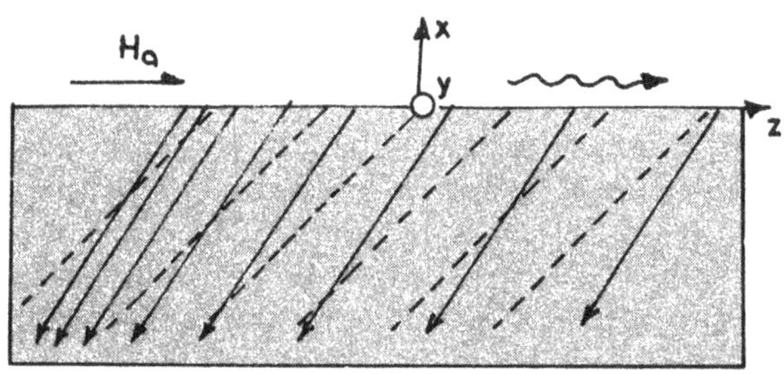

Figure 5.16b. Qualitative diagram of equi-phase and equi-amplitude planes for a backward-shedding leaky wave. The direction of the arrows represents the direction of energy flow (after Parekh and Bertoni, 1974a).

It is seen in Figure 5.15 that at $(f_0 - f_\sigma) < f = f_{11} < f_2'$, the K_4 branch of the refractive index diagram extends to infinity. Consequently, for any real k_z, the quasi-magnetic component is a homogeneous plane wave. Again, the magnetoelastic Rayleigh wave is leaky in this frequency range. Since the z-component of $\mathbf{v}_g$ is negative, the leaky wave is said to be of the backward-shedding type. Figure 5.16b displays the equi-phase surfaces and the direction of energy flow (equi-amplitude surface) for a typical backward-shedding wave. Thus, as the wave frequency increases from f_L to $(f_0 - f_\sigma)$, the direction in which energy leaks away rotates from the direction of surface wave propagation to the inward normal to the ferrimagnetic surface (forward-shedding wave). However, as the wave frequency increases from $(f_0 - f_\sigma)$ to f_2', the direction of energy leakage rotates from the inward surface normal to the direction opposite to that of surface wave propagation (backward-shedding wave).

Finally, when the wave frequency moves past f_2', all the component waves are inhomogeneous. Hence a bound surface wave results, there being no mechanism for energy leakage.

5.2.4. *Effect of Magnetic Loss**

In the foregoing analysis, the ferrimagnetic substrate was assumed to be magnetically lossless. Nevertheless, a leaky wave mechanism causes absorption of the surface wave in the frequency range $f_L < f < f_2'$, which is essentially a nonresonant type of absorption. It is worth investigating the effect of finite magnetic linewidth on the dispersion and loss characteristics of surface waves. It is anticipated that, when the linewidth is finite, energy leakage would occur over the entire dispersion spectrum.

The foregoing procedure of analysis and computation (Section 5.2.1–5.2.3) is repeated after substituting $\omega_0 + \frac{1}{2} j\gamma\Delta H$ for ω_0. The frequency dependence of the complex wavenumber has been shown in Figure 5.17 for different values of the linewidth. It is seen that as linewidth increases the peaks in the dispersion curve (Figure 5.17a) reduce in height and broaden out on a frequency scale. Figure 5.17b indicates that the abrupt rise in loss factor at the frequency $f = f_L$ is replaced by a relatively smooth rise. Also the peak value of the loss decreases.

Considering the fact that the resonance linewidth of the ferrimagnetic samples used in magnetostatic wave experiments is almost invariably less than 2 Oe, we conclude that it is the leaky wave mechanism, rather than magnetic linewidth, that constitutes the major part of absorption of magnetoelastic Rayleigh waves.

5.2.5. *Other Studies*

The detailed analysis of attenuation of magnetoelastic surface waves has been carried out by Emtage (1976). Unlike in the preceding analysis, the direction of propagation now is regarded as a [111]-direction rather than a cube edge. The magnetoelastic and elastic anisotropies are accounted for and are found to lead to nonreciprocal attenuation.

* This section is based on a paper by Parekh (1976).

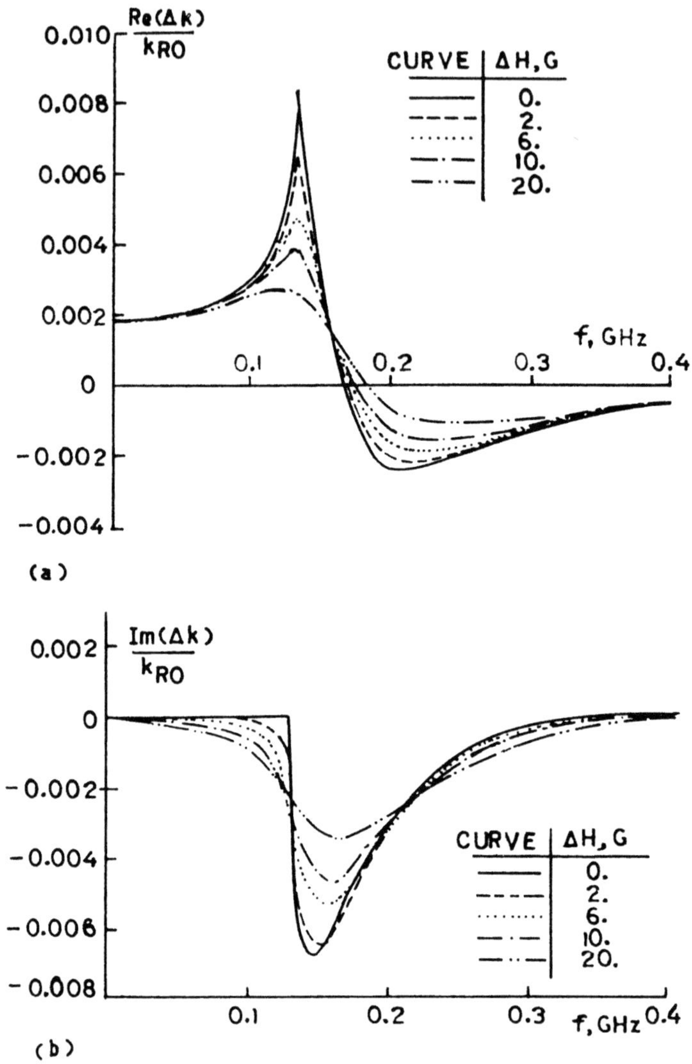

Figure 5.17. Computed frequency dependence of (a) $\mathrm{Re}(\Delta k/k_{R0})$, (b) $\mathrm{Im}(\Delta k/k_{R0})$ (after Parekh, 1976).

While the above discussion has been restricted to the case of propagation along the parallel bias field, the dispersion characteristics of magnetoelastic waves propagating at an oblique angle relative to the tangential bias field have been theoretically investigated by Scott and Mills (1977).

Ganguly et al. (1976a, b) investigated the magnetoelastic surface waves in a magnetic film deposited on a piezoelectric substrate. Accounting for the intrinsic stress developed on a deposited film, they have obtained very good agreement between theory and experiment.

5.3. *Normal Magnetization*

In this section we discuss magnetoelastic Rayleigh wave propagation on a ferrimagnetic substrate. The following is based on a paper by Parekh and Bertoni (1974b).

5.3.1. *The Dispersion Equation*

Figure 5.18 shows the configuration wherein the region $z \leq 0$ is occupied by an elastically as well as magnetically isotropic ferrimagnetic substrate, while the region $z > 0$ is free space. The dc magnetic field is along the z-axis, which is perpendicular to the substrate. The wave propagates along the x-axis and has a uniform field distribution along the y-axis.

The method of analysis is very similar to that discussed in Section 5.2. The wave field quantities vary as $\exp j(\omega t - \beta\hat{n} \cdot \mathbf{r})$, where $\beta\hat{n} = k_x\hat{x} + k_z\hat{z}$. For a given k_x, one obtains four values of k_z^2 designated as $[k_z^{(q)}]^2$, where $q = 1, 2, 3, 4$. As before, it can be argued that $-k_z^{(q)}$ represents the desired root. Thus, the elastic displacement, rf magnetization, and magnetostatic potential in the ferrimagnetic region, in the presence of the wave, are expressed as

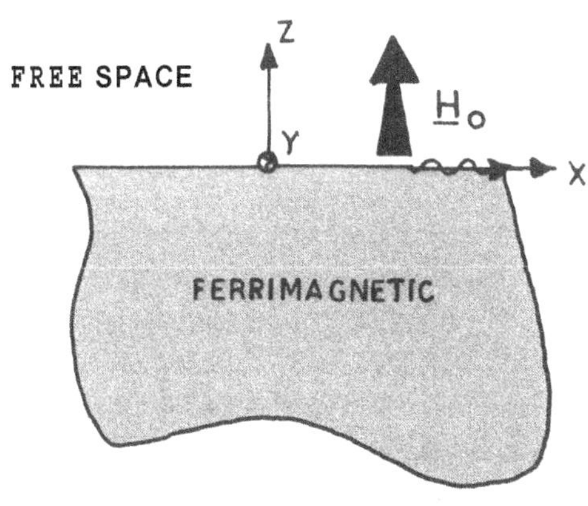

Figure 5.18. Ferrimagnetic half space: the dc magnetic field is along the z-axis, which is perpendicular to the substrate. The wave propagates along the x-axis and has a uniform field distribution along the y-axis (after Parekh and Bertoni, 1974b).

$$\left.\begin{aligned} \mathbf{u} &= \left[\sum_q Q_q \mathbf{u}_n^{(q)} \exp(jk_z^{(q)} z)\right] \exp j(\omega t - k_x x) \\ \mathbf{m} &= \left[\sum_q Q_q \mathbf{m}_n^{(q)} \exp(jk_z^{(q)} z)\right] \exp j(\omega t - k_x x) \\ \psi_{\mathrm{m}} &= \left[\sum_q \psi_{\mathrm{m}}^{(q)} Q_q \exp(jk_z^{(q)} z)\right] \exp j(\omega t - k_x x) \end{aligned}\right\} \tag{5.49}$$

where

$$\left.\begin{aligned} \mathbf{u}_n^{(q)} &= u_{nx}^{(q)}\hat{x} + u_{ny}^{(q)}\hat{y} + u_{nz}^{(q)}\hat{z} \\ \mathbf{m}_n^{(q)} &= m_{nx}^{(q)}\hat{x} + m_{ny}^{(q)}\hat{y} \end{aligned}\right\} \tag{5.50}$$

Using the condition $\nabla^2\psi_{\mathrm{m}} = -4\pi\nabla\cdot\mathbf{m}$, we obtain

$$\psi_{\mathrm{m}}^{(q)} = -\frac{4\pi jk_x m_{nx}^{(q)}}{k_x^2 + (k_z^{(q)})^2} \tag{5.51}$$

The magnetostatic potential in the free space region ($z > 0$) can be expressed as

$$\psi_{\mathrm{S}} = -Q_5 \exp(-k_z x) \exp j(\omega t - k_z z) \tag{5.52}$$

For the unmetallized configuration, the boundary conditions can be obtained as before and are expressed as

$$\left.\begin{aligned} \partial u_z/\partial z + \tfrac{1}{2}\partial u_x/\partial x &= 0 \\ \partial u_y/\partial z + (\sigma/\gamma b_2) m_y &= 0 \\ \psi_{\mathrm{m}} &= \psi_{\mathrm{S}} \\ \partial\psi_{\mathrm{m}}/\partial x + 4\pi m_x &= \partial\psi_{\mathrm{S}}/\partial x \end{aligned}\right\} \tag{5.53}$$

For the metallized configuration, the fourth of these conditions is unnecessary, and the right-hand side of the last equation should be replaced by zero. Following the preceding analysis for a tangential bias, we obtain the dispersion relation as

$$\mathscr{D} \equiv [a_{\alpha\beta}] = 0 \tag{5.54}$$

where each of the α and β varies from 1 to 5 for the unmetallized configuration and 1 to 4 for the metallized configuration. The expressions for $a_{\alpha\beta}$ are given as follows:

$$\left.\begin{aligned}
a_{1\beta} &= -2k_z^{(\beta)}u_{nz}^{(\beta)} + k_x u_{nx}^{(\beta)}, && \beta \neq 5\\
a_{15} &= 0\\
a_{2\beta} &= jk_z^{(\beta)}u_{ny}^{(\beta)} + (\sigma/\gamma b_2)m_{ny}^{(\beta)}, && \beta \neq 5\\
a_{25} &= 0\\
a_{3\beta} &= jk_z^{(\beta)}u_{nx}^{(\beta)} - jk_x u_{nz}^{(\beta)} + (\sigma/\gamma b_2)m_{nx}^{(\beta)}, && \beta \neq 5\\
a_{35} &= 0\\
a_{4\beta} &= -4\pi k_x m_{nx}^{(\beta)}k_z^{(\beta)}/[(k_z^{(\beta)})^2 + k_x^2], && \beta \neq 5\\
a_{45} &= k_x\\
a_{5\beta} &= 4\pi jk_x m_{nx}^{(\beta)}/[k_x^2 + (k_z^{(\beta)})^2], && \beta \neq 5\\
a_{55} &= 1
\end{aligned}\right\} \quad (5.55)$$

5.3.2. *Dispersion Characteristics*

The procedure for obtaining the numerical solution to the dispersion relation is analogous to the preceding case of tangential bias configuration. It is found that the surface wave is leaky within the range $(f_0 - f_\sigma) < f < f_x$, where f_x is a numerically determined frequency slightly greater than f_2'. Outside this frequency range, all the component waves are inhomogeneous and hence a bound surface wave results.

The computed dispersion curves have been presented in Figure 5.19 for both unmetallized and metallized configurations. The real and imaginary parts of the complex wavenumber k_R have been normalized with respect to the wavenumber k_l of the purely elastic longitudinal bulk wave. When $f < (f_0 - f_\sigma)$, the wavenumber k_R on the low-frequency branch is real $[\mathrm{Im}(k_R) = 0]$. This means that the magnetoelastic Rayleigh wave is a bound surface wave. When $f > (f_0 - f_\sigma)$, $[\mathrm{Im}(k_R) \neq 0]$ and the magnetoelastic Rayleigh wave becomes a leaky wave. The leakage rate slowly increases and then decreases to zero at a particular frequency f_z, which is different for metallized and unmetallized configurations. When $f > f_z$, the leakage rate grows fast as f approaches f_x. At frequency $f = f_z$, there is no energy leakage because the quasi-magnetostatic plane wave (which is responsible for energy leakage) decouples from the surface wave at this frequency.

The mechanism for energy leakage can be understood with the help of the refractive index diagram shown in Figure 5.20 for a typical set of data. Only the outermost branch (K_4 branch; see Section 5.2.3) has been plotted. At a frequency $f_1 < (f_0 - f_\sigma)$, the K_4 branch crosses the k_x-axis at a point which lies to the left of the point at which $k_x = k_R(f_1)$. Thus, the component wave corresponding to the K_4 branch decays into the ferrimagnetic substrate. Other component waves for which the branches of the refractive index diagram lie inside the K_4 branch would all necessarily decay into the substrate. Thus, the resulting magnetoelastic surface wave is bound. When f lies between $(f_0 - f_\sigma)$ and f_2' (such as f_{II} and f_{III} in Figure 5.20), the K_4 branch extends to infinity. Consequently, for any real value of k_x, the quasi-magnetic plane wave is homogeneous. Thus, the magnetoelastic

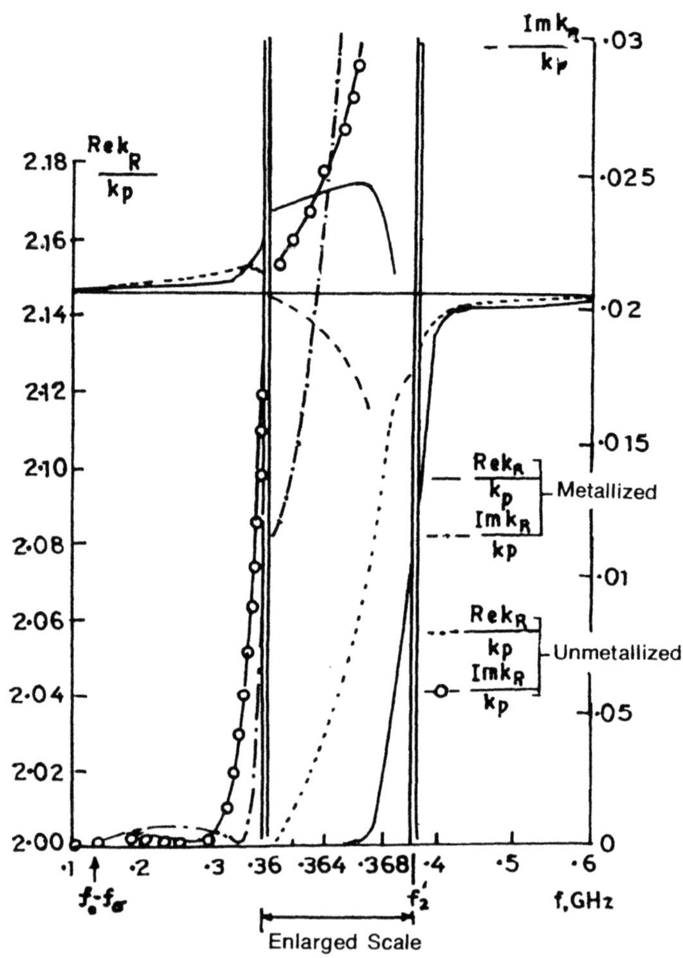

Figure 5.19. Dispersion curves for magnetoelastic Rayleigh waves in the configuration corresponding to Figure 5.18. For the metallized configuration, $\text{Re}(k_R/k_p)$ and $\text{Im}(k_R/k_p)$ are represented by a solid line (———) and a dot-dashed line (—·—), respectively. For the unmetallized configuration, $\text{Re}(k_R/k_p)$ and $\text{Im}(k_R/k_p)$ are represented by broken lines (- - - -) and a circle line -○-○-○-), respectively. For Ga–YIG, $k_p = 8100f\ \text{cm}^{-1}$ (after Parekh and Bertoni, 1974b).

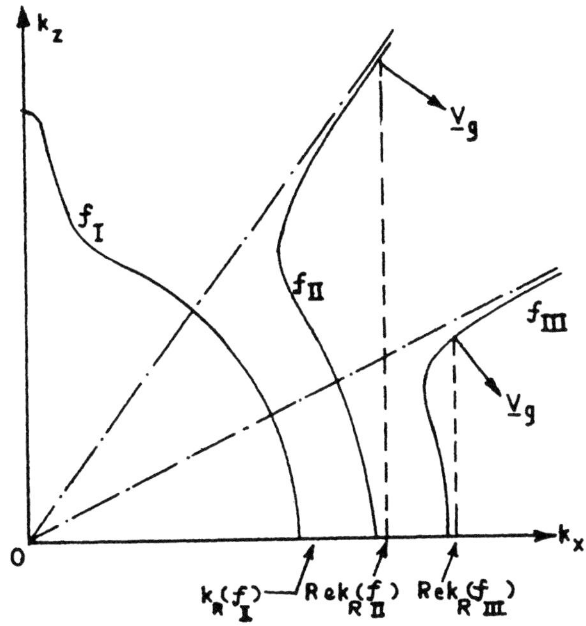

Figure 5.20. Qualitative frequency dependence of the outermost branch of the refractive index diagram for magnetoelastic plane waves. The various frequencies satisfy the inequality $f_{\rm I} < (f_0 - f_\sigma) < f_{\rm II} < f_{\rm III} < f_2'$ (after Parekh and Bertoni, 1974b).

Rayleigh wave is leaky in this frequency range and is of the forward-shedding type.

5.4. *Experimental Results and Devices*

As noted in the introduction to this chapter, this section presents some experimental results demonstrating the excitation and propagation of magnetoelastic waves on substrates and in films. Subsequently, some practical devices such as delay lines, isolators, oscillators, and convolvers will also be discussed.

5.4.1. *Excitation and Propagation Loss*

Magnetoelastic surface waves on a plane YIG plate were first excited by Voltmer et al. (1969). As shown in Figure 5.21, a YIG plate of dimensions $1.0 \times 0.5 \times 0.1\,\text{cm}^3$ was chosen. The [110]-crystallographic axis was perpendicular to the substrate whereas the [001]-axis was parallel to

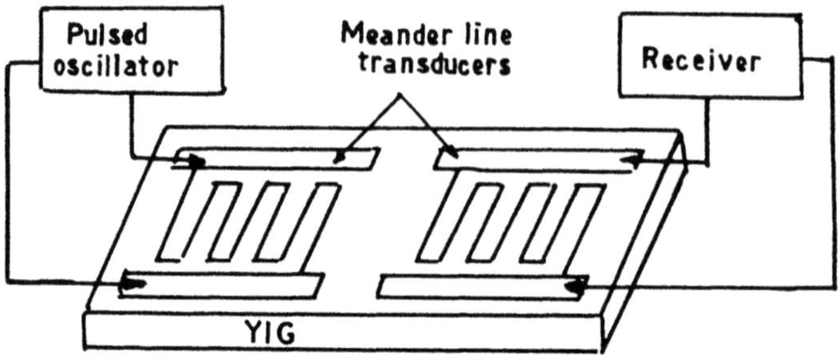

Figure 5.21. Experimental arrangement for surface wave excitation by a meander-line transducer (after Voltmer et al., 1969).

the long edge. The meander line transducers were fabricated by evaporating aluminum following standard photoresist techniques. The period of the meander lines was 59 μm, which would correspond to a 60-MHz surface wave. The surface wave was detected in a rather limited range of the magnetic field, specifically, from about 170 to 470 Oe. The surface waves disappeared when the dc field was in the plane of the substrate but transverse to the direction of propagation.

Daniel (1973) measured the absorption of magnetoelastic surface waves at a frequency around 30 MHz on a YIG substrate. He used ZnO film and $LiNbO_3$ platelets over interdigital electrodes deposited on the YIG substrate. The orientation of the substrate was (100) and the direction of propagation was a [100]-axis. The absorption was measured for different orientations of the dc magnetic field and was found to be smaller than the "magnetoelastic resonance" loss for bulk shear magnetoelastic waves (see Section 2.6.2). This clearly indicates that the propagation loss associated with magnetoelastic surface waves is due to the leaky nature rather than resonance absorption, which confirms the predictions by Parekh and Bertoni (1972, 1974a, b).

5.4.2. *Magnetoelastic Surface Wave Delay*

It was seen in Sections 5.1–5.3 that magnetoelastic waves are dispersive. Since the various resonance and cut-off frequencies corresponding to the magnetoelastic waves can be changed by varying the biasing field, it should be possible to switch from one branch of dispersion curves to another, thereby causing a change in the delay characteristics. This electronic tunability of delay has been experimentally investigated by Tsai et al. (1974).

The experiment was performed on a (111)-oriented YIG plate of size $1.5 \times 0.5 \times 0.1\,cm^3$ with the [110]-direction parallel to the direction of surface wave propagation. Thus, the configuration corresponds to that discussed in Section 5.2 with a specific crystallographic orientation. After polishing the crystal surface with aluminum oxide powder, meander lines consisting of 10 pairs of lines were fabricated using the standard photolithographic techniques. The lines were about 7 mm apart. This line spacing would correspond to a Rayleigh wave with a frequency of approximately 35.3 MHz in the parallel bias case. An exciting signal from a pulsed high-frequency oscillator was applied to one of the meander lines. The delayed signal was picked up on an oscilloscope after being passed through wide-band amplifiers. The delay time was measured from the beginning of the exciting pulse to the maximum of the delayed pulse.

Figure 5.22 presents the curves showing the variation of magnetoelastic surface wave delay time with a biasing field. The curves for eight distinct frequencies have been displayed. The magnetic tunability of group delay is evident, and the delay characteristics are obviously frequency-dependent.

Not all the features of the foregoing experiment have been explained by existing theories. Specifically, several distinct modes were observed whereas the existing theory predicts only one guided surface mode.

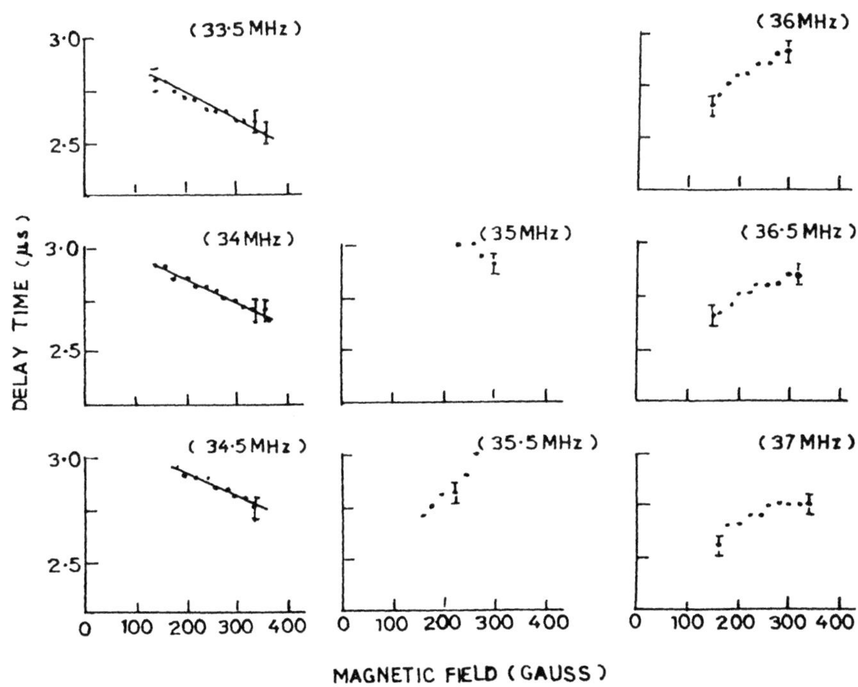

Figure 5.22. Variation of magnetoelastic surface wave delay time with dc magnetic field for different frequencies of operation (after Tsai et al., 1974).

5.4.3. *Magnetoelastic Surface Wave Isolator*

The nonreciprocal attenuation of magnetoelastic surface waves in different configurations has been experimentally and theoretically investigated by Lewis and Patterson (1972), Emtage (1976), Daniel (1977), and Volluet (1977a). In what follows we describe the experiment of Lewis and Patterson (1972).

Figure 5.23 shows the configuration wherein a 5-μm-thick Ga–YIG film is grown over a (112)-oriented GGG substrate. A 2-μm-thick ZnO film is deposited over the Ga–YIG, and interdigital transducers (IDT), with a period 16 μm, are formed for coupling. The dc magnetic field is applied in the plane of the film along a [111]-direction and the wave propagates along a [110]-direction. The excitation efficiency is a maximum at a frequency around 200 MHz.

Figure 5.24 shows the attenuation of oppositely propagating magnetoelastic surface waves. The attenuation shown in the figure is relative to a high field value. It is seen that the attenuation characteristics are approximately reciprocal at a high or zero bias field. There is appreciable nonreciprocity in the intermediate range of the biasing field. This nonreciprocity is maximum when the biasing field is around 40 Oe, in which case the wave propagating along the $+y$-direction exhibits an attenuation of the order of 30 dB/cm, which would lead to good isolator performance.

5.4.4. *Magnetoelastic Surface Wave Oscillator*

As in the case of magnetostatic waves, it is possible to fabricate a delay line oscillator based on magnetoelastic surface waves. Volluet (1977a, b) has reported such an oscillator.

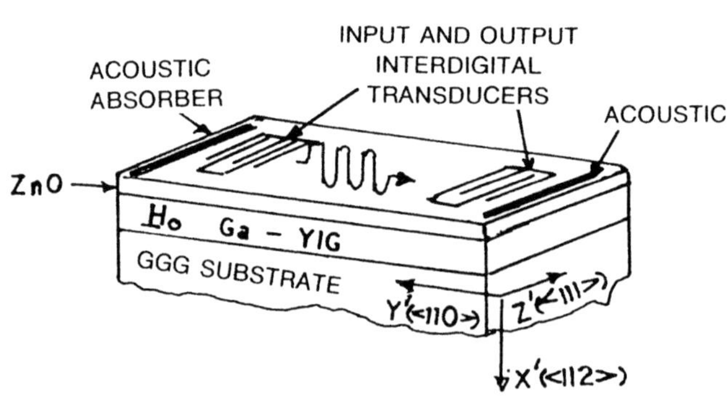

Figure 5.23. Magnetoelastic surface wave isolator: the Ga–YIG film grown by VPE has a thickness 5.0 μm. The sputtered film of ZnO is 2.0 μm thick (after Lewis and Patterson, 1972).

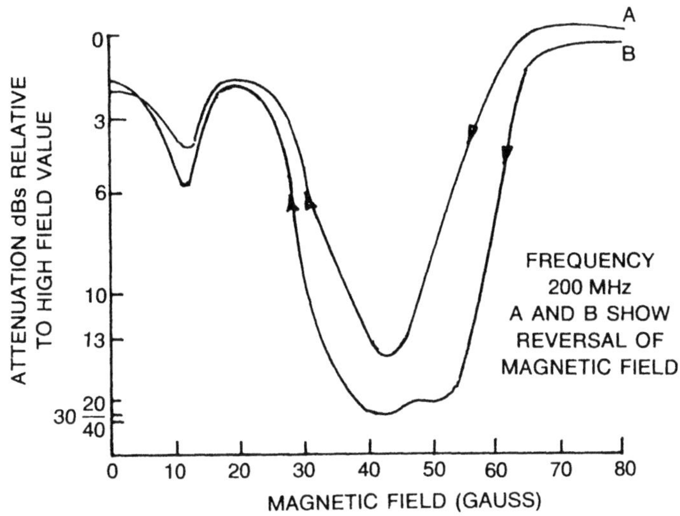

Figure 5.24. Attenuation of forward and backward propagating Rayleigh waves as a function of the biasing field for the isolator configuration shown in Figure 5.23 (after Lewis and Patterson, 1972).

A sketch of the experimental arrangement is shown in Figure 5.25 wherein a magnetoelastic surface wave propagates in a (111)-oriented Ga–Gd–YIG film grown on a gadolinium–gallium–garnet (GGG) substrate by liquid-phase epitaxy. The (111)-plane is the easy plane and thus only a small in-plane magnetic field is required for magnetic saturation. The Ga–Gd–YIG film thickness is about 9 μm. A 1-μm-thick layer of piezoelectric ZnO is deposited on the Ga–Gd–YIG by sputtering. Finally, interdigital transducers are fabricated to launch magnetoelastic surface waves, of frequency 339.9 MHz, propagating along the [110]-direction.

The variation of the phase velocity shift with an applied magnetic field as a function of the angle between the direction of propagation and the dc field is shown in Figure 5.26. It is seen that the phase velocity shift is maximum when the direction of the dc field coincides with that of propagation. The phase velocity shift is significantly large when the biasing field is around 20–80 Oe. The relative acoustic attenuation for $\mathbf{k} \| \mathbf{H}_0$ has been shown in Figure 5.27 in which the corresponding curve for the relative phase velocity shift has also been shown. It is seen that at $H_0 = 0$ the attenuation is rather large. This is presumably because of the presence of magnetic domains. Almost total absorption is observed when the biasing field is between 10 and 40 Oe. The relative attenuation is quite small when H_0 exceeds 50 Oe.

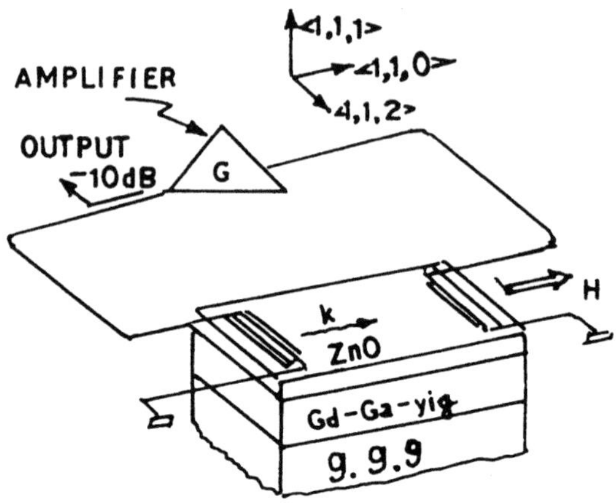

Figure 5.25. Experimental arrangement for the magnetoelastic surface wave oscillator (after Volluet, 1977a).

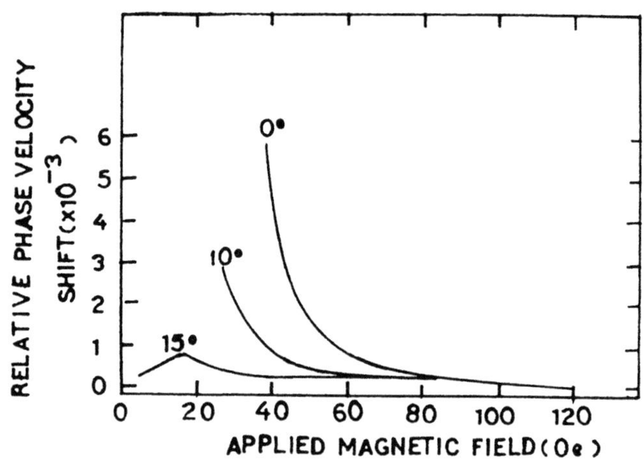

Figure 5.26. Variation of the relative phase velocity shift with dc magnetic field for different angles between $\mathbf{H}_0$ and $\mathbf{k}$ shown alongside each curve (after Volluet, 1977b).

Coming back to Figure 5.25, we see that the delayed signal is first amplified and then fed back to the delay line. If L, v, and φ_e, respectively, represent the path length between the transducers, wave velocity, and electronic phase shift introduced by the amplifier plus the transducers, the frequency of oscillation satisfies the following condition:

$$2\pi f_L/v + \varphi_e = 2n\pi \tag{5.56}$$

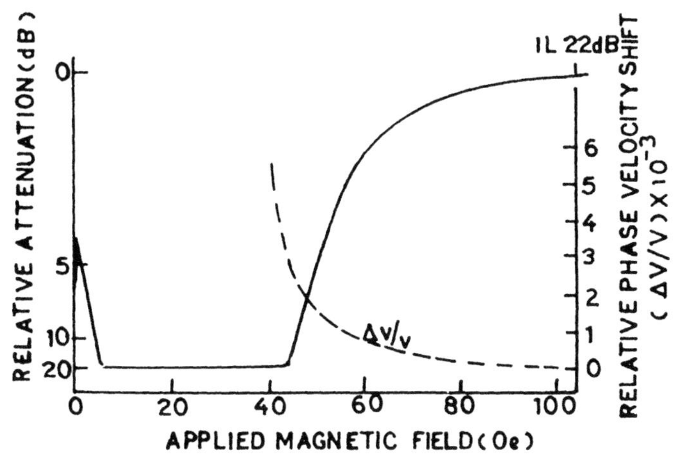

Figure 5.27. Variation of the relative phase velocity shift (dashed line) and elastic attenuation (solid line) with a dc magnetic field (after Volluet, 1977b).

where n is an integer. If Δv represents the change in velocity, the corresponding change in the oscillator frequency is given by

$$\Delta f / f = \Delta v / v \tag{5.57}$$

Figure 5.28 shows the performance of such a feedback oscillator. The amplifier gain is greater than the insertion loss (by 10 dB) to overcome the increasing attenuation. It is seen that a total frequency variation of about 620 KHz is obtained when the biasing field is swept between 50 and 120 Oe.

5.4.5. *Magnetoelastic Surface Wave Convolution*

Convolution of magnetoelastic surface waves was first demonstrated by Lundstrom and Robbins (1974). Subsequently, Robbins and Lundstrom (1975), Parekh et al. (1975), and Robbins (1976a, b) carried out further experimental and theoretical investigations to improve the convolution characteristics. The rationale (Robbins, 1976a) for magnetoelastic surface wave convolution can be understood in analogy with the corresponding case of electromagnetic (magnetostatic) waves.

The equation of motion for magnetization, i.e., equation (1.14) of Chapter 1, is intrinsically nonlinear. Thus, when a dc field $H_0\hat{z}$ and an rf field $\mathbf{h}(\omega)$ are applied, the magnetization contains the Fourier components at frequencies $0, \omega, 2\omega, 3\omega, \ldots$. Thus,

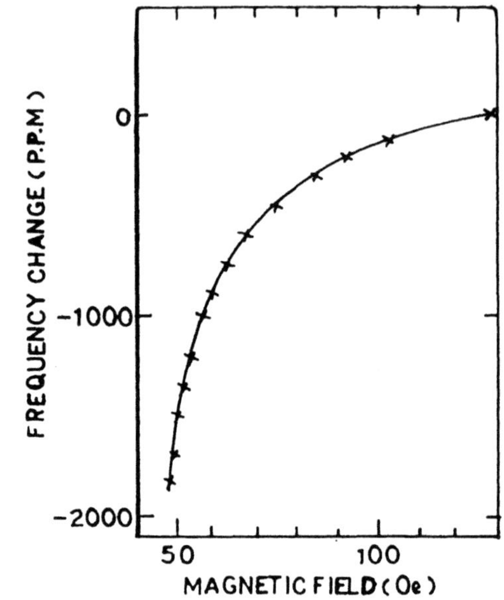

Figure 5.28. Variation of an oscillator frequency shift with a dc magnetic field (after Volluet, 1977b).

$$\mathbf{M} = M_0\hat{z} + \mathbf{m}(\omega) + \mathbf{m}_2(\omega) + \mathbf{m}_3(3\omega) + \cdots$$

Using equation (1.14) of Chapter 1, we have

$$(\dot{\mathbf{m}}_2)_z = -\gamma(m_x h_y - m_y h_x)$$

Substitution for m_x and m_y from equation (1.21) of Chapter 1 yields

$$\begin{aligned}(\dot{\mathbf{m}}_2)_z &\propto (h_x^2 + h_y^2)\\ &\propto |\mathbf{h}|^2\end{aligned}$$

Thus, the nonlinear (2ω) component of the magnetization varies as the square of the amplitude of the rf signal. While this result is based on the electromagnetic/magnetostatic interactions, its generalization to the analogous case of magnetoelastic interaction is straightforward.

Now consider the magnetoelastic surface wave signals $\mathcal{S}_1$ and $\mathcal{S}_2$, both at frequency ω. The nonlinear component at frequency 2ω can be expressed as (Robbins, 1976a)

$$m_{\text{nl}} = \eta(\mathcal{S}_1 + \mathcal{S}_2)^2$$

where the factor η depends on various parameters, such as wave frequency, dc field, strength of magnetoelastic coupling, etc. Let $\mathcal{S}_1$ and $\mathcal{S}_2$ correspond to oppositely propagating shear waves having velocity v_R (Rayleigh wave velocity). The signals can be expressed as

$$\left.\begin{aligned} \mathcal{S}_1(z,t) &= S_1(t - z/v_R)\cos(\omega t - \beta z) \\ \mathcal{S}_2(z,t) &= S_2(t + z/v_R)\cos(\omega t + \beta z) \end{aligned}\right\} \tag{5.58}$$

Figure 5.29 shows the experimental arrangement in which the total flux intercepted by the pick-up coil is expressed as

$$\Phi = W\int_0^L m_{nl}(z,t)dz \tag{5.59}$$

where W is the acoustic beamwidth, L is the coil width, and m_{nl} is regarded as uniform over the beamwidth. If the coil is connected to a detector which is tuned at frequency 2ω, the magnitude of the averaged detected voltage of the convolved signal can be obtained from equation (5.59) as

$$V_c = |d\Phi/dt| = 2\omega W\eta v_R \int_{-\infty}^{+\infty} S_1(\xi)S_2(2t - \xi)d\xi \tag{5.60}$$

where $\xi = t - z/v_R$ and the duration of each of S_1 and S_2 is less than L/v_R.

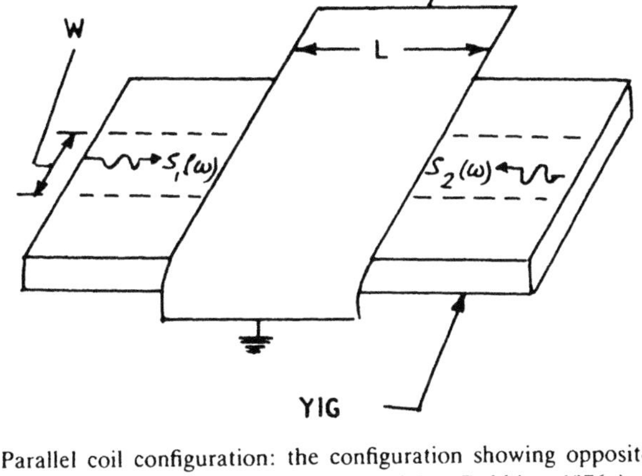

Figure 5.29. Parallel coil configuration: the configuration showing oppositely propagating signals and the pick-up coil for a convolved signal (after Robbins, 1976a).

If S_1 and S_2 are identical rectangular pulses of amplitude S_i and duration τ, equation (5.59) transforms into

$$V_c = 2\omega W \eta v_R \tau S_i^2$$

The average power delivered to a load R_L is obviously given by $P_c = V_c^2/2R_L$.

The foregoing semiempirical formulas and their derivation provide an understanding of the process of convolution. Now we shall describe some experimental results.

Figure 5.30 shows the experimental configuration of Lundstrom and Robbins (1974). The structure consists of the epoxy-bonded $LiNbO_3$–YIG–$LiNbO_3$ composite. The YIG part is a $0.5 \times 0.5 \times 0.125$ in. single crystal with a (110)-plane as propagating surface. The direction of propagation coincides with the [001]-axis. Interdigital transducers for coupling are fabricated over $LiNbO_3$. The central frequency of operation is 50 MHz and the bandwidth around this frequency is about 6 MHz. Both the input signals in the experiment were of 1.7 W magnitude and 1.5 μsec duration. The convolved signal was picked up both by parallel coil configuration (Figure 5.29) as well as by perpendicular coil configuration (Figure 5.31). The results have been summarized in Table 5.1. The best results were obtained when the biasing field was in the plane of the YIG crystal but oriented at an angle of 20° with respect to the direction of wave propagation. This clearly demonstrates the feasibility of the magnetostatic surface wave propagation.

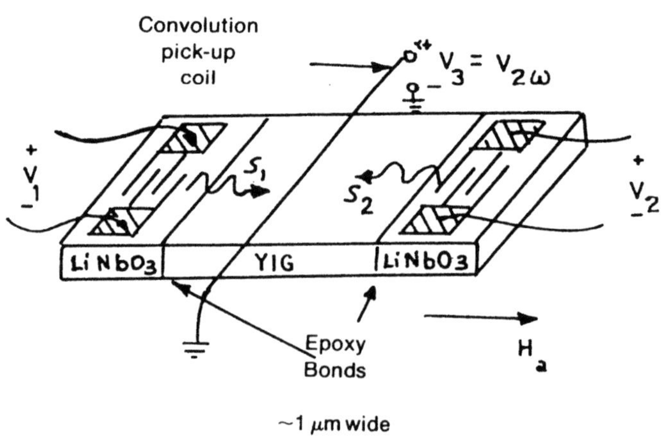

Figure 5.30. Experimental arrangement for magnetoelastic Rayleigh wave convolution (after Lundstrom and Robbins, 1974).

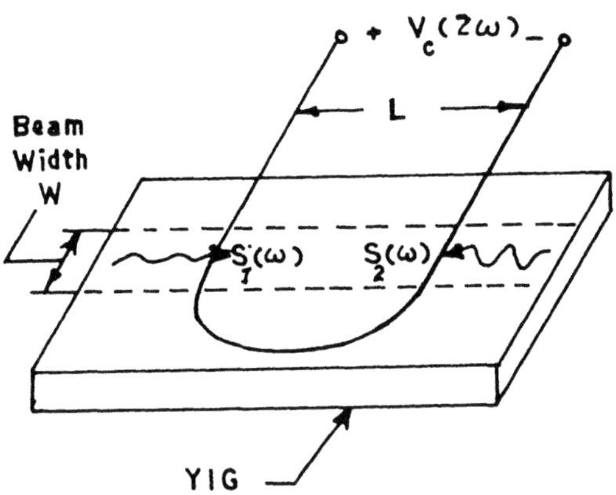

Figure 5.31. Perpendicular coil configuration: a full-turn wire loop is employed to sense the nonlinear magnetization component (after Robbins, 1976a).

Table 5.1. Convolver Performance for Various Field and Coil Orientations

Field direction	Coil orientation	Relative convolution magnitude P_3/P_1 (dB)	Magnitude of field (G)
$\mathbf{H}_a \parallel \mathbf{k}$	$\perp$	−70	263
	$\parallel$	−82	250
$\mathbf{H}_a \perp \mathbf{k}$	$\perp$	−96	700
	$\parallel$	−95	1450
$\mathbf{H}_a \perp \mathbf{k}$ and $\parallel$ YIG	$\perp$	−99	125
	$\parallel$	−87	150
$\mathbf{H}_a$ at 20° to $\mathbf{k}$ and $\parallel$ YIG	$\perp$	−59	250
	$\parallel$	−67	250

Input power $P_1 = P_2 = 1.7$ W for all configurations (after Lundstrom & Robbins, 1974).

Parekh et al. (1975) have used a ZnO transducer in fabricating a magnetoelastic Rayleigh wave convolver and obtained a significantly improved performance compared to the $LiNbO_3$ composite. Robbins (1976b) has employed magnetoelastic surface wave convolution to measure the demagnetizing factor of a plate YIG.

5.4.6. *Other Studies*

Ganguly et al. (1975, 1977) and Forester et al. (1978) have investigated the phase shift of elastic waves in films of different magnetostrictive materials and proposed phase shifters.

In spite of the numerous exhaustive theoretical analyses (some of which have been discussed in Sections 5.1–5.3), there are several experimental results which are as yet unexplained. A recent paper by Komoriya and Thomas (1979) describes experimental results pertaining to magnetoelastic surface waves on substrates as well as films of YIG. They have reported observing several magnetoelastic modes where only one was expected; this feature has been reported in the past by many other investigators. It is believed that future investigations will clarify some of these unexplained results.

APPENDIX A

Structure and Properties of Common Ferrimagnetics

Ferrimagnetic insulators are found to have a variety of crystal structures. Of these, there are three main classes, namely garnets, spinels, and hexagonal magnetoplumbites, which are frequently used in microwave devices. Therefore, we shall limit our discussion only to these materials.

A.1. *Ferrimagnetic Garnet*

The crystal structure of ferrimagnetic garnets is isomorphic with that of the classical garnet $Ca_3Fe_2(SiO_4)_3$. Yttrium–iron–garnet (YIG, perhaps the most important microwave magnetic material today) was the first ferrimagnetic garnet discovered, independently, by Bertaut and Forrat (1956) and by Geller and Gilleo (1957a, b). Since then, several pure, mixed, and substituted rare-earth iron garnets (RIG) have been prepared and studied. We shall briefly summarize the structure and basic properties of the various ferrimagnetic garnets. Details are available in several excellent books including those by Lax and Button (1962), von Aulock (1965), Tebble and Craik (1969), Smit (1971), Riches (1972), Standley (1972), and Heck (1974).

A.1.1. *Structure*

The present description follows Geller and Gilleo (1957a, b), Lax and Button (1962), and von Aulock (1965). The unit cell of an iron garnet (MIG, M is a trivalent metal cation) is very nearly cubic. The edge length of the cube edge is approximately 12.5 Å. The unit cell contains eight formula units of MIG, i.e., $8M_3Fe_2\,(FeO_4)_3$ or simply $8M_3Fe_5O_{12}$. Thus, in all, there are 160 ions per unit cell. The 96 nonmagnetic, divalent oxygen ions form a three-dimensional matrix which contains three types of "sites" for cations. There are 24 tetrahedral (d) sites, 16 octahedral (a) sites and 24 dodecahedral (c) sites.* All the 40 trivalent ferric ions are distributed among a- and d-sites while the 24 metal ions occupy c-sites which, incidentally, are the largest sites.

A.1.2. *Magnetization*

A.1.2.1. *Pure Garnet*

As noted above, all the ferric ions in an iron garnet are distributed among the a- and d-sites. The interaction between ions at a- and d-sites is very strong and is of an antiferromagnetic character. The interaction among the ions within a-sites, as well as among those within d-sites, is also antiferromagnetic. However, the a–a and d–d interactions are much weaker than the a–d interaction (Lax and Button, 1962). As a consequence, the magnetic moments of the ferric ions in a-sites are aligned opposite to those in d-sites, whereas all the magnetic dipoles within a- (or d-) sites are parallel to one another (Figure A.1). Hence, the ferric ions form two oppositely magnetized sublattices. Consequently, there will be a net magnetic moment, equivalent to that of one ferric ion per formula unit, which is $5\,\mu_B$ (see Appendix B). If the metal cation in the c-site is nonmagnetic (e.g., yttrium) there is no further contribution to the magnetic moment and, therefore, the total magnetic moment per unit cell is $40\,\mu_B$; this can be divided by the volume of the unit cell of YIG to obtain an estimate of spontaneous magnetization at absolute zero. The magnetization at room temperature would be somewhat less than that at absolute zero and can be best determined experimentally; it is found to be about 1.75 kG. The reduction of magnetization at higher temperatures results from the weakening of the forces responsible for magnetic order. As a consequence, the magnetization of each sublattice decreases as the temperature is raised; the rate of decrease is generally different for different sublattices. In the case of pure YIG, where the c-sublattice does not make any contribution toward the magnetic moment, the magnetization decreases monotonically with temperature (Figure A.2) and eventually vanishes at the Néel temperature (T_N), beyond which the garnet exhibits ordinary paramagnetism. For pure YIG, $T_N \simeq 550\,°K$ which is well above room temperature.

* A tetrahedral site is a vacancy in between four oxygen ions occupying the four corners of a tetrahedron. Likewise, octahedral and dodecahedral sites are surrounded by six and eight oxygen ions, respectively.

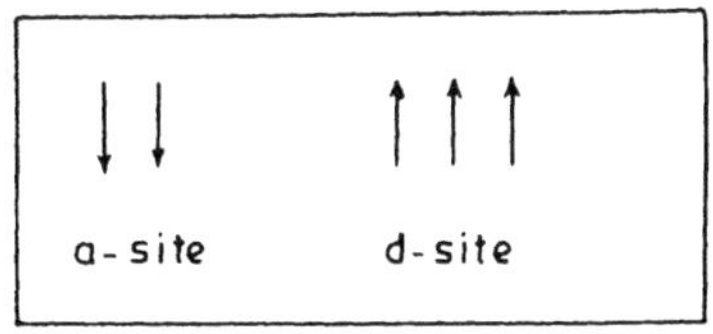

Figure A.1. Relative orientation of magnetic dipoles a- and d-sites.

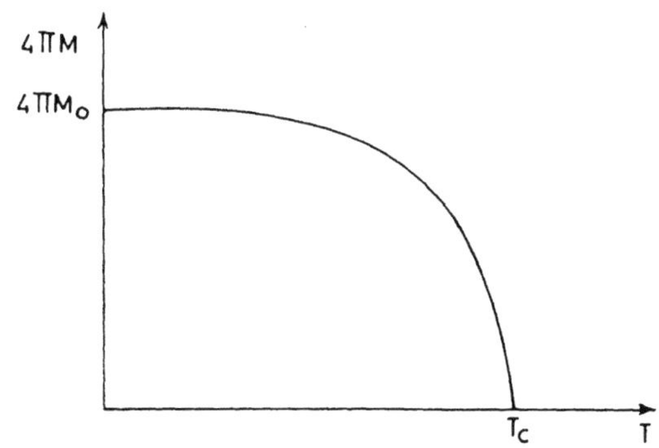

Figure A.2. Variation of magnetization with temperatures when c-sublattice magnetization is zero or small.

Next, we consider the case of a pure rare-earth iron garnet (RIG). Rare-earth ions are usually large and, therefore, they invariably go to c-sites. Since the rare-earth ions are magnetic, the c-sublattice also contributes to spontaneous magnetization of RIG. It is found that the d–c, a–c, and c–c interactions are all antiferromagnetic in character. The d–c interaction is the strongest among these, though it is usually much weaker than the a–d interaction (Lax and Button, 1962). As a result, the c-sublattice magnetization is oriented antiparallel to the d-sublattice magnetization (Figure A.3). It is also evident from the figure that the net magnetization would be the difference between the c-sublattice magnetization and the resultant of the a–d sublattice magnetizations. If the c-sublattice magnetization is less than that of the resultant of the a- and d-sublattices, such as in the case of EuIG, the ferric ion magnetic moment is dominant at all temperatures and the net magnetization decreases monotonically as the temperature increases. However, most of the rare-earth ions (e.g., gadolinium) have relatively large magnetic dipole moments due to which the rare-earth ion magnetic moment is dominant at low temperatures; the spontaneous magnetization is parallel to the magnetization of the c-sublattice. When the temperature is raised, the c-sublattice magnetization decreases much more rapidly than the net a–d magnetization.* At a certain temperature called the compensation point (T_{comp}), the

* This is so because the a–d interaction is much stronger than the d–c or a–c interaction.

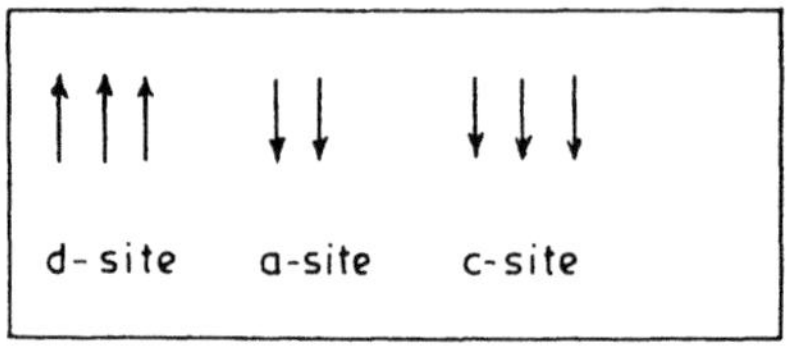

Figure A.3. Relative orientation of magnetic dipoles in a-, d-, and c-sublattices.

contributions from c- and a–d sublattices cancel out and zero magnetization results. With further increase in the temperature, the ferric ion moment dominates; the magnetization initially increases, attains a certain maximum, and once again starts decreasing, finally rendering the garnet paramagnetic at T_c (Figure A.4).

A.1.2.2. *Mixed Garnets*

In mixed garnets, the c-sites contain more than one type of rare-earth ion. In general, the c-sublattice magnetization of mixed garnets is almost equal to the weighted sum of the c-sublattice magnetizations of corresponding pure garnets. Thus, it is possible to control the compensation point (and room temperature magnetization) by appropriate mixing.

A.1.2.3. *Substituted Garnets*

Substituted garnets are obtained from pure garnets on replacing ferric ions in a- or d-sites by other trivalent ions, such as Al, Gd, etc. The substituted ion can be magnetic or nonmagnetic and may exhibit preference for either the a- or d-site. The magnetization depends on the degree of substitution in a complicated manner and can be greatly varied. The presence of foreign ions in a- or d-sites invariably leads to the weakening of the a–d interaction which, in turn, is responsible for the lowering of T_c.

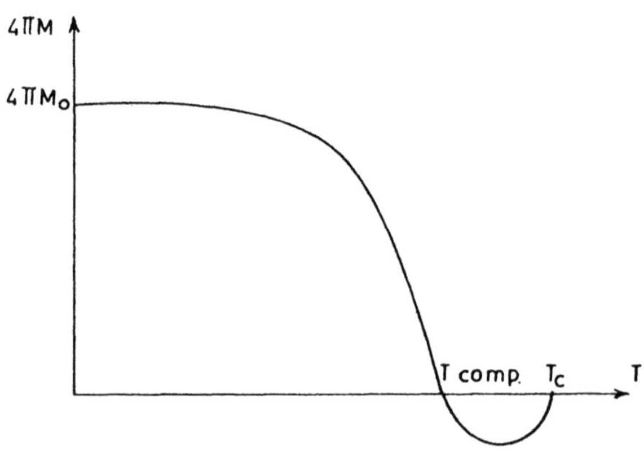

Figure A.4. Variation of magnetization with temperature for a typical RIG characterized by large c-sublattice magnetization.

A.1.3. *Magnetocrystalline Anisotropy*

The anisotropy energy for the cubic structure of garnet is given by equation (1.51). It is rather strongly dependent on temperature and decreases to zero at T_c. The first-order anisotropy constant k_1 is generally small and the second-order anisotropy constant k_2 is negligible. The anisotropy field ($2k_1/M_0$) for most garnets is around 50 Oe.

A.1.4. *Linewidth*

Garnets are low-loss materials at microwave frequencies and are characterized by narrow resonance linewidths. The X-band linewidth of polycrystalline YIG is about 50 Oe and depends mainly on porosity. Most other garnets have larger linewidths. A well-prepared single-crystal YIG has an extremely narrow linewidth (< 1 Oe), which can be further reduced by crystal surface polishing, since the mechanism responsible for the loss of microwave energy from a uniform-mode magnon to degenerate higher-order magnons is scattering due to surface imperfections.* Once again, the single crystals of garnets other than YIG have relatively large linewidths.

A.1.5. *Other Properties*

A ferrimagnetic garnet, specifically YIG, exhibits low losses for acoustic wave propagation at frequencies below 1 GHz. The acoustic loss factor for single-crystal YIG is about one tenth of that for quartz.

A.2. *Spinel Ferrites*

A.2.1. *Crystal Structure*

Spinel ferrites are the oldest known, and hence most extensively studied, ferrites (Smit and Wijn, 1959; Lax and Button, 1962; von Aulock, 1965; Tebble and Craik, 1969). The crystal structure of spinel ferrites is isomorphic with that of the classical spinel $MgAl_2O_4$. The unit cell is cubic with an edge length of approximately 8.4 Å for most spinels at room temperature. Thus, the unit cell of spinels is much smaller than that of garnets. It contains eight formula units of spinel, i.e., $8\,MFe_2O_4$, where M is a divalent metal cation. The 32 oxygen ions form a matrix which contains

* A well-grown single crystal of YIG is practically free from volume defects.

tetrahedral and octahedral sites for cations. There are 64 tetrahedral (a-) sites of which only 8 are occupied, while there are 32 octahedral (b-) sites of which only 16 are occupied. Thus, there is a lot of randomness in the distribution of cations which, incidently, contributes to the linewidth.

Oxygen ions are more tightly packed in spinels than in garnets. The exchange interaction is also stronger in spinels which results in: (i) their Curie temperature being higher and (ii) a variety of foreign ions being accepted without appreciably weakening the forces responsible for magnetic order, producing a large number of mixed and substituted spinels.

The metal ion in spinels is divalent while the ferric ions are trivalent. Hence, the charge transfer mechanism (Sparks, 1964) contributes to the electrical conductivity which is therefore not as low as for garnets.

Spinel ferrites can be classified according to the distribution of cations in a- and b-sites. In a "normal" spinel, the eight divalent metal ions go to a-sites while the 16 trivalent ferric ions occupy b-sites. In "inverted" spinels, the divalent ions prefer b-sites while the trivalent ferric ions are equally distributed in a- and b-sites. Also, many spinel ferrites are "partially inverted" in which divalent ions are distributed in some ratio among a- and b-sites. The degree of inversion depends on the nature of the ions and the quenching temperature (Smit and Wijn, 1959). Some examples of normal spinels are Mn and Zn ferrites while those of inverted spinels are Co, Fe, and Ni ferrites. Mg ferrites and many other spinels are partially inverted.

A.2.2. *Magnetization*

If the divalent ion M is nonmagnetic, the spontaneous magnetization is zero for a normal as well as an inverted spinel because the ferric ions, whether in a- or b-sublattices, are coupled by an antiferromagnetic interaction. In a partially inverted spinel, however, the trivalent ions are distributed unequally among the a- and b-sites and hence a small magnetization results.

When M is magnetic and occupies b-sites (i.e., it forms an inverted spinel, which is the case with most spinel ferrites), the ferric ions are equally distributed among the a- and b-sites and, therefore, the ferric ion moments do not contribute to spontaneous magnetization in completely inverted spinels. In this case, the magnetization results from the divalent metal ions. This can be compared with the case of garnets for which the c-sublattice moment is directed opposite to the nonzero ferric ion moment which makes the resultant magnetization of garnets, in general, smaller than that of pure spinel ferrites.

When some of the divalent ions are replaced by other divalent ions, a mixed spinel is formed. The prediction of the magnetization of mixed ferrites is rather difficult, since different divalent ions have different "affinities" for a- and b-sites. In general, spontaneous magnetization and the Néel temperature can be greatly varied by suitable mixing.

It is also possible to replace some ferric ions in a spinel by other trivalent ions, such as Al^{3+}, and form substituted spinels. Once again, the magnetization and Néel temperature can be controlled by appropriate substitution.

A.2.3. *Magnetocrystalline Anisotropy*

The magnetocrystalline anisotropy energy for the cubic structure of spinel ferrites has the same form as for garnets. The first-order anisotropy constant k_1 is, in general, larger for spinels than for garnets. Moreover, in some cases, e.g., Fe-rich Ni and Mn ferrites, the second-order anisotropy constant is also comparable to k_1.

A.2.4. *Linewidth*

In polycrystalline or single-crystal form, the resonance linewidths of most spinel ferrites are much larger than the corresponding linewidths of garnets. There are only a few exceptions, such as Li ferrite, the single crystals of which can have linewidths as low as 1 Oe. One of the reasons for the larger linewidth of spinels is the randomness in cation distribution. Moreover, the presence of divalent ions also enhances the linewidth on account of the charge transfer mechanism (Sparks, 1964). The volume defect density is also relatively high in spinels.

A.3. *Hexagonal Magnetoplumbites*

Hexagonal magnetoplumbites have extremely complicated structures (Smit and Wijn, 1959; Lax and Button, 1962; von Aulock, 1965) characterized by hexagonal symmetry. As discussed in Chapter 1, the main advantage of these materials in microwave devices is their large magnetocrystalline anisotropy (uniaxial or planar), the expression for which is given by equation (1.52). Hexagonal magnetoplumbites are found to have different chemical compositions and hence different structures, which are briefly discussed below.

A.3.1. *M-Type*

Ferroxdure ($Ba^{2+}Fe^{3+}_{12}O_{19}$) is a typical M-type hexagonal magnetoplumbite. This and other M-type materials are uniaxial. Ba ions can be partially (totally) replaced by Pb^{2+}, Sr^{2+}, and other divalent ions to form mixed (pure) magnetoplumbites. Moreover, the trivalent Fe ion can be replaced by other trivalent ions (e.g., Al^{3+}, Cr^{3+}, etc.) to prepare substituted M-type magnetoplumbites. The saturation magnetization of pure M-type materials is large (~ 5 kG), which is, however, reduced in the case of substitution for ferric ions. The Néel temperature is large (~ 700 °K). Finally, the uniaxial anisotropy field is normally extremely large (> 10 kG).

A.3.2. *W-Type*

These materials have the chemical formula $M_2BaFe_{16}O_{27}$, and are often represented as M_2W, where $W = BaFe_{16}O_{27}$, and M is a divalent metal ion, such as Fe, Zn, Ni, etc. The W-type hexagonal magnetoplumbites generally have large spontaneous magnetization (> 4 kG) and a high Néel temperature (~ 700 °K), which is usually reduced on substitution. Most of the W-type materials are uniaxial and have rather large anisotropy fields (> 10 kG). However, the replacement of an appropriate fraction of divalent ions by Co can cause planar anisotropy (von Aulock, 1965).

A.3.3. *Y-Type*

The hexagonal magnetoplumbites having the chemical formula $M_2Ba_2Fe_{12}O_{22}$ are called Y-type materials and are represented as M_2Y, where Y is a divalent ion.* The saturation magnetization of Y-type magnetoplumbites is intermediate (~ 2 kG) and the Curie temperature is often in the range 450–600 °K. The distinguishing feature of Y-type hexagonal ferrimagnetics is their large planar anisotropy; the maximum observed planar anisotropy field is about 40 kG in the case of properly oriented Co_2Y (Braden et al., 1966). The linewidths of well-prepared crystals of Y-type materials are generally lower than those of other hexagonal magnetoplumbites. Specifically, Zn_2Y has been found to have a very narrow linewidth (< 10 Oe), at the X-band, in single-crystal form (Stinson and Green, 1965).

A.3.4. *Z-Type*

The Z-type hexagonal ferrimagnetics have a chemical formula M_2Z, where $Z = Ba_3Fe_{24}O_{41}$. Their magnetization at room temperature normally exceeds 3 kG while the Curie temperature is about 650 °K. The Z-type materials are uniaxial with the exception of Co_2Z, which has planar anisotropy.

* Here Y represents the group $Ba_2Fe_{12}O_{22}$ and should not be confused with yttrium.

Table A.1. Ranges of Electrical Parameters of Ferrimagnetics

Parameter	Approximate range
Resistivity	$10^6\,\Omega\,m$–$10^{12}\,\Omega\,m$
Dielectric constant at microwave frequencies	9–15
Dielectric loss factor ($\tan\delta$)	0.00025–0.001

A.3.5. *U-Type*

The U-type hexagonal magnetoplumbites have a chemical formula M_2U, where $U = Ba_4Fe_{36}O_{60}$. These materials have a high Curie temperature and a moderately large uniaxial anisotropy field.

Table A.1 summarizes the approximate ranges of the basic electrical parameters of typical ferrimagnetics. The magnetic properties of the various microwave ferrimagnetics are available in the literature; see, for example, Lax and Button (1962), von Aulock (1965), Tebble and Craik (1969), Smit (1971), Riches (1972), Standley (1972), and Heck (1974).

APPENDIX B

Magnetic Moment of Atoms and Ions

To understand the magnetic properties of atoms and ions one requires some knowledge of quantum mechanics. We shall only summarize useful results, and the reader interested in details is referred to White (1934), Shore and Menzel (1968), and Schiff (1968), etc.

The concepts of orbital and spin angular momenta were introduced in Chapter 1. In order to relate the net angular momentum and magnetic moment of an atom (or ion) to the angular momenta of the constituent electrons, we must know the quantum state of the atom which, in turn, is obtained from the solution of the Schrödinger equation (Schiff, 1968).

In the case of the spherically symmetric potential experienced by a single electron,* it is found that the orbital angular momentum $\mathbf{l}$ has an absolute magnitude $(l(l+1))^{1/2}\hbar$, where $\hbar$ is Plank's constant divided by 2π and $l = 0, 1, 2, \ldots, (n-1)$, where n is the principal quantum number (White, 1934). The state $l = 0$ corresponds to zero orbital angular momentum and hence zero (orbital) moment, and this is called the s-state. The states with $l = 1, 2$, and 3 are called respectively p-, d-, and f-states.

It also follows from quantum mechanics (Schiff, 1968) that the projection of $\mathbf{l}$ on the external magnetic field is quantized. If the magnetic field is directed along the z-axis then $l_z = m_l\hbar$, where m_l can assume $(2l+1)$ values given by $m_l = -l, -(l+1), \ldots, (l-1), l$. As a consequence of this quantization, the orbital magnetic moment (which is related to the orbital angular momentum through the gyromagnetic ratio, as discussed in Chapter 1) is also quantized. The maximum observable orbital magnetic

* Apart from the H atom, $He^+, Li^+, Be^{3+}, \ldots$ etc., are also one-electron ions. Under certain approximations, alkali metal atoms, namely, Na, K, Rb, etc., also behave like one-electron atoms.

moment is therefore given by

$$\mu_l = \gamma_l l\hbar \equiv \mu_B l \tag{B.1}$$

where $\gamma_l = e/2mc_0$ is the orbital gyromagnetic ratio and $\mu_B = \gamma_l\hbar$ is the so-called "Bohr magneton," which is numerically equal to 9.21×10^{-21} erg/Oe and is the unit of atomic magnetic dipole moments.

As noted in Chapter 1, the electron possesses an intrinsic angular momentum called spin in addition to the usual orbital angular momentum. The spin $\mathcal{S}$ has a maximum observable value $\mathcal{S}\hbar$, where $\mathcal{S} = 1/2$. The projection of the spin angular momentum on the magnetic field axis is quantized; $\mathcal{S}_z = m_s\hbar$, where $m_s = +1/2$ or $-1/2$. It also follows from quantum mechanics that the spin gyromagnetic ratio γ_s is just the double of the orbital gyromagnetic ratio γ_l. Accordingly, the maximum observable spin magnetic moment corresponding to a spin $\mathcal{S}$ is, therefore, given by

$$\mu_s = \gamma_{\mathcal{S}}\mathcal{S}\hbar = 2\mu_B\mathcal{S} \tag{B.2}$$

It is now clear that a one-electron atom is characterized by quantum numbers n, l, m_l, and $m_{\mathcal{S}}$.

A multielectron atom can be treated under the self-consistent field approximation (Shore and Menzel, 1968), in which case the individual electrons are still described by the same quantum numbers, n, l, m_l, and $m_{\mathcal{S}}$, as for the one-electron atom. However, the filling of various shells and orbitals* is restricted by Pauli's exclusion principle, according to which no two electrons can have same quantum state,† i.e., have identical sets of quantum numbers. Therefore, the total number of electrons that can occupy a shell is the same as the total number of different quantum states (for that n), namely

$$\sum_{l=0}^{(n-1)} 2(2l+1) = 2n^2 \tag{B.3}$$

The scheme of the filling of shells has been discussed by White (1934), Shore and Menzel (1968), etc. For a complete description of the quantum state of a shell in a multielectron atom, one must specify (i) electronic configuration, (ii) total orbital angular momentum **L**, and (iii) total spin **S**.‡

* The electron states corresponding to a given n are said to belong to a shell. The shells with $n = 1, 2, 3, \ldots$ are, respectively, called K-, L-, M-, ... shells. On the other hand, the electron states corresponding to a given l (n prespecified) are said to belong to an orbital. The orbitals with $l = 0, 1$, and 2 are called s-, p-, and d-orbitals.

† This statement is true for all Fermi particles (Schiff, 1968).

‡ The scheme that we are now describing is only an approximation, called the Russell–Saunders or L–S coupling scheme. Another approximation is the so-called j–j coupling, in which the orbital and spin angular momenta are first added to get the net

If we have two electrons with orbital quantum numbers l_1 and l_2, the total orbital quantum number L can assume values from $(l_1 + l_2)$ to $|l_1 - l_2|$ in decreasing steps of unity (White, 1934). When there are three or more electrons, L can be obtained by the process of pairwise addition. In an external field, the projections of **L** in the unit of $\hbar$ are from $+L$ to $-L$, decreasing in steps of unity. Similar coupling and projection rules hold for the total spin S. The total angular momentum **J** of the shell is equal to the vector sum of **L** and **S**:

$$\mathbf{J} = \mathbf{L} + \mathbf{S} \tag{B.4}$$

The quantum number J can assume values from $(L + S)$ to $(L - S)$ in steps of unity. The projection of **J** on the external field is also quantized:

$$m_J = J, (J-1), \ldots, (-J+1), J$$

The total observable magnetic moment depends on J through the g-factor as

$$\mu = gJ\mu_B \tag{B.5}$$

where $g = 1$ for pure orbital motion $(S = 0)$ and $g = 2$ for pure spin $(L = 0)$. When L and S are both different from zero, the g-factor is influenced by the strength of spin–orbit interaction and is given by (Wagner, 1972; Shore and Menzel, 1968)

$$g = 1 + \frac{J(J+1) + S(S+1) - L(L+1)}{2J(J+1)} \tag{B.6}$$

It is evident from the preceding discussion that for a given electron shell, L, S, and J can assume several sets of values. Due to the various interactions, the energies corresponding to different sets are in general different. The ground state is determined from the empirical rules, called Hund's rules (White, 1934; Shore and Menzel, 1968), according to which: (i) the total spin S assumes the maximum value permitted by the exclusion principle, (ii) for this S, the total orbital quantum number L is maximum, and (iii) $J = L + S$ or $|L - S|$ according to whether the orbital is more or less than half filled. It is worth noting that $L = 0$ and $S = 0$ for a completely

angular momentum $\mathbf{j}_i$ of the ith electron. This is followed by the vector addition of all the **j**'s of the various electrons to obtain the total angular momentum $\mathbf{J} = \Sigma_i \mathbf{j}_i$. The L–S coupling is approximately valid when the electrostatic interaction between electrons is much stronger than the spin–orbit interaction, while the j–j coupling is applicable to the reverse situation (White, 1934).

filled shell. Accordingly, $J = 0$ and hence a completely filled shell does not contribute to the magnetic moment.

To facilitate the discussion of specific examples, the general scheme of filling of electron shells is shown in Table B.1. The electronic configuration of a given atom can be found by filling various orbitals and shells in the order shown in the table. While the filling procedure up to 3p-states ($n = 3, l = 1$) is in perfect order, it is found that the 4s-states ($n = 4, l = 0$) are filled before the 3d-states ($n = 3, l = 2$); this happens because the energy of the 4s-states is lower than that of the 3d-states. This effect is more pronounced for atoms with a still larger number of electrons. For example, in rare-earth atoms, the 5s-, 5p-, and 6s-states are filled before the 4f-states.

Now consider the case of an iron atom which has 26 electrons in all. According to the preceding discussion, 18 electrons occupy the states up to the 3p-states. After this, two electrons fill the 4s-states. The remaining 6 electrons occupy the 3d-states. It is evident that for the neutral Fe atom, only 3d-electrons ($l = 2$) contribute to the angular momentum and hence to the magnetic moment. According to Table B.1, the 3d-orbital can accommodate, in all, 10 electrons. Therefore, a maximum of 5 electrons can have their respective spins oriented parallel ($m_S = 1/2$), whereas the spin of the remaining electron should be oriented antiparallel ($m_S = -1/2$), which makes $S = 2$. For this S, Hund's rule (ii) requires that $L = 2$ for $(\Sigma_l m_l)_{\max} = 2$. Since the orbital (3d) is more than half filled, $J = L + S = 4$. It is found from equation (B.6) that $g = 3/2$. Thus, the magnetic moment $\mu = 3/2 \times 4\mu_B = 6\mu_B$. In most cases of interest, however, the iron atom or ion would be subjected to the crystalline fields, in which case the orbital motion contributes insignificantly (Appendix A) to the magnetic moment.

Table B.1

Principal quantum number n	Orbital quantum number l	Orbital magnetic quantum number m_l	Spin magnetic quantum number m_s	Maximum number of electrons	
1	0	0	$\pm 1/2$	2	
2	0	0	$\pm 1/2$	2	8
	1	$0, \pm 1$	$\pm 1/2$	6	
3	0	0	$\pm 1/2$	2	18
	1	$0, \pm 1$	$\pm 1/2$	6	
	2	$0, \pm 1, \pm 2$	$\pm 1/2$	10	
4	0	0	$\pm 1/2$	2	32
	1	$0, \pm 1$	$\pm 1/2$	6	
	2	$0, \pm 1, \pm 2$	$\pm 1/2$	10	
	3	$0, \pm 1, \pm 2, \pm 3$	$\pm 1/2$	14	

In this case $g = 2$ and $J = S = 2$, implying that $\mu = 4\mu_B$. For Fe^{2+} ions, the situation is not different, because the two removed electrons are the 4s-electrons, which would not contribute to the magnetic moment anyway. For ferric ions (Fe^{3+}), there are only five 3d-electrons, the spins of all of which are parallel in the ground state, i.e., $S = 5/2$ and hence $\mu = 5\mu_B$.

We can now discuss the magnetic moment of magnetite. One formula unit of magnetite (Fe_3O_4) contains two ferric ions and one ferrous ion apart from four nonmagnetic oxygen ions. Had the magnetic moment of all iron ions been parallel, the net magnetic moment per formula unit would have been $4\mu_B + 2 \times 5\mu_B = 14\mu_B$. However, the observed magnetic moment per formula unit is only $4\mu_B$, which can be explained only by assuming the existence of oppositely magnetized sublattices and inverted spinal structure (Appendix A), whereby the ferric ion contribution cancels out. In fact, this was one of the main arguments in support of the sublattice model of ferrimagnets in the past (Lax and Button, 1962).

APPENDIX C

Coordinate Transformations

C.1. *Eulerian Angles and Rotation Matrices*

The study of wave propagation in a given configuration becomes more convenient by making a transformation to a coordinate system which suits the geometry of the configuration. Figure C.1 elucidates this transformation. The new primed coordinate system is obtained from the unprimed system by carrying out rotations in the following sequence:

(i) a rotation by angle φ about the z-axis,
(ii) a rotation by angle θ about the transformed x-axis, and
(iii) a rotation by angle ψ about the transformed z-axis, i.e., the z'-axis.

The angles (φ, θ, ψ) are called Eulerian angles.* It is evident from Figure C.1 that, as a consequence of the three Eulerian rotations, the axes of the primed coordinate system are arbitrarily oriented with respect to those of the unprimed system. In what follows, the aforementioned rotations will be represented by matrices.

Refer to Figure C.2, which shows the first of the three Eulerian rotations. If the coordinates of a point P in the original (SY) and transformed (SY_1) coordinate system are (x, y, z) and (X_1, Y_1, Z_1), respectively, we have

$$X_1\hat{X}_1 + Y_1\hat{Y}_1 + Z_1\hat{Z}_1 = x\hat{x} + y\hat{y} + z\hat{z} \tag{C.1}$$

where $\hat{x}, \hat{y}, \hat{z}$ $(\hat{X}_1, \hat{Y}_1, \hat{Z}_1)$ are the basic unit vectors of the original (trans-

* See, for example, Goldstein (1950).

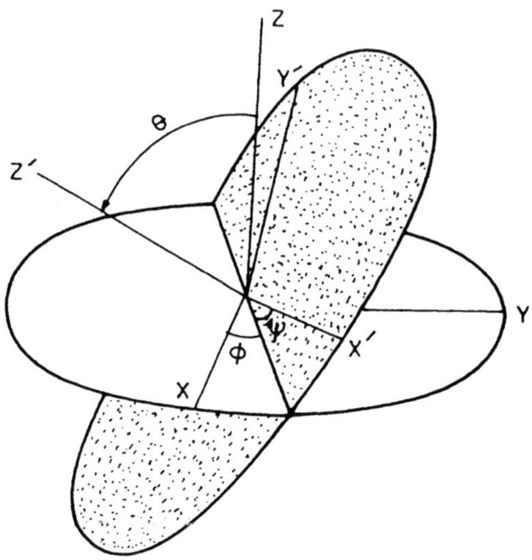

Figure C.1. The unprimed and primed coordinate systems. The relative orientation is characterized by the angles (φ, θ, ψ) which are called the Eulerian angles.

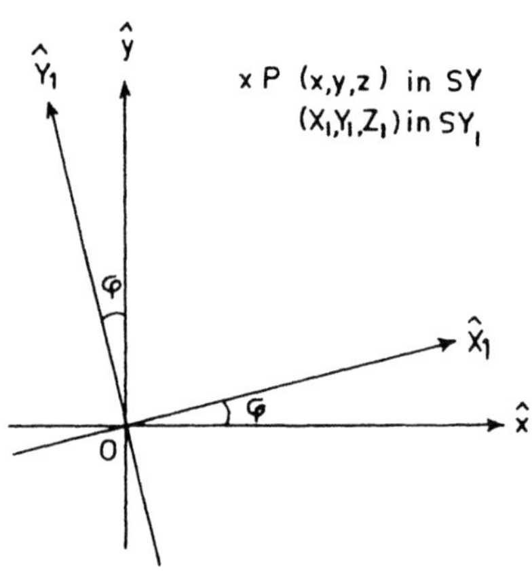

Figure C.2. The coordinate system SY_1 is obtained from the coordinate system SY by subjecting it to a rotation by angle φ about the z-axis.

formed) coordinate system. It follows from Figure C.2 that

$$\left.\begin{aligned}\hat{z} &= \hat{Z}_1\\ \hat{x} &= (\hat{x}\cdot\hat{X}_1)\hat{X}_1 + (\hat{x}\cdot\hat{Y}_1)\hat{Y}_1\\ &= \cos\varphi\hat{X}_1 - \sin\varphi\hat{Y}_1\\ \hat{y} &= (\hat{y}\cdot\hat{X}_1)\hat{X}_1 + (\hat{y}\cdot\hat{Y}_1)\hat{Y}_1\\ &= \sin\varphi\hat{X}_1 + \cos\varphi\hat{Y}_1\end{aligned}\right\}\qquad(C.2)$$

Substitution from equation (C.2) into equation (C.1) and comparison of the coefficients of $\hat{X}_1$, $\hat{Y}_1$, and $\hat{Z}_1$ on both sides yields

$$X_1 = x\cos\varphi + y\sin\varphi$$

$$Y_1 = -x\sin\varphi + y\cos\varphi$$

$$Z_1 = z$$

These equations can be expressed in matrix form as

$$\begin{pmatrix} X_1 \\ Y_1 \\ Z_1 \end{pmatrix} = \begin{pmatrix} \cos\varphi & \sin\varphi & 0 \\ -\sin\varphi & \cos\varphi & 0 \\ 0 & 0 & 1 \end{pmatrix} \begin{pmatrix} x \\ y \\ z \end{pmatrix} \tag{C.3}$$

Evidently, the 3×3 matrix R_1 on the right-hand side of this equation relates the coordinates of the same point P in coordinate systems SY and SY_1. It is seen that R_1 is an orthogonal matrix, i.e., $R_1\tilde{R}_1 = \tilde{R}_1 R_1 = 1$, where $\tilde{R}_1$ is the transpose of the matrix R_1, obtained by interchanging the rows and columns of R_1.

Next consider the second Eulerian rotation which is shown in Figure C.3. If (X_2, Y_2, Z_2) represent the coordinates of the point P in system SY_2, the aforementioned procedure can be followed to show that

$$\begin{pmatrix} X_2 \\ Y_2 \\ Z_2 \end{pmatrix} = \begin{pmatrix} 1 & 0 & 0 \\ 0 & \cos\theta & \sin\theta \\ 0 & -\sin\theta & \cos\theta \end{pmatrix} \begin{pmatrix} X_1 \\ Y_1 \\ Z_1 \end{pmatrix} \tag{C.4}$$

Finally, the third Eulerian rotation, which leads to the primed coordinate system of Figure C.1, can be represented as

$$\begin{pmatrix} x' \\ y' \\ z' \end{pmatrix} = \begin{pmatrix} \cos\psi & \sin\psi & 0 \\ -\sin\psi & \cos\psi & 0 \\ 0 & 0 & 1 \end{pmatrix} \begin{pmatrix} X_2 \\ Y_2 \\ Z_2 \end{pmatrix} \tag{C.5}$$

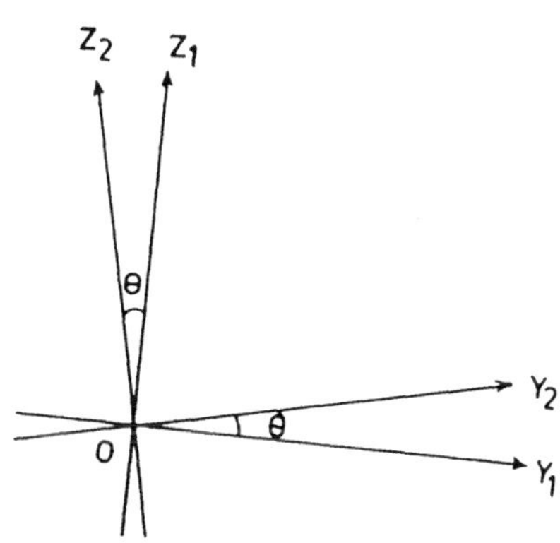

Figure C.3. The coordinate system SY_2 is obtained from the system SY_1 by subjecting it to a rotation by angle θ about the X_1-axis.

It follows from equations (C.3)–(C.5) that

$$\begin{pmatrix} x' \\ y' \\ z' \end{pmatrix} = \begin{pmatrix} R_{11} & R_{12} & R_{13} \\ R_{21} & R_{22} & R_{23} \\ R_{31} & R_{32} & R_{33} \end{pmatrix} \begin{pmatrix} x \\ y \\ z \end{pmatrix} \tag{C.6}$$

where the matrix $R = [R_{ij}]$ is given by

$$R = \begin{pmatrix} \cos\psi & \sin\psi & 0 \\ -\sin\psi & \cos\psi & 0 \\ 0 & 0 & 1 \end{pmatrix} \begin{pmatrix} 1 & 0 & 0 \\ 0 & \cos\theta & \sin\theta \\ 0 & -\sin\theta & \cos\theta \end{pmatrix} \begin{pmatrix} \cos\varphi & \sin\varphi & 0 \\ -\sin\varphi & \cos\varphi & 0 \\ 0 & 0 & 1 \end{pmatrix}$$

$$= \begin{pmatrix} \cos\psi\cos\varphi - \cos\theta\sin\varphi\sin\psi & \cos\psi\sin\varphi + \cos\theta\sin\psi\cos\varphi & \sin\psi\sin\theta \\ -\sin\psi\cos\varphi - \cos\theta\sin\varphi\cos\psi & -\sin\psi\sin\varphi + \cos\theta\cos\varphi\cos\psi & \cos\psi\sin\theta \\ \sin\theta\sin\varphi & -\sin\theta\cos\varphi & \cos\theta \end{pmatrix} \tag{C.7}$$

This matrix R, which is in fact the product of three orthogonal matrices, can also be shown to be orthogonal. Writing $x = x_1$, $y = x_2$, and $z = x_3$ and similarly for primed coordinates, we can express equation (C.6) as

$$x'_i = \sum_j R_{ij} x_j \tag{C.8}$$

where $i, j = 1, 2, 3$. The orthogonality condition $R\tilde{R} = \tilde{R}R = 1$ can be expressed as

$$\left.\begin{aligned} \sum_i \tilde{R}_{ki} R_{ij} &= \sum_i R_{ij} R_{ik} = \delta_{jk} \\ \sum_i R_{ki} \tilde{R}_{ij} &= \sum_i R_{ki} R_{ji} = \delta_{jk} \end{aligned}\right\} \tag{C.9}$$

where δ_{jk} is the Kronecker delta function, which is unity for $j = k$ and zero for $j \neq k$. Multiplying both sides of equation (C.8) by R_{ik}, summing over i and using equation (C.9), we see that

$$x_k = \sum_i R_{ik} x'_i$$

or, equivalently,

$$x_i = \sum_j R_{ji} x'_j \tag{C.10}$$

C.2. Transformation of Maxwell's Equations

It is noted that the transformation relations for the field vectors **E**, **b**, **h**, etc., are similar to those for **r**, i.e., analogous to equations (C.8) and (C.10). The second-rank tensor may be shown to transform under rotation according to the following matrix equation:

$$\boldsymbol{\mu}' = R\boldsymbol{\mu}\tilde{R} \tag{C.11}$$

Equivalently, in terms of the matrix elements, we have*

$$\mu'_{ij} = R_{il}\mu_{lm}\tilde{R}_{mj} = R_{il}R_{jm}\mu_{lm}$$

This follows since

$$\begin{aligned} b'_i &= R_{il}b_l \\ &= R_{il}\mu_{lm}h_m \qquad (\text{since } \mathbf{b} = \boldsymbol{\mu}\cdot\mathbf{h}) \\ &= R_{il}\mu_{lm}R_{jm}h'_j \equiv \mu'_{ij}h'_j \end{aligned}$$

The third rank, antisymmetric unimodular tensor ε_{ijk} (the Levi-Civita tensor[†]) transforms as

$$\begin{aligned} \varepsilon'_{pqr} &= R_{pi}R_{qj}R_{rk}\varepsilon_{ijk} \\ &= \varepsilon_{pqr} \end{aligned} \tag{C.12}$$

Now consider the transformation of Maxwell's equations under coordinate rotation. Maxwell's equations for a neutral, nonconducting and dielectrically isotropic ferrimagnetic material can be rewritten from Section 2.1 as

$$\left.\begin{aligned} \nabla\cdot\mathbf{b} &= 0 \\ \nabla\cdot\mathbf{D} &= 0 \\ \nabla\times\mathbf{E} &= -\frac{1}{c_0}\frac{\partial\mathbf{b}}{\partial t} \\ \nabla\times\mathbf{h} &= \frac{\varepsilon_f}{c_0}\frac{\partial\mathbf{E}}{\partial t} \end{aligned}\right\} \tag{C.13}$$

* Unless otherwise specified, summation over repeated suffixes is implied.

† See Jeffreys (1965) for an introduction to the Levi–Civita tensor.

where

$$\mathbf{b} = \boldsymbol{\mu} \cdot \mathbf{h}$$

Consider the third of these equations. The lth component can be expressed as

$$\varepsilon_{lpq} \frac{\partial E_q}{\partial x_p} = -\frac{1}{c_0} \frac{\partial}{\partial t} (\mu_{lm} h_m)$$

Transforming to primed coordinates, we have

$$\varepsilon_{lpq} R_{rp} R_{sq} \frac{\partial}{\partial x'_r} E'_s = -\frac{1}{c_0} R_{il} R_{jm} R_{nm} \frac{\partial}{\partial t} (\mu'_{ij} h'_n)$$

If we multiply both sides by $R_{i'l}$, sum over l, and use the relations (C.9), this transforms to

$$\varepsilon_{lpq} R_{i'l} R_{rp} R_{sq} \frac{\partial E'_s}{\partial x'_r} = -\frac{1}{c_0} \delta_{i'i} \delta_{jn} \frac{\partial}{\partial t} (\mu'_{ij} h'_n)$$

which, on using equation (C.12), reduces to

$$\varepsilon'_{krs} \frac{\partial}{\partial x'_r} E'_s = -\frac{1}{c_0} \frac{\partial}{\partial t} (\mu'_{kj} h'_j)$$

This equation can be expressed in vector notation as

$$\nabla' \times \mathbf{E}' = -\frac{1}{c_0} \frac{\partial}{\partial t} (\boldsymbol{\mu}' \cdot \mathbf{h}') \tag{C.14}$$

Similarly, the remaining equations in equation (C.13) can be expressed in a primed coordinate system as

$$\left.\begin{aligned} \nabla' \times \mathbf{h}' &= \frac{\varepsilon_f}{c_0} \frac{\partial E'}{\partial t} \\ \nabla' \cdot \mathbf{b}' &= 0 \\ \nabla' \cdot \mathbf{D}' &= 0 \end{aligned}\right\} \tag{C.15}$$

Evidently, Maxwell's equations are form-invariant under rotation; however, the permeability tensor $\boldsymbol{\mu}$ is transformed to $\boldsymbol{\mu}'$, which is given by equation (C.11).

C.3. The Transformed Permeability Tensor

The transformed permeability tensor* can be obtained from equation (C.11) on substituting for R from equation (C.7) and the appropriate form of $\boldsymbol{\mu}$ from Chapter 1. However, in most cases of practical interest, it is sufficient to reorient only one of the axes in a preferred direction, thus leaving the orientation of the remaining two axes arbitrary. To achieve this, it is sufficient to perform only two, instead of three, independent rotations; these can be the first two Eulerian rotations. Thus, substituting $\psi = 0$ in equation (C.7), the rotation matrix R reduces to

$$R = \begin{pmatrix} \cos\varphi & \sin\varphi & 0 \\ -\cos\theta\sin\varphi & \cos\theta\cos\varphi & \sin\theta \\ \sin\theta\sin\varphi & -\sin\theta\cos\varphi & \cos\theta \end{pmatrix} \tag{C.16}$$

where $-\pi \leqslant \varphi \leqslant \pi$ and $-\pi/2 \leqslant \theta \leqslant \pi/2$.† The transformed permeability tensor can be obtained from equations (C.11) and (C.16) as (Srivastava, 1977):

$$\boldsymbol{\mu}' = \begin{pmatrix} \mu'_{11} & \mu'_{12} & \mu'_{13} \\ \mu'_{21} & \mu'_{22} & \mu'_{23} \\ \mu'_{31} & \mu'_{32} & \mu'_{33} \end{pmatrix} \tag{C.17}$$

where

$$\left.\begin{aligned}
\mu'_{11} &= \mu' = \mu_{xx}\cos^2\varphi + \mu_{yy}\sin^2\varphi \\
\mu'_{12} &= [(\mu_{yy} - \mu_{xx})\sin\varphi\cos\varphi + j\kappa]\cos\theta \\
\mu'_{13} &= [(\mu_{xx} - \mu_{yy})\sin\varphi\cos\varphi - j\kappa]\sin\theta \\
\mu'_{21} &= [(\mu_{yy} - \mu_{xx})\sin\varphi\cos\varphi - j\kappa]\cos\theta \\
\mu'_{22} &= \sin^2\theta + \mu''\cos^2\theta \\
\mu'_{23} &= \mu'_{32} = (1 - \mu'')\sin\theta\cos\theta \\
\mu'_{31} &= [(\mu_{xx} - \mu_{yy})\sin\varphi\cos\varphi + j\kappa]\sin\theta \\
\mu'_{33} &= \cos^2\theta + \mu''\sin^2\theta \\
\mu'' &= \mu_{xx}\sin^2\varphi + \mu_{yy}\cos^2\varphi
\end{aligned}\right\} \tag{C.18}$$

* The transformed permeability tensor was first obtained by Tyres (1959).

† In the usual definition of Eulerian angles (Goldstein, 1950), the ranges of φ and θ are from 0 to 2π and 0 to π, respectively. We have chosen the range of φ as from $-\pi$ to $+\pi$ in order to have symmetry about $\varphi = 0$. The same is true for the range of θ. However, when θ lies between $-\pi/2$ and $+\pi/2$, the z'-axis cannot be oriented in an arbitrary direction relative to the biasing field. Nevertheless, the y'-axis can still be arbitrarily oriented by a proper choice of φ and θ in these ranges. Hence, we shall consider wave propagation along the y'-axis (see Chapter 2, Part A). Note that negative angles correspond to the motion of a left-handed screw.

C.4. Transformation of Anisotropy Field*

The anisotropy field for a cubic ferrite in the unprimed coordinate system is given by equation (1.56) of Chapter 1; it can be rewritten as

$$\mathbf{H}_A = -\frac{2k_1}{M_0}\left(\sum_i \alpha_i \hat{x}_i - \mathbf{P}\right)$$
$$\mathbf{P} = \sum_i \alpha_i^3 \hat{x}_i \tag{C.19}$$

It is required to transform this to a primed coordinate system. It is seen that the unit vectors and the components of $\boldsymbol{\alpha}$ transform as

$$\hat{x}'_s = \sum_j R_{sj}\hat{x}_j, \qquad \hat{x}_j = \sum_s R_{sj}\hat{x}'_s$$
$$\alpha'_p = \sum_i R_{pi}\alpha_i, \qquad \alpha_i = \sum_p R_{pi}\alpha'_p \tag{C.20}$$

It is now easy to show that

$$\sum_i \alpha_i \hat{x}_i = \sum_i \alpha'_i \hat{x}'_i \tag{C.21}$$

whereas

$$\mathbf{P} \equiv \sum_i \alpha_i^3 \hat{x}_i = \sum_{p,q,r,s} T_{pqrs}\alpha'_p\alpha'_q\alpha'_r\hat{x}'_s \tag{C.22}$$

where

$$T_{pqrs} = \sum_i R_{pi}R_{qi}R_{ri}R_{si} \tag{C.23}$$

Since each of p, q, r, and s can assume values from 1 to 3, there will be 81 terms in the right-hand side of equation (C.22). However, several of these terms are identical. For example, the term $\alpha'_1\alpha'_2\alpha'_3\hat{x}'_3$ will occur six times† with the same value of T_{pqrs}. Thus, the coefficient of the term $6\alpha'_1\alpha'_2\alpha'_3\hat{x}'_3$ is obtained from equation (C.23) as

* The only difference between the present treatment and that of Bajpai et al. (1979) is that the primed and unprimed coordinates are interchanged.

† The combinations (p, q, r) which will lead to this term are $(1, 2, 3)$, $(1, 3, 2)$, $(2, 3, 1)$, $(2, 1, 3)$, $(3, 1, 2)$, and $(3, 2, 1)$.

$$\sum_i R_{12}R_{2i}R_{3i}^2 = R_{11}R_{21}R_{31}^2 + R_{12}R_{22}R_{32}^2 + R_{13}R_{23}R_{33}^2$$

Substituting for R_{ij} from equation (C.16), this coefficient is seen to be equal to $\frac{1}{8}\sin\theta\sin 2\theta\sin 4\varphi$. As another example, it is seen that the term $\alpha_3'^3\hat{x}_1'$ will occur only once in the right-hand side of equation (C.22), for $p = q = r = 3$ and $s = 1$. The coefficient of this term is then equal to

$$\sum_i R_{3i}^3 R_{1i} = \cos^4\theta + (\sin^4\varphi + \cos^4\varphi)\sin^4\theta$$

Similarly, the coefficients of all the terms in equation (C.22) can be obtained. In the linear approximation, we have $\alpha_3' \simeq 1$ whereas α_1' and α_2' are small enough to neglect their products, second and higher powers.* Thus, **P** is obtained as follows (Bajpai et al., 1979):

$$\left.\begin{aligned} \mathbf{P} = & -\tfrac{1}{4}\sin^3\theta\sin 4\varphi(\hat{x}_1' + 3\alpha_1'\hat{x}_3') + \tfrac{3}{2}\sin^2\theta\sin^2 2\varphi\,\alpha_1'\hat{x}_1' \\ & + \tfrac{3}{8}\sin\theta\sin 2\theta\sin 4\varphi\,(\alpha_1'\hat{x}_2' + \alpha_2'\hat{x}_1') \\ & + [\cos^3\theta\sin\theta - \sin^3\theta\cos\theta(\sin^4\varphi + \cos^4\varphi)](\hat{x}_2' + 3\alpha_2'\hat{x}_3') \\ & + 3\sin^2\theta\cos^2\theta(1 + \sin^4\varphi + \cos^4\varphi)\alpha_2'\hat{x}_2' \\ & + [\cos^4\theta + (\sin^4\varphi + \cos^4\varphi)\sin^4\theta]\hat{x}_3' \end{aligned}\right\} \tag{C.24}$$

Substitution from equations (C.21), (C.22), and (C.24) into (C.19) yields the expression for the anisotropy field in the primed coordinate system.

C.5. Transformation of Magnetoelastic Interaction Energy†

The magnetoelastic interaction energy for a cubic ferrite is given by equation (1.108) of Chapter 1. It can be rewritten as

$$U_{me} = b_2 \sum_{i,j} \alpha_i\alpha_j S_{ij} + (b_1 - b_2)\sum_i \alpha_i^2 S_{ii} \tag{C.25}$$

It is seen that the first term is invariant under transformation to primed coordinates, whereas the second term is not. Specifically:

* In the examples discussed here, the term $\alpha_1'\alpha_2'\alpha_3'\hat{x}_3' \simeq \alpha_1'\alpha_2'\hat{x}_3'$ is neglected whereas the term $\alpha_3'^3\hat{x}_1' \simeq \hat{x}_1'$ is retained.

† This section is based on a paper by Rattan and Srivastava (1978).

$$\sum_{i} \alpha_i \alpha_j S_{ij} = \sum_{i,j} \alpha'_i \alpha'_j S'_{ij} \tag{C.26}$$

$$\sum_{i} \alpha_i^2 S_{ii} = \sum_{p,q,r,s} T_{pqrs} \alpha'_p \alpha'_q S'_{rs} \tag{C.27}$$

where the coefficients T_{pqrs} are defined by equations (C.23) and can be evaluated by the procedure described in Section C.4. Retaining only those terms which contribute first-order terms to $\mathbf{H}_{me}$ and $\mathbf{T}^{me}$, we can express the magnetoelastic interaction energy in primed coordinates as

$$\begin{aligned} U_{me} = {} & b_2(\alpha_3'^2 S_3' + \alpha_2'\alpha_3' S_4' + \alpha_3'\alpha_1' S_5') + (b_1 - b_2)[T_{1133}(\alpha_3'^2 S_1' + 2\alpha_1'\alpha_3' S_5') \\ & + 2T_{1123}(\alpha_2'\alpha_3' S_1' + \alpha_1'\alpha_3' S_6') + 2T_{1113}\alpha_1'\alpha_3' S_1' \\ & + T_{2233}(\alpha_3'^2 S_2' + 2\alpha_2'\alpha_3' S_4') + 2T_{2223}\alpha_2'\alpha_3' S_2' + 2T_{2213}[\alpha_1'\alpha_3' S_2' + \alpha_2'\alpha_3' S_6') \\ & + T_{3333}\alpha_3'^2 S_3' + T_{3312}(\alpha_3'^2 S_6' + 2\alpha_3'\alpha_1' S_4' + 2\alpha_2'\alpha_3' S_5') \\ & + T_{3323}(\alpha_3'^2 S_4' + 2\alpha_2'\alpha_3' S_3') + T_{3313}(\alpha_3'^2 S_5' + 2\alpha_1'\alpha_3' S_3')] \end{aligned} \tag{C.28}$$

where the coefficients T_{pqrs} are obtained as follows:

$$\left.\begin{aligned} T_{1133} &= \tfrac{1}{2}\sin^2\theta \sin^2 2\varphi \\ T_{1123} &= -\tfrac{1}{4}\sin 2\theta \sin^2 2\varphi \\ T_{1113} &= \tfrac{1}{4}\sin\theta \sin 4\varphi \\ T_{2233} &= \tfrac{1}{4}\sin^2 2\theta\,(1 + \sin^4\varphi + \cos^4\varphi) \\ T_{2223} &= \tfrac{1}{2}\sin 2\theta[\sin^2\theta - (\sin^4\varphi + \cos^4\varphi)\cos^2\theta] \\ \mathrm{T}_{2213} &= -\tfrac{1}{8}\cos\theta \sin 2\theta \sin 4\varphi \\ T_{3312} &= \tfrac{1}{8}\sin\theta \sin 2\theta \sin 4\varphi \\ T_{3323} &= \tfrac{1}{2}\sin 2\theta[\cos^2\theta - (\sin^4\varphi + \cos^4\varphi)\sin^2\theta] \\ T_{3313} &= -\tfrac{1}{4}\sin^3\theta \sin 4\varphi \\ T_{3333} &= \cos^4\theta + (\sin^4\varphi + \cos^4\varphi)\sin^4\theta \end{aligned}\right\} \tag{C.29}$$

APPENDIX D

Basic Elastic Wave Theory

In this appendix we summarize the main results of elastic wave theory. The background of elastic field theory is provided by a number of books and reviews, including those by Sokolnikoff (1946), Mason (1958), Brekhvoskikh (1960), Nadeau (1964), Nye (1964), Thurston (1964), Kittel (1968), Chou and Pagano (1967), Fedorov (1968), Landau and Lifshitz (1970), Musgrave (1970), Auld (1973). We shall, however, follow in our notation and conventions the relatively recent and excellent text by Auld (1973). Our treatment will be brief and the reader interested in details is referred to Auld (1973) or other references cited above.

D.1. *Strain*

The particle displacement field $\mathbf{U}(\mathbf{L}^c, t)$ at time t is defined as

$$\mathbf{u}(\mathbf{L}^c, t) = \mathscr{L}(\mathbf{L}^c, t) - \mathbf{L}^c \tag{D.1}$$

where $\mathscr{L}(\mathbf{L}^c, t)$ and $\mathbf{L}^c$ represent the instantaneous and equilibrium positions of particles. The strain tensor $\boldsymbol{S}$ is defined as follows:

$$\boldsymbol{S} = d\mathscr{L}^2 - dL^2 = 2S_{ij}dL_i^c\, dL_j^c \tag{D.2}$$

where summation over repeated suffixes is implied. Using equations (D.1) and (D.2), we obtain the strain coefficient S_{ij} as

$$S_{ij} = \frac{1}{2}\left(\frac{\partial u_i}{\partial L_j^c} + \frac{\partial u_j}{\partial L_i^c} + \frac{\partial u_k}{\partial L_i^c}\frac{\partial u_k}{\partial L_j^c}\right) \tag{D.3}$$

In linear theory (small displacement theory) and cartesian coordinates, S_{ij} is given by

$$S_{ij} = \frac{1}{2}\left(\frac{\partial u_i}{\partial x_j} + \frac{\partial u_j}{\partial x_i}\right) \tag{D.4}$$

It follows from equations (D.3) and (D.4) that the strain tensor is symmetric ($S_{ij} = S_{ji}$) and, therefore, it has 6 rather than 9 independent components. In matrix representation, we write

$$\mathbf{S} = \begin{pmatrix} S_{xx} & S_{xy} & S_{xz} \\ S_{yx} & S_{yy} & S_{yz} \\ S_{zx} & S_{zy} & S_{zz} \end{pmatrix} \equiv \begin{pmatrix} S_1 & \frac{1}{2}S_6 & \frac{1}{2}S_5 \\ \frac{1}{2}S_6 & S_2 & \frac{1}{2}S_4 \\ \frac{1}{2}S_5 & \frac{1}{2}S_4 & S_3 \end{pmatrix} \tag{D.5}$$

Hence, in abbreviated subscript notation, strain can be represented by a 6-element column matrix, i.e.,

$$\mathbf{S} = \begin{pmatrix} S_1 \\ S_2 \\ S_3 \\ S_4 \\ S_5 \\ S_6 \end{pmatrix} = [S_I] \tag{D.6}$$

The relation (D.4) between $\mathbf{S}$ and $\mathbf{u}$ is expressed in operator form as (summation over repeated suffixes is implied)

$$S_I = \nabla_{Ij} u_j \tag{D.7}$$

where S_I is given by equation (D.6), u_j represents the components of $\mathbf{u}$, and, in cartesian coordinates, the operator ∇_{Ij} has the following matrix representation*:

$$\nabla_{Ij} = \begin{bmatrix} \partial/\partial x & 0 & 0 \\ 0 & \partial/\partial y & 0 \\ 0 & 0 & \partial/\partial z \\ 0 & \partial/\partial z & \partial/\partial y \\ \partial/\partial z & 0 & \partial/\partial x \\ \partial/\partial y & \partial/\partial x & 0 \end{bmatrix} \tag{D.8}$$

* By convention, capital letters appearing in suffixes assume values from 1 to 6 in contrast to the lower-case letters, which vary from 1 to 3.

D.2. *Stress*

Stress is a measure of traction force acting on a unit area. Quantitatively, if

$$\mathbf{T}_i = \hat{x}_j T_{ji} \tag{D.9}$$

represents the force per unit area on an area element normal to the $+i$-direction, then T_{ji} are said to form the stress tensor $\boldsymbol{T}$. In general, in linearized theory, the stress tensor is also symmetric. As such, it can also be expressed in abbreviated subscript notation

$$\boldsymbol{T} = \begin{pmatrix} T_{xx} & T_{xy} & T_{xz} \\ T_{yx} & T_{yy} & T_{yz} \\ T_{zx} & T_{zy} & T_{zz} \end{pmatrix} = \begin{pmatrix} T_1 & T_6 & T_5 \\ T_6 & T_2 & T_4 \\ T_5 & T_4 & T_3 \end{pmatrix} \tag{D.10}$$

Note the difference in the abbreviated subscript notation for $\boldsymbol{S}$ and $\boldsymbol{T}$. The generalization of Newton's second law of motion to elastic fields leads to the following equation of motion in cartesian coordinates:

$$\rho \frac{\partial^2 u_i}{\partial t^2} = F_i^b + \frac{\partial T_{ij}}{\partial x_j} \tag{D.11}$$

where ρ is the density of the solid and $\mathbf{F}^b$ represents the body force per unit volume. Equation (D.11) can be rewritten in abbreviated subscript notation as

$$\rho \frac{\partial^2 u_i}{\partial t^2} = F_i^b + \nabla_{iJ} T_J \tag{D.12}$$

where the operator ∇_{iJ} has the following matrix representation:

$$\nabla_{iJ} = \begin{bmatrix} \partial/\partial x & 0 & 0 & 0 & \partial/\partial z & \partial/\partial y \\ 0 & \partial/\partial y & 0 & \partial/\partial z & 0 & \partial/\partial x \\ 0 & 0 & \partial/\partial z & \partial/\partial y & \partial/\partial x & 0 \end{bmatrix} \tag{D.13}$$

Note that the matrix in equation (D.13) is the transpose of the matrix in equation (D.8).

D.3. *Hooke's Law: Stiffness and Compliance*

It is an experimental fact that stress is proportional to strain when the latter is small (Hooke's law). Thus, in linearized theory, we have

$$T_{ij} = C_{ijkl}S_{kl} \tag{D.14}$$

where C_{ijkl} are called elastic stiffness coefficients. The symmetry of $\boldsymbol{S}$ and $\boldsymbol{T}$ can be explored to prove that C_{ijkl} are symmetric with respect to interchange between i and j and also between k and l. This makes it possible to rewrite equation (D.14) in abbreviated notation as

$$T_I = C_{IJ}S_J \tag{D.15}$$

It can be shown on the basis of energy considerations that $C_{IJ} = C_{JI}$. In matrix notation, equation (D.15) can be expressed as

$$\boldsymbol{T} = \boldsymbol{C} \cdot \boldsymbol{S} \tag{D.16}$$

The inverse of the stiffness matrix $\boldsymbol{C}$ is called the compliance matrix ($\boldsymbol{s}$). Hence, by definition, $\boldsymbol{s}$ is such that

$$\boldsymbol{C} \cdot \boldsymbol{s} = \boldsymbol{s} \cdot \boldsymbol{C} = 1 \tag{D.17}$$

Therefore, we have

$$\boldsymbol{S} = \boldsymbol{s} \cdot \boldsymbol{T} \tag{D.18}$$

Equivalently,

$$S_I = s_{IJ}T_J \tag{D.19}$$

It can be shown that $s_{IJ} = s_{JI}$. There are at the most 21 constants C_{IJ} (or s_{IJ}) for any medium. The crystal symmetry imposes further restrictions on these constants and the actual number of independent constants for usual crystal structures is, in fact, much less than 21. For example, the stiffness matrix for a cubic lattice is given by (Auld, 1973)

$$\boldsymbol{C} = \begin{bmatrix} c_{11} & c_{12} & c_{12} & 0 & 0 & 0 \\ c_{12} & c_{11} & c_{12} & 0 & 0 & 0 \\ c_{12} & c_{12} & c_{11} & 0 & 0 & 0 \\ 0 & 0 & 0 & c_{44} & 0 & 0 \\ 0 & 0 & 0 & 0 & c_{44} & 0 \\ 0 & 0 & 0 & 0 & 0 & c_{44} \end{bmatrix} \tag{D.20}$$

It is seen that only three independent constants are required to characterize the stiffness of a cubic crystal. The compliance matrix for a cubic lattice is also of the same form as $\boldsymbol{C}$. The relation between C_{IJ} and s_{IJ} is as follows:

$$\left.\begin{aligned} s_{11} &= (c_{11}+c_{12})/[(c_{11}-c_{12})(c_{11}+2c_{12})] \\ s_{12} &= -c_{12}/[(c_{11}-c_{12})(c_{11}+2c_{12})] \\ s_{44} &= 1/c_{44} \end{aligned}\right\} \tag{D.21}$$

D.4. *Damping*

The effect of damping is incorporated by adding a viscosity term to Hooke's law as follows:

$$T_I = C_{IJ}S_J + \eta_{IJ}\,\partial S_J/\partial t \tag{D.22}$$

where $\boldsymbol{\eta}$ is called the viscosity tensor and has a form similar to that of $\boldsymbol{C}$. It follows that, for harmonic time variation, the effect of damping can be taken into account effectively by making C_{IJ} complex.

D.5. *Boundary Conditions*

A realistic problem involves finite media and hence boundaries. The acoustic field quantities are required to satisfy certain restrictions, the so-called boundary conditions, at the interfaces. Let $\mathbf{v}$ and $\mathbf{T}_n$ represent the particle velocity and normal stress fields, respectively, on one side of a boundary and $\mathbf{v}'$ and $\mathbf{T}_n'$ represent the same quantities on the other side. The boundary conditions are

$$\begin{aligned} \mathbf{v} &= \mathbf{v}' \\ \mathbf{T}_n &= \mathbf{T}_n' \end{aligned} \tag{D.23}$$

D.6. *Elastic Energy*

The differential elastic energy density is given by

$$dU_e = C_{IJ}S_J dS_I = T_I dS_I \tag{D.24}$$

from which it follows that

$$T_I = \partial U_e/\partial S_I \tag{D.25}$$

The total elastic energy density is obtained from equation (D.24) as

$$U_e = \tfrac{1}{2}S_I C_{IJ} S_J = \tfrac{1}{2}T_I S_I \tag{D.26}$$

D.7. *Elastic Field Equations*

The elastic field equations along with the constitutive relations are summarized as follows:

$$\left.\begin{aligned} \rho\partial^2 u_i/\partial t^2 &= \nabla_{iJ}T_J + F_i \\ S_I &= \nabla_{Ij}u_j \\ T_J &= (C_{JI} + j\eta_{JI})S_I \end{aligned}\right\} \qquad \text{(D.27)}$$

where ρ represents the density of the medium. These equations can be combined to obtain the following equation, which is in fact the analogue of a wave equation in electromagnetics and is called the Christoffel equation:

$$\rho(\partial^2 u_i/\partial t^2) = F_i + \nabla_{iJ}(C_{JI} + j\eta_{JI})\nabla_{Ij}u_j \qquad \text{(D.28)}$$

The Christoffel operator, $\Gamma_{ij} \equiv \nabla_{iJ}(C_{JI} + j\eta_{JI})\nabla_{Ij}$, can be expressed in matrix form by making use of equations (D.8) and (D.13) along with the appropriate stiffness matrix. The solution of an elastic wave problem requires the solution of equation (D.27) or, equivalently, of equation (D.28), subject to the appropriate boundary conditions.

D.8. *Elastic Plane Waves in an Infinite Medium*

Consider the propagation of a plane elastic wave in a medium, the stiffness matrix of which is given by equation (D.20). If $\mathbf{k} = \beta\hat{n}$ represents the wavenumber, where $\hat{n}$ is a unit vector along the direction of phase propagation, the derivative $\partial/\partial x_i$ appearing in matrices (D.8) and (D.13) can be replaced by $-j\beta\hat{n}\cdot\hat{x}_i$. Thus, in the absence of the body force ($\mathbf{F} = 0$), the Christoffel matrix $\boldsymbol{\Gamma}$ can be obtained as

$$\boldsymbol{\Gamma} = \beta^2 \begin{bmatrix} c_{11}n_x^2 + c_{44}(n_y^2 + n_z^2) & (c_{12} + c_{44})n_x n_y & (c_{12} + c_{44})n_x n_z \\ (c_{12} + c_{44})n_x n_y & c_{11}n_y^2 + c_{44}(n_x^2 + n_z^2) & (c_{12} + c_{44})n_y n_z \\ (c_{12} + c_{44})n_x n_z & (c_{12} + c_{44})n_y n_z & c_{11}n_z^2 + c_{44}(n_x^2 + n_y^2) \end{bmatrix} \qquad \text{(D.29)}$$

Thus, the Christoffel equation (D.28) can be expressed as

$$\boldsymbol{A}\cdot\mathbf{u} = 0 \qquad \text{(D.30)}$$

where

$$\mathbf{u} = \begin{pmatrix} u_x \\ u_y \\ u_z \end{pmatrix} \tag{D.31}$$

and the elements of the matrix $\boldsymbol{A} = [A_{ij}]$ can be expressed as

$$\left.\begin{aligned} A_{11} &= c_{11}n_x^2 + c_{44}(n_y^2 + n_z^2) - \rho\omega^2/\beta^2 \\ A_{22} &= c_{11}n_y^2 + c_{44}(n_z^2 + n_x^2) - \rho\omega^2/\beta^2 \\ A_{33} &= c_{11}n_z^2 + c_{44}(n_x^2 + n_y^2) - \rho\omega^2/\beta^2 \\ A_{12} &= A_{21} = (c_{12} + c_{44})n_x n_y \\ A_{23} &= A_{32} = (c_{12} + c_{44})n_y n_z \\ A_{31} &= A_{13} = (c_{12} + c_{44})n_x n_z \end{aligned}\right\} \tag{D.32}$$

It follows from equation (D.30) that the dispersion relation can be obtained by equating the determinant of $\boldsymbol{A}$ to zero. We shall discuss the special cases of propagation in the xy-plane.

For propagation in the xy-plane, we have $n_z = 0$, $n_x = \cos\theta_x$ and $n_y = \sin\theta_x$, where θ_x is the angle between $\hat{n}$ and the x-axis. The solution of equations (D.30)–(D.32) shows that there is a pure shear wave polarized along the z-axis for which we have

$$(\omega/\beta)_{\mathrm{I}} = (c_{44}/\rho)^{1/2} \tag{D.33}$$

In addition to this, there is a quasi-shear and a quasi-longitudinal wave with phase velocities given by

$$(\omega/\beta)_{\mathrm{II,III}} = (2\rho)^{-1/2}\{(c_{11}+c_{44}) \mp [(c_{11} - c_{44})^2\cos^2 2\theta_x + (c_{12} + c_{44})^2\sin^2 2\theta_x]^{1/2}\}^{1/2} \tag{D.34}$$

The medium will be elastically isotropic if the velocity of the quasi-shear wave,* i.e., $(\omega/\beta)_{\mathrm{II}}$, as given by the upper sign in equation (D.34), is the same as $(\omega/\beta)_{\mathrm{I}}$ as given by equation (D.33), for all values of θ_x. Equating $(\omega/\beta)_{\mathrm{I}}$ and $(\omega/\beta)_{\mathrm{II}}$ for all θ_x, the elastic isotropy condition is obtained as

$$c_{44} = (c_{11} - c_{12})/2 \tag{D.35}$$

It may be noted that, for a YIG single crystal, this condition is approximately satisfied.

* When the elastic isotropy condition holds, the wave with velocity $(\omega/\beta)_{\mathrm{II}}$ will become a pure shear wave.

APPENDIX E

Uniform Precessional Mode Frequency for Small Ellipsoids

If an ellipsoid is placed in a uniform dc magnetic field $\mathbf{H}_0$, the internal dc field can be inferred from equations (1.96) and (1.102) of Chapter 1 and is expressed as

$$\mathbf{H}_i = \mathbf{H}_0 - \boldsymbol{N} \cdot (4\pi \mathbf{M}_0) \tag{E.1}$$

If an rf field $\mathbf{h}$ is superimposed on $\mathbf{H}_0$, it will give rise to an rf magnetization $\mathbf{m}$. When the ellipsoid is sufficiently small, $\mathbf{h}$ can be regarded as uniform and the rf magnetic field inside the ellipsoid can be expressed as*

$$\mathbf{h}_i = \mathbf{h} - \boldsymbol{N} \cdot (4\pi \mathbf{m}) \tag{E.2}$$

Replacing $\mathbf{H}_0$ and $\mathbf{h}$ by $\mathbf{H}_i$ and $\mathbf{h}_i$, respectively, in equation (1.19) of Chapter 1, we have

$$j\omega \mathbf{m} = -\gamma M_0 \hat{z} \times (\mathbf{h} - 4\pi \boldsymbol{N} \cdot \mathbf{m}) - \gamma \mathbf{m} \times \hat{z}(H_0 - 4\pi N_z M_0)$$

Considering $\boldsymbol{N}$ to be a diagonal tensor and following the procedure described in Section 1.4, one obtains the so-called "effective susceptibility tensor" of the ellipsoid. Finally, the effective permeability tensor is obtained as

$$\boldsymbol{\mu}^{\text{eff}} = \begin{pmatrix} \mu_{xx}^{\text{eff}} & j\kappa^{\text{eff}} & 0 \\ -j\kappa^{\text{eff}} & \mu_{yy}^{\text{eff}} & 0 \\ 0 & 0 & 1 \end{pmatrix} \tag{E.3}$$

* This assumes the rf field $\mathbf{h}$ to be quasi-static.

where

$$
\left.
\begin{aligned}
\mu_{xx}^{\text{eff}} &= 1 + \frac{\omega_0 + (N_y - N_z)\omega_{\text{m}}}{\omega_{\text{r}}^2 - \omega^2} \\
\mu_{yy}^{\text{eff}} &= 1 + \frac{\omega_0 + (N_x - N_z)\omega_{\text{m}}}{\omega_{\text{r}}^2 - \omega^2} \\
\kappa^{\text{eff}} &= \frac{\omega\omega_{\text{m}}}{\omega_{\text{r}}^2 - \omega^2} \\
\omega_{\text{r}}^2 &= \{[\omega_0 + (N_x - N_z)\omega_{\text{m}}][\omega_0 + (N_y - N_z)\omega_{\text{m}}]\}
\end{aligned}
\right\} \qquad \text{(E.4)}
$$

In the above expressions, ω_{r} is the desired uniform precessional mode frequency for the ellipsoid (at this frequency μ_{xx}^{eff}, $\mu_{yy}^{\text{eff}} \rightarrow \infty$).

APPENDIX F

Poynting's Theorem

The electromagnetic field in any medium is governed by equations (2.1) of Chapter 2. The electromagnetic energy and power relations can be obtained as follows*:

$$\nabla \cdot (\mathscr{E} \times \mathscr{H}) = \mathscr{H} \cdot \nabla \times \mathscr{E} - \mathscr{E} \cdot \nabla \times \mathscr{H}$$

Using the third and fourth of equations (2.1), we have (in real representation)

$$\nabla \cdot (\mathscr{E} \times \mathscr{H}) = -\frac{1}{c_0}\frac{\partial}{\partial t}(\mathscr{H} \cdot \mathscr{B} + \mathscr{D} \cdot \mathscr{E}) - \frac{4\pi}{c_0}\mathscr{J} \cdot \mathscr{E}$$

Integrating over a volume V, we have

$$\frac{c_0}{4\pi}\int_V \nabla \cdot (\mathscr{E} \times \mathscr{H})dV = -\frac{1}{4\pi}\int_V (\mathscr{H} \cdot \mathscr{B} + \mathscr{D} \cdot \mathscr{E})dV - \int_V \mathscr{J} \cdot \mathscr{E}dV$$

Using Gauss' divergence theorem and assuming the medium to be temporally and spatially nondispersive, we have

$$-\frac{1}{4\pi}\frac{d}{dt}\int_V (\mathscr{H} \cdot \mathscr{B} + \mathscr{D} \cdot \mathscr{E})dV = \int_V \mathscr{J} \cdot \mathscr{E}dV + \frac{c_0}{4\pi}\int_S \mathscr{E} \times \mathscr{H} \cdot \hat{n}dS \quad \text{(F.1)}$$

where $\hat{n}$ is the outward normal to S. The left-hand side of this equation is interpreted as the rate of decrease of energy stored in the volume V. Thus, the right-hand side should represent the power dissipated in the volume

* The script form of the symbols is used to indicate the field vectors in real representation.

plus the power flowing out of the volume V. Since the first term on the right-hand side represents the power absorbed, it follows that the second term represents the outward power flow across the surface S. Thus, the instantaneous power flow per unit area is given by*

$$\tilde{\mathbf{P}} = \frac{c_0}{4\pi} \mathscr{E} \times \mathscr{H}$$

In complex representation, we have **E** and **H** such that

$$\mathscr{E} = \mathrm{Re}(\mathbf{E}) = \tfrac{1}{2}(\mathbf{E} + \mathbf{E}^*)$$

and similarly for $\mathscr{H}$. Thus

$$\begin{aligned} \tilde{\mathbf{P}} &= \frac{c_0}{4\pi} \cdot \frac{1}{2}(\mathbf{E} + \mathbf{E}^*) \cdot \frac{1}{2}(\mathbf{H} + \mathbf{H}^*) \\ &= \frac{c_0}{8\pi} \mathrm{Re}(\mathbf{E} \times \mathbf{H} + \mathbf{E} \times \mathbf{H}^*) \end{aligned}$$

For harmonic fields, the $\mathbf{E} \times \mathbf{H}$ term will vary at frequency 2ω while the $\mathbf{E} \times \mathbf{H}^*$ term will be time-independent. The first term will average out to zero over a cycle and thus the time-averaged power flow per unit area (**P**) is given by

$$\mathbf{P} = \frac{c_0}{8\pi} \mathrm{Re}(\mathbf{E} \times \mathbf{H}^*) \qquad \text{(F.2)}$$

It also follows from equation (F.1) that the energy density (for harmonic fields) is given by

$$\begin{aligned} \tilde{U} &= \frac{1}{4\pi} (\mathscr{H} \cdot \mathscr{B} + \mathscr{D} \cdot \mathscr{E}) \\ &= \frac{1}{8\pi} \mathrm{Re}[(\mathbf{H} \cdot \mathbf{B} + \mathbf{H}^* \cdot \mathbf{B}) + (\mathbf{E} \cdot \mathbf{D} + \mathbf{E}^* \cdot \mathbf{D})] \end{aligned}$$

Again the time-averaged energy density is seen to be

$$U = \frac{1}{8\pi} \mathrm{Re}(\mathbf{H}^* \cdot \mathbf{B} + \mathbf{E}^* \cdot \mathbf{D}) \qquad \text{(F.3)}$$

* See, for example, Jordan and Balmain (1971), for the interpretation of a Poynting vector as power flow per unit area, and further discussion.

When the medium is dispersive, the foregoing analysis is to be modified. We shall only summarize the results, and the reader interested in details should refer to Landau and Lifshitz (1960) and Yeh and Liu (1972). Whereas the expression for power flow is still given by equation (F.2), the energy density given by equation (F.3) is modified (for a biased ferrimagnetic medium) as follows:

$$U = \frac{1}{8\pi}\mathrm{Re}\left\{\mathbf{h}^* \cdot \left[\frac{\partial}{\partial\omega}(\omega\boldsymbol{\mu})\cdot\mathbf{h}\right] + \mathbf{D}\cdot\mathbf{E}^*\right\} \tag{F.4}$$

It is worth obtaining the limiting forms of equations (F.2) and (F.4) under the magnetostatic approximation. Equation (F.4) immediately reduces to

$$U = \frac{1}{8\pi}\mathrm{Re}\left\{\mathbf{h}^* \cdot \left[\frac{\partial}{\partial\omega}(\omega\boldsymbol{\mu})\cdot\mathbf{h}\right]\right\} \tag{F.5}$$

In order to transform* equation (F.2), it is seen that

$$\nabla\cdot(\mathbf{E}\times\mathbf{h}^*) = \mathbf{h}^*\cdot\nabla\times\mathbf{E} - \mathbf{E}\cdot\nabla\times\mathbf{h}^*$$

Using the magnetostatic approximation and harmonic variation, we have

$$\begin{aligned}\nabla\cdot(\mathbf{E}\times\mathbf{h}^*) &= -j(\omega/c_0)\mathbf{h}^*\cdot\mathbf{b}\\ &= -j(\omega/c_0)(\nabla\psi^*\cdot\mathbf{b})\\ &= -j(\omega/c_0)\nabla\cdot(\psi^*\mathbf{b})\end{aligned}$$

The last step follows since $\nabla\cdot\mathbf{b} = 0$. Thus, the Poynting vector may be expressed as:

$$P = \mathrm{Re}(-j(\omega/8\pi)\psi^*\mathbf{b}) \tag{F.6}$$

* The subsequent procedure follows Auld (1973).

APPENDIX G

Partially Magnetized Ferrimagnetics

In general, the macroscopic magnetization of a ferro- or ferrimagnetic sample is much smaller than its saturation magnetization.* An external magnetic field is required to magnetically saturate the sample. This behavior is explained by the domain theory according to which a macroscopic sample is composed of a large number of regions called domains, each of which is magnetically saturated. The direction of magnetization in different domains may be different. Consequently, the net magnetization of the sample will be less than the saturation magnetization; it can even be zero. When such a sample is subjected to a dc magnetic field, the domains with magnetization parallel to the field grow in size at the expense of the unfavorably oriented domains. Thus, the magnetization of the sample increases (Figure G.1). When the magnetic field is sufficiently large, the magnetization of each domain is oriented parallel to the dc field and the sample is magnetically saturated.

When the magnetic field is reduced to zero, the magnetization does not retrace its path; it relaxes to some finite value M_r, called the remanent magnetization (Figure G.2). In fact, a dc magnetic field in the direction opposite to that of the magnetization is required to reduce the magnetic induction to zero. This magnetic field, called the coercive force, depends on a number of parameters including magnetocrystalline anisotropy, impurities, porosity, and internal strains. The full variation of the magnetic induction with magnetic field is shown in Figure G.3; the resulting curve is called the hysteresis loop.

The remanence ratio R, defined as the ratio of the remanent magnetization (M_r) to the saturation magnetization (M_0), is an important parame-

* This is true at temperatures well below the critical temperature.

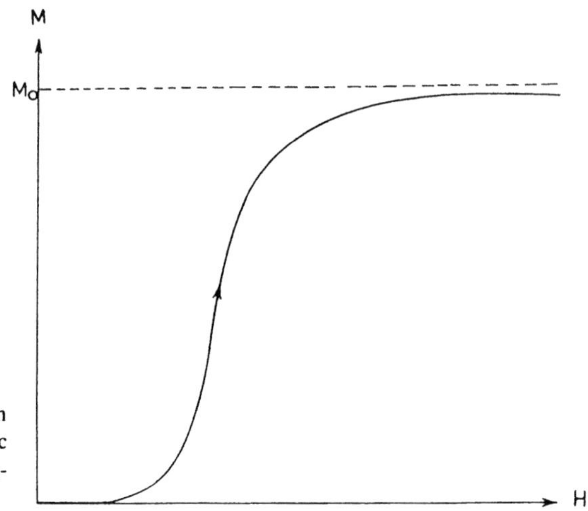

Figure G.1. Magnetization curves for a ferro-(ferri)magnetic sample which is initially unmagnetized.

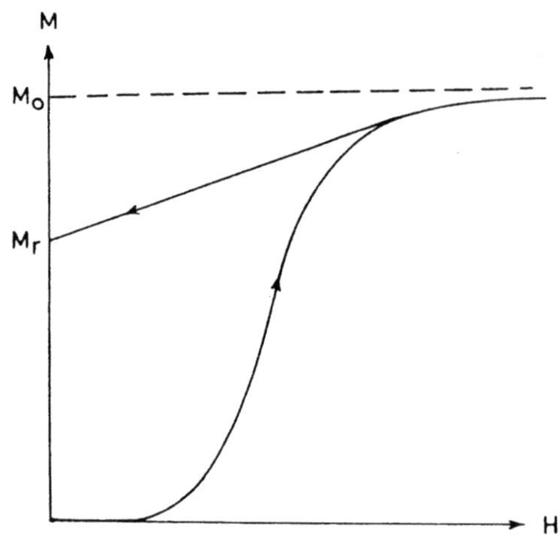

Figure G.2. Magnetization curve, where the dc magnetic field is first increased to magnetically saturate the sample and is then gradually reduced to zero. The magnetization does not retrace its path, but relaxes to a finite value called the remanence magnetization.

ter relevant to the fabrication of latching devices; it is required to be as close to unity as possible. The remanence ratio depends in a complicated manner on magnetocrystalline anisotropy, magnetostriction, grain size, porosity, etc. The highest achieved value of R is approximately 0.8 in the case of certain Mg–Mn and Li ferrites. Most garnets have lower values of R (up to 0.6). The remanence ratio of a ferrimagnetic toroid* can be generally increased by subjecting it to hydrostatic pressure (Wijn et al., 1954/55).

* The ferrimagnetic samples used in latching devices are usually in the form of toroids.

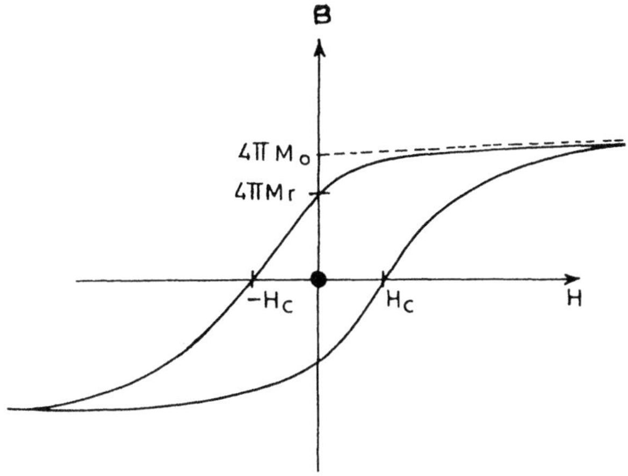

Figure G.3. $B - H$ loop: one complete cycle of variation of magnetic induction with magnetic field.

It is evident from the preceding discussion that when the biasing field is reduced to zero, the magnetization of a ferrimagnetic sample reduces to some value $4\pi M_r$ which lies between zero and the saturation magnetization $4\pi M_0$. In order to analyze the microwave behavior and applications of partially magnetized ferrimagnetics, it is necessary to characterize their microwave permeability. The microwave permeability tensor of demagnetized and partially magnetized ferrimagnetics has been investigated by Rado (1953, 1956), Green et al. (1969), Schlömann (1970), and Green and Sandy (1974a, b). Ignoring the magnetocrystalline anisotropy, the permeability tensor can be expressed as

$$\boldsymbol{\mu} = \begin{pmatrix} \mu & j\kappa & 0 \\ -j\kappa & \mu & 0 \\ 0 & 0 & \mu_z \end{pmatrix} \tag{G.1}$$

where, for a magnetically saturated medium, we have $\mu_z = 1$ while $\mu = \mu' - j\mu''$ and $\kappa = \kappa' - j\kappa''$ are given by equations (1.31) and (1.32) of Chapter 1.

For a partially magnetized ferrimagnetic medium (with magnetization $4\pi M$), Rado (1953, 1956) showed, using averaging techniques, that

$$\kappa' = \gamma 4\pi M/\omega \tag{G.2}$$

$$\mu' = 1 \tag{G.3}$$

Whereas equation (G.2) agrees well with the experimental results, equation (G.3) is found to be at variance with experimental data. Moreover, μ_z'

is also found to deviate from unity. Thus, while κ' can still be regarded as given by equation (G.2), it is necessary to obtain more accurate expressions for μ and μ_z.

Schlömann (1970) carried out an approximate analysis for a completely demagnetized ferrimagnetic sample assuming oppositely magnetized cylindrical domains. He obtained $(\mu')_{\text{dem}}$ as

$$(\mu')_{\text{dem}} = \frac{1}{3} + \frac{2}{3}\left[1 - \left(\gamma \frac{4\pi M_0}{\omega}\right)^2\right]^{1/2} \tag{G.4}$$

This expression is in good agreement with experimental results (Green and Sandy, 1974a, b).

In the case of a partially magnetized ferrimagnetic sample, Green and Sandy (1974a) have presented the following empirical formulas for the dependence of μ' and μ'_z on the degree of magnetization M/M_0:

$$\mu' = \mu'_{\text{dem}} + (1 - \mu'_{\text{dem}})(M/M_0)^{3/2} \tag{G.5}$$

$$\left.\begin{aligned} \mu'_z &= (\mu'_{\text{dem}})^x \\ x &= (1 - M/M_0)^{5/2} \end{aligned}\right\} \tag{G.6}$$

Furthermore, when $\gamma 4\pi M_0/\omega \leqslant 0.75$, which is generally true, one is justified in neglecting κ'' for computational purposes (Green and Sandy, 1974a). The remaining quantities, μ'' and μ''_z, are best determined experimentally for each sample; Green and Sandy (1974b) have tabulated them for a variety of material compositions.

References

References for Chapter 1

Akhiezer, A.I., Bar'yakhtar, V.G., and Peletminskiĭ, S.V., 1968, *Spin Waves*, North-Holland Publishing Co., Amsterdam.

Anderson, P.W., 1963a, in: *Solid State Physics*, Vol. 14 (F. Seitz and D. Turnbull, eds.), Academic Press, New York.

Anderson, P.W., 1963b, in: *Magnetism*, Vol. I (G.T. Rado and H. Suhl, eds.), Academic Press, New York.

Artmann, J.O., 1956, Microwave resonance relations in anisotropic single crystal ferrites, *Proc. IRE*, **44**, 1284.

Auld. B.A., 1965, Geometrical optics of magnetostatic wave propagation in a uniform magnetic field, *Bell Syst. Tech. J.*, **44**, 495.

Auld, B.A., 1968, in: *Advances in Microwaves*, Vol. 3 (L. Young, ed.), Academic Press, New York.

Bady, I., 1961, Ferrites with planar anisotropy at microwave frequencies, *IRE Trans. Microwave Theory Tech.*, **MTT-9**, 52.

Berk, A.D., and Lengyel, B.A., 1955, Magnetic fields in small ferrite bodies with application to microwave cavities containing such bodies, *Proc. IRE*, **43**, 1587.

Berlincourt, D.A., Curran, D.R., and Jaffa, H., 1964, in: *Physical Acoustics*, Vol. IA (W.P. Mason, ed.), Academic Press, New York.

Brown, W.F., Jr., 1965, Theory of magnetoelastic effects in ferromagnetism, *J. Appl. Phys.*, **36**, 994.

Darby, M.I., and Issac, E.D., 1974, Magnetocrystalline anisotropy of ferro- and ferrimagnetic materials, *IEEE Trans. Magnt.*, **MAG-10**, 259.

Feynman, R.P., 1972, *Statistical Mechanics*, W.A. Benjamin, Inc., Reading, Mass.

Gardiol, F.E., 1967, On the thermodynamic paradox in ferrite loaded waveguides, *Proc. IEEE*, **55**, 1616.

Gardiol, F.E., 1972, Evaluate ferrite tensor components fast, *Microwaves* (May Issue), 52.

Gilbert, T.A., 1955, Armour Research Foundation Rept. No. 11, ARF, Chicago, Ill., unpublished.

Haas, C.W., and Callen, H.B., 1963, in: *Magnetism*, Vol. I (G.T. Rado and H. Suhl, eds.), Academic Press, New York.

Helszajn, J., and McStay, J., 1969, External permeability tensor of magnetised ferrite ellipsoid in terms of uniform mode ellipticity, *Proc. IEEE.*, **116**, 2088.

Joseph, R.I., 1966, Ballistic demagnetizing factors in uniformly magnetised cylinder, *J. Appl. Phys.*, **37**, 4639.

Joseph, R.I., 1967, Ballistic demagnetizing factors in a uniformly magnetized rectangular prism, *J. Appl. Phys.*, **38**, 2405.

Joseph, R.I., and Schlömann, E., 1965, Demagnetizing fields in non-ellipsoidal bodies, *J. Appl. Phys.*, **36**, 1579.

Kanamori, J., 1963, in: *Magnetism*, Vol. I (G.T. Rado and H. Suhl, eds.), Academic Press, New York.

Kittel, C., 1947, Interpretation of anomalous Larmor frequencies in ferromagnetic resonance experiment, *Phys. Rev.*, **71**, 270.

Kittel, C., 1948, On the theory of ferromagnetic resonance absorption, *Phys. Rev.*, **73**, 155.

Kittel, C., 1971, *Introduction to Solid State Physics*, 4th edition, John Wiley & Sons, Inc., New York.

Landau, L.D., and Lifshitz, E.M., 1935, On the theory of dispersion of magnetic permeability in ferrimagnetic bodies, *Phys. Z. Sowjetunion.* (English), **8**, 153.

Lax, B., and Button, K.J., 1962, *Microwave Ferrites and Ferrimagnetics*, McGraw-Hill Book Co., New York.

Lewandowsky, S.J., 1964, Ferrite ellipsoid in parallel field, *Br. J. Appl. Phys.*, **15**, 193.

Osborne, J.A., 1945, Demagnetizing factors of the general ellipsoid, *Phys. Rev.*, **67**, 351.

Polder, D., 1949, On the theory of ferromagnetic resonance, *Philos. Mag.*, **40**, 99.

Schiff, L.I., 1968, *Quantum Mechanics*, 3rd edition, McGraw-Hill Book Co., New York.

Schlömann, E., 1970, Demagnetizing fields in thin magnetic films due to surface roughness, *J. Appl. Phys.*, **41**, 1617.

Schneider, B., 1972, Effect of magnetocrystalline anisotropy on magnetostatic spin wave modes in plates I: Theoretical results for infinite plates, *Phys. Status Solidi*, (b)**51**, 325.

Smit, J., and Wijn, H.P.J., 1959, *Ferrites*, John Wiley & Sons, Inc., New York.

Sommerfeld, A., 1952, *Electrodynamics*, Academic Press, New York.

Sparks, M., 1964, *Ferromagnetic Relaxation Theory*, McGraw-Hill Book Co., New York.

Stratton, J., 1941, *Electromagnetic Theory*, McGraw-Hill Book. Co., New York.

Suhl, H., 1955, Ferromagnetic resonance in nickel ferrite between one and two kilomegacycles, *Phys. Rev.*, **97**, 555.

Tiersten, H.F., 1964, Coupled magnetomechanical equations for magnetically saturated insulators, *J. Math. Phys.*, **5**, 1298.

Vittoria, C., Bailey, G.C., Barker, R.C., and Yelon, A., 1973, Ferromagnetic resonance field and linewidth in an anisotropic magnetic metallic medium, *Phys. Rev.*, **B7**, 2112.

Vittoria, C., Craig, J.N., and Bailey, G.C., 1974, General dispersion law in a ferrimagnetic cubic magnetoelastic conductor, *Phys. Rev.*, **B10**, 3945.

von Aulock, W.H., 1965, *Handbook of Microwave Ferrite Materials*, Academic Press, New York.

von Aulock, W.H., and Rowan, J.H., 1957, Measurement of dielectric and magnetic properties of ferromagnetic materials at microwave frequencies, *Bell. Syst. Tech. J.*, **36**, 427.

Wagner, D., 1972, *Introduction to the Theory of Magnetism*, Pergamon Press, Oxford.

Waldron, R.A., 1957, Theory of measurements of the permeability tensor of a ferrite by means of a resonance cavity, *Proc. Inst. Electr. Eng.*, **104B**, 307.

Waldron, R.A., 1959, Electromagnetic fields in ferrite ellipsoid, *Br. J. Appl. Phys.*, **10**, 20.

White, R.M., 1970, *Quantum Theory of Magnetism*, McGraw-Hill Book Co., New York.

References for Chapter 2

Auld, B.A., 1960, Walker modes in large ferrite samples, *J. Appl. Phys.*, **31**, 1642.

Auld, B.A., 1965, Geometrical optics of magnetoelastic wave propagation in a non-uniform magnetic field, *Bell Syst. Tech. J.*, **44**, 495.

Auld, B.A., 1971, in: *Applied Solid State Science*, Vol. II (R. Wolfe, ed.), Academic Press, New York.

Bolle, D.M., and Lewin, L., 1973, On the definitions of parameters in ferrite-electromagnetic wave interactions, *IEEE Trans. Microwave Theory Tech.*, **MTT-21**, 118.

Bömmel, H.E., and Dransfeld, K., 1960, Spin phonon interactions in yttrium–iron–garnet, *Bull. Am. Phys. Soc.*, **5**, 357.

Brillouin, L., 1960, *Wave Propagation and Group Velocity*, Academic Press, New York.

Clark, A.E., DeSavage, B., Coleman, W., Callen, E.R., and Callen, H.B., 1963, Saturation magnetostriction of single-crystal YIG, *J. Appl. Phys.*, **34**, 1296.

Clogston, A.M., Suhl, H., Walker, L.R., and Anderson, P.W., 1956, Ferromagnetic resonance linewidth in insulating materials, *J. Phys. Chem. Solids*, **1**, 129.

Eberhardt, N., Horvarth, V.V., and Knerr, R.H., 1970, On plane and quasi-optical wave propagation in gyromagnetic media, *IEEE Trans. Microwave Theory Tech.*, **MTT-18**, 554.

Engineer, M.H., and Nag, B.R., 1965, Propagation of electromagnetic waves in rectangular guides filled with a semiconductor in the presence of a transverse magnetic field, *IEEE Trans. Microwave Theory Tech.*, **MTT-13**, 641.

Freeman, A.J., and Schmid, H., 1975, *Magnetoelastic Interaction Phenomena in Crystals*, Gordon & Breach Science Publishers, London.

Gabriel, G.J., and Brodwin, M.E., 1966, Distinction between gyroelectric and gyromagnetic medium, *IEEE Trans. Microwave Theory Tech.*, **MTT-14**, 292.

Gurevich, A., 1965, *Ferrites at Microwave Frequencies*, Boston Technical Publishers, Inc., Boston, Mass.

Hogan, C.L., 1952, The ferromagnetic Faraday effect at microwave frequency and its application, *Bell. Syst. Tech. J.*, **31**, 1.

Hogan, C.L., 1953, Ferromagnetic Faraday effect at microwave frequencies and its applications, *Rev. Mod. Phys.*, **25**, 253.

Kittel, C., 1958, Interaction of spin waves and ultrasonic waves in ferrimagnetic crystals, *Phys. Rev.*, **110**, 836.

Kong, J.A., 1975, *Theory of Electromagnetic Waves*, John Wiley & Sons, Inc., New York.

Lax, B., and Button, K.J., 1962, *Microwave Ferrites and Ferrimagnetics*, McGraw-Hill Book Co., New York.

Lighthill, M.J., 1964. *Fourier Analysis and Generalized Functions*, Cambridge University Press, Cambridge.

Matthews, H., and LeCraw, R.C., 1962, Acoustic wave rotation by magnon–phonon interaction, *Phys. Rev. Lett.*, **8**, 397.

Mercereau, J.E., and Feynman, R.P., 1956, Physical conditions for ferromagnetic resonance, *Phys. Rev.*, **104**, 63.

Parekh, J.P., and Bertoni, H.L., 1973, Exchange free magnetoelastic plane waves, *J. Appl. Phys.*, **44**, 2866.

Rosenbaum, F.J., 1964, Electromagnetic wave propagation in lossy ferrites, *IEEE Trans. Microwave Theory Tech.*, **MTT-12**, 517.

Schlömann, E., 1960, Generation of phonons in high power ferromagnetic resonance experiments, *J. Appl. Phys.*, **31**, 1647.

Sodha, M.S., and Srivastva, N.C., 1976, Differential phase-shift at microwave frequencies using planar ferrites, *IEEE Trans. Microwave Theory Tech.*, **MTT-24**, 215.

Srivastava, N.C., 1976a, Electromagnetic wave propagation in a rectangular waveguide partially loaded with planar ferrite, *Appl. Phys.*, **11**, 277.
Srivastava, N.C., 1976b, Magnetostatic modes of a slab of hexagonal planar ferrite, *J. Appl. Phys.*, **47**, 5447.
Srivastava, N.C., 1977, Electromagnetic Wave Propagation in Magnetized Ferrites, Ph.D. Thesis, unpublished, Indian Institute of Technology, New Delhi.
Srivastava, N.C., 1978a, Non-reciprocal reflection of electromagnetic waves from magnetized ferrites, *J. Appl. Phys.*, **49**, 3181.
Srivastava, N.C., 1978b, Propagation of surface waves through gap between oppositely magnetized ferrite substrates, *IEEE Trans. Microwave Theory Tech.*, **MTT-26**, 213.
Steele, M.C., and Vural, B., 1969, *Wave Interactions in Solid State Plasmas*, McGraw-Hill Book Co., New York.
Suhl, H., and Walker, L.R., 1954, Topics in guided wave propagation through gyrotropic media: I. Completely filled cylindrical guide; II. Transverse magnetization and non-reciprocal helix; III. Perturbation theory and miscellaneous results, *Bell Syst. Tech. J.*, **33**, 519; 939; 1133.
Vittoria, C., Craig, J.N., and Bailey, G.C., 1974, General dispersion law in a ferromagnetic cubic magnetoelastic conductor, *Phys. Rev.*, **B10**, 3945.
Vlasov, K.B., and Ishmukhametov, B.Kh., 1960, Rotation of plane of polarization of elastic waves in magnetically polarized magnetoelastic media, *Soviet Phys. — JETP*, **37**, 531.
von Aulock, W.H., and Fay, C.E., 1968, *Linear Ferrite Devices for Microwave Applications*, Academic Press, New York.
Walker, L.R., 1957, Magnetoelastic modes in ferromagnetic resonance, *Phys. Rev.*, **105**, 390.
White, R.L., and Solt, I.H., Jr., Multiple ferromagnetic resonance in ferrite spheres, *Phys. Rev.*, **104**, 56.
Waldron, R.A., 1961, *Ferrites*, D. Van Nostrand Co., Toronto.

References for Chapter 3

Brillouin, L., 1960, *Wave Propagation and Group Velocity*, Academic Press, New York.
Culshaw, W., 1961, in: *Advances in Electronics and Electron Physics*, Vol. 15 (L. Marton, ed.), Academic Press, New York.
Gupta, S.S., and Srivastava, N.C., 1979, Physics of microwave reflection at a dielectric–ferrite interface, *Phys. Rev.*, **B19**, 5403.
Heinrich, B., and Meshcheryakov, V.F., 1969, Passage of electromagnetic wave through a ferromagnetic metal in the antiresonance region, *JETP Lett.*, **9**, 378.
Ince, W.J., and Temme, D.H., 1969, in: *Advances in Microwaves*, Vol. 4 (L. Young, ed.), Academic Press, New York.
Kanda, M., and May, W.G., 1975, A millimeter-wave reflection beam isolator, *IEEE Trans. Microwave Theory Tech.*, **MTT-23**, 506.
Kogelnik, H., 1975, in: *Integrated Optics* (T. Tamir, ed.), Springer-Verlag, Berlin, New York.
Landau, L.D., and Lifshitz, E.M., 1960, *Electrodynamics of Continuous Media*, Pergamon Press, London.
Lieu, O.L.S., Alexandrakis, G.C., and Huerta, M.A., 1977, Theory of nonlinear phenomena in ferromagnetic transmission resonance, *Phys. Rev.*, **B16**, 476.
Lotsch, H.K.V., 1970, Beam displacement at total reflection: the Goos–Hanchen effect: I; II, *Optik*, **32**, 116; 189.
Lotsch, H.K.V., 1971, Beam displacement at total reflection: the Goos–Hanchen effect: III; IV, *Optik*, **32**, 299; 553.
Mueller, R.S., 1971, Reflection and refraction of plane waves from plate ferrite surfaces, *J. Appl. Phys.*, **42**, 2264.

Mueller, R.S., 1974, Ferrite prisms and lenses, *J. Appl. Phys.*, **45**, 3587.
Mueller, R.S., 1975, Transmission of electromagnetic waves through stratified ferrite, *J. Appl. Phys.*, **46**, 2295.
Mueller, R.S., 1976, Interface effects in microwave ferrite structures, *J. Appl. Phys.*, **47**, 4164.
Philliphs, T.G., 1970, Ferromagnetic resonance transmission in iron group metals, *J. Appl. Phys.*, **41**, 1109.
Renard, R.H., 1964, Total reflection : a new evaluation of Goos–Hanchen shift, *J. Opt. Soc. Am.*, **54**, 1190.
Seavey, M.H., and Tannenwald, P.E., 1957, Electromagnetic propagation effects in ferromagnetic resonance, Tech. Rept. 143, MIT Lincoln Lab., MIT, Boston, Mass.
Sodha, M.S., and Srivastva, N.C., 1975, Modulation of electromagnetic waves by dependent magnetic fields in a ferrite, *Appl. Phys.*, **8**, 23.
Srivastava, N.C., 1977, Electromagnetic Wave Propagation in Magnetized Ferrites, Ph.D. Thesis, unpublished, Indian Institute of Technology, New Delhi.
Srivastava, N.C., 1978, Nonreciprocal reflection of electromagnetic waves from magnetized ferrite, *J. Appl. Phys.*, **49**, 3181.
Suhl, H., and Walker, L.R., 1954, Topics in guided wave propagation through gyromagnetic media: I. The completely filled cylindrical guide; II. Transverse magnetization and the nonreciprocal helix; III. Perturbation theory and miscellaneous results, *Bell Syst. Tech. J.*, **33**, 579; 939; 1133.
Tien, P.K., 1977, Integrated optics and new wave phenomena and optical waveguides, *Rev. Mod. Phys.*, **49**, 361.
Vittoria, C., Bailey, G.C., Barker, R.C., and Yelon, A., 1973, Ferromagnetic resonance field and linewidth in an anisotropic magnetic metallic medium, *Phys. Rev.*, **B7**, 2112.
Vittoria, C., Rubinstein, M., and Lubitz, P., 1975, Transmission and magnetoacoustic efficiencies in ferromagnetic conducting film, *Phys. Rev.*, **B12**, 5150.
von Aulock, W.H., 1965, *Handbook of Microwave Ferrite Materials*, Academic Press, New York.
von Aulock, W.H., and Fay, C.E., 1968, *Linear Ferrite Devices for Microwave Application*, Academic Press, New York.

References for Chapter 4

Adam, J.D., 1970, Delay of magnetostatic surface waves in YIG, *Electron Lett.*, **6**, 718.
Adam, J.D., and Collins, J.H., 1976, Microwave magnetostatic delay devices based on epitaxial YIG, *Proc. IEEE*, **64**, 794.
Adam, J.D., Bennett, G.A., and Wilkinson, J., 1970, Experimental observation of magnetostatic modes in a YIG slab, *Electron. Lett.*, **6**, 434.
Adam, J.D., Collins, J.H., and Owens, J.M., 1973, Magnetostatic surface wave group delay equaliser, *Electron. Lett.*, **9**, 537.
Adam, J.D., Owens, J.M., and Collins, J.H., 1974, Magnetostatic delay lines for group delay equalization in millimetric waveguide communicating system, *IEEE Trans. Magnt.*, **MAG-10**, 783.
Adam, J.D., Collins, J.H., and Owens, J.M., 1975, Microwave device applications of epitaxial magnetic garnets, *Radio Electron Eng.*, **45**, 738.
Adam, J.D., Patterson, R.W., and O'Keefe, T.W., 1978, Magnetostatic wave in interdigital transducers, *J. Appl. Phys.*, **49**, 1797.
Akhiezer, A.I., Bar'yakhtar, V.G., and Peletminskiĭ, S.V., 1963, Coherent amplification of spin waves, *Phys. Lett.*, **4**, 129.

Auld, B.A., 1960, Walker modes in large ferrite samples, *J. Appl. Phys.*, **31**, 1642.

Awai, I., Ohtsuki, K., and Ikenoue, J., 1976, Interaction of magnetic surface waves with drifting carriers, *Jpn. J. Appl. Phys.*, **15**, 1297.

Bajpai, S.N., Rattan, I., and Srivastava, N.C., 1979, Magnetostatic volume waves in dielectric layered structure: effect of magnetocrystalline anisotropy, *J. Appl. Phys.*, **50**, 2887.

Bajpai, S.N., and Srivastava, N.C., 1980a, Magnetostatic bulk waves in arbitrarily magnetized dielectric layered structure, *Phys. Status Solidi*, (a)**57**, 307.

Bajpai, S.N., and Srivastava, N.C., 1980b, Magnetostatic bulk wave propagation in a multi-layered structure, *Electron. Lett.*, **16**, 269.

Bardai, Z.M., Adam, J.D., Collins, J.H., and Parekh, J.P., 1976, Delay lines based on magnetostatic volume waves in epitaxial YIG, AIP Conf. Proc. No. 34, 268.

Bardati, F., and Lampariello, P., 1979, The model spectrum of a lossy ferrimagnetic slab, *IEEE Trans. Microwave Theory Tech.*, **MTT-27**, 679.

Basterfield, J., 1969, Chemical polishing of yttrium iron garnet, *J. Phys. D. Appl. Phys.*, **2**, 115.

Bennett, G.A., and Adam, J.D., 1970, Identification of surface wave resonances on a metal backed YIG slab, *Electron. Lett.*, **6**, 789.

Benson, H., and Mills, D.L., 1969, Variation principle in spin wave theory: application to the theory of magnetostatic surface waves, *Phys. Rev.*, **188**, 849.

Bini, M., Filleti, P.L., Millanta, L., and Rubio, N., 1976, Energetic derivation of the amplification of magnetic waves interacting with a flow of charges in a semiconductor, *J. Appl. Phys.*, **47**, 3209.

Bini, M., Millanta, L., and Rubio, N., 1977, Thin film magnetostatic amplifier; analytical expression of dispersion and gain properties, *Electron Lett.*, **13**, 114.

Bini, M., Millanta, L., and Rubio, N., 1978a, Interaction of magnetic waves with drifting charges, *IEEE Trans. Magnt.*, **MAG-14**, 811.

Bini, M., Filetti, P.L., Millanta, L., and Rubio, N., 1978b, Amplification of surface magnetic waves in transversely magnetized ferrite slabs, *J. Appl. Phys.*, **49**, 3554.

Bongianni, W.L., Collins, J.H., Pizzarello, F.A., and Wilson, D.A., 1969, Propagating magnetic waves in epitaxial YIG, IEEE Int. MW Symp. Digest, Dallas, 376.

Bongianni, W.L., 1972, Magnetostatic propagation in a dielectric layered structure, *J. Appl. Phys.*, **43**, 2541.

Bongianni. W.L., 1974, X-band signal processing using magnetic waves, *Microwave J.*, **17**, 49.

Bresler, A.D., 1959, TE_{n0} surface wave at ferrite–air interface, Polytech. Inst. Brooklyn Microwave Res. Inst. Memo 48, R 723–59, PIB–651.

Brekhovskikh, L.M., 1960, *Waves in Layered Media*, Academic Press, New York.

Briggs, R.J., 1964, *Electron-Stream Interactions with Plasmas*, The MIT Press, Massachusetts.

Brundle, L.K., and Freedman, N.J., 1968a, Nonlinear behaviour of magnetostatic surface waves, *Electron. Lett.*, **4**, 427.

Brundle, L.K., and Freedman, N.J., 1968b, Magnetoelastic surface waves on YIG slab, *Electron. Lett.*, **4**, 132.

Castéra, J.P., 1978, Tunable magnetostatic surface wave oscillator, *IEEE Trans. Magnt.*, **MAG-14**, 826.

Castéra, J.P., and Hartemann, P., 1978, Magnetoelastic surface wave oscillators and resonators, Proc. VIII European Microwave Conf., 658.

Chang, N.S., and Matsuo, Y., 1968, Possibility of utilizing the coupling between a backward wave in YIG and waves associated with drift carrier stream in semiconductor, *Proc. IEEE.*, **56**, 765.

Chang, N.S., and Matsuo, Y., 1975, Characteristics of wave propagation in a composite system consisting of ferrite and semiconductor, *Trans. Inst. Electron. Commun. Engr. Jpn.*, **B58**, 315.

Chang, N.S., and Matsuo, Y., 1977, Ferromagnetic loss effect on magnetoelastic surface wave amplification by YIG-semiconductor coupled system, *IEEE Trans. Magnt.*, **MAG-13**, 1308.

Chang, N.S., Yamada, S., and Matsuo, Y., 1975, Characteristics of magnetoelastic surface wave propagation in a layered structure consisting of metals, dielectrics, a semiconductor and YIG, *Electron. Lett.*, **11**, 83.

Chang, N.S., Yamada, S., and Matsuo, Y., 1976a, Amplification of magnetostatic surface waves in a layered structure consisting of metals, dielectrics, a semiconductor and YIG, *J. Appl. Phys.*, **47**, 385.

Chang, N.S., Yamada, S., and Matsuo, Y., 1976b, Amplification characteristics of magnetostatic surface and volume waves in semiconductor-dielectric-YIG-metal system, *Wave Electron.*, **2**, 341.

Collins, J.H., Adam, J.D., and Owens, J.M., 1972, Microwave device applications of epitaxial ferrimagnetic films, Proc. European Solid State Devices Conf., 83.

Collins, J.H., and Pizzarello, F.A., 1973. Propagating magnetic waves in thick films. A contemporary technology to surface wave acoustics, *Int. J. Electron.*, **34**, 319.

Collins, J.H., Owens, J.M., and Smith, C.V., Jr., 1977, Magnetostatic wave signal processing, Proc. Ultrasonics Symposium, 541.

Courtois, L., Declercq, G., and Purichard, M., 1971, On the nonreciprocal aspect of gyromagnetic surface wave, AIP Conf. Proc. No. 5, 1541.

Damon, R.W., and Eshbach, J.R., 1961, Magnetostatic modes of a ferromagnetic slab, *J. Phys. Chem. Solids*, **19**, 308.

Damon, R.W., and van de Varrt, H., 1965, Propagation of magnetostatic spin waves at microwave frequencies in a normally magnetized disc, *J. Appl. Phys.*, **36**, 3453.

De Wames, R.E., and Wolfram, T., 1976, Characteristics of magnetostatic surface waves for a metallized ferrite slab, *J. Appl. Phys.*, **41**, 5243.

Elachi, C., 1975, Electromagnetic wave propagation in periodic media, *IEEE Trans. Magnt.*, **MAG-11**, 36.

Eshbach, J.R., and Damon, R.W., 1960, Surface magnetoelastic modes and surface spin waves, *Phys. Rev.*, **118**, 1208.

Ganguly, A.K., and Vittoria, C., 1974, Magnetostatic wave propagation in double layers of magnetically anisotropic slab, *J. Appl. Phys.*, **45**, 4665.

Ganguly, A.K., and Webb, D.C., 1975, Microstrip excitation of magnetostatic surface waves: Theory and experiment, *IEEE Trans. Microwave Theory Tech.*, **MTT-23**, 998.

Ganguly, A.K., Vittoria, C., and Webb, D., 1974, Interaction of surface magnetic waves in anisotropic magnetic slabs, AIP Conf. Proc. Magnetism and Magnetic Materials, 495.

Ganguly, A.K., Webb, D.C., and Banks, C., 1978, Complex radiation impedence of microstrip excited magnetostatic surface waves, *IEEE Trans. Microwave Theory Tech.*, **MTT-26**, 444.

Gardiol, F.E., 1967, On the thermodynamic paradox in ferrite loaded waveguides, *Proc. IEEE*, **55**, 1616.

Gerson, T.J., and Nadan, J.S., 1974, Surface electromagnetic modes of a ferrite slab, *IEEE Trans. Microwave Theory Tech.*, **MTT-22**, 757.

Glass, H.L., and Elliot, M.T., 1975, Attainment of intrinsic linewidth in yttrium iron garnet films grown by liquid phase epataxy, X Int. Conf. Crystallography, Amsterdam, paper 08.3–8.

Grant, P.M., Adam, J.D., and Collins, J.H., 1974, Surface wave device applications in microwave communication system, *IEEE Trans. Commun.*, **22**, 1410.

Gupta, S.S., and Srivastava, N.C., 1979, Power flow and energy distribution of magnetostatic bulk waves in dielectric layered structure, *J. Appl. Phys.*, **50**, 6697.

Gupta, S.S., and Srivastava, N.C., 1980, Theory of magnetic surface wave propagation in a thick YIG slab, *J. Appl. Phys.*, **51**, 4618.

Gupta, S.S., and Srivastava, N.C., 1980b, Ray optics approach to magnetostatic bulk wave propagation in a YIG slab, *IEEE Trans. Microwave Theory Tech.*, **MTT-28**, 915.

Howarth, J., 1975, A magnetoelastic delay line equaliser, Proc. IEEE MTTS Int. MW Symposium, 371.

Hurd, R.A., 1970, Surface waves at ferrite-metal boundaries, *Electron. Lett.*, **6**, 262.

Kawasaki, K., Takagi, H., and Umeno, M., 1974a, Passband control of surface magnetostatic waves by spacing a metal plate apart from the ferrite surface, *IEEE Trans. Microwave Theory Tech.*, **MTT-22**, 924.

Kawasaki, K., Takagi, H., and Umeno, M., 1974b, The interaction of surface magnetostatic waves with drifting carriers in semiconductors, *IEEE Trans. Microwave Theory Tech.*, **MTT-22**, 918.

Kogelnik, H., and Weber, H.P., 1974, Rays, stored energy and power flow in dielectric waveguides, *J. Opt. Soc. Am.*, **64**, 174.

Lax, B., and Button, K.J., 1956, Theory of ferrites in rectangular waveguides, *IRE Trans. Antennas Propagt.*, **AP-4**, 531.

Levinstein, H.J., Licht, S., Landorf, R.W., and Blank, S.L., 1971, Growth of high quality garnet thin films from supercooled melt, *Appl. Phys. Lett.*, **19**, 486.

Masuda, M., Chang, N.S., and Matsuo, Y., 1974, Magnetostatic surface waves in ferrite slab adjacent to semiconductor, *IEEE Trans. Microwave Theory Tech.*, **MTT-22**, 132.

Mee, J.E., Pullian, G.R., Archer, A.L., and Besser, P.J., 1969, Magnetic oxide films, *IEEE Trans. Magnt.*, **MAG-5**, 717.

Mercereau, J.E., and Feynman, R.P., 1956, Physical conditions for ferromagnetic resonance, *Phys. Rev.*, **104**, 63.

Merry, J.B., and Sethares, J.C., 1973, Low loss magnetostatic surface waves at frequencies up to 15 GHz, *IEEE Trans. Magnt.*, **MAG-9**, 527.

Miller, N.D.J., 1976a, Magnetostatic volume wave propagation in a dielectric layered structure, *Phys. Status Solidi*, (a)**37**, 83.

Miller, N.D.J., 1976b, Nondispersive magnetostatic volume wave delay line, *Electron. Lett.*, **12**, 466.

Miller, N.D.J., 1977, Nonreciprocal propagation of magnetostatic volume waves, *Phys. Status Solidi* (a)**43**, 593.

Miller, N.D.J., 1978, Nonreciprocal magnetostatic volume waves, *IEEE Trans. Magnt.*, **MAG-14**, 829.

Miller, N.D.J., and Brown, D., 1976, Tunable magnetoelastic surface wave oscillator, *Electron. Lett.*, **12**, 209.

Morgenthaler, F.R., 1970, Nonreciprocal magnetostatic surface waves with independently controllable propagation and decay constant, *J. Appl. Phys.*, **41**, 1014.

Morgenthaler, F.R., 1977, Magnetostatic waves bound to a dc field gradient, *IEEE Trans. Magnt.*, **MAG-13**, 1252.

Neviere, M., Petit, R., and Cadilhac, M., 1973, About the theory of optical grating complex waveguide system, *Opt. Commun.*, **8**, 113.

Newburgh, R.G., Blacksmith, P., Budreau, A.J., and Sethares, J.C., 1974, Acoustic and magnetic surface wave ring interferometers for rotation sensing, *Proc. IEEE*, **62**, 1621.

O'Keeffe, T.W., and Patterson, R.W., 1978, Magnetostatic surface wave propagation in finite samples, *J. Appl. Phys.*, **49**, 4886.

Olson, F.A., and Yaeger, J.R., 1965, Microwave delay techniques using YIG, *IEEE Trans. Microwave Theory Tech.*, **MTT-13**, 63.

Owens, J.M., Collins, J.H., and Adam, J.D., 1975, Planar microwave multipole filters using LPE–YIG, AIP Conf. Proc. No. 24, 497.

Parekh, J.P., 1973, Magnetostatic surface waves on a partially metallized YIG plate, *Proc. IEEE*, **61**, 1371.

Parekh, J.P., 1975, Dielectrically induced surface waves and the magnetodynamic modes of a YIG plate, *J. Appl. Phys.*, **46**, 5040.

Parekh, J.P., 1979, Theory for magnetostatic forward volume wave excitation, *J. Appl. Phys.*, **50**, 2452.

Parekh, J.P., and Ponamgi, S.R., 1973, Dielectrically induced surface wave on a YIG substrate, *J. Appl. Phys.*, **44**, 1384, errt. 4791.

Parekh, J.P., and Tuan, H.S., 1979a, Excitation of magnetostatic surface waves in an arbitrary direction on tangentially YIG film, *IEEE Trans. Magnt.*, **MAG-15**, 1747.

Parekh, J.P., and Tuan, H.S., 1979b, Meander line excitation of magnetostatic surface waves, *Proc. IEEE*, **67**, 182.

Peng, S.T., Tamir, T., and Bertoni, H.L., 1975, Theory of periodic waveguides, *IEEE Trans. Microwave Theory Tech.*, **MTT-23**, 123.

Pizzarello, F.A., Coerver, L.E., and Collins, J.H., 1970, Magnetic steering of magnetostatic waves in epitaxial YIG films, *J. Appl. Phys.*, **41**, 1016.

Renard, R.H., 1964, Total reflection: a new evaluation of the Goos–Hanchen shift, *J. Opt. Soc. Am.*, **54**, 1190.

Robinson, B.B., Vural, B., and Parekh, J.P., 1970, Spin-wave/carrier wave interaction, *IEEE Trans. Electron Devices*, **ED-17**, 224.

Schilz, W., 1973, Spin wave propagation in epitaxial YIG films, *Phillips Res. Rep.*, **28**, 50.

Schlömann, E., 1969, Amplification of magnetoelastic surface waves by interaction with drifting carriers in crossed electric and magnetic fields, *J. Appl. Phys.*, **40**, 1422.

Schneider, B., 1972, Effect of crystalline anisotropy on the magnetostatic spin wave modes in ferromagnetic plates: I. Theoretical results for infinite plates, *Phys. Status Solidi*, (b)**51**, 325.

Seidel, H., and Fletcher, R.C., 1959, Gyromagnetic modes in waveguides partially loaded with ferrite, *Bell Syst. Tech. J.*, **38**, 1427.

Seshadri, S.R., 1970, Surface magnetostatic modes of ferrite slab, *Proc. IEEE*, **58**, 506.

Seshadri, S.R., 1978, Theory of a YIG film filter, *J. Appl. Phys.*, **49**, 6079.

Sethares, J.C., 1975, Magnetostatic surface waves on a cylinder, Air Force Cambridge Res. Rept. AFCRL-TR-0380.

Sethares, J.C., 1978, Magnetostatic surface wave transducer design, Int. Microwave Symp. Digest, IEEE Cat. #78 CH 1335–7 MTT.

Sethares, J.C., and Merry, J.B., 1974, Magnetostatic surface waves in ferrimagnetics above 4 GHz, Air Force Cambridge Res. Rept. AFCRL-TR-74-0112, Phys. Sc. Res. Paper No. 587.

Sethares, J.C., and Stiglitz, M.R., 1974, Propagation loss and MSSW delay lines, *IEEE Trans. Magnt.*, **MAG-10**, 787.

Sethares, J.C., and Weinberg, I.J., 1979a, Apodization of variable coupling magnetoelastic surface wave transducers, *J. Appl. Phys.*, **50**, 2458.

Sethares, J.C., and Weinberg, I.J., 1979b, Insertion loss of apodized-weighted and nonuniform magnetostatic surface wave transducers, Joint Inter MAG–MMM Conf., New York, paper 6–C.

Sethares, J.C., Tsai, T., and Koltunov, I., 1978, Periodic magnetostatic surface wave transducers, Rome Air Force Dev. Centre Res. Rept., RADC–TR–78–78.

Sittig, E.K., and Coquin, G.A., 1968, Filters and dispersive delay lines using repetitively mismatched ultrasonic transmission line, *IEEE Trans. Sonics Ultrason.*, **SU-15**, 111.

Sparks, M., 1964, *Ferromagnetic Relaxation Theory*, McGraw-Hill Book Co., New York.

Sparks, M., 1969, Magnetostatic surface modes of a YIG slab, *Electron. Lett.*, **5**, 618.

Srivastava, N.C., 1976, Magnetostatic modes of a slab of hexagonal planar ferrite, *J. Appl. Phys.*, **47**, 5447.

Srivastava, N.C., 1978b, Propagation of magnetostatic waves along curved ferrite surfaces, *IEEE Trans. Microwave Theory Tech.*, **MTT-26**, 252.

Srivastava, N.C., 1978c, Propagation of surface waves through the gap between oppositely magnetized ferrite substrates, *IEEE Trans. Microwave Theory Tech.*, **MTT-26**, 213.

Steele, M.C., and Vural, B., 1969, *Wave Interaction in Solid State Plasmas*, McGraw-Hill Book Co., New York.

Sykes, C.G., Adam, J.D., and Collins, J.H., 1976, Magnetostatic wave propagation in periodic structures, *Appl. Phys. Lett.*, **29**, 388.

Szustakowski, M., and Wecki, B., 1973, Amplification of magnetostatic surface waves in YIG–Ge hybrid system, *Proc. Vib. Probl.*, **14**, 155.

Tamir, T., 1973, Inhomogeneous wave types at planar interface: II. Surface waves, *Optik*, **37**, 204.

Tien, P.K., 1971, Light waves in thin films and integrated optics, *Appl. Opt.*, **10**, 2395.

Tien, P.K., 1977, Integrated optics and new wave phenomena and optical waveguides, *Rev. Mod. Phys.*, **49**, 361.

Trivelpiece, A.W., Ignatius, A., and Holscher, P.C., 1961, Amplification of magnetostatic waves by interaction with charged carriers drifting through a semiconductor, *J. Appl. Phys.*, **32**, 259.

Tsai, M.C., Wu, H.J., Owens, J.M., and Smith, C.V., Jr., 1976, Magnetostatic propagation for uniform normally magnetized multilayer planar structures, *AIP Conf. Proc.*, No. 34, 280.

Tsutsumi, M., 1974, Magnetostatic surface wave propagation through air gap between adjacent magnetic substrates, *Proc. IEEE*, **62**, 541.

Tsutsumi, M., and Yuki, Y., 1975, Magnetostatic wave propagation in periodically magnetized ferrites, *Electron. Comm. Jpn.*, **58**, 74.

Tsutsumi, M., Bhattacharya, T., and Kumagai, N., 1976, Effect of magnetic perturbation on magnetic surface wave propagation, *IEEE Trans. Microwave Theory Tech.*, **MTT-24**, 591.

Tsutsumi, M., Sakaguchi, Y., and Kumagai, N., 1977a, Behaviour of magnetostatic waves in a periodically corrugated YIG slab, *IEEE Trans. Microwave Theory Tech.*, **MTT-25**, 224.

Tsutsumi, M., Sakaguchi, Y., and Kumagai, N., 1977b, The magnetostatic surface wave propagation in a corrugated YIG slab, *Appl. Phys. Lett.*, **31**, 779.

Vaskovskii, A.V., Zubkov, V.I., Ki'ldishev, V.N., and Murmuzev, B.A., 1972, Interaction of surface magnetostatic waves with carriers on a ferrite-semiconductor interface, *JETP Lett.*, **16**, 2.

Vaslow, D.F., 1973, Group delay time for a surface wave on a YIG film backed by a grounded dielectric, *Proc. IEEE*, **61**, 142.

Vaslow, D.F., 1974, Surface waves on a ferrite magnetized perpendicular to the interface, *IEEE Trans. Microwave Theory Tech.*, **MTT-22**, 743.

Vittoria, C., and Wilsey, N.D., 1974, Magnetostatic wave propagation loss in an anisotropic insulator, *J. Appl. Phys.*, **45**, 414.

Volluet, G., 1980, Unidirectional magnetostatic forward volume wave transducers, *IEEE Trans. Magnt.*, **MAG-16**, 1162.

Vural, B., 1966, Interaction of spin waves with drifted carriers in solids, *J. Appl. Phys.*, **37**, 1030.

Vural, B., and Thomas, E., 1968, Helicon-spin wave interaction in the magnetic semiconductor $Ag_xCd_{1-x}Cr_2Se_4$, *Appl. Phys. Lett.*, **12**, 14.

Walker, L.R., 1957, Magnetoelastic modes in ferromagnetic resonance, *Phys. Rev.*, **105**, 390.

Walker, L.R., 1963, in: *Magnetism*, Vol. 1, (G.T. Rado and H. Suhl, eds.), Academic Press, New York.

Webb, D.C., Vittoria, C., Lubitz, P., and Lesoff, H., 1975, Magnetostatic propagation in thin films of liquid phase epitaxy YIG, *IEEE Trans. Magnt.*, **MAG-11**, 1259.

Weinberg, I.J., and Sethares, J.C., 1978, Magnetostatic wave transducers with variable coupling, Rome Air Force Dev. Centre Res. Rept., RADC-TR-78-205.

White, R.L., and Solt, I.H., Jr., 1956, Multiple ferromagnetic resonance in ferrite spheres, *Phys. Rev.*, **104**, 56.

Wolfram, T., 1970, Magnetostatic surface waves in layered magnetic structures, *J. Appl. Phys.*, **41**, 4748.

Wu, H.J., Smith, C.V., Jr., Collins, J.H., and Owens, J.M., 1977, Bandpass filtering with multibar magnetostatic surface wave transducers, *Electron. Lett.*, **13**, 610.
Yamada, S., Chang, N.S., and Matsuo, Y., 1977, Energy analysis for the amplification phenomena of magnetostatic surface waves in a YIG-semiconductor coupled system, *IEEE Trans. Microwave Theory Tech.*, **MTT-25**, 600.
Young, P., 1969, Effect of boundary conditions on the propagation of surface magnetostatic waves in a transversely magnetized thin slab of YIG, *Electron. Lett.*, **5**, 429.
Yukawa, T.., Yamada, S., Abe, K., and Ikenoue, J., 1977. Effect of metal on dispersion relation of magnetostatic surface waves, *Jpn. J. Appl. Phys.*, **16**, 2187.
Yukawa, T., Ikenoue, J., Yamada, S., and Abe, K., 1978a, Effect of metal on dispersion relation of magnetostatic volume waves, *J. Appl. Phys.*, **49**, 346.
Yukawa, T., Maeda, T., Fujimoto, H., Abe, K., and Ikenoue, J., 1978b, Absolute instability of magnetostatic surface waves interacting with drifting carriers in closed ferrite-semiconductors structure, *IEEE Trans. Magnt.*, **MAG-14**, 817.

References for Chapter 5

Auld, B.A., 1971, in: *Applied Solid State Science*, Vol. II (R. Wolfe, ed.), Academic Press, New York.
Auld, B.A., 1973, *Acoustic Fields and Waves in Solids*, Vol. II, John Wiley and Sons, Inc., New York.
Bhattacharya, T., Tsutsumi, M., and Kumagai, N., 1976, The piezoelectric magnetoelastic wave propagation through the conducting plate in a composite medium, *IEEE Trans. Microwave Theory Tech.*, **MTT-24**, 226.
Daniel, M.R., 1973, Ferroacoustic interaction in yttrium–iron–garnet with Rayleigh waves, *J. Appl. Phys.*, **44**, 1404.
Daniel, M.R., 1977, Experimental observation of non-reciprocal attenuation of surface waves in yttrium–iron–garnet, *J. Appl. Phys.*, **48**, 1732.
Emtage, P.R., 1976, Nonreciprocal attenuation of magnetoelastic surface waves, *Phys. Rev.*, **B13**, 3063.
Forester, D.W., Vittoria, C., Webb, D.C., and Davis, K.L., 1978, Variable delay lines using magnetostrictive metallic glass overlays, *J. Appl. Phys.*, **49**, 1794.
Ganguly, A.K., Davis, K.L., Webb, D.C., and Vittoria, C., 1976a, Magnetoelastic surface waves in a magnetic film-piezoelectric substrate configuration, *J. Appl. Phys.*, **47**, 2696.
Ganguly. A.K., Davis, K.L., Webb, D.C., and Vittoria, C., 1976b, Magnetic control of surface elastic wave in a novel layered structure, *AIP Conf. Proc.*, No. 34, 259.
Ganguly, A.K., Davis, K.L., Webb, D.C., Vittoria, C., and Forester, D.W., 1975, Magnetically tuned surface acoustic wave phase shifter, *Electron. Lett.*, **11**, 610.
Ganguly, A.K., Webb, D.C., Davis, K.L., Koon, N.C., and Milstein, J.B., 1977, Magnetically variable surface wave velocity in a highly magnetostrictive rare-earth–iron compound, Ultrasonic Symposium Proc., IEEE Cat. # 77 CH1264–1SU.
Komoriya, G., and Thomas, G., 1979, Magnetoelastic surface waves on YIG substrate, *J. Appl. Phys.*, **50**, 6459.
LeCraw, R.C., and Comstock, R.L., 1965, in: *Physical Acoustics*, Vol. IIIB (W.P. Mason, ed.), Academic Press, New York.
Lewis, M.F., and Patterson, E., 1972, Acoustic surface wave isolator, *Appl. Phys. Lett.*, **20**, 276.
Lundstrom, M.S., and Robbins, W.P., 1974, Characteristics of magnetoelastic Rayleigh wave convolution on YIG, Ultrasonic Symposium Proc., IEEE Cat. #CHO 896–1SU.

Matthews, H., and van de Varrt, H., 1969, Magnetoelastic Love waves, *Appl. Phys. Lett.*, **15**, 373.
Parekh, J.P., 1969a, Magnetoelastic surface waves in ferrites, *Electron. Lett.*, **5**, 322.
Parekh, J.P., 1969b, Propagation characteristics of magnetoelastic surface waves, *Electron. Lett.*, **5**, 540.
Parekh, J.P., 1970, Influence of surface metallisation on a magnetoelastic surface wave, *Electron. Lett.*, **6**, 47.
Parekh, J.P., 1976, Effect of magnetic linewidth of YIG on the propagation characteristics of magnetoelastic Rayleigh waves, *J. Appl. Phys.*, **47**, 2228.
Parekh, J.P., and Bertoni, H.L., 1972, Magnetoelastic Rayleigh type surface wave on a tangentially magnetized YIG substrate, *Appl. Phys. Lett.*, **20**, 362.
Parekh, J.P., and Bertoni, H.L., 1974a, Magnetoelastic Rayleigh waves propagating along a tangential bias field on a YIG substrate, *J. Appl. Phys.*, **45**, 434.
Parekh, J.P., and Bertoni, H.L., 1974b, Magnetoelastic Rayleigh waves on a YIG substrate magnetized normal to its substrate, *J. Appl. Phys.*, **45**, 1860.
Parekh, J.P., Shen, S., and Thomas, G., 1975, Magnetoelastic Rayleigh wave delay line and convolver utilizing ZnO transducers, Ultrasonic Symposium Proc., IEEE Cat. # 75 CHO994–4SU.
Robbins, W.P., and Lundstrom, M.S., 1975, Magnetoelastic Rayleigh wave convolver, *Appl. Phys. Lett.*, **26**, 73.
Robbins, W.P., 1976a, Approximate theory of magnetoelastic surface wave convolution, Ultrasonic Symposium Proc., IEEE Cat. # 76 CH1120–5SU.
Robbins, W.P., 1976b, Measurement of demagnetizing factors using magnetoelastic Rayleigh wave convolution, *J. Appl. Phys.*, **47**, 5116.
Scott, R.Q., and Mills, D.L., 1977, Propagation of surface magnetoelastic waves on ferromagnetic crystal substrate, *Phys. Rev.*, **B15**, 3545.
Strauss, W., 1966, in: *Physical Acoustics*, Vol. IVB (W.P. Mason, ed.), Academic Press, New York.
Tsai, T.L., Komoriya, G., Thomas, G., and Parekh, J.P., 1974, Variation of magnetoelastic delay time with frequency and magnetic field in YIG, Ultrasonic Symposium Proc., IEEE Cat. #74 CHO 896–1SU.
Tsutsumi, M., Bhattacharya, T., and Kumagai, N., 1975, Piezoelectric–magnetoelastic surface wave guided by interface between semi-infinite piezoelectric and magnetoelastic media, *J. Appl. Phys.*, **46**, 5072.
van de Varrt, H., and Matthews, H., 1970a, Propagation magnetoelastic waves in an infinite plate, *Appl. Phys. Lett.*, **16**, 153.
van de Varrt, H., and Matthews, H., 1970b, Magnetoelastic Love waves in a magnetic layered nonmagnetic substrate, *Appl. Phys. Lett.*, **16**, 222.
van de Varrt, H., 1971, Magnetoelastic Love wave propagation in metal coated layered substrates, *J. Appl. Phys.*, **42**, 5305.
Volluet, G., 1977a, Surface acoustic wave oscillator tuned by magnetoelastic effect, *Electron. Lett.*, **13**, 588.
Volluet, G., 1977b, Applications of magnetoelastic effects to acoustic surface wave devices, Ultrasonic Symposium Proc., IEEE Cat. #77 CH1264 1SU.
Voltmer, F.W., White, R.M., and Turner, W., 1969, Magnetostrictive generation of surface elastic waves, *Appl. Phys. Lett.*, **15**, 153.
White, R.M., 1970, Surface elastic waves, *Proc. IEEE*, **58**, 1238.

References for Appendices

Auld, B.A., 1973, *Acoustic Fields and Waves in Solids*, Vol. I & II, John Wiley and Sons, Inc., New York.

Bajpai, S.N., Rattan, I., and Srivastava, N.C., 1979, Magnetic volume waves in dielectric layered structure: effect of magnetocrystalline anisotropy, *J. Appl. Phys.*, **50**, 2887.

Bertaut, F., and Forrat, F., 1956, Structure of ferrimagnetic ferrites of rare-earths, Compt. Rend., **242**, 382.

Braden, R.A., Gordon, I., and Harvey, R.L., 1966, Microwave properties of planar hexagonal ferrites, *IEEE Trans. Magnt.*, **MAG-2**, 43.

Brekhovskikh, L.M., 1960, *Waves in Layered Media*, Academic Press, New York.

Chou, P.C., and Pagano, N.J., 1967, *Elasticity Tensor, Dyadic and Engineering Approaches*, D. van Nostrand Co., New York.

Fedorov, F.I., 1968, *Theory of Elastic Waves in Crystals*, Plenum Press, New York.

Green, J.J., and Sandy, F., 1974a, Microwave characterization of partially magnetized ferrites, *IEEE Trans. Microwave Theory Tech.*, **MTT-22**, 641.

Green, J.J., and Sandy, F., 1974b, A catalog of low power loss parameters and high power thresholds for magnetized ferrites, *IEEE Trans. Microwave Theory Tech.*, **MTT-22**, 645.

Green, J.J., Schlömann, E., Sandy, F., and Saunders, J., 1969, Characterization of the microwave tensor permeability of partially magnetized materials, Semiannual Rept. RADC–TR–69–73.

Geller, S., and Gilleo, M.A., 1957a, Crystal structure and ferrimagnetism of yttrium–iron–garnet, $Y_3Fe_3(FeO_4)_3$, *J. Phys. Chem. Solids*, **3**, 30.

Geller, S., and Gilleo, M.A., 1957b, Structure and ferrimagnetism of yttrium rare-earth iron garnets, *Acta Crystallogr.*, **10**, 239.

Goldstein, H., 1950, *Classical Mechanics*, Addison Wesley Publishing Co., Inc., New York.

Heck, C., 1974, *Magnetic Materials and Their Applications*, Butterworths, London.

Jeffreys, H., 1965, *Cartesian Tensors*, The Cambridge University Press, Cambridge.

Jordan, E.C., and Balmain, K.G., 1971, *Electromagnetic Waves and Radiating Systems*, Prentice Hall India Pvt. Ltd., New Delhi.

Kittel, C., 1968, *Introduction to Solid State Physics*, 3rd edition, John Wiley and Sons, Inc., New York.

Landau, L.D., and Lifshitz, E.M., 1960, *Electrodynamics of Continuous Media*, Pergamon Press, London.

Landau, L.D., and Lifshitz, E.M., 1970, *Theory of Elasticity*, Pergamon Press, New York.

Lax, B., and Button, K.J., 1962, *Microwave Ferrites and Ferrimagnetics*, McGraw-Hill Book Co., New York.

Mason, W.P., 1958, *Physical Acoustics and the Properties of Solids*, D. van Nostrand, New York.

Musgrave, M.J.P., 1970, *Crystal Acoustics*, Holden-Day, San Francisco.

Nadeau, G., 1964, *Introduction to Elasticity*, Holt, Rinehart and Winston, New York.

Nye, J.F., 1964, *Physical Properties of Crystals*, Oxford Press, England.

Rado, G.T., 1953, Theory of microwave permeability tensor and Faraday effect in nonsaturated ferromagnetic materials, *Phys. Rev.*, **89**, 529.

Rado, G.T., 1956, On the electromagnetic characterization of ferromagnetic media: Permeability tensor and spin wave equations, *IRE Trans. Antennas Propagat.*, **AP-4**, 512.

Rattan, I., and Srivastava, N.C., 1978, Magnetoelastic interaction energy of a cubic ferromagnet in an arbitrary coordinate system, *Phys. Status Solidi*, (a)**46**, 475.

Riches, E.E., 1972, *Ferrites: A Review of Materials and Applications*, Mills and Boon, London.

Schiff, L.I., 1968, *Quantum Mechanics*, 3rd edition, McGraw-Hill Book Co., New York.
Schlömann, E., 1970, Microwave behaviour of partially magnetized ferrites, *J. Appl. Phys.*, **41**, 204.
Smit, J., and Wijn, H.P.J., 1959, *Ferrites*, John Wiley and Sons, Inc., New York.
Shore, B.W., and Menzel, D.H., 1968, *Principles of Atomic Spectra*, John Wiley and Sons, Inc., New York.
Smit, J., 1971, ed., *Magnetic Properties of Materials*, McGraw-Hill Book Co., New York.
Sokolnikoff, I.S., 1946, *Mathematical Theory of Elasticity*, McGraw-Hill Book Co., New York.
Sparks, M., 1964, *Ferromagnetic Relaxation Theory*, McGraw-Hill Book Co., New York.
Srivastava, N.C., 1977, Electromagnetic wave propagation in magnetized ferrites, Ph.D. Thesis, unpublished, Indian Institute of Technology, New Delhi.
Standley, K.J., 1972, *Oxide Magnetic Materials*, 2nd edition, Oxford Press, New York.
Stinson, D.C., and Green, M.A. Jr., 1965, Resonance properties of single-crystal hexagonal ferrites over x-band frequencies, *IEEE Trans. Magnt.*, **MAG-1**, 414.
Tebble, R.S., and Craik, D.J., 1969, *Magnetic Materials*, Wiley Interscience, London.
Thurston, R.N., 1964, in: *Physical Acoustics*, Vol. IA, (W.P. Mason, ed.), Academic Press, New York.
Tyras, G., 1959, The permeability matrix for a ferrite medium magnetized in an arbitrary direction and its eigenvalues, *IRE Trans. Microwave Theory Tech.*, **MTT-7**, 176.
von Aulock, W.H., 1965, *Handbook of Microwave Ferrite Materials*, Academic Press, New York.
Wagner, D., 1972, *Introduction to the Theory of Magnetism*, Pergamon Press, New York.
White, H.E., 1934, *Introduction to Atomic Spectra*, McGraw-Hill Book Co., New York.
Wijn, H.P.J., Gorter, E.W., Esveldt, C.J., and Geldermans, P., 1954/55, Conditions for square hysteresis loop in ferrites, *Phillips Tech. Rev.*, **16**, 49.
Yeh, K.C., and Liu, C.H., 1972, *Theory of Ionospheric Waves*, Academic Press, New York.

Author Index

Abe, K., 235, 238, 240–41
Adam, J.D., 145, 152, 154, 156, 158, 160, 174, 180, 201-202, 206, 217, 219, 222, 249, 281, 289
Akhiezer, A.I., 18, 266
Alexandrakis, G.C., 127
Anderson, P.W., 1, 78
Artmann, J.O., 24
Auld, B.A., 44, 72, 80, 83, 143, 291, 298, 302, 363, 375
Awai, I., 266, 268, 270, 273

Bady, I., 32
Bailey, G.C., 24, 96, 127
Bajpai, S.N., 227-28, 254, 286–87, 289, 360–61
Balmain, K.G., 374
Banks, C., 289
Bardai, Z.M., 221-22
Bardati, F., 145
Barker, R.C., 24
Bar'yakhtar, V.G., 18, 226
Basterfield, J., 156
Bennett, G.A., 145, 178, 180, 201-202, 206
Benson, H., 153
Berk, A.D., 43
Berlincourt, D.A., 21
Bertaut, F., 337
Bertoni, H.L., 80, 84, 250, 308, 312, 314, 321, 326
Bhattacharya, T., 242, 254, 297
Bini, M., 145, 266, 278
Blacksmith, P., 166, 174–75
Blank, S.L., 160
Bolle, D.M., 58
Bömmel, H.E., 84
Bongianni, W.L., 158, 174, 209, 211-13
Braden, R.A., 344
Brekhovskikh, L.M., 188, 363
Bresler, A.D., 145
Briggs, R.J., 272
Brillouin, L., 127
Brodwin, M.E., 71
Brown, D., 278
Brown, W.F., 44
Brundle, L.K., 143, 145, 155–56, 177-78
Button, K.J., 9, 59, 67, 245, 337-38, 341, 343, 345, 351

Cadilhac, M., 250
Callen, E.R., 85
Callen, H.B., 11, 85
Castéra, J.P., 278–80
Chang, N.S., 266, 268, 274, 278
Chou, P.C., 363
Clark, A.E., 85
Clogston, A.M., 78
Coerver, L.E., 172, 174, 176
Coleman, W., 85
Collins, J.H., 158, 160, 172, 174, 176, 219, 221-22, 249, 279–81, 289
Comstock, R.L., 291
Coquin, G.A., 249
Courtois, L., 145, 147
Craig, J.N., 29, 96, 127
Craik, D.J., 337, 341, 345
Culshaw, W., 111
Curran, D.R., 21

Damon, R.W., 143, 145, 150, 152, 170, 174, 181, 204, 206, 270–71
Daniel, M.R., 326
Darby, M.I., 22
Davis, K.L., 291, 320, 335
Declercq, G., 145, 147, 150
DeSavage, B., 85
DeWames, R.E., 203, 206
Dransfeld, K., 84

Eberhardt, N., 52
Elachi, C., 249
Elliot, M.T., 160
Emtage, P.R., 306, 319, 328
Engineer, M.H., 70
Eshbach, J.R., 143, 145, 150, 152, 170, 174, 204, 206, 270–71
Esveldt, C.J., 378

Fay, C.E., 59, 111
Fedorov, F.I., 363
Feynman, R.P., 18, 73, 143
Filleti, P.L., 266, 278
Fletcher, R.C., 196
Forat, F., 337
Forester, D.W., 335
Freedman, N.J., 143, 145, 155–56, 177–78
Freeman, A.J., 49

Magnetostatic wave propagation *continued*
in metal-backed ferrimagnetic slab *continued*
nonreciprocal delay, 194
nonreciprocal dispersion, 194
nonreciprocal propagation, 195, 206
surface modes, 195, 198, 200, 203
Maxwell's equations, 47, 129, 203

Néel temperature, 1, 338, 344
Nondispersive delay
band of, 256
frequency range of, 257, 261, 263–65
region of, 211, 219, 223, 227, 233–34

Partially magnetized ferrimagnetics, 377ff
Particle displacement, 91
field, 363
Pauli's exclusion principle, 348
Permeability tensor, 8ff, 29, 32, 50, 122, 146, 228, 256, 379
effective, 371
transformed, 151
Phase angle, 130
Phase shift at total reflection, 188
Phase shift of elastic waves, 335
Phase shifter, 335
Phase velocity, 53, 58
Planar magnetoplumbites, 31ff, 108
Planck's constant, 347
Power
absorption coefficient, 123
flow, 116, 118, 287, 289
incident, 117
reflected, 111, 117
total, 285
Poynting
theorem, 184, 373ff
vector, 285, 375
Propagation loss, 266, 281

Rare-earth iron garnets, 337
Ray model for attenuated total reflection, 122ff
Rayleigh wave velocity, 333
Reflection and transmission of electromagnetic waves by a ferrimagnetic slab, 127ff
Reflection beam
isolator, 113
phase shifter, 111
Reflection coefficient, 190
Reflection loss, 112
forward, 112–13
reverse, 112–13
Refractive index diagram, 82, 86, 88ff, 93, 311, 316, 318, 323–24
Relative permittivity of drifting carriers, 269

Seshadri mode, 272
Spin waves, 74
Spinel ferrites, 341ff
crystal structure, 341ff
line width, 343
magnetization, 342ff
magnetocrystalline anisotropy, 343
Strain, 363ff
tensor, 363–64
Strength of modulation, 250, 252
Stress, 365ff
tensor, 365
Surface elastic waves, 158
Susceptibility tensor, 14

Transducers
interdigital, 328–29, 334
meander line, 326–27
zinc oxide, 337
Transformation of
anisotropy field, 360ff
coordinates, 353ff
magnetoelastic energy, 361ff
Maxwell's equations, 357ff
permeability tensor, 359ff

Uniaxial magnetoplumbites, 30ff
Uniform precessional mode, 7
frequency of, 371ff

Waves in magnetoplasmas, 70ff
Weiss field, 19

Subject Index

Acoustic
beam width, 333
loss factor, 341
relative attenuation, 329
Angular momentum
orbital, 2, 347
quenching of orbital, 3
spin, 3, 347
total, 3, 348–49
Anisotropy
energy, 341, 343
energy of a cubic ferrimagnet, 22
energy of a hexagonal magnetoplumbite, 23, 30
field, 23–26, 32, 80, 154, 344
first-order constant, 23, 341, 343
magnetocrystalline, 22ff, 256, 377–79
planar, 344
second-order constant, 23–24, 341, 343
uniaxial, 344
Approximation
linear, 10
magnetostatic, 71ff, 79, 235, 240, 268, 375
small-signal, 8, 14
Attenuation coefficient, 68–69

Bandwidth, 108, 111
for amplification, 268
Bidirectional propagation, 246
Bohr magneton, 348
Boltzmann constant, 21
Brigg's criteria, 273

C-axis, 30–31
Christoffel
equation, 368
Christoffel *continued*
matrix, 368
operator, 368
Classical garnet, 337
Coercive force, 377
Collision frequency, 70, 269
Conservation of energy, 47
Constitutive relations, 49
Coupling loss, 266, 281
Critical angle, 103
Cubic ferrimagnet, 24ff
Curie temperature, 1

Damon–Eshbach mode, 272
Damping, 10ff
Demagnetization, 33ff, 160, 179
Demagnetizing
factors for
a circular cylinder, 42
an ellipse, 37
a sphere, 38
a thin disk, 39
a YIG plate, 335
tensor, 37, 39
Dielectric
anisotropy, 49
loading, 108, 111
tensor, 49

Effective
absorption coefficient, 122–23
boundary condition model, 203, 206
mass, 70
Elastic
displacement, 135–36, 294, 297, 299, 300, 309, 321

Elastic *continued*
- energy, 367ff
- energy density, 367
- field equation, 368
- losses, 141
- plane waves in infinite medium, 366ff
  - dispersion relation of, 369
- polarization, 134
- stiffness coefficient, 366
- stress, 80
- waves, 82–83, 297, 304, 335

Electromagnetic interaction, 332
Electromagnetic plane waves, 47ff
- dispersion relation, 52, 55, 61–62, 71, 74, 76
- effect of damping, 67ff
- exchange-dominated, 74ff
- exchange-free, 50ff
- in hexagonal planar magnetoplumbites, 61ff
- in lossless ferrimagnetics, 52ff
- mode classification, 63ff
- propagating and nonpropagating regions, 64ff

Electromagnetic wave propagation across interfaces, 97ff
- anomalous reflection, 103
- causality condition, 100
- differential phase shift, 106, 108–10
- frequency range of OR/TR, 101–2
- nonreciprocal phase shift, 106
- phase shift at total reflection, 105ff
- reflection and transmission coefficients, 98ff
- reflection from lossy ferrimagnetics, 111ff
- regions of OR/TR, 100ff
- Snell's law, 99–100
- total reflection, 108–9

Electromagnetic wave propagation in periodic structures, 133
Energy density in magnetostatic field, 185
Energy flow, 115ff
Eulerian
- angles, 227, 353
- rotations, 353, 355, 359

Evanescent field, 117
Exchange
- effects, 74
- energy, 18
- integral, 18–19

Exchange *continued*
- interaction, 18ff, 74
- torque, 20–21

Excitation efficiency, 286, 289
Exciting current, 284

Faraday
- ellipticity, 61
- rotation, 58, 84, 128

Frequency of oscillator, 330
- change in, 331

Ferrimagnetic garnets, 337ff
- crystal structure, 337ff
- line width, 341
- magnetization of
  - mixed garnets, 340
  - pure garnets, 338ff
  - substituted garnets, 340ff
- magnetocrystalline anisotropy, 341

Ferrite–air mode, 196, 203, 209–10
Ferrite–metal mode, 196, 203, 209–10

g-factor, 3, 349
Gauss's theorem, 373
Group velocity, 54, 58, 90

Hall effect, 267
Hard axis, 114–15, 254, 263, 265
Hexagonal magnetoplumbites, 30ff, 343ff
- crystal structure of
  - M-type, 343–44
  - U-type, 345
  - W-type, 344
  - Y-type, 344
  - Z-type, 344

Hook's law, 365ff
- compliance matrix, 366
- effect of damping, 367
- stiffness matrix, 366, 368

Hund's rule, 350

Input resistance, 287, 289
Insertion loss, 281

Kronecker delta function, 356

Landau–Lifshitz damping, 80
Laplace equation, 137, 145
Lateral shift, 115ff, 188
- nonreciprocal, 118
- reciprocal, 120

Levi–Civita tensor, 357

Fujimoto, H., 266

Gabriel, G.J., 71
Ganguly, A.K., 254, 282, 284, 287, 289, 291, 320, 325
Gardiol, F.E., 10, 15, 246
Geldermans, P., 378
Geller, S., 337
Gerson, T.J., 168, 197–98
Gilleo, M.A., 337
Glass, M.L., 160
Goldstein, H., 7, 353, 359
Gordon, I., 344
Gorter, E.W., 378
Grant, P.M., 158
Green, J.J., 379–80
Green, M.A. Jr., 344
Gupta, S.S., 69, 98, 145, 154, 181, 187, 193, 221, 224, 227
Gurevich, A., 59

Haas, C.W., 11
Hartemann, P., 278, 280
Harvey, R.L., 344
Heck, C., 337, 345
Heinrich, B., 127
Helszajn, J., 43
Hogan, C.L., 59
Holscher, P.C., 266
Hovrath, V.V., 52
Howarth, J., 278
Huerta, M.A., 127
Hurd, R.A., 242

Ignatius, A., 266
Ikonue, J., 235, 238, 240–41, 266, 268, 270, 273
Ince, W.J., 111
Ishmukhametov, B.Kh., 84
Issac, E.D., 22

Jaffe, H., 21
Jeffreys, H., 337
Jordan, E.C., 374
Joseph, R.I., 35, 36, 39, 42

Kanamori, J., 44
Kanda, M., 113
Kawasaki, K., 220–21, 266, 278
Ki'ldishev, V.N., 266, 278
Kittel, C., 4, 19, 22, 24, 43, 84, 363
Knerr, R.H., 52
Kogelnik, H., 116, 187
Koltunov, I., 282, 284, 287, 289
Komoriya, G., 336
Kong, J.A., 49
Koon, N.C., 335
Kumagai, N., 242, 249–50, 254, 297

Lampariello, P., 145
Landau, L.D., 11, 123, 363, 375
Landorf, R.W., 160
Lax, B., 9, 59, 67, 245, 337–38, 341, 343, 345, 351
LeCraw, R.C., 84–85, 291
Lengyel, B.A., 43
Levinstein, H.J., 160
Lewandowsky, S.J., 43
Lewin, L., 58
Lewis, M.F., 328
Licht, S., 160
Lieu, O.L.S., 127
Lifshitz, E.M., 11, 123, 363, 375
Lighthill, M., 48
Liu, C.H., 375
Lotsch, H.K.V., 117
Lubitz, P., 127
Lundstrom, M.S., 331, 334

Maeda, T., 266
Mason, W.P., 363
Masuda, M., 266, 268
Matsuo, Y., 266, 268, 274, 278
Matthews, H., 84–85, 298, 303–304
May, W.G., 113
McStay, J., 266, 278
Mee, J.E., 160
Menzel, D.H., 347–49
Mercereau, J.E., 73, 143
Merry, J.B., 145, 155–56, 158
Meshcheryakov, V.F., 127
Millanta, L., 266, 278
Miller, N.D.J., 221–22, 227, 230, 234, 278
Mills, D.L., 153, 320
Milstein, J.B., 335
Morganthaler, F.R., 145
Mueller, R.S., 97, 127, 132–33
Murmuzev, B.A., 266, 278
Musgrave, M.J.P., 363

Nadan, J.S., 167, 197–98
Nadeau, G., 363
Nag, B.R., 70
Neviere, M., 250
Newburgh, R.G., 166, 174–75
Nye, J.F., 363

Ohtsuki, K., 266, 268, 270, 273
O'Keeffe, T.W., 221, 281, 289
Olson, F.A., 143
Osborne, J.A., 37
Owens, J.M., 158, 160, 174, 219, 223, 235, 240, 279, 281, 289

Pagano, N.J., 363
Parekh, J.P., 80, 84, 145, 147, 150, 197–98, 201–203, 221–22, 266, 282, 289, 292, 308, 312, 314, 319, 321, 326, 331, 336
Patterson, E., 328
Patterson, R.W., 221, 281, 289
Peletminskiĭ, S.V., 18, 266
Peng, S.T., 250
Petit, R., 250
Phillips, T.G., 127
Pizzarello, F.A., 158, 172, 174, 176
Polder, D., 9
Ponamgi, S.R., 145, 147, 150
Purichard, M., 145, 147, 150

Rado, G.T., 379
Rattan, I., 254, 360–61
Renard, H., 117, 121, 188, 192
Riches, E.E., 337, 345
Robbins, W.P., 331–32, 334–35
Robinson, B.B., 266

Rosenbaum, F.J., 69
Rowan, J.H., 43
Rubinstein, M., 127
Rubio, N., 266, 268

Sakeguchi, Y., 249–50
Sandey, F., 379–80
Saunders, J., 379
Schiff, L.I., 3, 317
Schilz, W., 234, 281, 289
Schlömann, E., 35–36, 39, 42, 80, 82, 266, 379–80
Schmid, H., 49
Schneider, B., 24, 254
Seavey, M.H., 320
Seidel, H., 196
Seshadri, S.R., 195, 197, 204, 249, 270–72
Sethares, J.C., 145, 155–56, 158, 166, 174–75, 282, 284, 287, 289
Shen, S., 331, 335
Shore, B.W., 347–49
Sittig, E.K., 249
Smit, J., 44, 337, 342–43, 345
Smith, C.V. Jr., 223, 235, 240, 279–81, 289
Sodha, M.S., 62, 127, 133
Sommerfeld, A., 36
Sparks, M., 11, 145, 158, 177–78, 180, 342–43
Srivastava, N.C., 62, 65, 69, 72–73, 97–98, 101, 115, 127, 133, 145, 154, 166, 181, 187, 193, 221, 224, 227–28, 242, 244, 254, 286–87, 289, 359–61
Standley, K.J., 337, 345
Steele, M.C., 70, 72, 266, 269
Stiglitz, M.R., 145, 158
Stinson, D.C., 344
Solt, I.H., 73, 143, 155
Stratton, J., 36
Strauss, W., 291
Suhl, H., 24, 70, 78, 127–28
Sykes, C.G., 249
Szutakowski, M., 266

Tagaki, H., 220–21, 266, 268
Tamir, T., 143, 150
Tannenwald, P.E., 97, 127, 132
Tebble, R.S., 337, 341, 345
Temme, D.H., 111
Thomas, E., 266, 278
Thomas, G., 331, 335–36
Thurston, R.N., 363
Tien, P.K., 116, 190
Tiersten, H.F., 44
Trivelpiece, A.W., 266
Tsai, M.C., 223, 235, 240
Tsai, T., 282, 284, 287, 289
Tsai, T.L., 326
Tsutsumi, M., 242, 247, 249–50, 254, 297
Tuan, H.S., 282, 289
Turner, W., 291, 325

Umeno, M., 220–21, 266, 278

van de Vaart, H., 281, 298, 303–304
Vashkowskii, A.V., 266, 278
Vaslow, D.F., 145, 212
Vittoria, C., 24, 96, 127, 215, 217, 242, 254, 291, 320, 335
Vlasov, K.B., 84
Volluet, G., 289, 328
Voltmer, F.W., 325
von Aulock, W.H., 23, 59, 108, 111, 335, 341, 343–45
Vural, B., 70, 72, 266, 269, 278

Wagner, D., 3, 349
Waldron, R.A., 43, 59
Walker, L.R., 70, 73, 127–28, 143
Webb, D.C., 242, 254, 289, 291, 320, 325
Weber, H.P., 187
Wecki, B., 266
Weinberg, I.J., 289
White, R.L., 3, 73, 143, 155, 291, 347, 349
White, R.M., 291, 325
Wijn, H.P.J., 44, 342–43, 378
Wilkinson, J., 145, 178, 180
Wilsey, N.D., 215, 217, 254
Wilson, D.A., 178
Wolfram, T., 203, 206, 242, 246
Wu, H.J., 223, 235, 240, 279, 281, 289

Yamada, S., 235, 238, 240–41, 266, 274, 278
Yeh, K.C., 375
Yelon, A., 24
Young, P., 201
Yukawa, T., 235, 238, 240, 266
Yuki, Y., 249

Zubkov, V.I., 266, 278

Line width, 14, 67, 78, 113, 126, 215, 266, 277, 319
Loss factor, 217
Love waves, 298, 302–3

Magnetic
induction, 138, 146, 152, 161, 181, 189, 195, 221, 236, 242, 251, 282, 293–94, 310, 377.
losses, 141
moment, 338–39, 347, 350–51
spin moment, 348
Magnetic dipole
moment of, 1, 2, 5, 348
potential energy of, 4
Magnetic field
biasing, 7
demagnetizing, 35–36
effective, 19, 26
external, 4
internal, 34
total instantaneous, 48
uniform, 7
Magnetic wave
interaction with drifting carriers, 266ff
amplification factor, 273
amplification of, 266ff
dispersion relation, 270, 274
gain parameter, 277
on air–ferrimagnetic–semiconductor composite, 267ff
on metal–dielectric–semiconductor–YIG, 273ff
propagation in a ferrimagnetic half space, 144ff
attenuation rate, 158
dispersion relation, 146, 152
group delay time, 154, 156
group velocity, 153–54
surface waves, 150, 152, 154–55
Magnetization, 2ff
dc, 25
instantaneous, 8
nonuniform, 20–21, 35
remanent, 337
spontaneous, 1, 304, 315, 333
Magnetoelastic
coupling, 45, 297, 302, 304, 315, 333
coupling parameter, 83–84
energy, 44
field equations, 79ff
Magnetoelastic *continued*
first constant, 45
interaction, 44ff, 79
magnetic field, 45–46
second constant, 45
stress, 45–46, 81
Magnetoelastic plane waves
dispersion relation, 81ff, 83, 85–86
effect of anisotropy, 95ff
effect of exchange, 93ff
exchange-free, 80ff
polarization of modes, 91ff
Magnetoelastic Rayleigh wave, 308ff, 321ff
convolver, 335
dispersion relation, 308ff, 321ff
effect of magnetic loss, 319ff
energy loss for leaky waves, 316ff
leaky waves, 311, 323
numerical solution of dispersion relation, 312ff, 323ff
Magnetoelastic surface wave
absorption of, 326
attenuation of, 319ff, 328
nonreciprocal, 328
reciprocal, 328
convolution, 331ff
delay, 326
isolator, 328
oscillator, 328ff
propagation in ferrimagnetic half space, 292ff
dispersion relation, 295
effect of metallization, 295
propagation in ferrimagnetic layered nonmagnetic structure, 304ff
propagation in ferrimagnetic slab, 303ff
propagation in nonmagnetic layer over ferrimagnetic substrate, 297ff
dispersion relation, 299–301
effect of metallization, 300
Magnetoelastic waves
nonreciprocal reflection of, 141
propagation in layered planar structures, 291ff
bulk, 291
excitation and propagation, 325ff
excitation and propagation loss, 325ff
excitation efficiency, 328
surface, 291

Magnetoelastic waves *continued*
reflection and transmission coefficients, 137
reflection and transmission of, 133ff
total reflection of, 141
Magnetostatic
condition, 135, 143
field equation, 73
interaction, 322
potential, 75, 78, 137–38, 145, 151, 161, 166, 171, 181, 189, 195, 201, 208, 221, 235, 242, 251, 268, 282, 294, 299, 309, 321–22
region, 74
Magnetostatic surface wave, 297, 302–4
equalizers, 174
resonances, 177, 206
ring interferometer, 176
rotation rate sensor, 174–75
Magnetostatic surface wave propagation
on a dielectric layered structure
dispersion relation, 254
effect of anisotropy, 252ff
effect of cubic anisotropy, 254, 256ff
effect of hexagonal planar anisotropy, 254, 266ff
group delay time, 255–56, 262–63, 265
group velocity, 254
magnetostatic analysis, 254
on a double-layered structure, 242ff
backward waves, 245, 248
dispersion relation, 243, 246
electromagnetic analysis, 245
forward waves, 244–45, 248
group velocity, 245
magnetostatic analysis, 245–46
reciprocal dispersion, 243
on ferrimagnetic dielectric metal structure, 209ff
attenuation factor, 245
dispersion relation, 212, 228
electromagnetic analysis, 212
energy and power, 223ff
group delay time, 212–13
group velocity, 215, 222, 229
magnetostatic analysis, 212, 228
nonreciprocal propagation, 212, 233
reciprocal delay, 232
Magnetostatic surface wave propagation *continued*
on ferrimagnetic layer over another ferrimagnetic substrate, 246ff
on metal–ferrimagnetic–metal structure, 235ff
dispersion relation, 236, 240
effect of anisotropy, 238
effect of demagnetization, 238
reciprocal propagation, 240
on periodic structures, 248ff
dispersion relation, 252
Magnetostatic waves, 74, 83–84, 93–94
amplification of, 266, 274
bulk, 127, 297, 302–4
delay line, 278–80
dispersion relation, 284
excitation of, 282
forward excitation of, 286
in layered structures, 143ff
oscillator, 278ff
transduction of, 281ff
Magnetostatic wave propagation
in isotropic ferrimagnetic slab, 160ff
backward waves, 172
dispersion relation, 160–61, 163, 174
effect of cylindrical curvature, 166
electromagnetic analysis, 167ff
energy distribution and power flow, 184ff
forward waves, 183
group delay time, 164, 188, 192
group velocity, 164, 182–83
magnetostatic analysis, 169ff, 181ff, 188
steering of surface waves, 172, 174
surface waves, 160–61, 163, 174
zig-zag ray model, 187ff, 208, 277
in metal-backed ferrimagnetic slab, 194ff
bulk modes, 198
dispersion relation, 195, 198–99, 202, 204, 208.
effect of finite conductivity, 203ff
electromagnetic analysis, 194, 197ff
group delay, 196
magnetostatic analysis, 194, 195ff, 201ff

GPSR Compliance

The European Union's (EU) General Product Safety Regulation (GPSR) is a set of rules that requires consumer products to be safe and our obligations to ensure this.

If you have any concerns about our products, you can contact us on ProductSafety@springernature.com

In case Publisher is established outside the EU, the EU authorized representative is:

Springer Nature Customer Service Center GmbH
Europaplatz 3
69115 Heidelberg, Germany

Batch number: 10620143

Printed by Printforce, the Netherlands